Second European and Mediterranean Structural Engineering and Construction Conference (EURO-MED-SEC-2)

Responsible Design and Delivery of the Constructed Project

Beirut, Lebanon
23-28 July 2018

Editors:

Mohamed-Asem Abdul-Malak
Hiam Khoury
Amarjit Singh
Siamak Yazdani

ISBN: 978-1-7138-0001-9

Printed from e-media with permission by:

Curran Associates, Inc.
57 Morehouse Lane
Red Hook, NY 12571

Some format issues inherent in the e-media version may also appear in this print version.

Printed with permission by Curran Associates, Inc. (2020)

For permission requests, please contact The International Structural Engineering and Construction Society (ISEC) at the address below.

ISEC PRESS
PO Box 1334
Fargo, North Dakota 58107
USA

isec.sec@gmail.com

Additional copies of this publication are available from:

Curran Associates, Inc.
57 Morehouse Lane
Red Hook, NY 12571 USA
Phone: 845-758-0400
Fax: 845-758-2633
Email: curran@proceedings.com
Web: www.proceedings.com

TABLE OF CONTENTS

KEYNOTE PAPER

ENGINEERING/CONSTRUCTION EDUCATION AND KNOWLEDGE, RESOURCES, AND RISK MANAGEMENT

INNOVATIVE ARCHITECTURAL SOLUTIONS AND ENVIRONMENTAL AND SAFETY ISSUES

EXPERIMENTAL RESEARCH AND IMPROVED DESIGNS IN STRUCTURAL ENGINEERING

THE USE OF SUSTAINABLE FILLERS AND SPECIAL TECHNIQUES FOR IMPROVING CONCRETE BEHAVIOR

Preface

Under the ever-increasing complexity that continues to characterize the architecture, engineering, and construction (AEC) industry, professionals of all specialties are to be expected to act responsibly as they assume an array of varying roles pertaining to the planning, design, and delivery of today's construction projects. "Acting responsibly" involves a lot more than clearly assessing the needs, carefully identifying the problems, and skillfully providing design and construction-related services. It rather entails embracing a more conscious approach that steadily aims at the adoption of innovative solutions throughout the projects' implementation stages. Such intended innovations shall be meant to serve not only the interests and constraints of concerned development entities and bodies, as it may typically be the case, but also to better serve contemporary development criteria including those concerned with the social acceptability of planned projects as well as their short- and long-term environmental friendliness and economic viability.

It is with the above in mind that the theme of the EURO-MED-SEC-2 conference was chosen to be: *"The Responsible Design and Delivery of the Constructed Project"*. In this regard, the conference organizers are grateful for the overwhelming interest and response by the many prominent researchers and industry practitioners from all around the world. All their valuable contributions have made of these proceedings a credible reference that distinctively promotes the advancement of knowledge in a number of engineering (architectural, civil infrastructure, structural, materials, geotechnical, environmental, etc.) and construction fields. This set of proceedings has 62 manuscripts from 16 countries.

Nothing is more germane than for the International Structural Engineering and Construction (ISEC) Society to have strategically opted to hold the EURO-MED-SEC-2 conference in Beirut (Lebanon), a city that has been experiencing a massive program of rebirth, following one of the most fiercely destructive wars that had spanned over close to two decades. To this end, the American University of Beirut, the "leading" university in the Middle East region is honored to be hosting the activities of this conference, and the members of the Civil and Environmental Engineering Department at the Maroun Seeman Faculty of Engineering and Architecture are privileged to have partnered with the ISEC Society, and the renowned members of its administration and its competent support staff, for the successful organization and delivery of this conference.

Mohamed-Asem Abdul-Malak, *American University of Beirut, Lebanon*
Hiam Khoury, *American University of Beirut, Lebanon*
Amarjit Singh, *University of Hawaii at Manoa, USA*
Siamak Yazdani, *Heyer Engineering, USA*
Editors

Acknowledgements

Host
American University of Beirut, Beirut, Lebanon

Co-Sponsor
American Concrete Institute

Committees

Ruveyda Komurlu, *Kocaeli University, Turkey*
Mounir Mabsout, *American Unversity of Beirut, Lebanon*
Adnan Masri, *Beirut Arab University, Lebanon*
Carol Menassa, *University of Michigan at Ann Arbor, USA*
Fuad Okay, *Kocaeli University, Turkey*
Mehmet Ozbek, *Colorado State University, USA*

Local Organizing Committee
Mohamed-Asem Abdul-Malak, *American University of Beirut, Lebanon* ***(Conference Chair)***
Hiam Khoury, *American University of Beirut, Lebanon* ***(Conference Co-Chair)***
Ibrahim Alameddine, *American University of Beirut, Lebanon*
Ghassan Chehab, *American University of Beirut, Lebanon*
Mayssa Dabaghi, *American University of Beirut, Lebanon*
Farook Hamzeh, *American University of Beirut, Lebanon*
Shadi Najjar, *American University of Beirut, Lebanon*
George Saad, *American University of Beirut, Lebanon*
Issam Srour, *American University of Beirut, Lebanon*

List of Reviewers

Khairy Abdelkareem
Mohamed-Asem Abdul-Malak
Ramzi Abed
Grace Abou Jaoude Estephan
Majdi Abou Najm
Zaher Abou Saleh
Maya Abou Zeid
Hisham Abou-Ibrahim
Simaan Abourizk
Vitor Abrantes
Sameer Abu Eisheh
Munir Ahmed
Ibrahim Alameddine
Mohammed Algahtany
Mohammad Alhassan
John Aliu
Wael Alnahhal
Maha Alqam
Salem Alsanusi
Sana Amir
Faisal Ananda
David Arditi
Benazouz Chikh
Mayssa Dabaghi
Farah Demachkieh
Mounir El Asmar
Islam El-Adaway
Kamel Elamrousy
Samer Ezeldin
Mohamed Ezzeldin
Amine Ghanem
Murat Gunduz
Bilal Hamad
Samer Hamdar
Farook Hamzeh
Boštjan Harl
Amr Hassanein
Daniel Hastenpflug
Kyu Hong Hwang
Daniela Ionescu
Vineet Kamat
Rana Khallaf
Hiam Khoury
Paula Lazzari
Carol Menassa
George Morcous
Chady Najjar
Shadi Najjar
Fuad Okay
Mehmet Ozbek
Alexandre Pacheco
Shuvendu Patel
George Saad
Ezzeldin Yazeed Sayed-Ahmed
Awni Shaaban
Amira Shalaby
Amarjit Singh
John Smallwood
Issam Srour
Faith Jordan Srour
Afshin Turk
Mumtaz Usmen
Yiannis Vacanas
Enrique Villacis
Frank Yazdani

Ousama Baalbaki
Pablo Ballesteros-Pèrez
Mohammad Barakat
Abdulaziz Bubshait
Ghassan Chehab

Mounir Mabsout
Bruna Manica Lazzari
Paula Manica Lazzari
Adnan Masri
Manuela Melo

Responsible Design and Delivery of the Constructed Project
Edited by Abdul-Malak, M., Khoury, H., Singh, A., and Yazdani, S.

SIMULATION-BASED ANALYTICS: ADVANCING DECISION SUPPORT IN CONSTRUCTION

SIMAAN ABOURIZK

Dept of Civil and Environmental Engineering, University of Alberta, Edmonton, Canada

Construction projects are predominantly managed with a heavy reliance on the knowledge and experience of construction professionals and supporting enterprise resource planning systems. The construction sector continues to struggle with the management, analysis, and transformation of data into useful information for improved decision-making. While development of data-driven decision support systems for construction would improve the accuracy and relevancy of decision-making processes, several challenges currently limiting the incorporation of dynamic project data into prediction models must first be addressed. An envisioned solution for advancing data-driven decision making in construction using a simulation-based analytics framework capable of overcoming such limitations is presented and discussed. Concept feasibility is demonstrated through the successful completion of a prototype for quality-associated decision support that has been developed using the proposed conceptual framework.

Keywords: Data-driven decision-making, Predictive analytics, Quality management, Construction management.

1 DECISION SUPPORT IN CONSTRUCTION

Competitiveness of the construction sector depends on numerous improvements across multiple areas of project delivery as documented by many researchers and industrial organizations (Construction Industry Institute 2008, Construction Owners Association of Alberta 2015, Hanna 2016). Yet in spite of these advances, the construction industry remains challenged by a slow adoption of innovation and technological developments, high capital costs, decreased labor productivity, and reduced operational efficiency. While the construction industry has adopted a variety of information technology solutions (e.g. commercial enterprise resource planning systems, scheduling and contract management software, or in-house computer systems) for improved decision support, these systems lack the quantitative predictive capabilities required to produce desired results (Oracle 2010, Pritchard *et al.* 2012). Accordingly, construction management continues to rely heavily on the subjective knowledge and experience of construction professionals whose ability to capture intricate details of various processes, resources, and uncertainties associated with project execution are often limited (Manyika *et al.* 2011).

2 DATA-DRIVEN DECISION SUPPORT SYSTEMS: THE IDEAL

To remain competitive, the construction industry must take advantage of newly emerging technologies and tools to facilitate improvements in production efficiency, product quality, and safety management practices. A new generation of decision support systems capable of (1) providing interpretable, up-to-date project information, (2) generating reliable project predictions throughout both the project planning and execution phases of construction, and (3) allowing for analysis and investigation of various potential project scenarios would represent a notable advancement in how construction is managed in practice. Despite the desire for such tools, however, analytical systems capable of comprehensively and automatically generating reliable decision support outputs from existing project data have yet to be successfully implemented in construction practice.

3 ENVISIONED SOLUTION: SIMULATION-BASED ANALYTICS

Achievement of the ideal will require the development and implementation of systems that can integrate a variety of datasets and incorporate associated project uncertainty to dynamically reflect actual project conditions in an automated manner. Simulation modeling has emerged as a technological means of addressing the dynamicity and uncertainty associated with construction operations for improved system predictability (AbouRizk 2010) and represents a technique capable of providing the functionalities required to create this ideal system.

However, as with many decision support systems (Darema 2004), a lack of (1) appropriate approaches at the application level for enabling dynamic feedback and measurement coupling, (2) techniques for interfacing applications with measurement devices and instruments, (3) application algorithms adaptable to dynamic data inputs, (4) methods for handling uncertainty in input data, and (5) computer programs capable of supporting such environments have considerably limited the integration of real construction project data into simulation models, in turn, limiting the predictive capabilities of current systems. Indeed, models stemming from this work have remained limited to specific construction applications (e.g., module yard schedule, crane selection and allocation). The inability of simulation methods to achieve full, effective integration into organizational decision-making (Akhavian and Behzadan 2015) is due, primarily, to the inability of current process interaction-based simulation modeling methods to automatically recalibrate themselves as new project data are collected, greatly limiting their applicability during project delivery. Successful implementation of a simulation-based analytics solution will require:

- Advances in dynamic, data-driven simulation applications for decision support in construction engineering and management
- Investigation of methods for integrating (1) data that are currently collected, (2) simulation-derived data to supplement existing data when required, (3) measurement processes that are steered, as required, to enhance decision-support when possible, and (4) decision support applications
- Examination of structured methods for integrating analytics into organizational decision-making to improve effectiveness
- Transitioning the role of simulation beyond traditional analysis into an essential component of decision support

4 SIMULATION-BASED ANALYTICS FOR CONSTRUCTION

We have explored and built upon recent advances in data analytics applications (Barton and Court 2012), dynamic data-driven application systems (Darema 2004), and simulation-based analytics (Dube *et al.* 2014) to develop a conceptual, simulation-based analytics framework for construction (SAC) that we believe possesses the attributes and characteristics required to support data-driven decision-making in construction. The conceptual SAC framework, illustrated in Figure 1, is comprised of *data adapters*, which are involved in data transformation and preparation, and *analysis modules*, which are composed of a suite of algorithms that interact with *simulation components* to facilitate production of required *decision support metrics*. Rationale and functionalities of each component are detailed as follows.

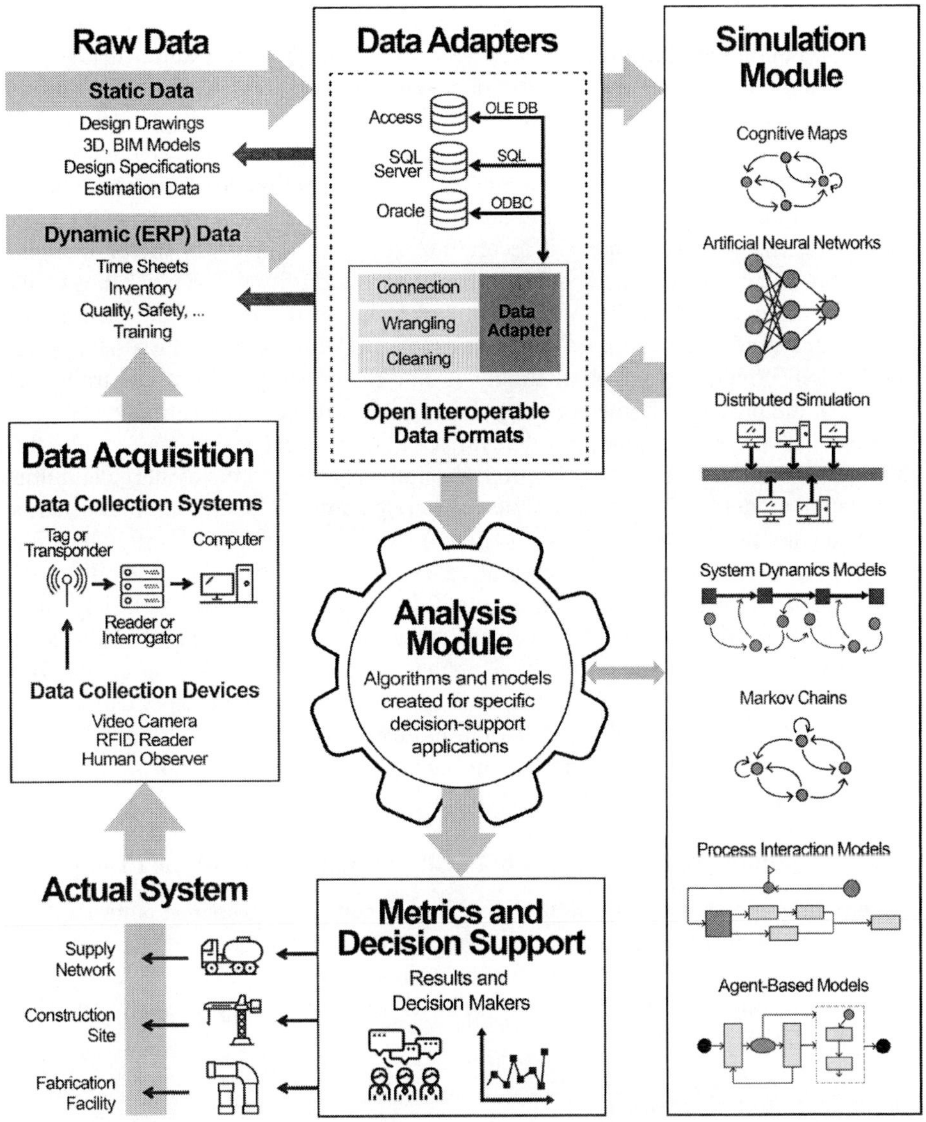

Figure 1. Proposed simulation-based analytics for construction framework.

4.1 SAC Framework Components

Construction companies handle vast amounts of raw data, which can be classified into two main types, namely dynamic data stored in data management systems (e.g., time sheets, safety incidences, quantities) and static data passed on from engineering (e.g., 2D, 3D, and BIM models). Raw data are generally not effective for decision support, as they are not always reliable, aggregated at the required level, or transformed into useful information. The *data adapter* portion of the framework is required to transform raw data into useful information. Data adapters apply a variety of analytical methods (e.g., data mining) to identify anomalies and missing data as well as simulation models to populate and validate data.

While the data adapter is responsible for collecting, organizing, and cleaning raw data, transformed data are not often stored in an easily interpretable format. As such, the conceptual SAC framework contains an *analysis module*, which is comprised of a suite of algorithms that facilitate production of required metrics for decision support. The primary function of this module is to drive the simulation process, invoking the appropriate analytic (e.g., descriptive analytic from a database query or predictive analytic derived from a data-driven simulation) and cycling through the process until decisions are made. The simulation component module includes (1) process interaction models, which simulate production processes, supply market, logistical issues, external factors (e.g., weather and economic conditions) and (2) factor models, which estimate the states of various measures, such as facility productivity based on values of identified influencing factors. Factor models, which can take the form of artificial neural networks, structural equations models, cognitive maps, and tree models, follow a three-stage process of factor identification, Principal Component Analysis, and factor analysis or model training.

At the core of the proposed framework is a dynamic data-driven application system (DDDAS) concept (Darema 2004). This concept refers to a paradigm that strives to seamlessly couple simulation and measurement (i.e., data collection) disciplines, enabling the dynamic addition of new data into simulation models. The addition of new data (e.g., archival data, real-time generated data, or sensor-detected measurements from actual systems) triggers an automated, internal calibration of the original model, increasing the representativeness and, in turn, the reliability of system predictions. DDDAS concepts have been successfully applied to model a variety of systems, including smart cities (Fujimoto *et al.* 2016), expressway traffic (Sunderrajan *et al.* 2016), dynamic manufacturing (Kück *et al.* 2016), and for crisis management analysis (Badr *et al.* 2015). Recent works in the construction domain include the development of a tunneling application, where data are collected in real-time and applied to simulation models to enhance predictions (Bi *et al.* 2015), and a framework for model-driven acquisition and analytics of visual data using unmanned aerial vehicles for construction progress monitoring (Lin *et al.* 2015).

4.2 Simulation Methods and Techniques to Facilitate Deployment of the Framework

The virtual environment is responsible for enabling dynamic data-driven simulation and for supporting the integration of various forms of simulation. It can be built using extensions of COSYE (AbouRizk and Hague 2009), Simphony (AbouRizk *et al.* 2016), and/or a variety of other information technology and modeling tools as required. Key elements of the virtual environment are detailed as follows.

4.2.1 *Data preparation algorithms*

Data adaptors are built to transform, clean, and prepare data for use in the analytics. Prepared data are warehoused using data incubators similar to those described by Fan et al. (2008) and Hammad (2009). Warehouse models consist of an object-oriented, integrated, non-volatile, and time-variant collection of data capable of supporting management decisions (Inmon 2005) and of providing a variety of decision support capabilities that cannot be attained using relational databases. Such capabilities include (1) validation, transformation, and scrubbing of data to ensure quality prior to storage, (2) collection and integration of data from various sources, and (3) use of multi-dimensional data models to organize data by subject allowing decision-makers to perform data analysis from various perspectives at various levels of detail. Concurrently, online analytical processing of data warehouses facilitates user-directed information retrieval and interactive analysis through visual operations of drill-down, roll-up, slicing and dicing, and pivoting.

Notably, data adapters should be designed specifically for particular datasets. For example, data adapters for unconventional data, such as images, differ from those designed to extract and transform data retrieved from enterprise resource planning systems, as image data must be inferred from geometric entities as opposed to being selected from an information system's database. Data adapters, therefore, may be designed to extract geometric model information from 3D models (Han *et al.* 2017) or to contain a variety of computer vision algorithms capable of reconstructing as-built 3D point cloud models from photographs (Martens 2017). The overall objective of this step is the development of intelligent data adapters that can identify a variety of information for planning various construction activities and processes.

4.2.2 *Simulation modeling infrastructure*

To function successfully, the framework must combine and integrate various subsystems that execute in a harmonious manner. This will require advancing construction simulation methods to facilitate real-time communication, enable dynamic data-driven simulation approaches, and accommodate hybrid simulations (e.g., system dynamics, agent-based modeling, discrete-event simulation, construction/time-stepped simulation, and fuzzy cognitive maps).

Simulation models often use statistical distributions or Markov chains to represent uncertainties that exist in the systems they are analyzing. Although useful for assessing static data, the ability to recalibrate these models following the incorporation of new data remains challenging. Bayes' theorem, which describes a method for updating probabilities when provided with new evidence, can be used for updating statistical distributions. The use of Bayes' theorem reported in literature, however, remains limited to normal distribution updating (Lynch 2007, Chung *et al.* 2004) due to the low dimensionality associated with normal distribution parameters and to the computational challenges associated with evaluating Bayes' mathematical formulation for updating probabilistic models. Various numerical techniques, including appropriate conjugate selection (Gelman *et al.* 2013) and numeric integration, will allow updating capabilities to be extended to other distribution types commonly used in construction simulation. Markov chain updating will be achieved using a variety of approaches, including application of Bayes' theorem variants and techniques such the Baum-Welch (Rabiner 1989), Ensemble Learning (MacKay 1997), and Viterbi (Rabiner 1989) algorithms.

5 SIMULATION-BASED ANALYTICS FOR CONSTRUCTION

Feasibility of the conceptual framework is being examined through the completion of specialized decision support prototypes, including an earned value management prototype, which captures project schedule while considering the historical performance of individual activities, and a safety management prototype, which allows users to test and predict the impact of various safety management strategies on safety performance.

Concept feasibility has been demonstrated through the successful completion of a prototype for quality-associated decision support developed using SAC concepts. This prototype uses quality management, engineering design, and cost management data to determine operator quality performance, product complexity, and quality performance for project quality forecasting and quality-induced rework cost management. The specialized framework underlying the prototype contains several SAC framework components, namely a data adapter, data analysis module, simulation module, and decision support module, which have been customized to facilitate system application for quality-associated decision support. R (R Core Team 2017) programming software for statistical computing and graphics was utilized to drive framework components.

5.1 Data Sources and Data Adapter

Data sources available for analysis include dynamic quality management data (e.g., quality inspection results, rework time, and operator information) and static engineering design data (e.g., product design attributes, such as pipe material, thickness, and diameter). A data adapter specific to this dataset was built to transform raw data, through data connection (i.e., integration of multiple sources of data), wrangling (i.e., conversion of data into a tabular format), and cleaning (i.e., noise reduction and omission of missing values) into a compatible and interpretable dataset. Example of data adapter inputs and outputs are detailed in Figure 2.

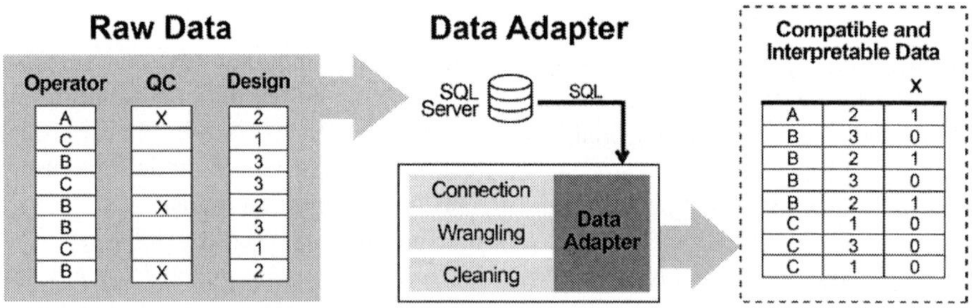

Figure 2. Example of real-time updating mechanism of quality performance.

5.2 Analysis Module

The analysis module functions to convert transformed data into useful decision support information or into a format that is suitable for simulation input. In this specialized framework, the analysis module is responsible for determining quality performance (i.e., failure rates) of individual operators and of particular product designs.

5.3 Simulation Module and Decision-Support Outputs

The simulation module generates data for desired prediction metrics for a given decision support application using transformed data from the data adapter and/or analysis module. The specialized framework uses a variety of simulation or simulation-associated components. Specifically, a Metropolis-Hasting Algorithm component is used to convert binary quality data (i.e., failure rates) into distributions for simulation purposes. A Monte Carlo simulation module (Ji and AbouRizk 2016) and an absorbing Markov chain (Ji and AbouRizk 2018b) are used to incorporate uncertainty and to model the rework process for forecasting purposes, respectively. Specific functionalities of the simulation module are described as follows.

5.3.1 *Bayesian-based quality performance modeling*

To estimate product quality performance, a Bayesian-based solution was developed to drive a distribution for incorporating uncertainty (Ji and AbouRizk 2017). In addition to providing more accurate, reliable, and interpretable estimation of product quality performance, the proposed solution addresses some of aforementioned challenges associated with simulation model updating. Application of these methods allows the simulation models to realign with dynamic, real-time data generated by the actual system. As demonstrated in Figure 3 the quality performance distribution can be updated using real-time quality inspection data.

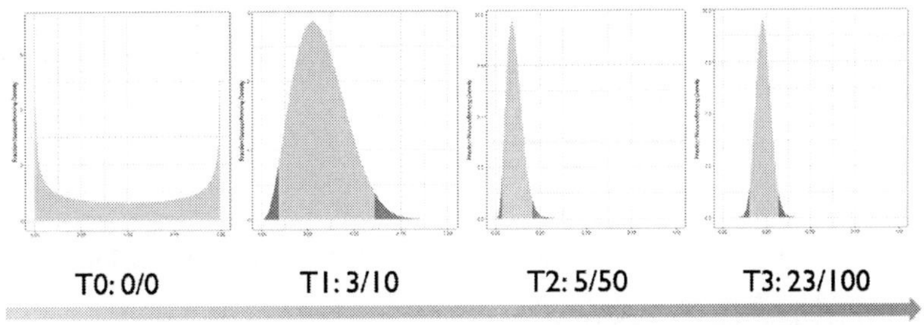

Figure 3. Example of real-time updating mechanism of quality performance.

The components comprising the specialized simulation-based analytics environment allows simulation models to be adjusted by real-time data and measurements. The approach also develops descriptive and predictive analytical metrics, namely operator quality performance measurements and project quality performance forecasts, for supporting and improving decision-making processes (Ji and AbouRizk 2018a). For instance, by using historical product quality performance data as inputs, quality performance for a new project can be simulated to achieve a quality performance distribution at a project-level, supporting decision making in consideration of particular risk attitudes (Figure 4).

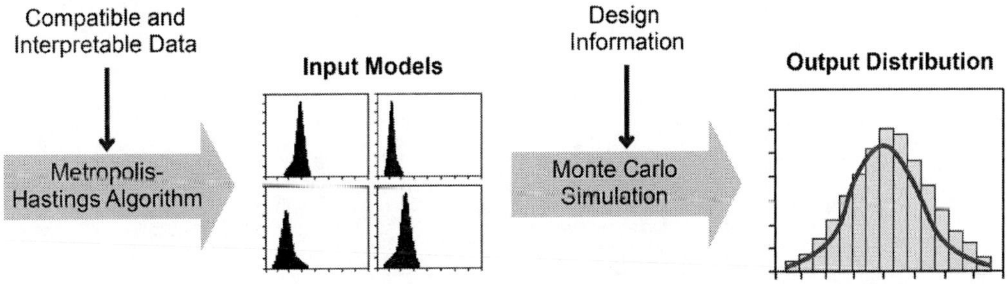

Figure 4. Workflow of the simulation model for project quality performance measurement.

5.3.2 *Product complexity clustering*

To further extend the functionality of the proposed system, a hybrid data mining approach for quantitatively analyzing product complexity from product quality performance data (i.e., failure rate) was developed (Ji *et al.* 2018). The proposed model is comprised of three steps, which (1) measure product complexity by introducing a Bayesian-based nonconforming quality performance indicator, (2) score each type of product complexity by developing a Hellinger distance-based distribution similarity measurement, and (3) cluster products into homogeneous complexity groups by using the agglomerative hierarchical clustering technique. Practitioners can implement this approach to enhance their product complexity management practices from the perspectives of (1) strategic bidding, (2) complexity-driven production planning, and (3) customized training. As demonstrated in Figure 5, practitioners can directly obtain product design information, complexity score, total business percentage, and its corresponding complexity level from the outputs.

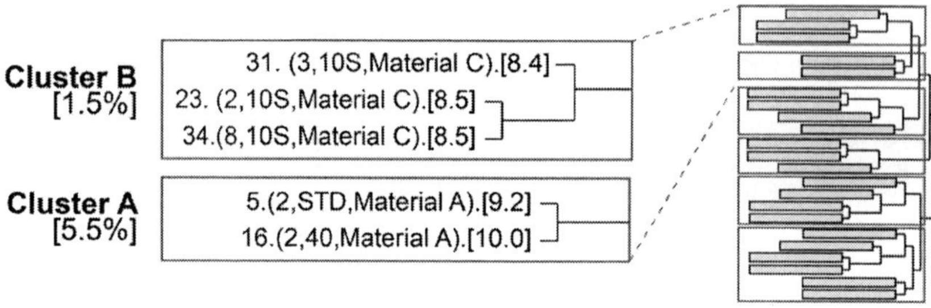

Figure 5. Example of outcomes of product complexity analysis.

5.3.3 *Rework cost estimation and control*

A novel functionality for supporting quality-induced rework cost estimation and control for construction product prefabrication was also developed using absorbing Markov chains (Ji and AbouRizk 2018b). Two types of decision support metrics are established to support decision-making processes, namely (1) rework cost estimation during the project planning phase and (2) rework cost control during the project execution phase. As shown in Figure 6, a distribution of rework cost estimation is simulated in project planning phase and is utilized to construct a control chart for monitoring purposes. At each time point, the simulated rework cost is updated by incorporating real-time quality and rework cost information. Abnormal patterns can be detected for practitioners to analyze root causes and improve their operation processes.

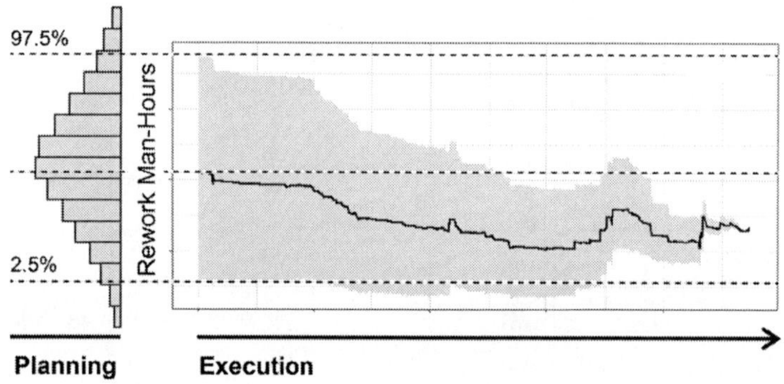

Figure 6. Decision support metrics for quality-induced rework cost estimation and control.

6 CONCLUSIONS

The use of predictive analytics and other advanced data analytics methods capable of extracting value from existing data is becoming increasingly common and employed across all types of industries and sectors (Manyika *et al.* 2011). While the construction industry is advancing the way they collect and store data, methods capable of transforming raw data into valuable and interpretable information for decision-support remains limited (Dean 2014). The simulation-based analytics framework for construction proposed here represents a new approach for decision support that takes advantage of emerging concepts and technologies in data analytics and computing sciences.

While demonstration projects and prototypes developed using a simulation-based analytics approach demonstrate the feasibility of the proposed framework, transformation and refinement of demonstration projects into automated, fully-functional decision support systems will require additional research and development in the areas of distributed simulation, information integration, and dynamic data updating. For such systems to become implemented and universally-accessible in industry, research efforts must also focus on creating systems that are easy-to-use and relevant to decision makers. Collaborations with industry will be essential for integrating analytics into organizational effectiveness and for enhancing sector efficiency and competitiveness.

Acknowledgments

This research is funded by a Collaborative Research and Development Grant (CRDPJ 492657) from the Natural Sciences and Engineering Council of Canada. Section 5 of this paper is adapted from Dr. Wenying Ji's doctoral thesis (Ji 2017). Simaan AbouRizk would like to thank Graham Industrial Services LP, JV Driver Projects Inc., Ledcor Constructors Inc., and PCL Constructors Inc. for their continued support of this project and Dr. Catherine Pretzlaw for her assistance with manuscript composition.

References

AbouRizk S. M. and Hague, S., An Overview of the COSYE Environment for Construction Simulation, in *Proceedings of the 2009 Winter Simulation Conference*, Rossetti, M. D., Hill, R. R., Johansson, B., Dunkin, A., and Ingalls, R. G. (eds.), 2624-2634, IEEE, Piscataway, USA, 2009.

AbouRizk, S. M., Role of Simulation in Construction Engineering and Management, *Journal of Construction Engineering and Management*, ASCE, 136(10), 1140-1153, 2010.

AbouRizk, S. M., Hague, S., Ekyalimpa R., and Newstead, S., Simphony: A Next Generation Simulation Modelling Environment for the Construction Domain, *Journal of Simulation*, Springer, 10(3), 207-215, 2016.

Akhavian, R. and Behzadan, A. H., Construction Equipment Activity Recognition for Simulation Input Modeling Using Mobile Sensors and Machine Learning Classifiers, *Advanced Engineering Informatics*, Elsevier, 29(4), 867-877, 2015.

Badr, Y., Hariri, S., Al-Nashif, Y., and Blasch, E., Resilient and Trustworthy Dynamic Data-Driven Application Systems (DDDAS) Services for Crisis Management Environments, *Procedia Computer Science*, Elsevier, 51, 2623-2627, 2015.

Barton, D. and Court, D., Making Advanced Analytics Work for You, *Harvard Business Review,* Harvard Business Publishing, 90(10), 78-83, October 2012.

Bi, L., Ren, B., Zhong, D., and Hu, L., Real-Time Construction Schedule Analysis of Long-Distance Diversion Tunnels Based on Lithological Predictions using a Markov Process, *Journal of Construction Engineering and Management*, ASCE, 141(2), 4014076, 2015.

Chung, T. H., Mohamed, Y., and AbouRizk, S. M., Simulation Input Updating using Bayesian Techniques, in *Proceedings of the 2004 Winter Simulation Conference*, Ingalls R. G., Rossetti, M. D., Smith, J. S., and Peters, B. A. (eds.), 1238-1243, IEEE, Piscataway, USA, 2004.

Construction Industry Institute, *Leveraging Technology to Improve Construction Productivity (RS240-1)*, Construction Industry Institute, Austin, USA, 2008.

Construction Owners Association of Alberta, *COAA Major Projects Performance Assessment System (Alberta Report #2)*, Construction Owners Association of Alberta, Edmonton, Canada, 2015.

Darema, F., Dynamic Data Driven Applications Systems: A New Paradigm for Application Simulations and Measurements, in *Computational Science – ICCS 2004, Part III*, Bubak, M., van Albada, G. D., Sloot, P. M. A., and Dongarra, J. (eds.), 662-669, Springer-Verlag Berlin Heidelberg, Berlin, Germany, 2004.

Dean, J., *Big Data, Data Mining, and Machine Learning: Value Creation for Business Leaders and Practitioners*, John Wiley & Sons, Inc., Hoboken, USA, 2014.

Dube, P., Gonçalves, J. P. M., Mahatma, S., Barahona, F., Naphade, M., and Bedeman, M., Simulation Based Analytics for Efficient Planning and Management in Multimodal Freight Transportation Industry, in *Proceedings of the 2014 Winter Simulation Conference,* Tolk, A., Diallo, S. Y., Ryzhov, I. O., Yilmaz, L., Buckley, S., and Miller, J. A. (eds.), 1943-1954, IEEE, Piscataway, USA, 2014.

Fan, H., Kim H., AbouRizk, S., and Han, S. H., Decision Support in Construction Equipment Management using a Nonparametric Outlier Mining Algorithm, *Expert Systems with Applications*, Elsevier, 34(3), 1974-1982, 2008.

Fujimoto, R. M., Celik, N., Damgacioglu, H., Hunter, M., Jin, D., Son, Y. J., and Xu, J., Dynamic Data Driven Application Systems for Smart Cities and Urban Infrastructures, in *Proceedings of the 2016*

Winter Simulation Conference, Roeder, T. M. K., Frazier, P. I., Szechtman, R., Zhou, E., Huschka, T., and Chick S. E. (eds.), 1143-1157, IEEE, Piscataway, USA, 2016.

Gelman, A., Carlin, J. B., Stern, H. S., Dunson D. B., Vehtari, A., and Rubin, D. B., *Bayesian Data Analysis, 3rd Ed.*, CRC Press Taylor & Francis Group, Boca Raton, USA, 2013.

Hammad, A. M., *An Integrated Framework for Managing Labour Resources Data in Industrial Construction Projects: A Knowledge Discovery in Data (KSS) Approach*, Doctoral Thesis, University of Alberta, Edmonton, Canada, 2009.

Han, P. A., Siu, M. F. F., AbouRizk, S., Hu, D., and Hermann, U., 3D Model-Based Quantity Take-Off for Construction Estimates, in *Computing in Civil Engineering 2017*, Lin, K. Y., El-Gohary, N., and Tang, P. (eds.), 118-124, ASCE, Reston, USA, 2017.

Hanna A. S., Benchmark Performance Metrics for Integrated Project Delivery, *Journal of Construction Engineering and Management*, ASCE, 142(9), 04016040, 2016.

Inmon, W. H., *Building the Data Warehouse, 4th Ed.*, John Wiley & Sons, Inc., Hoboken, USA, 2005.

Ji, W. and AbouRizk, S., A Bayesian Inference Based Simulation Approach for Estimating Fraction Nonconforming of Pipe Spool Welding Processes, in *Proceedings of the 2016 Winter Simulation Conference*, Roeder, T. M. K., Frazier, P. I., Szechtman, R., Zhou, E., Huschka, T., and Chick S. E. (eds.), 2935-2946, IEEE, Piscataway, USA, 2016.

Ji, W. and AbouRizk, S., Credible interval estimation for fraction nonconforming: analytical and numerical solutions, *Automation in Construction*, Elsevier, 83, 56-67, 2017.

Ji, W., Simulation-Based Analytics for Fabrication Quality-Associated Decision Support, Doctoral Thesis, University of Alberta, Edmonton, Canada, 2017.

Ji, W., AbouRizk, S., Zaïane, O., and Li, Y., Complexity Analysis Approach for Prefabricated Construction Products using Uncertain Data Clustering, *Journal of Construction Engineering and Management*, ASCE [in press], 2018.

Ji, W. and AbouRizk, S., Simulation-Based Analytics for Quality Control Decision Support: Pipe Welding Case Study, *Journal of Computing in Civil Engineering,* ASCE, 32(3), 05018002, 2018a.

Ji, W. and AbouRizk, S., Data-Driven Simulation Model for Quality-Induced Rework Cost Estimation and Control using Absorbing Markov Chain, *Journal of Construction Engineering and Management*, ASCE [in press], 2018b.

Kück, M., Ehm, J., Hildebrandt, T., Freitag, M., and Frazzon, E. M., Potential of Data-Driven Simulation-Based Optimization for Adaptive Scheduling and Control of Dynamic Manufacturing Systems, in *Proceedings of the 2016 Winter Simulation Conference*, Roeder, T. M. K., Frazier, P. I., Szechtman, R., Zhou, E., Huschka, T., and Chick S. E. (eds.), 2820-2831, IEEE, Piscataway, USA, 2016.

Lin, J. J., Han, K. K., and Golparvar-Fard, M., A Framework for Model-Driven Acquisition and Analytics of Visual Data using UAVs for Automated Construction Progress Monitoring, in *Computing in Civil Engineering 2015*, O'Brien, W. J. and Ponticelli, S. (eds.), ASCE, Reston, USA, 156-164, 2015.

Lynch, S. M., *Introduction to Applied Bayesian Statistics and Estimation for Social Scientists*, Springer-Verlag New York, New York, USA, 2007.

MacKay D. J. C., Ensemble Learning for Hidden Markov Models, 1997. Retrieved from www.citeseerx.ist.psu.edu/viewdoc/download?doi=10.1.1.52.9627&rep=rep1&type=pdf on March 20, 2018.

Manyika, J., Chui, M., Brown, B., Bughin, J., Dobbs, R., Roxburgh, C., and Hung Byers, A., *Big Data: The Next Frontier for Innovation, Competition, and Productivity*, McKinsey Global Institute, New York, USA, 2011.

Martens, O., *Automated Tool of Visual Progress Monitoring in Construction*, Master of Science Thesis, University of Alberta, Edmonton, Canada, 2017.

Oracle, *Predictive Analytics: Bringing the Tools to the Data*, Oracle Corporation, Redwood Shores, USA, 2010.

Pritchard, R. D., Weaver, S. J., and Ashwood, E., *Evidenced-Based Productivity Improvement: A Practical Guide to the Productivity Measurement and Enhancement System (ProMES)*, Routledge Taylor & Francis Group, New York, USA, 2012.

R Core Team, R: A Language and Environment for Statistical Computing, 2017. Retrieved from www.R-project.org on January 19, 2018.

Rabiner, L. R., A Tutorial on Hidden Markov Models and Selected Applications in Speech Recognition, in *Proceedings of the IEEE*, IEEE, 77(2), 257-286, 1989.

Sunderrajan A., Viswanathan, V., and Cai, W., Data Driven Adaptive Traffic Simulation of an Expressway, in *Proceedings of the 2016 Winter Simulation Conference*, Roeder, T. M. K., Frazier, P. I., Szechtman, R., Zhou, E., Huschka, T., and Chick S. E. (eds.), 1194-1205, IEEE, Piscataway, USA, 2016.

Responsible Design and Delivery of the Constructed Project
Edited by Abdul-Malak, M., Khoury, H., Singh, A., and Yazdani, S.

REFORMING UNDERGRADUATE STRUCTURAL ENGINEERING EDUCATION: A LEAP OF FAITH

EZZELDIN SAYED-AHMED

Dept of Construction Engineering, American University in Cairo, Cairo, Egypt

Do we need to reform civil engineering education? The notable remark about any listing of top universities is that more than 60% of these universities are located in the G8 countries with GDP of more than 60% of the world total. This fact confirms that economic growth is strictly tied to social and cultural development, which is linked to the intellectual and scientific development presented by higher education. This also reveals that the answer to the concern of reforming engineering education is YES. Thus, teaching paradigms must change to fit needs and modes of today's students; they are not interested in passive learning. Active learning must be the path for today's generation of ubiquitous-learning. Different innovative teaching techniques for high-level structural engineering courses have been adopted by the author and proved to be rewarding. Among these paradigms are blended learning and flipped classes, which stretch the utilization of technology in classes. Other new techniques such as student-generated examinations, student-generated classes, design competitions and project-based learning are reflection of student-centered education. Here, the said innovative paradigms, which were recently applied by the authors will be presented.

Keywords: Active learning, Blended learning, Flipped classroom, Teaching pedagogy.

1 INTRODUCTION

The end product of higher education is the graduates; a mirror reflecting the level of success in achieving any program's objectives. Recently a serious and continuous decline in graduates' level of knowledge and skills is noticed; a strong evidence that something is going wrong within the process. Consequently, higher education components (students, faculty, facilities, curricula and teaching/learning paradigms) must be carefully inspected. In Egypt, scrutinizing the students reveals a shocking conclusion: there is no problem with the quality of in-take engineering students. We had, and still have the best students even compared to students worldwide. The second pillar of higher education is faculty. Current faculty rewards are way below the level, which can safely support their social needs and thus, they are involved in other non-academic activities to raise their income. Every day passes without resolving this issue turns this secondary non-academic activity to be the faculty's primary job. Naturally, it takes a valuable share of the faculty's time and the ones who end up paying the penalty are the students. We also have a clear shortage in our facilities; the third pillar. Almost all our facilities starting from our classrooms, laboratories, service areas and even faculty offices need a tremendous upgrade.

Talking about engineering curricula, we always get the famous comment saying: "*we had the best engineering education; when we continued our post-graduate study abroad we high-ranked our colleagues from other universities*". The first part of this statement is true if we modified to: "*we had the best engineering education; then*". At the time of our education, both the curricula and teaching paradigms served the need of this era. However, this does not mean that we must fix both for decades; whatever fitted us does not fit today's students. We must continuously update and improve the curriculum and more important the teaching/learning techniques. We should accept the fact that what we have learned, mastered and excelled in for some time, will always be surpassed by something newer and better; we should not only be willing but eager to accept that it is time to replace what we know with what works better. As such, this article will mainly focus on teaching/learning paradigms, which are also related to faculty to some extent.

If we always think that "*the true purpose of education is to stimulate creativity and passion of the students and the students' body*", we will definitely come to a conclusion that whatever we are doing now should be reformed: it is a leap of faith.

2 OBJECTIVES-ORIENTED CURRICULUM AND OUTCOMES-BASED COURSES

Each engineering program must have its own set of educational objectives, which serve the university and school mission and vision, which in turn serve the overall national higher education mission and vision. Furthermore, the program's educational objectives should reflect the needs of the different program's constituents: students, faculty, and hiring agencies.

Each course within the curriculum should be designed to introduce a certain set of outcomes. When combined, the matrix of the courses' outcomes should form the overall targeted program's outcomes, or attributes of program's graduates (knowledge, skills, values) reflecting the degree to which the program has met its educational objectives. It is worth noting here that curriculum design implementation and evaluation were subjects for many research and discussions (e.g., Walkington 2002, Yeomans and Atrens 2001, Yokomoto and Bostwick 1999). The program's objectives and outcomes significantly contribute toward the curriculum planning, and renewal: they give the ability to change or modify it based on measurable indicators not anecdotes.

Currently, we lack a clear definition of the educational objectives of our structural engineering programs, which is known and familiar to faculty. The only objective of each faculty is the transfer of his/her knowledge within his/her inherited course description to his/her students. Every faculty does this on his/her own with his/her personally defined teaching technique and strategy. In addition, curriculum development is based on an ad-hoc announcement, which came down to faculty individually and everyone/department division has their input which is then collected into a "*new*" curriculum without following any strategy/procedures. To generate an effective objective-oriented curriculum and outcome-based courses, ABET proposed the Continuous Quality Improvement procedures (Besterfield-Sacre *et al.* 2000, Naji *et al.* 2003). Another important aspect is the contribution of all the program's constituencies in defining and assessing the program's objectives and the courses' outcomes. Currently, engineering programs strive to achieve quality. One way of doing this is to obtain national/international accreditation. However, it is imperative that the quality/accreditation procedures become performance-based and not process-based: in the latter, everyone is working to document the quality procedures while in the former everyone is keen to reach into the heart and soul of quality itself.

3 TEACHING/LEARNING PARADIGMS

A brief description of programs' objectives and outcomes were discussed above. However, we must keep in mind that courses of the best-designed curriculum can be delivered in a spontaneous manner with no acquaintance between its courses. Only faculty and department collective body may weave the curriculum's courses to form a solid engineering program. The relationship between the courses presents one significant issue; another vital issue is how each course is delivered. Effective course offering is the only way to benefit the students from a course as opposed to exposing them to a number of sporadic topics, indubitably forgotten as soon as they complete the final exam. Currently, there are pressures on engineering educators everywhere to teach effectively and on the students to learn efficiently in order to meet today's challenges.

It is very interesting to note that most engineering university professors were not taught anything about teaching: they just follow the footsteps of their professors. Although some are gifted teachers, the majority gains it by trial and error over the years. Today, a lot is known about effective teaching via cumulative professors' experience (Kauchak and Eggen 2007, Van Rensburg 2003, Wankat 2002). Three aspects control effective teaching and course offering:

1. Learning styles: different ways our students take in and process information.
2. Course planning: the course learning objectives definition course outcomes' assessment.
3. Instructional approaches: active, cooperative, and problem-based learning techniques.

The first two items were discussed elsewhere (Sayed-Ahmed 2009). In the following sections, different innovative teaching/learning techniques for structural engineering courses adopted by the author and proved to be rewarding will be presented. Among these paradigms are blended learning and flipped classrooms, which stretch the utilization of technology in classes. Other new techniques such as student-generated examinations, student-generated classes, design competitions and project-based learning are direct reflections of student-centered education.

3.1 Flipped Classroom

I have applied this technique to some topics in one of the advanced reinforced concrete design courses (CENG 4158: Structural Systems and Advanced Design) in multiple semesters.

In this paradigm, lectures are uploaded online in form of mini-videos (less than 12 minutes each) along with the common lecture notes in PDF format. Online assessments are added after each part in order to enhance the students' perception of the online materials. Students are asked to watch the mini-videos, read the notes and go through the online assessment prior to the class. Furthermore, an activity is assigned to the students and they were asked to work on it and bring it to the class. During the normal face-to-face (F2F) class, the time is used to perform advanced practical application related to the online materials: active/collaborative learning is adopted based on the knowledge, which the students gained prior the class. Group work and discussions between the students actually engage almost all the students in an enthusiastic environment. In other words, class time is used for concept engagement and collaboration with peers in higher order practical applications under my direct supervision and guidance as a mentor. Figure 1 compares F2F class time for both traditional and flipped classroom paradigms.

Traditional Class Time	
Activity	Time
Warm-up & previous info	10 min
Lecture new content	45-50 min
Independent activity	10-15 min

Flipped Class Time	
Activity	Time
Warm-up activity and previous info	5 min
Q&A time on mini-videos	10-15 min
Guided teamwork practice activity	50-55 min

Figure 1. Comparison between class time in flipped and normal classrooms.

3.2 Blended Learning

Recently, I have designed and offered a blended learning course (CENG 592/5292: Advanced Topics in Construction Engineering: Design and Construction of Steel Structures). AUC defines blended learning as a combination of F2F and online instruction, which entails replacing 30% of the class time with online instruction. This BL course used a model of a sequential F2F/online.

BL proved to be very effective particularly for such graduate level course where students may have different levels of entry skills and knowledge. It provides them with self-paced coverage of materials and allows them to perform self-assessment. F2F class time is saved for discussion of content, and practical application after bringing the student to the same level of knowledge.

The online part of the course was designed to include mini-videos, lecture notes, e-assessment, active e-discussions, etc. The main purpose of the online part is to bring graduate students with different backgrounds to the same level before each face-to-face (F2F) class, which includes normal lecturing, students' presentations, assignments, mini-projects and case studies.

3.3 Student-Generated Examinations

As John Dewey said, "*if we teach today's students as we taught yesterday's, we rob them of tomorrow.*" Hence, the idea of starting student-generated examinations. A mid-term exam of one of my courses (CENG 4113: Structural Mechanics) has now regularly become a student-generated examination. Each student (or groups of two students) submit a mid-term exam on pre-specified topics. Each exam is accompanied by a model answer and a mark scheme. This mark scheme defines both the grade distribution and all possible alternatives for the solutions and mistakes. In class, exams are randomized and distributed to students; each student solves one of his/her colleague's exam. The student's grading of this mid-term exam is distributed as 50% on preparing the exam, its model answer and mark scheme and 50% on solving other student's exam. Recently, I have added a 5% grade on the difficulty level of the exam, which I assigned on each of the submitted and solved exams. The assigned time for the exam is 60 minutes with no limit on the number or distribution of exam's questions. It is, of course, an open book exam.

I have superb outcomes out of this pedagogy such as

- The range of items covered by students' questions is astonishing.
- The exam defines part(s), which students see as "the most important content" and other over-looked items in my classes.
- Most students study beyond the content and the lecture notes in order to be able not only to assign the question but also to prepare its mark scheme.
- The exam also reveals to me the students' perception of "fair" and "useful" questions and how well they answer the questions they created.

3.4 Student-Generated Classes

At the last three weeks of the semester, I leave topics of one of my courses (CENG 4113: Structural Mechanics) for the students to generate classes; these topics are introduction to fracture mechanics, beams/rafts on elastic foundation and theory of plates and shells.

Students are arranged in groups of two-to-three students each and specify the topic for each group. Then, I ask them to prepare one (or two) full class(es) on their topic in any form of lecturing and, in addition, they must present practical applications (in forms of solved examples, or any other form) and give their colleague an assignment on this topic to solve.

I have recorded multiple outcomes of this teaching/learning pedagogy. Students really appreciate the need for life-long learning, which is indeed my primary teaching goal. Each group performs a research-like activity and presents their findings to their colleagues; they also feel responsible for getting these findings to their colleagues during these lectures.

3.5 Project Based Learning (PBL)

PBL is a learner-centered teaching practice, where students collaboratively engage in solving complex and challenging real-world problems. The motivation to solve a problem drives them to learn the required/intended content. The problem actually becomes a vehicle for stimulating the cognitive process and developing problem-solving skills. I have implemented PBL in one of my classes (CENG 4911: Selected Topics in Construction Engineering – Design of Steel Structures) in form of a case study on a collapsed steel building. This case study presented a real engineering problem related to a building which partially failed during construction, and thus, needed rehabilitation and strengthening. The LMS discussion board was adopted to post (and discuss) alternatives for the reason(s) creating this engineering problem and for alternative solution(s) to rehabilitate/strengthen the structure presented in the case study. The first part (cause of failure) was performed as a "guided" discussion. Based on the first topic outcomes, the second topic (repair) was opened for discussion. Figure 2 shows the problem statement. It was accompanied by a PP presentation, drawings, photos and a video.

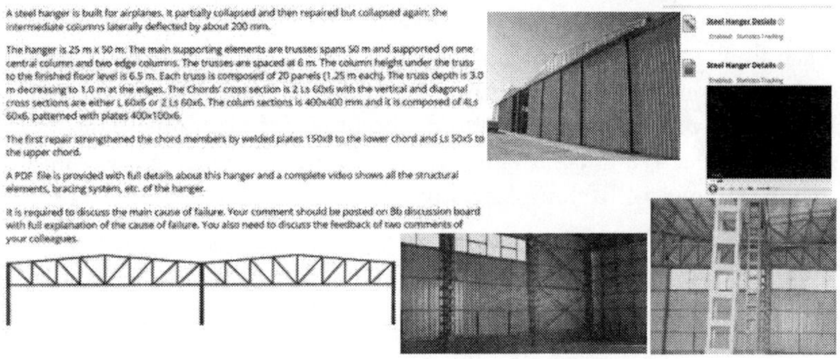

Figure 2. Problem statement and details adopted for BPL.

Students were so enthusiastic to analyze the structure in order to find its cause of failure by using all what they have learned in this course and others. Then, once we agreed on the main cause of failure, they endeavored to find the best method of repair in a very active atmosphere.

3.6 Course Competitions

I have generated two design competitions in some of my design courses. The first one is building and testing columns using Spaghetti and while the second includes building, analyzing and testing a bridge model created using a set of tongue pressers. The students compete to reach the best design where the intended learning objectives of these competitions are; motivating students to learn design concepts using non-traditional materials, enhancing the teamwork capabilities of the students, adopting design/build techniques, enhancing the students' engagement in my class. I follow a simple methodology where students' groups are formed and materials are distributed with a definition of design constraints. In the design phase, students choose an appropriate design and construct their models. For the bridge competition, students create a numerical simulation for their models in order to predict the failure load. This stage needs material testing as well in order to estimate the mechanical properties of the tongue pressers. The last stage is testing the model in order to experimentally predict the failure load. Ratios between this failure load to the predicted one and the failure load to the model weight were established. The winning team (100% of the grade) is the one achieving the first ratio closest to 1.0 and maximum value for the second ratio.

Figure 3. Design competitions: spaghetti columns and tongue presser bridges.

4 SUMMARY

Our teaching/learning paradigm in structural engineering courses must be updated to fit the needs and the challenges facing today's students, and more important, to get them engaged and interested to what they are learning. This article presents new paradigms such as flipped classrooms, blended learning, student-generated examinations and classes, project-based learning and design competitions which were applied to different senior or graduate design courses. They are revealed to be very effective in achieving the intended learning outcomes and in emphasizing to the students the importance of lifelong learning.

References

Besterfield-Sacre, M., Shuman, L. J., Wolfe, H., Atman, C. J., McGourty, J., Miller, R. L., Olds, B. M., and Roger, G. M., Defining the Outcomes: A Framework for EC 2000, *IEEE Transactions on Education*, 43(2), 100-110, 2000.

Kauchak, D. and Eggen, P., *Learning and Teaching: Research-based Methods*, Pearson, New York: Allyn Bacon, 2007.

Naji, K. K., Sayed-Ahmed, E. Y., and Elnawawy, O. A., 2003, Preparing the Civil Engineering Program at The Qatar University for Accreditation Review Under ABET EC2000. *3rd Forum Engineering Education: Building Partnership with Government, Industry and Society*, Sharjah, UAE, 2003.

Sayed-Ahmed, E. Y., Reforming Undergraduate Engineering Education: A Leap of Fate, *13th International Colloquium on Structural and Geotechnical Engineering*, Cairo, Egypt, 2009.

Van Rensburg, B. W. J., Structural Engineering Education: Effective Teaching and Learning, *ASEE/WFEO International Colloquium*, ASEE, USA, 2003.

Walkington, J., Curriculum Change in Engineering, *European Journal of Engineering Education,* 27(2), 133-148, 2002.

Wankat, P. C., *The Effective, Efficient Professor: Teaching, Scholarship and Service*, Pearson, Boston: Allyn and Bacon, 2002.

Yeomans, S. R., and Atrens, A., 2001. A Methodology for Discipline-Specific Curriculum Development, *International Journal of Engineering Education*, 17(6), 518-524, 2001.

Yokomoto, C. F., and Bostwick, W. D., Modelling the Process of Writing Measureable Outcomes for EC2000, *Proceedings Frontier in Education Conference*, 2,11b1, 18-22, 1999.

Responsible Design and Delivery of the Constructed Project
Edited by Abdul-Malak, M., Khoury, H., Singh, A., and Yazdani, S.

IMPROVING CONSTRUCTION EDUCATION: THE ROLE OF COLLABORATIONS BETWEEN HIGHER INSTITUTIONS AND THE CONSTRUCTION INDUSTRY

JOHN ALIU and CLINTON AIGBAVBOA

Sustainable Human Settlement and Construction Research Centre, University of Johannesburg, Johannesburg, South Africa

Higher Education Institutions (HEIs) are developmental hubs that contributes to a country's economic growth through knowledge generation, dissemination and skills diffusion among students. For HEIs to maintain their roles in fostering construction education to equip students for the industry, the current traditional pedagogical approaches are simply not enough to deliver on its mandate. One of the many ways through which HEIs can improve the quality of construction education today is through collaborations with the construction industry. Through a structured questionnaire survey, this paper identified the possible roles that collaboration between HEIs and the construction industry will play in fostering construction education. The sample for this study consisted of one hundred and twenty-six (126) respondents drawn from professionals in the Nigerian construction industry. The study revealed that exposing construction students to the world of work, strengthening universities in conducting quality and relevant research and creating and training students with key skills industry require, are among the benefits of collaboration between the HEIs and the construction industry. The study found that there is an increased need for HEIs to collaborate with the construction industry to be progressive, relevant and produce skilled graduates who will spur the activities of the construction industry. This study recommends that HEIs seek collaboration with the industry so as to address the shortcomings of higher education in Nigeria and Africa in general.

Keywords: Higher education institutions, Higher education, Survey design, Mean item score, Construction professionals, Nigeria.

1 INTRODUCTION

The concept of collaborations between higher education and the construction industry has existed for quite a long time, principally in technology-driven regions. Ramakrishnan and Yasin (2011) state that collaborations were first made known in 1903 by the Sunderland Technical Knowledge in Northern England and was earlier known as the Sandwich Programme. Since then, collaboration all over the world has gained ground and has become one of the vital agencies of policy making in HEIs in a bid to sustain the future of the construction industry. In fostering construction education as well as strengthening graduate skills, collaboration plays a key role and its importance cannot be underestimated (Turhan and Akman 2013). Feng *et al.* (2011) state that these collaborations play a critical role in the development of any economy. Nowadays, driven by the influence of evolving technology, it is imperative for HEIs to establish a cordial relationship with the construction industry in a bid to exchange ideas as well as create developmental strategies. Esham (2008) asserts that the construction industry looks forward to flourishing collaborations with HEIs to look out for adequately skilled students as they plan. This working relationship provides a constant flow of information and innovative ideas between both parties and its integration into the HEIs' curricula helps to meet the needs of the construction industry (Ishengoma and Vaaland 2016).

Advanced nations of the world today have also benefited immensely from the collaboration between HEIs and the construction industry. The linkages have been key in research and development coordination (R&D) agendas, and enhancement of relations between private and public sectors, amongst others (Guimón 2013). Its benefits are also recognizable in developing countries. According to Marotta *et al.* (2007), collaborations between HEIs and industry increases the tendency of industry establishments to create and introduce new innovations and inventions which in terms of patents can expand their exploration base. This collaboration stems from the fact that most industry firms constantly require skilled graduates because, the role of HEIs in technology impartation and diffusion makes it an important driver in economic development. Hence, it can be stated in this context that effective collaborations can offer solutions to the rapid increase of unemployment in developing countries (Guimón 2013).

According to Ramakrishnan and Yasin (2011), the results of collaborations can encourage more benefits which include generating new knowledge where industry professionals share their innovations and experiences by conducting lectures and seminars at HEIs; providing educational resources as students benefit from the hands-on and practical experience they are exposed to; providing students with up-to-date information on the industry as industry professionals can act as consultants to students; providing career planning tips to students which is key in defining their futures and providing students with the needs and requirements for industry employment (Ramakrishnan & Yasin 2011). From these benefits, it can be construed that collaborations can occur in various forms which include research and development, teaching and curriculum enhancement, as well as consultancy.

Furthermore, with the ever-growing demand for graduates to be equipped with more than just an academic degree, HEI-industry collaboration is fundamental. These collaborations help in the development of certain skills among graduates such as soft skills which include communications, critical thinking, leadership, time management skills, teamwork, interpersonal skills, management skills, and problem-solving skills (Aliu and Aigbavboa 2016). Hence, this study is significant because construction professionals of today are saddled with the responsibility of developing the built environment which improves the quality of life for any society. From the foregoing, the roles of collaboration between HEIs and the construction industry bodes well for the improvement of construction education and it is against this backdrop that this paper examines the benefits from a Nigerian perspective.

2 RESEARCH METHODOLOGY

This study adopted a descriptive survey design. This design was deemed appropriate because it was effective in providing both quantitative and numeric description of the respondents on the issue of interest in the study. For this current study, the target population was construction professionals in the Nigerian construction industry, namely architects, builders, engineers, quantity surveyors, estate surveyors and valuers, land surveyors, and town planners. The selection of these construction professionals was made because they were likely to have a considerable amount of knowledge in contributing to the objectives of this study. This study adopted the random sampling technique and it was preferred to the cluster sampling, stratified sampling or the sampling using multiple probability techniques. Random sampling gives all the participants an equal chance to be selected for the study with the same criteria which prompted this study to adopt this technique. The criterion for participation in this study was that each participant had to be a construction professional in the Nigerian construction industry. Most times, this method is used when the target population presents the same characteristics, or the sampling size is too large to represent the entire population efficiently and each member of the entire population has an equal chance of being selected as a sampling respondent. Hence, a total of 126 respondents took part in this study. The main instrument of data collection was a structured questionnaire which was designed by the researcher following a review of extant literatures. The instrument was validated by handing it to experts in the construction industry before using it for the study. In addition, focused group discussions were held with key stakeholders in HEIs including construction lecturers/educators and curricula planners. The data collected was analysed using the descriptive statistics. Also, closed-ended questions were used for the purposes of this study. This is because close-ended questions provided participants with a multiple of options to choose from without allowing them to put their opinions in their own words. The main advantage of using close-ended questions is the simplicity for data collection and analysis, thus less time consuming.

2.1 Mean Item Score

For the purpose of this study, mean item score (MIS) was used to analyse the data. MIS indicate the average level of agreement with an item. The Likert scales were transformed to a Mean Item Score (MIS) for each of the research objectives as applicable. The indices were then used to determine the rank of each item. These rankings made it possible to cross compare the relative importance of the items as perceived by the respondents. Following the mathematical computations, the criteria were then ranked in descending order of their relative importance.

3 FINDINGS AND DISCUSSIONS

3.1 Background Information of Participants

Respondents were requested to indicate the degree of importance of each of the roles of collaboration between the HEIs and the construction industry in upscaling construction education based on a five-point Likert scale (strongly disagree = 1, disagree = 2, neutral = 3, agree = 4, strongly agree = 5). One hundred and twenty-six complete questionnaires were received signifying an 84% response rate. Findings from the 126 usable questionnaires revealed that a total of 98 males took part in the study which represents 77.8 percent of the total population. A total of 22 females took part in the study which represents a total of 22.2 percent.

Also, 3.2 percent of the respondents were in the age group of 21-25 years old, 10.3 percent of the respondents were in the age group 26-30 years, 13.5 percent were in the age group 31-35 years, 17.5 percent were in the age group 36-40 years, 23.0 percent of the respondents were in the age group 41-45 years, 11.1 percent were in the age group of 46-50 years, 11.9 percent were in the age group of 51-55 years and 9.5 percent of the respondents were above 56 years old.

It revealed that 0.8 percent of the total respondents had less than a year's work experience and 4 percent had less than two years of work experience. It also revealed that 23.8 percent had experience that ranged from three to five years, 27 percent had experience in the range of six to ten years, 12.7 percent had experience that ranged from 11 to 15 years, 17.5 percent had experience that ranged from 16 to 20 years, 12.7 percent had experience that ranged from 21 to 25 years and 1.6 percent of the respondents had more than 25 years of industry experience. The years of experience of respondents were sufficient to provide useful responses to achieve the purpose of the study as 95.3 percent of the respondents for this study had over three years of work experience in the construction industry.

3.2 Mean Item Score for Ranking of Roles of Collaboration between HEIs and the Construction Industry

The mean ranking of each attribute was presented to provide a clearer picture of the agreement reached by the respondents. A summary of the test result is shown in the table below. The mean for each variable included the standard deviation.

Table 1. Results of mean item score.

ROLES OF COLLABORATION	$\overline{x}$	σX	R
Exposing construction students to the world of work	4.66	0.635	1
Strengthening universities in conducting quality and relevant research	4.49	0.629	2
Creating and training students with the key skills industry require	4.48	0.629	3
Providing equipment and services to facilitate learning for students	4.48	0.641	3
Providing students access to updated technical information	4.43	0.650	4
Encouraging construction site visits and field trips for students	4.43	0.686	4
Enhancing job opportunities for graduates	4.42	0.663	5
Creating an entrepreneurship culture among graduates	4.41	0.719	6
Publicizing university activities relevant to industry	4.38	0.725	7
Setting-up of schemes to enable students to attend industry events	4.35	0.696	8
Incorporating open-days in universities to further sensitize future graduates	4.32	0.734	9
Visiting universities to organize workshop on products by industry players	4.30	0.648	10
Integrating some industry-specific courses into the university curriculum	4.29	0.633	11
Incorporating industry/academic mentoring plan for students	4.27	0.686	12
Establishing student chapters of professional institutions	4.26	0.683	13
Improving the overall university curriculum	4.23	0.609	14
Introducing construction project exercises to classrooms	4.17	0.538	15
Involving industry in university curricula design, planning and evaluation	4.14	0.576	16

$\overline{x}$ = Mean item score; σX = Standard deviation; R = Rank

Table 1 reveals the respondents' rankings of the possible benefits of collaborations between HEIs and the industry in improving construction education. The table shows that, with a mean score (M) of 4.66 and standard deviation of (SD) = 0.635, 'exposing construction students to the world of work' was ranked the most important role of collaborations between HEI and the

Nigerian construction industry. ‘Strengthening universities in conducting quality and relevant research’ was ranked second with (M = 4.49; SD = 0.629); ‘creating and training students with the key skills industry require’ and ‘providing equipment and services to facilitate learning for students’ were ranked third with (M = 4.48; SD = 0.629) and (M = 4.48; SD = 0.641) respectively. ‘Providing students access to updated technical information’ and ‘encouraging construction site visits and field trips for students’ were both ranked fourth with (M = 4.43; SD = 0.650) and (M = 4.43; SD = 0.686) respectively; ‘enhancing job opportunities for graduates’ was ranked fifth with (M = 4.42; SD = 0.663); ‘creating an entrepreneurship culture among graduates’ was ranked sixth with (M = 4.41; SD = 0.719); ‘publicizing university activities relevant to industry’ was ranked seventh with (M = 4.38; SD = 0.725); ‘setting up of schemes to enable students to attend industry events’ was ranked eighth with (M = 4.35; SD = 0.696); ‘incorporating open-days in universities to further sensitize future graduates’ was ranked ninth with (M = 4.32; SD = 0.734); ‘visiting universities to organize workshop on products by industry players’ was ranked tenth with (M = 4.30; SD = 0.648); ‘integrating some industry-specific courses into the university curriculum’ was ranked eleventh with (M = 4.29; SD = 0.633) and ‘incorporating industry/academic mentoring plan for students’ was ranked twelfth with (M = 4.27; SD = 0.686). Furthermore, the table revealed that ‘establishing student chapters of professional institutions’ was ranked thirteenth (M = 4.26; SD = 0.683); ‘improving the overall university curriculum’ was ranked fourteenth with (M = 4.23; SD = 0.609); ‘introducing construction project exercises to classrooms’ was ranked fifteenth with (M = 4.17; SD = 0.538) and ‘involving industry in university curricula design, planning and evaluation’ was ranked the least with (M = 4.14; SD = 0.576).

4 IMPLICATION OF FINDINGS

From the empirical study, it was revealed that collaborations between the HEIs and the construction industry will enable students to attend industry events, thereby exposing them to the world of work. The ever-growing demand for graduates to be equipped with more than just an academic degree makes collaboration between HEI and the construction industry significant. Through collaborations, the integration of some industry-specific courses into the university curriculum is possible and aimed at preparing students for the world of work. Furthermore, findings also reveal that collaborations should be encouraged as they lead to the generation of new knowledge in cases where industry professionals are willing to share their innovations and experiences by conducting lectures and seminars at HEIs. These lectures and seminars help develop non-academic skills among students such as communication skills, leadership skills, teamwork, interpersonal skills and problem-solving skills.

5 LESSONS LEARNED AND CONCLUSIONS

HEIs act as an important driver of economic development, considering its roles in education and technology absorption, adaptation, and diffusion. It has also become a reliable source of knowledge for innovation which has stimulated the interests of the construction industry as firms and companies seek ties with HEIs to tap into the knowledge generated by academic research. In improving construction education, one of the effective approaches widely practiced is the collaborations between the HEIs and the construction industry. Various findings from literature describe it as a powerful tool in strengthening the ability of HEIs in conducting relevant and quality research which improves graduate employability for the construction industry. It also provides construction students with industry exposure and experience which acts as a foundation for their future professional careers. It is through collaborations that HEIs have developed into catalysts for economic growth as their roles in improving construction education for the future has become more significant. Future research can ensure the selection of a larger sample that cuts across more states and involves more construction industry employers in Nigeria. This will provide a better insight into the perceptions of HEI-industry collaboration. This will further allow for the generalisation of results to the greater population.

Acknowledgments

This research was supported by grants received from the Global Excellence and Stature (GES) Postgraduate Scholarships and supported by the Faculty of Engineering and the Built Environment, University of Johannesburg, South Africa.

References

Aliu, J., and Aigbavboa, C., Equipping 21st Century Construction Graduates: A Review of Key Skills in Fostering Infrastructural Development in Africa: *Proceedings of the 3rd International Conference on Infrastructure Development and Investment Strategies for Africa*, 164-179, Livingstone, Zambia, 2016, retrieved from http://diiconference.org/wp-content/uploads/2016/10/DII-2016-PROCEEDINGS-1.pdf on December 2017.

Esham, M., Strategies to Develop University-Industry Linkages in Sri Lanka. *National Education Commission Sri Lanka Study Series No.4 (2007),* 2008.

Feng, C., Ding, M., and Sun, B., A., Comparison Research on Industry-University-Research Strategic Alliances in Countries, *Asian Social Science,* 7(1), 102-105, 2011.

Guimón, J., Promoting University-Industry Collaboration in Developing Countries, Policy Brief, *The Innovation Policy Platform,* 1(3), 1-12, 2013.

Ishengoma, E., and Vaaland, T. I., Can University-Industry Linkages Stimulate Student Employability? *Education Training,* 58(1), 18-44, 2016. Retrieved from http://dx.doi.org/10.1108/ET-11-2014-0137 on December 2017.

Marotta, D., Mark, M., Blom, A., and Thorn, K., *Human Capital and University-Industry Linkages' Role in Fostering Firm Innovation: An Empirical Study of Chile and Colombia,* 2007. Retrieved from https: openknowledge.worldbank.org/bitstream/handle/10986/7558 on December 2017.

Ramakrishnan, K., and Yasin, N. M., Higher Learning Institution — Industry Collaboration: A Necessity to Improve Teaching and Learning Process, *6th International Conference on Computer Science & Education (ICCSE),* 1445-1449, 2011.

Turhan, C., and Akman, I., Employability of IT Graduates from The Industry's Perspective: A Case Study in Turkey. *Asia Pacific Education Review,* 14(4), 523-536. 10.1007/s12564013-9278-5, 2013.

Responsible Design and Delivery of the Constructed Project
Edited by Abdul-Malak, M., Khoury, H., Singh, A., and Yazdani, S.

MANAGING DESIGN PROJECTS: METHODS AND POSSIBLE IMPROVEMENTS USING BIM

HISHAM ABOU-IBRAHIM and FAROOK HAMZEH

Dept of Civil and Environmental Engineering, American University of Beirut, Beirut, Lebanon

The planning and monitoring of the design phase is crucial to ensure the aimed time, cost, and quality performances of the project are met. In this context, monitoring the progress of design is tightly related to the way design is being planned. In other words, the measures taken into account while planning the development of design; whether time, cost or quality measures, are themselves used to monitor the progress of design throughout the project. In this regard, this study sheds light on the methods used to manage design projects, highlights their shortcomings, and suggests new ways to properly manage the progress of design using BIM as a data repository of design information and management decisions. The study reviews current methods used to plan and monitor design projects and suggests improvements in needed areas. Results show that the reviewed methods differ in the way they approach design management and they do not fully represent all design characteristics. Nonetheless, measures used in these methods can mislead managers in detecting the actual progress of design. In this regard, this study suggests a comprehensive framework that builds on existing methods to enhance the management of design and to better control its development.

Keywords: Design management, Planning, Control, LOD, DSM, PPC, Man-hours.

1 INTRODUCTION

Managing the design phase of construction projects can be seen as the most challenging form of management in the Architecture, Engineering and Construction (AEC) industry. While the management of construction can be approached from a linear production perspective, the management of design deals with managing iterative and creative ideas in a social and multidisciplinary environment. Because the difference between the two processes is structural, the use of construction-based measures has failed to properly address design characteristics. Thus, the development of properly tailored design measures is crucial to enhance the management of design generation and product development.

Many aspects complicate the management of the design phase. Some of these aspects are related to the nature of design being an ill-structured and creative process that is basically based upon iterative loops of analysis, synthesis and evaluation (Simon 1984). Other aspects are caused by the involvement of several stakeholders from different backgrounds and mentalities with relative attitudes towards project's value (Buenano 1999). Therefore, proper design management is expected to consider these aspects while planning for the design phase as well as while monitoring and controlling design progress throughout the process.

Researchers have recognized the complexity of managing design tasks due to their interdependent relation. To solve for complex dependencies, the Design Structure Matrix (DSM) has been developed to help arrange design tasks in a way to reduce rework. Accordingly, better flow of information among design parties is expected (Austin *et al.* 1999). However, scheduling the design program resulting from the DSM requires not only the sequence of activities, but also the start/end dates, durations, and resources requirements of each activity. Accordingly, the lookahead planning technique is suggested to further detail design activities, identify constraints, allocate resources and release work packages (Hammond *et al.* 2000, Hamzeh *et al.* 2009).

Apart from the complexities of managing the design phase, the industry is currently witnessing an increasing use of Building Information Modeling (BIM) as a platform to run design projects. In this context, there are continuous efforts to reap full potentials from BIM and to enhance its implementation in construction projects (Al Hattab and Hamzeh 2016, Eastman *et al.* 2009). Amongst the notable efforts to formalize the development of BIM is the work on the Level of Development (LOD) concept (Hooper 2015, Abou-Ibrahim and Hamzeh 2016). LOD is initially introduced by the American Institute of Architects (AIA) and several parties have worked to define and develop LOD worldwide (AEC UK BIM Protocol 2012, BIMForum 2015, The American Institute of Architects 2013).

The proper implementation of BIM as a platform to run design projects can be beneficial to enhance overall project's value (Barlish and Sullivan 2012, Tribelsky and Sacks 2010). In this context, this study investigates the possibility of enhancing design management by introducing a new framework that uses BIM as data repository of both design and management information. The framework is developed to tackle the major shortcomings witnessed while using conventional management tools to run design projects. It uses real time data related to product and process measures enabling the design manager to constantly check the real progress of design and to better navigate the corresponding project.

2 RESEARCH METHODOLOGY

The research starts by investigating the major methods and corresponding measures used to manage design projects. Then, the pros and cons of each method are analyzed according to certain criteria. Finally, a new framework is developed to tackle the shortcomings witnessed in each of the reviewed methods. Figure 1 illustrates the steps followed in the study.

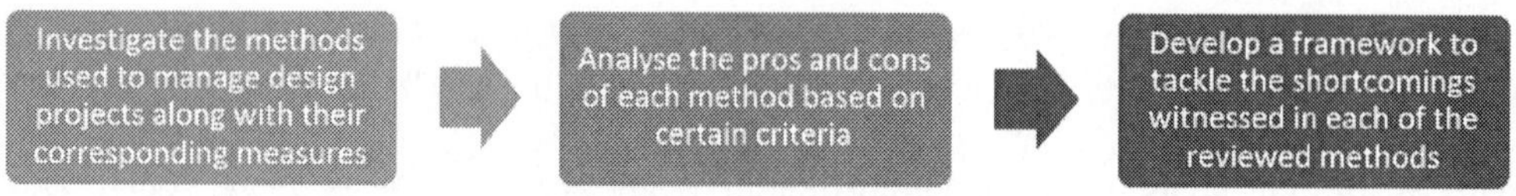

Figure 1. Research methodology.

3 MAJOR METHODS USED FOR MANAGING DESIGN PROJECTS

This section explores the major methods and their corresponding measures used to plan and monitor design projects. The three main methods considered in this study are: (1) the Top-Down planning approach using Work Breakdown Structure (WBS), (2) DSM combined with Lookahead Planning (also known as DePlan from Hammond *et al.* 2000), and (3) the use of LOD to plan design development in BIM-based projects.

3.1 Top-Down Planning using Work Breakdown Structure (WBS)

In the top-down management approach, managers break down the project into smaller activities and subjectively estimate required man-hours and needed resources to finish each of these sub-activities. In this scenario, managers build on their experience and make analogies to similar projects to subjectively estimate the required time and effort needed to finish each activity. Once the WBS is ready, the planner constructs an activity network and estimate total project's duration by identifying the critical path (Bashir and Thomson 1999).

This approach is very important to simplify the design undertaken and divide the project into manageable portions while assigning responsibilities to involved designers. However, problems in this approach reside in the following: (1) the reliance on subjective assessment of needed man-hours for each design activity, (2) the assumption that all needed activities are included in the WBS which is not always the case where new design tasks might emerge during design execution significantly affecting project's progress, (3) the use of the CPM method that does not capture the iterative nature of design where much more complicated interdependencies are present, and (4) the use of man-hours and Earned Value techniques to reflect on design progress. For instance, to assess progress using earned value techniques, the design manager would compare actually produced submittals or models with the planned submittals schedule. In this regard, finishing 25 plans out of 100 scheduled ones does not directly reflect a 25% completion of the design project where nothing is revealed about the actual information embedded in those submittals. In design, tracking only the produced submittals (whether plans, BIM models, or any other type of submittals) is not enough to assess actual design progress. In fact, the integrity and maturity of design information embedded in those submittals is what governs the actual progress of design. The same analysis applies when using the man-hours technique, where consuming 50% of man-hours does not reflect the completion of 50% of design.

3.2 DSM Combined with Lookahead Planning (DePlan)

DePlan follows two major steps to plan and control the production of design. The first step transforms the design process model into a DSM based on information dependencies among design tasks. The second step generates Weekly Work Plans (WWP) using the Last Planner system and Lookahead planning. Deplan uses Percent Plan Complete (PPC) to measure the progress of design by calculating the percent of completed activities out of scheduled activities (Hammond *et al.* 2000).

DePlan plays a major role in transforming design from an ill-structured and chaotic process to a more systematic and formalized process. Although DePlan enhances the management of iterative activities by using DSM and Lookahead planning to make sure constraints are removed and information flow is achieved, it does have some shortcomings. At the level of value generation, DePlan does not directly reflect the value of generated design in the eyes of involved stakeholders. At the level of design progress, it can only reflect progress in terms of activities completed out of scheduled activities, not in terms of actual design maturity. So although the PPC value might be high, it does not necessarily reflect a matured design. Finally at the level of design deliverables, DePlan is not attached to the quality of design deliverables whether they are drawings or BIM models.

3.3 LOD Plan for Design Development in BIM-Based Projects

With the emergence of BIM technologies in the AEC industry, a new way to plan the development of design, using the LOD concept, appears. Accordingly, the development of design is staged by defining which elements need to be modeled, at which phase, at what LOD level, and by whom. Accordingly, and as part of the BIM execution plan (BIM-XP), designers are required to model their design using the elements that appear in the BIM-XP according to pre-defined LOD levels. In this regard, designers are expected to follow the LOD guidelines to know how to graphically model these elements and which data to include in them (The American Institute of Architects 2013).

The LOD concept is essential to cope with BIM-based design where actual parametric objects are used in the model not sketches as in traditional 2D-CAD. LOD serves as a communication language between designers to highlight the actual maturity of their models and this is essential in a BIM environment for information sharing and coordination. However, the current concept and use of LOD shows some weaknesses. By definition, the current notion of LOD includes only six levels ranging from LOD 100 to LOD 500 (including LOD 350) without defining partial LOD levels witnessed throughout design. So, although elements are expected to reach their planned LOD levels at each design milestone according to the BIM-XP, these elements witness several maturity statuses throughout design due to iterations and coordination among stakeholders. Therefore, LOD can help design managers detect design maturity only at design milestones, not throughout design iterations and solution development. Nonetheless, being pre-defined, the use of LOD may not be as flexible as needed by designers where project specific LOD levels might be wanted. Moreover, LOD reflects the development of one model elements and therefore the development of the entire BIM model is not reflected.

4 PROS AND CONS OF EACH METHOD

This section presents the pros and cons of each of the three methods and their corresponding measures based on a set of criteria as highlighted in Tables 1 and 2. Table 1 reflects the ability of each planning approach to satisfactory manage the design process. The Top-Down approach is good at dividing the design undertaken into manageable design activities, but it fails in meeting other criteria. The combined DSM and Last Planner system is good at dynamically capturing design iterations while ensuring streamlined information flow among stakeholder; however, it is not directly linked to design deliverables (in other words, dynamics of the design process are not linked to the development of deliverables being models or drawings). The LOD Planning method fails to capture the iterative nature of design and it is relatively rigid to handle design dynamics since it is related to a pre-defined set of LOD levels; however, assigning adequate LOD levels to each discipline, at the right time, can streamline the flow of information among parties. Most importantly, the LOD planning method is the only method that relates design activities to modeling activities and therefore to product's development.

Table 1. Pros and cons of each method.

Criteria	Top-Down using WBS	DSM & Last Planner	LOD Plan
Divides design to manageable entities	Good	Good	Good
Ensures streamlined information flow	Unsatisfactory	Good	Satisfactory
Manages design process dynamically	Unsatisfactory	Good	Unsatisfactory
Captures design iterations	Unsatisfactory	Good	Unsatisfactory
Links design activities to design deliverables	Unsatisfactory	Unsatisfactory	Good

Table 2 highlights the ability of corresponding measures to reflect product and process characteristics. Man-Hours and Earned Value techniques can serve as a good indicator for incurred costs and consumed time. They can also reflect designers' productivity and planning quality if weighed against actual design progress assessed by the design manager. However, man-hours and EV techniques alone do not reflect actual development of the design or the fulfillment of owner's value. As for the PPC measure, it gives an indication about the time and cost performance of the project by reflecting the compliance to the plan; but, it cannot reflect actual design development or the fulfillment of owner's value. However, PPC is good at reflecting designers' productivity and planning quality. Finally, the LOD level is satisfactory to reflect design development at major milestones, but it cannot continuously track actual design progress occurring in a multidisciplinary environment. LOD can be used to reflect planning quality and designers' productivity if the manager compares planned and actual LOD levels at a given time, but this gives just an indication. Moreover, LOD alone does not reflect the time and cost performance of the project or the realization of owner's value.

Table 2. Pros and cons of corresponding measures.

Criteria	Man-Hours/ Earned Value	PPC	LOD Level
Reflects time and cost performance	Good	Satisfactory	Unsatisfactory
Reflects designers' productivity	Satisfactory	Good	Satisfactory
Reflects planning quality	Satisfactory	Good	Satisfactory
Reflects design development	Unsatisfactory	Unsatisfactory	Satisfactory
Reflects owner's value	Unsatisfactory	Unsatisfactory	Unsatisfactory

5 BIM-BASED DESIGN MANAGEMENT FRAMEWORK

The framework builds upon these three methods and it is suggested to be a comprehensive approach to manage the design using BIM. The framework targets the development of design at the level of BIM models where actual objects are being drafted to represent design intents. The framework is divided into two major parts: Part (1) that targets collaborative planning before the start of design information generation, and Part (2) that targets the scheduling and controlling of design activities during the design information generation process. Figure 2 presents the major steps followed in each part of the framework.

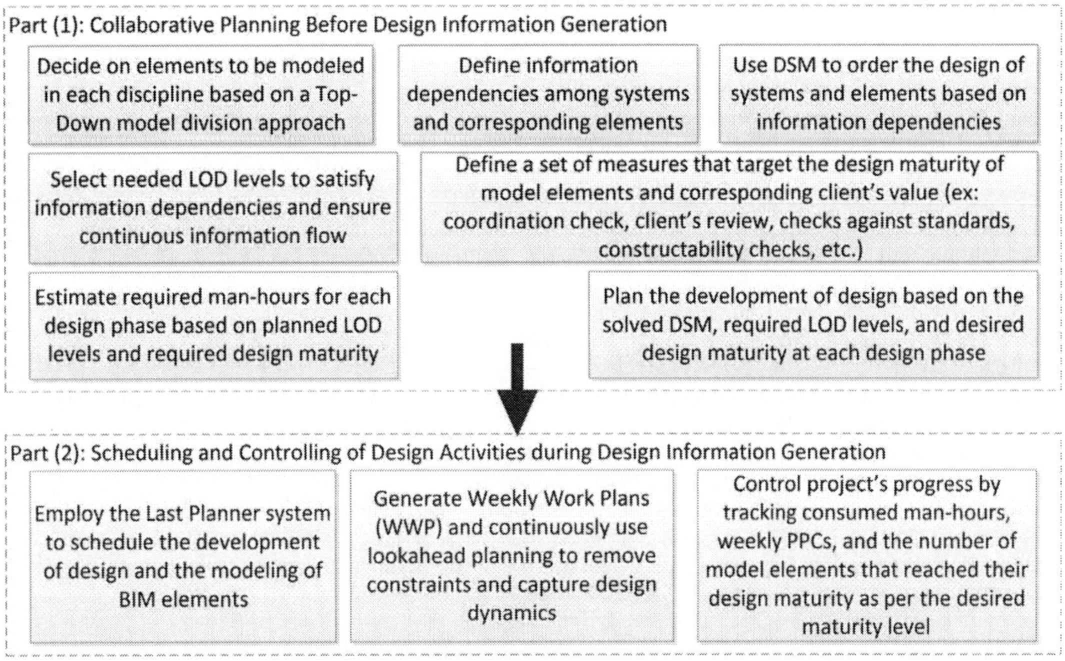

Figure 2. Suggested framework.

6 CONCLUSION

Different techniques are currently used to plan and monitor design projects. While each method has its strengths and weaknesses, design projects need new planning and control methods to overcome current shortcomings. Nonetheless, the increasing use of Building Information Modelling (BIM) in the construction industry, especially at the design phase, calls for the development of new measures that suits the object-oriented nature of BIM. This study investigates current methods used to manage design projects and analyses their adequacy in reflecting several design management aspects. Afterwards, a framework is suggested to enhance the management of design projects. The framework builds on the strengths of each of the investigated methods and provides a comprehensive approach to manage the process of generating design information. The framework is divided into two major parts. Part 1 includes several steps that aim to enhance pre-project planning using collaborative planning techniques, while the second part provides several steps to enable design manager to cope with the dynamic nature of design activities and to enhance the monitoring and controlling of design progress.

References

Abou-Ibrahim, H., and Hamzeh, F., Enabling Lean Design Management: An LOD-Based Framework, *Lean Construction Journal*, 12-24, 2016.

AEC UK BIM Protocol Version 2.0., 2012. Retrieved from https://aecuk.files.wordpress.com/2012/09/aecukbimprotocol-v2-0.pdf in October 2014.

Al Hattab, M., and Hamzeh, F., Using Social Network Theory and Simulation to Compare Traditional Versus BIM-Lean Practice for Design Error Management, *Automation in Construction*, 52, 59-69, 2016.

Austin, S., Baldwin, A., Baizhan, L., and Waskett, P., Analytical Design Planning Technique: a Model of the Detailed Building Design Process, *Design Studies,* 20, 279-296, 1999.

Barlish, K., and Sullivan, K., How to Measure the Benefits of BIM-A Case Study Approach. *Elsevier, Automation in Construction,* 24, 149-159, 2012.

Bashir , H., and Thomson, V., Metrics for Design Projects: A Review, *Design Studies, 20*, 263-277, 1999.

BIMForum, *Level of Development Specification,* 2015.

Buenano, G., *The Becoming of Problems in Design: Knowledge in Action to Frame Wicked Problems,* PhD Thesis, Department of Architecture, University of California, Berkeley, 1999.

Eastman, C., Lee, J., Jeong , Y. and Lee, J., Automatic Rule-Based Checking of Building Designs, *Automation in Construction, Elsevier,* 18, 1011-1033, 2009.

Hammond, J., Choo, H., Austin, S., Tommelein, I., and Ballard, G., Integrating Design Planning, Scheduling, and Control With DePlan, *8th Annual Conference of the International Group for Lean Construction,* Brigthon, UK, 2000.

Hamzeh, F., Ballard, G., and Tommelein, I., Is the Last Planner System Applicable to Design? A Case Study, *17th Annual Conference of the International Group for Lean Construction*, Cuperus, Y., and Hirota, E. H. (ed.), 165-176, Taipei, Taiwan, 2009.

Hooper, M., Automated Model Progression Scheduling Using Level of Development, *Construction Innovation, 15*(4), 428-448, 2015.

Simon, H., *The Structure of Ill-Structured Problems,* in *Developments in Design Methodology,* Cross, N. (ed.), J. Wiley & Sons, Chichester, UK, 1984.

The American Institute of Architects, *Guide, Instructions and Commentary to the 2013 AIA Digital Practice Documents,* 2013.

Tribelsky, E., and Sacks, R., Measuring Information Flow in the Detailed Design of Construction Projects, *Research in Engineering Design,* 21, 189-206, 2010.

HUMAN RESOURCE ALLOCATION IN ENGINEERING DESIGN COMPANIES

RUBA RIZK and ISSAM SROUR

Dept of Civil and Environmental Engineering, American University of Beirut, Beirut, Lebanon

The human capital is a major asset for engineering companies. On-time completion of project tasks and cost effectiveness are contingent on proper allocation of human resources. In practice, most engineering mangers assign resources according to availability without considering the level of employee's expertise while making the assignment decision. This may result in lower quality of design production or a difficulty in meeting deadlines in case an employee requires training or coaching. To better attain project requirements, the assignment practice could be improved by including resource characteristics and critical project information. This paper is part of an ongoing study that aims to develop a reliable resource assignment mathematical model that takes into account engineers' years of experience, their level of familiarity with different software programs used by the company, and other task-related attributes. Examples of the latter include scheduled submission date, and range of estimated period required to complete a task. The output of the model consists of assignment variables through pairing a group of human resources to a group of tasks with the objective of minimizing project duration and cost while maximizing resource utilization. Extensions to the model can allow engineering managers to monitor and control the progress of the project or the lack thereof.

Keywords: Project duration, Project cost, Resource assignment, Employee attributes, Project complexity, Resource utilization.

1 INTRODUCTION

Despite the advancements made in technology, machinery, and software programming, large scale design companies are still characterized by their main asset: human resources. The latter play a critical role in determining the level of quality of the company's work, which in turn translates into project success or failure. Nonetheless, companies tend to focus more on the technical perspective of a project and less so on human resources. To be able to survive the competitive market, corporations need to continuously enhance their performance (Garavan *et al.* 1999, Hodgetts *et al.* 1999, Losey 1999) which can be achieved through proper human resource management (Ngwenya and Aigbavboa 2017).

Managing employees is thought to be a troublesome activity that differs from handling capital or technology (Pirzada *et al.* 2013). One form of managing human resources is the process of allocation whereby individuals are assigned to certain tasks within the chosen development projects. This process is peculiarly critical since several attributes relating to the individual need to be taken into account and current allocation practices usually neglect the aforementioned and distribute resources according to the managers' intuition and availability of the resource

(Yoshimura *et al.* 2006, André *et al.* 2011). An employee ends up working on tasks whose skillset requirements do not match the worker's profile, and this translates into inefficient performance throughout the execution of the project. In addition to that, the estimated duration for design tasks may be inaccurate due to the lack of quantifiable features and the uniqueness of projects that leads to imprecise estimation of the effort an individual would need to invest to complete a certain task. Evident consequences of these inaccuracies are schedule and cost overruns (Bashir and Thomson 2001).

The objective of this paper is to develop a robust human resource allocation tool for engineering design companies that bases its assignment decision on resource-related and project-related factors when nominating individuals to work on specific tasks. The tool also aims at (1) making sure the time required to complete the demand of tasks does not exceed the allocated time for each task and (2) minimizing the total cost of engineering work-hours devoted to finalizing the project. The paper is organized as follows. The next section covers the literature on human resource allocation in various disciplines and the tools used to carry out this process. This is followed by a description of the model for resource allocation in design companies followed by a case study that demonstrates how the model works. The final section offers conclusions, discusses limitations of the work, and provides suggestions for future studies.

2 LITERATURE REVIEW

Several studies were carried out to come up with reliable human resource allocation models in different work fields ranging from health-care organizations, software developing companies and contracting companies. These tools were also developed using different methods like goal programming, decision making support system, multi-objective multistage combinatorial optimization models, dependency structure matrix, Taguchi's parameter design, and constraint satisfaction problems, to name a few.

Some examples include studies done by Otero *et al.* (2009) who detected delay in delivery software development projects due to the need for training employees who do not have the required skill set for certain tasks. Such an issue may rise when qualified staff members are preoccupied with other projects/tasks. Thus, Otero *et al.* (2009) developed systematic personnel assignment methods that evaluate the skill sets of employees against required skills for certain tasks using the Best-Fitted Resource approach. Their study, however, only focuses on the skills acquired by each employee and does not include other factors that normally affect the performance of the designer.

Similarly, Silva and Costa (2013) and Tsai *et al.* (2003) analyzed possible human resource allocation alternatives depending on the capabilities of employees and skill requirements for tasks related to software development projects. However, the authors also added the complexity of each project as a factor to be considered to weigh the type and number of professionals needed. Their ultimate goal was to minimize, or at least decrease, the cost and amount of time necessary to complete a project since software design projects almost never meet their deadlines.

Albers *et al.* (2012) handled human resource allocation with the use of Dependency Structure Matrices (DSM), which are normally used for modeling the structure of complex systems for organizational and planning purposes. The used model presents a multi-domain framework of the activities required for product development and a number of human resources responsible for fulfilling these tasks. It elaborates on the effect of different resource allocation options on the flow of information between activities, and how it enhances the dynamic relationships between the objective of a project, the objects used for completing the work and the operation systems. Nevertheless, the model does not discuss the characteristics relating to resources, or whether each resource is considered as being unique. The main focus is on how dependent the activities are on one another and if they require common resources to work on them.

Kwak and Lee (1997) used a variation from linear programming model, referred to as goal programming (GP) that also aims at minimizing total payroll costs while addressing the needs of patients at a healthcare organization. This technique gives goal constraints in addition to system constraints like minimum payroll goal, physician utilization goal, and physician assignment goal.

The above-mentioned set of studies shows that human resource allocation may be achieved through various methods and requires careful assessment to ensure the success of the service provided by the resource. However, it is noticeable that limited research has been done to create a human resource assignment model specific to engineering design and consultancy companies. These companies generate and review work under various disciplines such as structures, architecture, environmental, transportation, mechanical, electrical, planning and urban development, and façade engineering divisions. As human resources are the underlying assets of these companies, their proper allocation should remarkably enhance the degree to which a project can be considered successful. Previously developed models have not accounted for all resource and project related factors simultaneously. In order to generate more accurate assignment decisions, this study proposes incorporating these factors into the resource allocation model that will be further discussed in the following section.

3 METHODOLOGY

The aim of the human resource allocation model is to assign a pool of resources to a pool of tasks while taking into account the attributes that are related to both. This assignment aims at minimizing the total cost of work-hours spent completing the design tasks and ensuring that the task duration does not exceed its scheduled time.

The individual resources, who will be part of the allocation process are those designers who have up to four years of work experience and who manually perform the design work. They are characterized by the following variables:

- The level of experience the designer has in his/her respective area of specialization. This may take the form of "skilled" or "unskilled";
- The level of familiarity of the designer with the different software programs used by the company to complete the tasks at hand. This may take the form of "novice" or "expert";
- A binary variable indicating the availability of each resource defined by whether or not this employee is already working on a task; and
- The wage corresponding to each employee that depends on the first two variables mentioned above. For illustration, Table 1 corresponds to a range of wages for civil engineers as estimated by the U.S. Bureau of Labor Statistics (BLS) in 2010.

Table 1. Estimated wage of civil engineers as published by the U.S. BLS.

Employee Characteristics	Employee Wage ($/hr)
Skilled & Expert	45
Skilled & Novice/Unskilled & Expert	30
Unskilled & Novice	20

The tasks, on the other hand, are characterized by the following two variables:

- The time period within which the tasks must be completed, which may take the form of a deadline

- The expected time period required by a specific resource to complete a particular task. Each task's duration follows a PERT distribution characterized by an average and a standard deviation. The graph is divided into three zones that mimic the grouping in Table 1. The first group has a time range close to the least amount of time needed to complete the task; the second requires a value close to the average time; and the third needs the greatest amount of time that is estimated based around the highest value in the graph.

Based on the listed variables, the primary assignment solution appoints human resources to specific tasks in a way that guarantees the minimum cost needed to complete all activities in a timely manner. Commercial optimization software programs (e.g., Python or MATLAB) may be used to code this model and reach an optimum solution.

The following assumptions were made when developing the model:

- Every working day is comprised of ten hours;
- Employees are not interrupted while working;
- Each employee is to work on only one task at any given time;
- The level of compatibility between individuals working on the same team does not affect the productivity of each designer; and
- The tasks at hand require only one software program to be completed
- The current version of the model is conceived for a single design discipline

4 CASE STUDY

To illustrate the proposed model, a case study was considered. The case relates to the assignment process of five tasks to ten resources, who works in the same company (a prominent design firm in the Middle East), out of which six are available to perform the activities. Table 2 and Table 3 illustrate attributes relating to the human resources in this firm and those relating to the tasks to be completed, respectively. Table 4 is a result of the two input tables. It shows the expected time required for the resources to complete the tasks at hand. The only resources that are qualified to be part of the model formulation are those who are available. Thus, Table 4 only contains the first, second, fourth, seventh, eighth and tenth resources. From this available pool of resources, the allocation model analyzes which employees are best suitable for completing tasks within the preset timeframe for each activity. For example, it is evident that Task 4 (T4) is very critical since it is due in 30 working hours and only human resource 1 (HR1) can perform it on time.

Table 2. Human resource related attributes.

Human Resource	Resource Availability	Level of Experience	Level of Software Program Familiarity	Wage per Resource ($/hr)
1	1	Skilled	Expert	45
2	1	Unskilled	Novice	20
3	0	Unskilled	Novice	20
4	1	Unskilled	Expert	30
5	0	Skilled	Expert	45
6	0	Skilled	Expert	45
7	1	Unskilled	Expert	30
8	1	Skilled	Novice	30
9	0	Skilled	Novice	30
10	1	Unskilled	Novice	20

Table 3. Task related attributes.

Task	Average Duration (hrs)	Standard deviation (hrs)	Deadline (hrs)
1	30	10	40
2	20	15	40
3	40	13	50
4	35	5	30
5	50	10	50

Table 4. Expected time (hrs) required by available human resource (HR) to complete task (T).

HR/Task	T1	T2	T3	T4	T5
HR1	22	8	30	**30**	41
HR2	**37**	30	52	38	57
HR4	28	18	35	35	50
HR7	32	**13**	40	36	47
HR8	26	15	36	37	**47**
HR10	39	29	**50**	40	57

The optimal solution provided by the model is one where HR1 works on T4, HR2 works on T1, HR4 does not work on any task, HR7 works on T2, HR8 works on T5, and HR10 works on T3 (results are highlighted in Table 4). This assignment combination ensures that all tasks are done on-time and allocation is formed in a way that minimizes the cost of engineering hours spent working on these tasks. The optimum cost to complete all of the activities is $4,890.

5 CONCLUSIONS AND FUTURE WORK

Human resource assignment is a serious process that may have detrimental effects on projects if not implemented properly. Reliable resource allocation in engineering design companies increases the likelihood of meeting project deadlines while maintaining the lowest possible cost of engineering work-hours. This paper builds the case for formulated mathematical model that allows information, related to the company's assets and projects, to be stored making it possible to further enhance this system.

The model may be expanded to look into more than one discipline and therefore account for the interaction between different departments. Other potential extensions to the model may include allowing managers to monitor and control the progress of the project by asking employees to update the status of the initially assigned tasks. The model would then rerun the optimization program to make sure constraints are still satisfied or else, designers would be reallocated to maintain the model's goals.

References

Albers Albers, A., Breitschuh, J., and Lohmeyer, Q., Dependency Structure Matrix Modeling for Human Resource Allocation, *14th International Dependency and Structure Modeling Conference*, DSM 12. Kyoto, Japan, 2012.

André, M., Baldoquín, M. G., and Acuña, S. T., Formal Model for Assigning Human Resources to Teams in Software Projects, *Information and Software Technology*, 53(3), 259–275. https://doi.org/10.1016/j.infsof.2010.11.011, 2011.

Bashir, H. A. and Thomson, V., Models for Estimating Design Effort and Time, *Design Studies*, 22(2), 141–155, 2001.

Garavan, T. N., Bamicle, B., and O'Sulleabhain, F., Management Development: Contemporary Trends, Issues and Strategies, *J. Eur. Ind. Training,* 23(4/5), 3–12, 1999.

Hodgetts, R. M., Luthans, F., and Slocum, J. W., Strategy & IHRM Initiatives for The 00s Environment: Redefining Roles & Boundaries, Linking Competencies & Resources, *Organ. Dyn.*, 28(2), 7–21, 1999.
Kwak, N. K., and Lee, C., A Linear Goal Programming Model for Human Resource Allocation in a Health-Care Organization, *Journal of Medical Systems*, 21(3), 129–140, 1997.
Losey, M. R., Mastering the Competencies of HR Management, *Hum. Res. Manag.*, 38(2), 99–111, 1999.
Ngwenya, L. and Aigbavboa, C., Improvement of Productivity and Employee Performance Through an Efficient Human Resource Management Practices, In J. I. Kantola, T. Barath, S. Nazir, & T. Andre (Eds.), Advances in Human Factors, Business Management, Training and Education, 498, 727–737, 2017.
Otero, L. D., Centeno, G., Ruiz-Torres, A. J., and Otero, C. E., A Systematic Approach for Resource Allocation in Software Projects, *Computers & Industrial Engineering*, 56(4), 1333–1339. https://doi.org/10.1016/j.cie.2008.08.002, 2009.
Pirzada, D. S., Hayat, F., Ikram, A. A., Ayub, M., and Waheed, K. Impact of Human Resources Management Practices on Turnover, Productivity and Corporate Financial Performance, *Eur.J. Bus. Manag.*, 5(10), 163, 2013.
Silva, L. C., and Costa, A. P. C. S., Decision Model for Allocating Human Resources in Information System Projects, *International Journal of Project Management*, 31(1), 100–108. https://doi.org/10.1016/j.ijproman.2012.06.008, 2013.
Tsai, H.-T., Moskowitz, H., and Lee, L.-H., Human Resource Selection for Software Development Projects Using Taguchi's Parameter Design, *European Journal of Operational Research*, 151(1), 167–180. https://doi.org/10.1016/S0377-2217(02)00600-8, 2003.
Yoshimura, M., Fujimi, Y., Izui, K., and Nishiwaki, S., Decision-Making Support System for Human Resource Allocation in Product Development Projects, *International Journal of Production Research*, 44(5), 831–848. https://doi.org/10.1080/00207540500272519, 2006.

Responsible Design and Delivery of the Constructed Project
Edited by Abdul-Malak, M., Khoury, H., Singh, A., and Yazdani, S.

ASSESSMENT OF CONTRACTORS' PRACTICES TOWARDS RISKS OUT OF CONTRACTORS CONTROL IN CONSTRUCTION INDUSTRY

MOHAMMED ALGAHTANY[1], DEAN KASHIWAGI[2], KENNETH SULLIVAN[1], JACOB KASHIWAGI[1], and MAJED ALZARA[3]

[1]*School of Sustainable Engineering and the Built Environment, Arizona State University, Tempe, United States*
[2]*KSM Inc, Mesa, United States*
[3]*College of Engineering, Jouf University, Al-Jouf, Saudi Arabia*

Research has shown that public construction projects in Saudi Arabia have exhibited poor performance for the past three decades. Recent studies have identified the ownership of parties who cause risks and lead to low performance in the Saudi construction industry. These studies identified that contractors are not the main party that cause risks as owners, consultants, and other parties have the major share of causing risks in the industry. The aim of this study is to assess the current risk management and performance measurement practices applied by contractors to minimize risk out of their control (caused by other parties) through conducting a questionnaire survey. The questionnaire survey was sent to contractors who work in public construction projects in Saudi Arabia. The assessment included the contractors' practices in the identification of projects activities and risks, contractors' practices in mitigating other parties' activities and risks, and contractors' practices in measuring the performance of all project parties. The study identified that contractors' current risk management and performance measurement practices are not effective in minimizing projects risks caused by other parties and ineffective in measuring performance of all parties. The study recommends a proactive risk management approach that can help contractors to minimize risks caused by other parties and can measure the performance of all parties in projects.

Keywords: Risk management, Performance measurement, Risk mitigation, Project parties, Risk identification, Accountability.

1 INTRODUCTION

Several studies conducted in Saudi Arabia through the past three decades indicated the critical issue of the construction industry with inefficiencies, non-performance, cost and time overruns. Delay is regarded as one of the most serious and common issues in the Saudi construction projects (Faridi and El-Sayegh 2006). It was revealed by three researchers in Saudi that between 60-70% of the public construction projects faced delays in the completion of projects' time (Al-Sultan 1989, Al-Khalil and Al-Ghafly 1999, Assaf and Al-Hejji 2006). One study reported that the average time overrun percentages varied from the original duration of contracts by 10% to 30% (Assaf and Al-Hejji 2006). Another study found that the average time overrun percentage in projects was 39% from their predicted original schedules (Elawi *et al.* 2016). Additionally, in the

Saudi construction industry (SCI), several researchers have studied the prevailing issue of cost overruns (Alhomidan 2013, Allahaim and Liu 2015, Alghonamy 2015). The above projects cost overruns and delays are results and measures of risk occurrence that cause a low performing construction.

Several parties are involved in construction projects such as owner, designer, consultant, contractor, suppliers, and other stakeholders and all of project parties inevitably carry certain risks (Peckiene *et al.*, 2013). In the context of the SCI, three authors identified risks in the industry and categorized the risks considering the responsible parties of generating these risks. Assaf and Al-Hejji (2006) identified and assessed delay risks in the SCI. The study conclusion showed that the most common delay risk is change orders by owners. The final combination of results showed that project delays are mostly created by owners, followed by contractors, designers, labors and consultants. Another study assessed risks in the aviation construction projects in Saudi Arabia (Baghdadi and Kishk 2017). This research identified that the most important group of risks affecting projects is designer related risks such as design errors. The owners related risks such as design changes were ranked second followed by consultant related risks and then contractor related risks. The third study identified the ownership percentages of parties causing risks in the SCI (Elawi *et al.*, 2016). This study identified that project owners caused 53% of the risks. Other stakeholders (excluding owners and contractors) were responsible for 20% of the risks. In addition, Algahtany *et al.*, (2016) analyzed 19 studies that identified and assessed risks in the SCI and identified that risks out of contractor's control (caused by other parties) such as change of scope, change orders, specifications changes, design changes by owner or consultant, and mistakes in design are frequent risks in the SCI. In traditional practices in managing risks in the SCI, most of the risks are allocated to contractors (Al-Salman 2004), as other project parties do not take accountability of their risks (Al-Sobiei *et al.*, 2005). However, in the literature discussed above, owners and project parties other than contractors cause the major share of risks in SCI. Albogamy and Dawood (2015) found that there is a clear lack of risk management practices in Saudi Arabia, which identify the impact of the risks for project parties in construction process. In addition, Alsulamy (2015) identified that public organizations in Saudi Arabia suffer from the absence of performance measurement systems in their construction projects that identify performance of projects and differentiate between project parties' performances. According to Kashiwagi (2016) the responsibility for projects' risks should be identified and measured during project phases to identify the project parties' impact on project risks and to differentiate the actual performance of project parties.

2 OBJECTIVE OF THE STUDY

The literature discussed above showed that project parties other than contractors cause the major share of risks, which lead to low performance in the SCI. The main aim of the study is to assess the current practices applied by contractors to minimize risks out of their control (caused by other parties) by evaluating the contractors' current risk mitigation and performance measurement practices. The assessment includes the contractors' practices in the identification of projects activities and risks, contractors' practices in mitigating other parties' activities and risks, and contractors' practices in measuring the performance of all project parties.

3 RESEARCH METHODOLOGY

The aim of the study will be achieved through conducting a questionnaire survey to contractors who work in public projects in Saudi Arabia. The first part of the survey assesses the contractors' current practices towards risks out of their control by evaluating the contractors' current risk mitigation and performance measurement practices. The second part of the survey identifies the contractors' perceptions towards the effectiveness of their current risk mitigation practices and the need for a new approach in mitigating and measuring all parties' activities and risks.

3.1 Data Analysis and Results

The targeted population in this study is contractors who work in public projects in Saudi Arabia. 252 contractors responded to the survey. The survey was formed in English, translated into Arabic, and then distributed through online survey. Of the 252 respondents, 29 contractors are classified in grade 1 in the contractors' classification system in Saudi Arabia. 28 classified in grade 2. Most of the participated contractors are classified in grades 3 and 4 with 72 in grade 3 and 76 in grade 4, 44 contractors in grade 5, and 3 non- classified contractors. The contractors' classification system is used by public organizations in Saudi as the basis for prequalifying contractors and awarding projects to ensure contractors' capabilities and performance (Bubshait and Al-Gobali, 1996). The Saudi contractors' classification system utilizes specific financial and technical criteria to classify contractors and functions within 5 grades and 29 fields and the grades levels determine the financial values of projects budgets that contractors can bid for within their fields of classification (MOMRA, 2017). A total of 103 respondents (41%) have more than 15 years of experience in the construction industry. 26% in the group of 10-15 years of experience, 28% have from 5 to 10 years of experience, and 5% have less than 5 years of experience. Considering the participants' academic qualifications, 84% of the participants have bachelor's degree as their minimum level of education, with 14% holding master's degree, and 5% holds PhD degrees. Considering the participants' working position at the company, the majority were the companies' owners (59%), 16% are projects managers, 5% are field engineers, 13% are administration managers, and 7% choose the final choice (others).

The survey questions were divided into two parts. The first part asks about the contractors' current practices in mitigating and measuring risks and activities of project parties. For this part the respondents were required to select one of three choices that reveal their agreement about each statement. The three choices are agree, don't know, and disagree. The second part of the survey asks about the contractors' perceptions towards the effectiveness of current risk mitigation practices. For this part the respondents were required to select one of five choices that reveal their agreement about each statement. The five choices are strongly agree, agree, not sure, disagree, and strongly disagree.

3.1.1 *Assessment of contractors' practices towards risks out of their control*

The contractors' practices in the identification of projects activities and risks are shown in Figure 1. The majority of the respondents (82.54%) agreed that at early stages of projects, they as contractors identify all of their activities and risks. In addition, Figure 1, shows the approach of the contractors in the identification of activities and risks of other parties in early stages of projects as 56.35% of the participants agreed that they as a contractor identify all of other parties' activities and risks. However, 29.76% of the respondents disagreed with the statement, as they don't identify other parties' activities and risks in projects.

Figure 2, shows the analysis of the contractors' practices in including other parties activates and risks in project plans and mitigating these activities and risks throughout project phases. Only 22.22% of the participated contractors include other parties activates and risks in project plans. Furthermore, the contractors' current risk management practices were investigated, considering the mitigation of other parties' activities and risks throughout the project phases by the contractors. As shown in Figure 2, 65.08% of the contractors do not mitigate other parties' activities and risks throughout the projects phases and only 15.48% of the respondents mitigate other parties' activities and risks in the project phases.

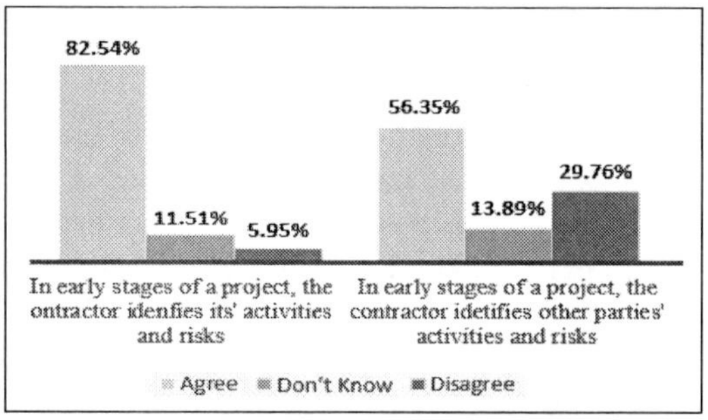

Figure 1. Contractors' practices in the identification of Projects' activities and risks.

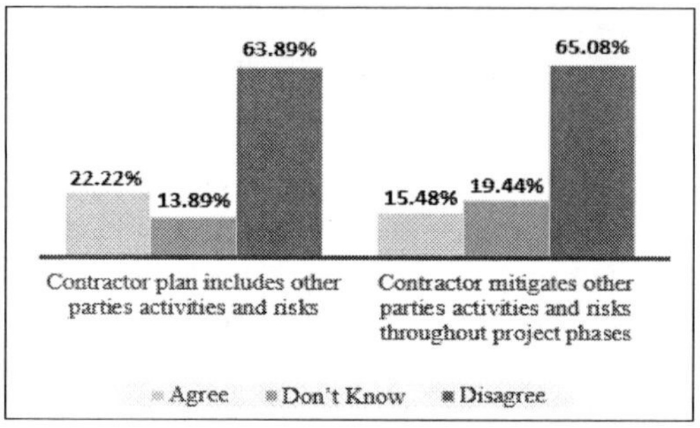

Figure 2. Contractors' practices in mitigating other parties' activities and risks.

The performance measurement practices for the participated contractors were investigated considering measuring their own performance and measuring other parties' performance during the project phases. As shown in Figure 3, most of the participated contractors (84.92%) measure and track their performance during projects phases based on time and cost deviations. However, their measurement practices considering other parties change as only 40.08% measure the performance of all parties in projects during projects phases and 43.65% do not measure the performance of all parties during project phases. A following question was asked for the participated contractors who measure all parties' performance to investigate if they share the measurement information of all parties' performance periodically with all parties involved in projects. Only 31% of those contractors share the measurement information with project parties.

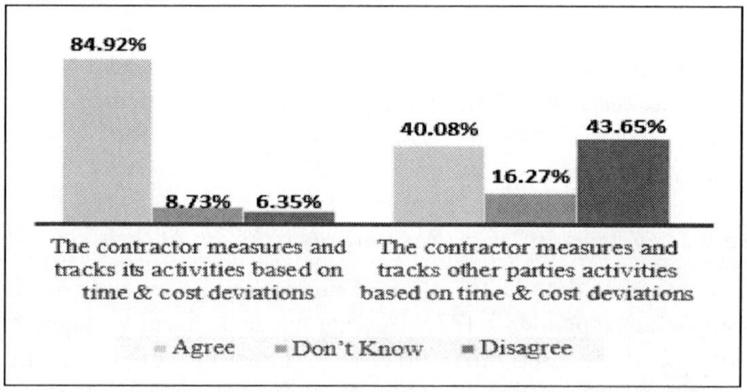

Figure 3. Contractors' practices in measuring activities performance.

3.1.2 *Contractors' perceptions of the current risk mitigation practices*

This section identifies the contractors' perceptions towards the effectiveness of the current risk mitigation practices applied by contractors and the need for a new approach in mitigating and measuring all parties' activities and risks. Table 1 provides the results of the descriptive statistics of the contractors' answers. The answers' scale ranges from 5 for strongly agree to 1 for strongly disagree. The table shows the analysis of sample response using the concept of weighted mean and standard deviation.

Table 1. Contractors' perceptions of the current risk mitigation practices

#	Question	%					Mean	Sd
		Strongly agree	Agree	Not sure	Disagree	Strongly disagree		
1	Activities of other parties in projects (activities out of contractor's control) are risks to the project if they are not done as expected.	46.43	44.84	5.95	1.98	0.79	4.34	0.75
2	Risks and activities that are out of contractor's control (generated by other parties) are a leading cause of low performance in projects in Saudi Arabia.	39.29	45.63	5.16	8.33	1.59	4.13	0.95
3	The contractors' current risk mitigation practices are not effective in minimizing risks because they do not identify and mitigate risks and activates of other parties (out of contractors control) in projects in Saudi Arabia.	37.30	46.43	9.92	5.16	1.19	4.13	0.88
4	There is a need for a proactive risk mitigation practice that identifies and mitigates activates and risks out of contractors control in projects in Saudi Arabia.	49.60	40.87	7.14	1.98	0.40	4.37	0.74
5	There is a need for a performance measurement system that identifies all parties' performance in projects to differentiate the actual contractors' performance from others' risks and performance.	54.37	37.30	6.35	1.19	0.79	4.43	0.73

4 CONCLUSIONS

The analysis of the collected data revealed the status of the current approach of mitigating risks applied by contractors in the SCI. The study results assured that the activities of projects parties other than contractors should be dealt with as risks as activities and risks out of contractors control are a leading cause of low performance in projects in Saudi Arabia. However, after the analysis of the contractors' current approach in mitigating these risks, results showed that most of the contractors do not include the risks and activities out of their control in projects plans and do not mitigate them throughout project phases. Furthermore, considerable percentage of contractors does not measure activities deviations and risks' effects of all parties in projects and thereby contractors are not able to differentiate their performance from the actual performance of other project parties. Transparency is not utilized by most of the contractors to minimize risks effects, as they do not share the measurement information periodically with all parties in projects.

The study results showed that the contractors' current risk mitigation practices are not effective in minimizing risks because they do not identify and mitigate risks and activates of other parties in projects. According to the majority of the participated contractors, there is a need for a proactive risk mitigation practice that identifies and mitigates activates and risks out of contractors control. Furthermore, there is a need for performance measurement system that identifies all parties' performance in projects to differentiate the actual contractors' performance from others' risks and performance. These results show the need to shift the focus of contractors' risk management and performance measurement practices from only managing and measuring their own activities and risks to be proactive by including all project parties' activities and risks.

Authors propose that this paradigm shift will increase project parties' accountability towards their activities and risks and lead to minimizing project risks. Project parties' accountability increases when their activities and risks are mitigated continuously by contractors with measuring the activities and risks' deviations from the planned schedule and cost with identifying sources of deviations. Transparency should be utilized in this approach through sharing performance measurements of project activities and risks with all project parties. This will differentiate the performance of each party by identifying the responsibility of causing risks or causing deviations in project planned schedule and budget.

References

Al-Khalil, M. I., and Al-Ghafly, M. A., Important Causes of Delay in Public Utility Projects in Saudi Arabia, *Construction Management & Economics*, 17(5), 647-655, 1999.

Alsulamy, S., Developing a Performance Measurement Framework for Municipal Construction Projects in Saudi Arabia, *PhD dissertation,* Edinburgh Napier University, 2015.

Albogamy, A., and Dawood, N., Development of a Client-Based Risk Management Methodology for The Early Design Stage of Construction Processes: Applied to the KSA, *Engineering, Construction and Architectural Management*, 22(5), 493-515, 2015.

Al-Salman, A. A., Assessment of Risk Management Perceptions and Practices of Construction Contractors in Saudi Arabia, *MSc Thesis,* King Fahd University of Petroleum and Minerals, 2004.

Al-Sobiei, O. S., Arditi, D., and Polat, G. Managing Owner's Risk of Contractor Default, *Journal of Construction Engineering and Management*, 131(9), 973-978, 2005.

Al-Sultan, A. S., Determination of Construction Contract Duration for Public Projects in Saudi Arabia, *MSc Thesis,* KFUPM, Dhahran, Saudi Arabia, 1989.

Alhomidan, A., Factors Affecting Cost Overrun in Road Construction Projects in Saudi Arabia, *International Journal of Civil & Environmental Engineering*, IJEE-IJENS, 13(3), 1-4, 2013.

Allahaim, F. S., and Liu, L., Causes of Cost Overruns on Infrastructure Projects in Saudi Arabia, *International Journal of Collaborative Enterprise*, 5(1-2), 32-57, 2015.

Alghonamy, A., Cost Overrun in Construction Projects in Saudi Arabia: Contractors' Perspective, *International Journal of Engineering & Technology*, IJET-IJENS, 15(4), 35-42, 2015.

Algahtany, M., Alhammadi, Y., and Kashiwagi, D., Introducing a New Risk Management Model to the Saudi Arabian Construction Industry, *Procedia Engineering*, 145, 940-947, 2016.

Assaf, S. A., and Al-Hejji, S., Causes of Delay in Large Construction Projects, *International Journal of Project Management*, 24(4), 349-357, 2006.

Baghdadi, A., and Kishk, M., Assessment of Risks Associated with Saudi Aviation Construction Projects and of the Risks' Importance, *International Journal of Innovation, Management and Technology*, 8(2), 123, 2017.

Bubshait, A. A., and Al-Gobali, K. H., Contractor Prequalification in Saudi Arabia, *Journal of Management in Engineering*, 12(2), 50-54, 1996.

Elawi, G., Algahtany, M., and Kashiwagi, D., Owners' Perspective of Factors Contributing to Project Delay: Case Studies of Road and Bridge Projects, *Procedia Engineering*, 145, 1402-1409, 2016.

Faridi, A. S., and El-Sayegh, S. M., Significant Factors Causing Delay in the UAE Construction Industry, *Construction Management and Economics*, 24(11), 1167-1176, 2006.

Kashiwagi, D., *Best Value Approach*, Tempe, AZ: KSM Inc, 2016.

MOMRA, Contractors Classification Agency, 2017. Retrieved from https://contractors.momra.gov.sa on October 12, 2017.

Peckiene, A., Komarovska, A., and Ustinovicius, L., Overview of Risk Allocation Between Construction Parties, *Procedia Engineering*, 57, 889-894, 2013.

Responsible Design and Delivery of the Constructed Project
Edited by Abdul-Malak, M., Khoury, H., Singh, A., and Yazdani, S.

DEVELOPMENT OF CONTRACTORS' MITIGATION PRACTICES TOWARDS RISKS OUT OF CONTRACTORS' CONTROL IN CONSTRUCTION INDUSTRY

MOHAMMED ALGAHTANY[1], KENNETH SULLIVAN[1], DEAN KASHIWAGI[2], and JACOB KASHIWAGI[1]

[1]School of Sustainable Engineering and the Built Environment, Arizona State University, Tempe, United States
[2]KSM Inc, Mesa, United States

Research has shown that public construction projects in Saudi Arabia have exhibited poor performance for the past three decades. Recent studies have identified that contractors are not the main party that cause risks as owners, designers, and other parties have the major share of causing risks in the industry. The traditional risk management practices have been ineffective at helping contractors deliver projects on time and within budget while meeting quality expectations. The aim of this study is to develop a risk mitigation model for the contractors to mitigate risks out of contractors' control (caused by other parties). The developed model is validated through identifying contractors' perceptions who work in public construction projects in Saudi of the new approach through conducting a questionnaire survey. The developed model focuses on increasing accountability of project parties through mitigating parties' activities and risks, measuring the activities and risks deviations (time and cost), and identifying sources of deviations. Transparency is utilized in the model through sharing weekly updates of the activities and risks combined with updated information of performance measurements of all project parties. The analysis of the study results showed that project risks will be minimized and performance of projects can be increased if contractors shift their focus using the new model from only managing their own activities and risks to managing all project parties' activities and risks.

Keywords: Risk management, Performance measurement, Accountability, Transparency, Project parties, Project delivery.

1 INTRODUCTION

Different parties are involved in construction projects such as owner, designer, consultant, contractor, suppliers, and other stakeholders and all of these parties certainly carry certain risks (Peckiene *et al.* 2013). In the context of the Saudi construction Industry (SCI), three researchers have identified the ownership of parties who cause risks and lead to low performance. Assaf and Al-Hejji (2006) identified that time overruns in construction are mostly created by owners, followed by vendors, designers, labors and consultants. Another study identified that designers are the most party that cause risks in the aviation projects in Saudi and owners were ranked second followed by consultants and then contractors (Baghdadi and Kishk 2017). Another study identified the ownership percentages of project parties who cause risks in the SCI (Elawi *et al.*

2016). This study identified that 53% of the risks are caused by owners and other stakeholders (excluding contractors) were responsible for 20% of the risks.

The reported data are perceptions of poor performance of projects in SCI as there is a lack of centralized systems that track and collect actual performance data to provide precise and quantitative data on projects performance in the governmental organizations in Saudi (Almutairi 2017, Al-Otaibi and Price 2009). Alsulamy (2015) identified that the construction projects in public organizations in Saudi suffer from the absence of performance measurement systems that identify the performance of projects and identify the performance of all project parties.

So far, in the SCI, there has been little consideration given to applying risk management practices to minimize risks' affects in the construction sector. According to Baghdadi and Kishk (2015) there is a lack of efficient risk management practices in the SCI. Ikediashi *et al.* (2014) identified that poor risk management practices in the SCI is the most critical factor for construction projects failure. In addition, according to Albogamy and Dawood (2015) there is a clear lack of applying risk management practices in the SCI that reveal the impact of the risks for project parties. In the traditional practices in managing risks in the SCI, most of the risks are allocated to contractors and none to owners (Al-Salman 2004) as owners do not take accountability and automatically transfer them to other parties (Al-Sobiei *et al.* 2005). However, in the literature discussed above, owners and project parties other than contractors cause a major share of project risks in the SCI. According to Kashiwagi *et al.* (2013), risk is defined as what contractor has no control over including risks and activities of other parties in projects. Assaf *et al.* (2015) stated that it is important to identify and manage the risks that impede contractors from delivering projects on specified cost, time and quality objectives. To minimize effects of project risks, according to Algahtany *et al.* (2016), contractors should identify risks out of their control (caused by other parties) and plan in advance how to reduce the effects of these risks.

2 OBJECTIVE OF THE STUDY

The literature discussed above showed that project parties other than contractors cause the major share of project risks, which lead to low performance in SCI. However, with the identification that risks out of contractors' control are a leading cause of low performance, there is a lack of efficient risk mitigation practices by contractors to manage these risks. The aim of this study is to develop a risk mitigation model for the contractors to mitigate risks out of contractors' control.

3 METHODOLOGY

The methodology of the study will include the following steps:

(i) Propose a risk management framework to manage risks out of contractors' control.

(ii) Conduct a questionnaire survey to identify the contractors' perceptions of the new approach in managing risks out of contractors' control through.

(iii) Identify the practical framework in mitigating risks out of contractors' control.

3.1 Proposed Risk Management Framework

The main components in the developed risk mitigation model are derived from a delivery structure called the Performance Information Procurement System/ Performance Information Risk Management System (PIPS/PIRMS). The PIPS\PIRMS model replaces insufficient decision making (expectations made by project parties) by the expert vendor plan that states a clear difference between expectations and reality and can identify the project scope and how it will be accomplished (Kashiwagi and Kashiwagi 2012). The expert vendor preplans the project from beginning to end in a clarification period with identifying activities and risks they don't control (Kashiwagi *et al.* 2010). The contractors minimize activities deviations and risks' effects by performing risk management and performance measurement through mitigating project activities and risks throughout the project execution stage (Kashiwagi 2016).

3.1.1 *The use of performance measurement in managing risks*

In the process of applying risk management in projects, identified and emerging risks should be measured during project phases to identify the areas, which require enhancement and to identify project team's performance in mitigating risks (Kashiwagi 2016). According to Hatry *et al.* (1990), performance measures are needed in planning, monitoring and evaluating project activities and risks. Performance measurement can be combined with risk management in order to measure and minimize risks affects (Kashiwagi and Kashiwagi 2012). According to Kim (2010) applying performance measurement methods in managing project risks is necessary to help in measuring risks influence on project time and cost. The use of performance measurement in projects will dramatically increase the performance level of vendors and the overall level of the industry productivity (El-Mashaleh *et al.* 2007). Measuring performance increases projects' performance because it increases accountability and provides a transparent environment. Measuring performance provides transparency to projects by giving insights into activities, responsibilities, costs, and outputs. Furthermore, it is a way of incentive or rewards for projects' activities outputs (De Bruijn 2002). Considering accountability, performance measurement helps in increasing the accountability of the responsible parties (Ammons 1995) and it is considered as an effective method of shaping accountability in projects (De Bruijn 2002).

3.2 Survey Analysis and Results

The targeted population in this study is contractors who work in public projects in Saudi Arabia. 252 contractors responded to the survey. The survey was formed in English and then translated into Arabic and then distributed through online survey. Of the 252 respondents, 29 contractors are classified in grade 1 in the contractors' classification system in Saudi Arabia. 28 classified in grade 2. Most of the participated contractors are classified in grades 3 and 4 with 72 in grade 3 and 76 in grade 4, 44 contractors in grade 5, and 3 non- classified contractors. A total of 103 respondents (41%) have more than 15 years of experience in the construction industry. 26% in the group of 10-15 years of experience, 28% have from 5 to 10 years of experience, and 5% have less than 5 years of experience. Considering the participants' academic qualifications, 84% of the participants have bachelor's degree as their minimum level of education, with 14% holding master's degree, and 5% holds PhD degrees. Considering the participants' working position at the company, the majority were the companies' owners (59%), 16% are projects managers, 5% are field engineers, 13% are administration managers, and 7% choose the final choice (others).

The survey questions were divided into two parts. The first part identifies the contractors' perceptions of the proposed risk mitigation framework. The second part of the survey identifies the contractors' perceptions towards the effectiveness of the new approach in managing risks out of contractors' control. The respondents were required to select one of five choices that reveal their agreement about each statement. The five choices are strongly agree, agree, not sure, disagree, and strongly disagree. Table 1 shows the descriptive statistics for the contractors' perceptions of the new approach in mitigating risks out of contractors' control. The answers' scale ranges from 5 for strongly agree to 1 for strongly disagree. It shows the analysis of sample response using the concept of weighted mean and standard deviation.

Table 1. Contractors' perceptions of the new approach in mitigating risks out of contractors' control.

#	Question	%					Mean	Sd
		Strongly agree	**Agree**	**Not sure**	**Disagree**	**Strongly disagree**		
1	The existence of an expert project manager with the contractor (who can plan project from start to end and can identify all activates out of contractor control) is essential to the success of project.	85.03	13.61	0.00	1.36	0.00	4.82	0.48
2	When contractors make a milestone schedule for a project, they should identify and include all risks and activities of client and other related parties in the plan.	53.74	36.05	6.80	2.72	0.68	4.39	0.79
3	Following up with risks and activities of project parties by reminding them what, when, and how to manage them, would help responsible parties to act better to perform their tasks.	53.74	38.10	5.44	2.72	0.00	4.43	0.72
4	Being transparent by sharing with project's parties a weekly update of all upcoming activities and risks that need to be managed will help and motivate those responsible to be more accountable for their activities.	54.42	38.78	4.76	1.36	0.68	4.45	0.71
5	Identifying activities' deviations from the planned schedule and cost, explaining why the deviation occurred and identifying the responsible party would help in measuring all parties' performance.	40.14	49.66	7.48	2.04	0.68	4.26	0.74
6	Being transparent by sharing with project's parties a weekly update of all parties' performance information will help and motivate them to be more accountable for their activities.	44.90	45.58	6.80	2.72	0.00	4.33	0.72
7	The addition of projects plans that include all parties' activities in the project contracts will help the parties to comply with their responsibilities.	62.59	32.65	4.08	0.68	0.00	4.57	0.61
8	Contractor who is doing the work is the best party that can prepare the weekly reports to mitigate and measure project activities and risks and client or consultant can review the report.	46.26	38.10	6.80	6.80	2.04	4.19	0.97
	Effectiveness of the new approach in mitigating risks out of contractors' control							
1	The identification and mitigation of project parties' risks by the contractor (by knowing how they will manage or minimize the risks) will help in reducing risks' affects.	45.58	48.98	4.08	1.36	0.00	4.39	0.63
2	If the contractor makes a milestone schedule for project plan from beginning to end that includes risks and activities of all stakeholders and measures all parties' performance during projects, this will increase the performance of the project.	54.42	36.73	7.48	1.36	0.00	4.44	0.69

The developed risk mitigation model shown in Figure 1, focuses on increasing accountability of project parties through mitigating parties' activities and risks with measuring their deviations (time and cost) and identifying sources of deviations. Transparency is utilized in the model through sharing weekly updates of the activities and risks combined with updated information of performance measurements of project parties. The practical framework in mitigating risks out of contractors' control consists of a Weekly Risky Report (WRR) that includes a project activities schedule & milestones table that identifies project activities with their responsible party, percentage of completion, and deviation from the planned schedule. Secondly, the WRR includes a project risks deviations table that identifies the plan to minimize risk deviation, effect of risk on the planned schedule and budget, explaining why the deviation occurred and identifying the responsible party. Finally, the WRR includes a performance measurement table, which identifies the performance measurements of project and project parties based on deviations from the planned schedule and budget. The contractor is responsible for maintaining the WRR and sending it out weekly with project parties. The client's project manager or consulting engineer would be used to ensure that contractor is keeping the dominant performance information of time and cost deviations along with explaining the risk, why it occurred, and identifying the responsible party.

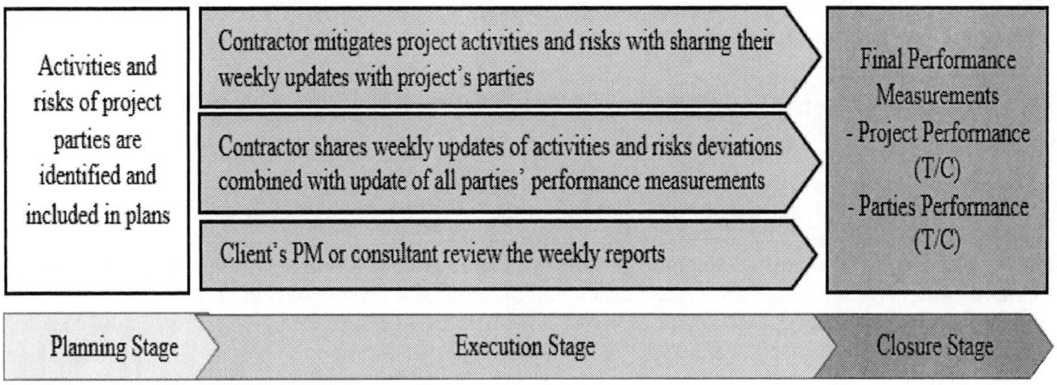

Figure 1. Risks out of contractors' control mitigation model.

4 CONCLUSION

The analysis of the collected data revealed the contractors' perceptions of the new approach in mitigating risks out of contractors' control. According to contractors' views, the activities and risks of all project parties should be identified, included in project plans, and added to project contracts to help project parties to comply with their responsibilities. The study results showed that accountability of project parties increases when contractors mitigate activities and risks and measure deviations with identifying sources of deviations. Furthermore, accountability also increases when transparency increases through sharing weekly updates of upcoming activities and risks with project parties combined with the updated information of performance measurements with updating project parties' performance information.

The study results identified that contractor is the best party that can maintain and distribute the weekly reports to mitigate and measure the project activities and risks and client or consultant can review the report. In this procedure, the client's project manager or consulting would be used to ensure that the contractor is keeping the dominant performance information of time and cost deviations along with explaining the risk, why it occurred, and identifying the responsible party. This assurance process is necessary to make sure contractors are performing the risk management and quality control process that can minimize deviations in project schedule and budget. The analysis of the study results showed that project risks will be minimized and performance of projects will be increased if contractors shift their focus using the new model from only managing their own activities and risks to managing all project parties' activities and risks.

References

Albogamy, A. and Dawood, N., Development of a Client-Based Risk Management Methodology for the Early Design Stage of Construction Processes: Applied to the KSA, *Engineering, Construction and Architectural Management*, 22(5), 493-515, 2015.

Al-Salman, A. A., Assessment of Risk Management Perceptions and Practices of Construction Contractors in Saudi Arabia, *MSc thesis,* King Fahd University of Petroleum and Minerals, 2004.

Al-Sobiei, O. S., Arditi, D., and Polat, G. Managing Owner's Risk of Contractor Default, *Journal of Construction Engineering and Management*, 131(9), 973-978, 2005.

Al-Otaibi, M. and Price, A. D., Analysis and Evaluation of Criteria for Pre-Selecting Contractors in the Saudi Arabian Construction Sector, *Procs 26th Annual ARCOM Conference. Leeds, UK, Association of Researchers in Construction Management*, (pp. 1141-1148), 2009.

Algahtany, M., Alhammadi, Y., and Kashiwagi, D., Introducing a New Risk Management Model to the Saudi Arabian Construction Industry, *Procedia Engineering*, 145, 940-947, 2016.

Almutairi, S., Assessment and Develop the Saudi's Contractors Classification System, *PhD dissertation,* Arizona State University, 2017.

Alsulamy, S., Developing a Performance Measurement Framework for Municipal Construction Projects in Saudi Arabia, *PhD dissertation,* Edinburgh Napier University, 2015.

Ammons, D. N., Overcoming the Inadequacies of Performance Measurement in Local Government: The Case of Libraries and Leisure Services, *Public Administration Review*, 37-47, 1995.

Assaf, S. A. and Al-Hejji, S., Causes of Delay in Large Construction Projects, *International Journal of Project Management*, 24(4), 349-357, 2006.

Assaf, S., Hassanain, M., and Al-Zahrani, S., Causes of Contractors' Failure in Industrial Projects in Saudi Arabia, *Research Journal of Applied Sciences, Engineering and Technology,* 9(3), 158-164, 2015.

Baghdadi, A. and Kishk, M., Assessment of Risks Associated with Saudi Aviation Construction Projects and of the Risks' Importance, *International Journal of Innovation, Management and Technology*, 8(2), 123, 2017.

Baghdadi, A. and Kishk, M., Saudi Arabian Aviation Construction Projects: Identification of Risks and Their Consequences, *Procedia Engineering*, 123, 32-40, 2015.

De Bruijn, H., Performance Measurement in the Public Sector: Strategies to Cope with the Risks of Performance Measurement, *International Journal of Public Sector Management*, 15(7), 578-594, 2002.

Elawi, G. S. A., Algahtany, M., and Kashiwagi, D., Owners' Perspective of Factors Contributing to Project Delay: Case Studies of Road and Bridge Projects, *Procedia Engineering*, 145, 1402-1409, 2016.

El-Mashaleh, M. S., Edward Minchin Jr, R., and O'Brien, W. J., Management of Construction Firm Performance Using Benchmarking, *Journal of Management in Engineering*, 23(1), 10-17, 2007.

Hatry, H. P. and Alexander, M., *Service Efforts and Accomplishments Reporting: Its Time Has Come*, Vol. 10, The Board, 1990.

Kashiwagi, D., Kashiwagi, J., Smithwick, J., Kashiwagi, I., and Kashiwagi, A., *Owner Controlled Delivery of Construction Degrades Quality and Value*, The Seventh International Structural Engineering and Construction Conference, ISEC-7, Honolulu, 2013.

Kashiwagi, D. and Kashiwagi, J., A New Risk Management Model, *Journal of Risk Analysis and Crisis Response*, 2(4), 233-251, 2012.

Kashiwagi, J., Sullivan, K., and Kashiwagi, D., New Contract Model for Project Management, *PM-05 Advancing Project Management for the 21st Century*, 228-335, 2010.

Kashiwagi, D., *Best Value Approach*, Tempe, AZ: KSM Inc, 2016.

Kim, S. G., Risk Performance Indexes and Measurement Systems for Mega Construction Projects, *Journal of Civil Engineering and Management*, 16(4), 586-594, 2010.

Ikediashi, D. I., Ogunlana, S. O., and Alotaibi, A., Analysis of Project Failure Factors for Infrastructure Projects in Saudi Arabia: A Multivariate Approach, *Journal of Construction in Developing Countries*, 19(1), 35-52, 2014.

Peckiene, A., Komarovska, A., and Ustinovicius, L., Overview of Risk Allocation between Construction Parties, *Procedia Engineering*, 57, 889-894, 2013.

Responsible Design and Delivery of the Constructed Project
Edited by Abdul-Malak, M., Khoury, H., Singh, A., and Yazdani, S.

DIVERSITY AND KNOWLEDGE SHARING IN CONSTRUCTION: A MATHEMATICAL MODEL

DAOUD KIOMJIAN[1], F. JORDAN SROUR[2], and ISSAM SROUR[1]

[1]*Dept of Civil and Environmental Engineering, American University of Beirut, Beirut, Lebanon*

[2]*Adnan Kassar School of Business, Lebanese American University, Beirut, Lebanon*

Conventional wisdom in the management literature holds that diversity is positively correlated with performance. Yet, the findings from the construction field indicate that this is not always the case. In an effort to study the interaction between diversity and performance in the construction industry, this paper presents the elements of a theoretical mathematical model to explore the relationship between diversity and knowledge sharing which is a precursor of performance. This model includes five dimensions of diversity: ethnicity, age, experience, language and education. At the heart of the mathematical model is a fuzzy based system that generates the probability of knowledge sharing among members with different demographic attributes. The presented fuzzy system will, in future work, become the foundation of an agent based model used to study the impact of worker interactions on productivity.

Keywords: Fuzzy system, Performance, Modeling, Labor management, Demographic.

1 INTRODUCTION AND BACKGROUND

The construction industry often assumes "rational behavior" by workers, and thus project managers tend to focus only on the technical and financial aspects of their projects (Dadfar and Gustavsson 1992). They often fail to acknowledge the socio-technical nature of these projects overlooking the fact that many of the technical and financial problems they face are rooted in the "people". This is unfortunate in a field, where the human element is a keystone to success. Construction projects are labor intensive environments, where work is accomplished by crews and rarely by individuals. Whenever crews are involved, the issues of coordination and cooperation arise; managers have to constantly question whether crew members are working in harmony or behaving as a dysfunctional unit. This facet of construction management becomes increasingly complex if the crew members are demographically diverse (Lau and Murnighan 1998). Diversity has to be properly managed, and sensitivities must be taken into account if effective work is to be expected (Ling *et al.* 2013).

Demographic differences can affect the rate of social contact among crew members (O'Reilly *et al.* 1989). Team members may want to communicate with those who share their demographic attributes as such communication has lower overhead (Haas and Cummings 2015). For example, two people of the same age are more likely to have successful communication since they share the same values because they have been through similar political and economic cycles (Ryder 1965). In addition, two people with the same educational background will communicate more eloquently since they have similar vocabularies (Haas and Cummings 2015). The effects of demographic diversity on communication and crew operation attracted the attention of multiple construction management scholars. Dadfar and Gustavsson (1992) studied the effect of cultural diversity on communication in Swedish contractors. Along the same lines, Loosemore and Muslmani (1999) investigated the problems arising due to the cultural difference between UK and Persian Gulf nationals. In more recent wok, Ling *et al.* (2013) studied the cultural differences among migrant workers in the Singapore construction industry and generated a set of recommendations for diversity management to avoid diminishing productivity and miscommunication.

These works, however, focus on a single dimension: either culture or age. Other work from the field of management suggests that diversity is a multidimensional construct (Lau and Murnighan 1998). Such a reductionist approach of studying diversity as one-dimensional may lead analysts or managers to neglect other demographic attributes which are actually significant (Lau and Muringhan 1998). When observed as a multidimensional phenomenon, diversity is not only limited to heterogeneity, it also yields faultlines or subgroups (Gibson and Vermeulen 2003). The concept of demographic faultlines is introduced via the example in Table 1. When viewed from the single dimension of cultural diversity, teams one and two are identical. However, when both dimension of age and culture are considered simultaneously, the members of Team 1 show greater cohesion as they belong to both a common ethnicity and age group. Members of Team 2 only share a similar nationality. Faultlines and subgroups have a significant impact on the success of communication among team members. Better communication usually means that there is a better flow of knowledge and information among team members (Gibson and Vermeulen 2003). An improved flow of knowledge means that the members will learn faster and thus will be able to attain the required level of performance faster. However, does this mean that all crew members have to belong to the same age group and ethnicity? The answer is no, since in that case younger members will never get the chance to benefit from the experiences of their older colleagues. What the concept of faultlines suggests is that crews have to be formed and managed while considering the sensitivities associated with such differences. If properly managed, differences could be used to improve communication and foster better performance.

Table 1. Example of faultlines and subgroups.

Team 1			**Team 2**		
Member	Age	Nationality	Member	Age	Nationality
A	20	Chinese	D	29	American
B	25	Chinese	E	40	American
C	27	Chinese	F	55	American

Demographic traits such as age group, ethnicity, and educational background can also be qualitative in nature. What does it mean if to say "Worker A is young" or "Workers A and B share a similar cultural background"? Being young or culturally similar depends significantly on the definition of age-groups and cultural similarity. As such, the two statements embedded in the question above could be both true and false at the same time; and, therefore, they are considered to be "fuzzy". Fuzzy theory allows us to deal with uncertainty and also allows for computing with words (Sivanandam *et al.* 2007). Fuzzy theory therefore permits representing linguistic constructs such as "young", "similar", "big", "late" (Sivanandam *et al.* 2007). Fuzzy logic is an attractive alternative for studying the impact of demographic attributes on communication, since there is a lack of consensus in the literature regarding the effects of these attributes on communication, and thus cooperation in work crews. There is no evidence, for instance, that different demographic attributes have a different impact on communication and learning (Gibson and Vermeulen 2003). Fuzzy logic is suitable to such a problem where there are incomplete or imprecise relationships between the variables of interest (Chan *et al.* 2009). The fuzzy system presented in this paper takes the worker's linguistic or demographic attributes as inputs, and outputs an array of probabilities representing the likelihood of communication with every other individual in the team. The outputs of the fuzzy system will go as inputs to an agent based model that aims at studying the impact of diversity on performance in construction crews. The next section provides a description of the fuzzy logic model in detail, and the last section provides a conclusion with a road map for future research directions.

2 THE MODEL

Figure 1 presents and overview of the fuzzy system.

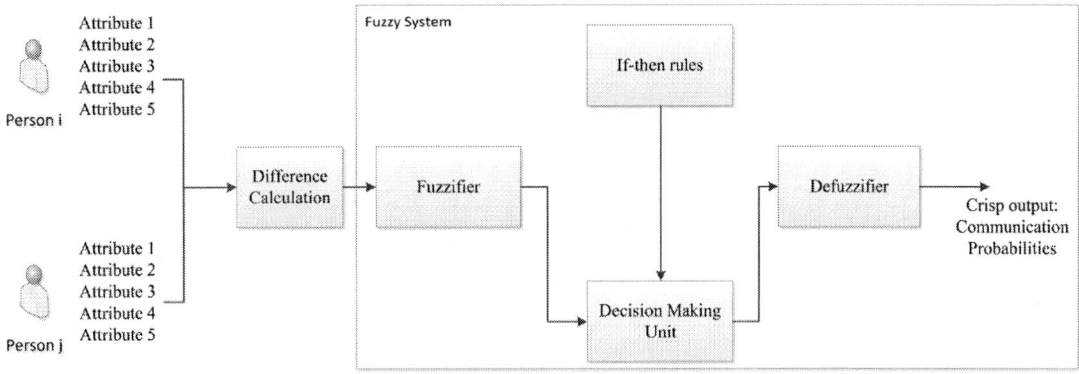

Figure 1. Fuzzy system design.

The fuzzifier takes in the crisp inputs which are the differences of demographic attributes among crew members. For example, if one crew member is 29 years old and the other is 40 years old, then the input to the fuzzier would be the age difference which is 11 in this case. The fuzzifier includes the membership function $\mu_A(x)$, which converts the inputs into a fuzzy number <0, 1>. If the fuzzy number obtained was 1, then the input belongs completely into a certain fuzzy set. Specifically, the fuzzifier takes as input the overlap among individual crew members which is in line with the methodology adopted in Gibson and Vermeulen (2003), who used the overlap among different crew members to calculate the heterogeneity and the strength of the subgroups or faultlines. Three of the five diversity dimensions used in this paper are qualitative or linguistic in nature, and these are nationality, language and education. Regarding the overlap in these dimensions, the following membership function was used:

Perfect Overlap [0.8-1.0]; Good Overlap [0.6-0.9]; Average Overlap [0.3-0.7]; Weak Overlap [0.1-0.4]; Opposition [0.0-0.2].

These measures (Perfect-Good-Average-Weak and Opposition), comprise the fuzzy sets to which the overlap between every pair of workers belongs. The numbers shown in parentheses offer a numerical measure of the overlap between every two workers. Thus, if two workers have the same nationality the overlap value would be one and the overlap between these two workers would belong entirely to the perfect overlap set. However, if two workers belong to two countries with similar cultures-one Lebanese and one Jordanian- then the value of the overlap would be in the range 0.8-0.9. The overlap between these two workers would then have 50% membership to the perfect overlap set and 50% membership to the good overlap set. The membership of the differences to the fuzzy sets shown in Figure 2 ranges between 0 and 1. If the membership value is 0, then the difference is entirely excluded from the set, whereas if it was found to be 1, then the difference belongs entirely to a set. Any number in between 0 and 1 indicates partial membership and thus the membership values to every set ranges between 0 and complete. The graphical representation of this membership function is shown in Figure 2. Unlike Gibson and Vermeulen (2003), who assumed a binary alignment among ethnicities, we allow for partial alignment. Thus, rather than assuming that an American and a German national would have zero alignment, we assume that an American would have "some" overlap with the German person due to the shared western culture and other aspects. Fuzzy systems allow the representation of this partial overlap and thus are a viable candidate for studying such complex realities. The same applies to language and education. A worker with a technical degree will have more overlap with a person with basic technical training, than with a worker who can't read or write.

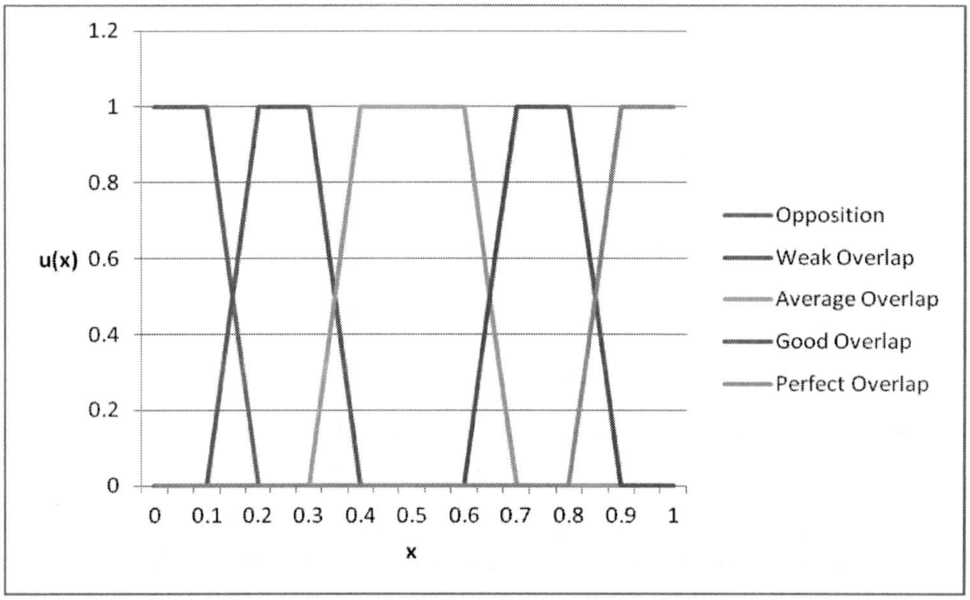

Figure 2. Membership function for linguistic inputs.

The age differences are based on a quantitative measure, and thus can be modeled by a mathematical function. The work of Zemke *et al.* (1999) identifies three generations at the current workplace. On average, a generation is the set of people with 16 years age difference. As such, people with an age difference of less than 16 years are considered to belong to the same generation, those with an age difference between 16-32 years are considered moderately different and those with an age difference of more than 32 years are considered to be very different. The membership function for the age difference is shown in Figure 3a.

The last measure for comparing crew members is the level of experience. Experienced workers are more familiar with their tasks and communicate about it easily. Overlap in experience is measured proportionally similarly to Gibson and Vermeulen (2003). The minimum value for experience overlap is zero, and the maximum is one. Thus, if one worker has three years of experience, and the other has five then the overlap is 0.6, the same value would be obtained when comparing a worker with six years of experience and a worker with ten years of experience. The membership function for experience is shown in Figure 3b.

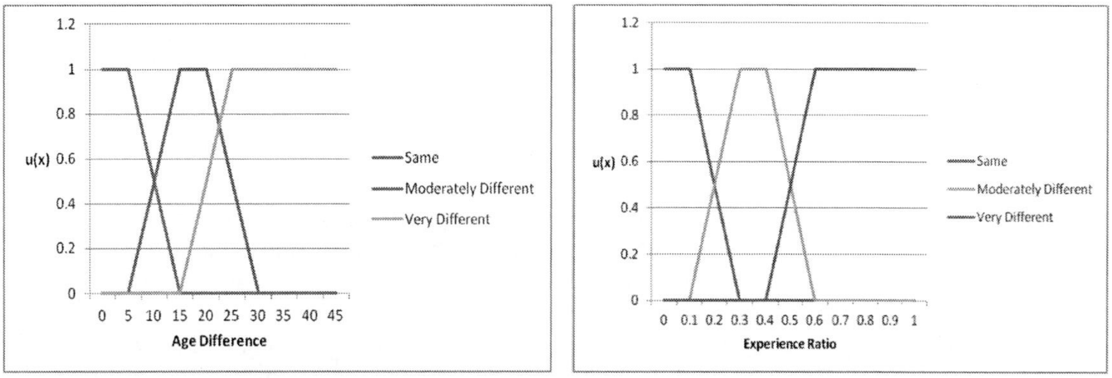

Figure 3. (a) Membership function for age difference, (b): Membership function for experience ratio.

With the fuzzifier component of the fuzzy based system defined, the next step is to define the if-then rules. These rules are logical statements that provide the intelligent component of the fuzzy system with a recipe to formulate decisions. The intelligent unit in any fuzzy system typically assumes one of two formats, Takagi-Sugeno or Mamdani. Since the input in our system is intuitive and is expected to be done by human experts, the Mamdani paradigm will be applied. Unlike Takagi-Sugeno, the output of a Mamdani fuzzy based system is also fuzzy and thus the need arises for a defuzzifier as depicted in Figure 1. A sample of the if-then rules applied in our system is shown in Table 2. It is worth mentioning that the proposed system has 243 rules; Table 2 shows an example illustrating these rules. What the first rule states is that if nationality, language and education overlap perfectly and the age and experience are the same then the output, or likelihood of communication, is highly probable.

Table 2. Sample fuzzy system if then rules.

Input					Output
Nationality	**Language**	**Education**	**Age**	**Experience**	**Probability**
Perfect Overlap	Perfect Overlap	Perfect Overlap	Same	Same	Highly Probable
Average Overlap	Average Overlap	Average Overlap	Moderately Different	Moderately Different	Average
Opposition	Opposition	Opposition	Very Different	Very Different	Improbable

However, what does an output of "Highly Probable" signify? This output is still fuzzy and needs to be converted into a numerical format, since our system is expected to generate numerical probabilities. This is when the defuzzifier comes into play, and it converts such fuzzy terms into numbers. The defuzzification method used in this system is the Centroid method. It was selected since it returns estimates which are at the center of the membership function curve. Therefore, the generated probabilities will not be too high or too low.

3 CONCLUSION

Demographic diversity could have a significant impact on communication and thus knowledge sharing in any industry. This impact could be more severe in the construction industry, where the work is done by crews and not individuals. Workers could be from different nationalities, age groups, experience levels and with varying educational backgrounds. All of these factors lead to the creation of faultlines or subgroups and workers are most likely to communicate with those belonging to the same subgroup. However, given the soft nature of the problem standard mathematical methods fail in quantifying the impacts of these attributes for a multitude of reasons. First, many demographic constructs are loosely defined. Moreover, there seems to be a disagreement on the impact of individual demographic attributes on communication and knowledge sharing. As such, a fuzzy based system was proposed to study this matter. Fuzzy based systems work well with systems that are represented linguistically and with systems that have inputs with a high degree of uncertainty.

The proposed fuzzy based system takes the demographic differences among crew members in as inputs and generates the probabilities of communication among these members. The system has three linguistic inputs, and two numerical or continuous inputs. We used the Mamdani approach while designing the system since it allows for intuitive estimates and for human estimates. Moreover, the defuzzifier uses the Centroid method to generate the outputs. This method was selected due to its popularity and its capability of generating moderate estimates. These probabilities will be used in future work as the theoretical foundation for an Agent Based Model that studies the impact of demographic diversity on knowledge sharing among construction crews.

References

Chan, A. P., Chan, D. W., and Yeung, J. F., Overview of The Application of "Fuzzy Techniques" in Construction Management Research, *Journal Of Construction Engineering And Management*, 135(11), 1241-52, 2009.

Dadfar, H, and Gustavsson, P., Competition by Effective Management of Cultural Diversity: The Case of International Construction Projects, *International Studies of Management & Organization*, 22(4), 81-92, 1992.

Gibson, C., and Vermeulen, F., A Healthy Divide: Subgroups as a Stimulus for Team Learning Behavior, *Administrative Science Quarterly*, 48(2), 202-39, 2003.

Haas, M. R., and Cummings, J. N., Barriers to Knowledge Seeking within MNC Teams: Which Differences Matter Most?, *Journal of International Business Studies*, 46(1), 36-62, 2015.

Lau, D. C., and Murnighan, J. K., Demographic Diversity and Faultlines: The Compositional Dynamics of Organizational Groups, *Academy of Management Review,* 23(2), 325-40, 1998.

Ling, F. Y., Dulaimi, M. F., and Chua, M., Strategies for Managing Migrant Construction Workers from China, India, and The Philippines. *Journal of Professional Issues in Engineering Education and Practice*, 139(1), 19-26, 2013.

Loosemore, M. and Muslmani, H. A., Construction Project Management in The Persian Gulf: Inter-Cultural Communication. *International Journal of Project Management*, 17(2), 95-100, 1999.

O'Reilly III, C. A., Caldwell, D. F., and Barnett, W. P., Work Group Demography, Social Integration, and Turnover, *Administrative Science Quarterly*, 1, 21-37, 1989.

Ryder, N. B., The Cohort as a Concept in the Study of Social Change, *American Sociological Review*, 1, 843-61, 1965.

Sivanandam, S. N., Sumathi S, and Deepa S. N., Introduction to Fuzzy Logic using MATLAB, Berlin: Springer, 2007.

Zemke, R., Raines, C., and Filipczak, B., Generations at Work: Managing The Clash of Veterans, Boomers, Xers, And Nexters in Your Workplace, Amacom, 1999.

Responsible Design and Delivery of the Constructed Project
Edited by Abdul-Malak, M., Khoury, H., Singh, A., and Yazdani, S.

PRODUCT VERSUS PROCESS: A JUNGLE COMMUNITY CASE STUDY FOUR YEARS LATER OF TECHNOLOGY TRANSFER

ENRIQUE VILLACIS[1], MARIA LORENA RODRIGUEZ[2], and CYNTHIA AYARZA[2]

[1]*School of Architecture, Design, and Arts, Pontifical Catholic University, Quito, Ecuador*
[2]*Ensusitio Arq., Quito, Ecuador*

More than four years have passed since the collaboration with the Kichwa jungle community of Santa Rita in Ecuador. While designing and building the Cocoa Cabin, we were able to see how technology transfer took place. Based on this case study we will refer to the negative and positive impacts of the technology transfer system over time, acknowledging the differences between a product versus process approach. Comparing three specific points in time where technology transfer was implemented in: the Cocoa City (2010-2015), the Cocoa Cabin (2013-2014), Bolivar′ Lookout (2017), demonstrating how imposed technology leads to misuse or disuse because of lack of participation and understanding, and how technology development, when escorted by people and collaborative processes, leads to successful technology transfer.

Keywords: Participatory design, Endogenous, Empowerment, Knowledge transfer, Ecuador.

1 WHY TECHNOLOGY TRANSFER?

From the beginning of time, technology transfer has benefited the human conditions. "Within groups, apprenticeship was the main vehicle for knowledge transmission across generations." (Gorman 2002). Though in contemporary times technology transfer has been focused on the product, the technology rather than the necessity, the real understanding of the idea or ideal behind it: knowledge transfer. "The cultural and political aspects of technology transfer are even more complex than the economic or technical ones. (…) Most often the introduction of a new technology will not meet with one kind of reaction, but with several at once; in other words, it will provoke conflicts within the importing society" (Headrick 1988).

Introducing a technology in a specific context if not understood may cause more problems than benefits, and probably cause dependence, in which case we should think about who is the real benefactor. "Economic hit men (EHM) are highly paid professionals who cheat countries around the globe out of trillions of dollars. They funnel money (...) from foreign "aid" organizations into the coffers of huge corporations and the pockets of a few wealthy families who control the planet's natural resources. They play a game as old as empire, but one that has taken on new and terrifying dimensions during this time of globalization. I should know; I was an EHM" (Perkins 2004).

This study will trace different periods and methods of technology transfer within a site-specific case study located in Santa Rita in Ecuador. Two approaches, product vs. process, will be confronted through time in the community. This comparison will be illustrated by the use of construction detailing with two different ideological approaches, dependences vs. empowerment.

2 WHY ENDOGENOUS TECHNOLOGY TRANSFER IN OUR TIMES?

Decreased oil production, energy crisis, and global warming are facts that have negative global impact. Although Ecuador "has 4.6 billion barrels of proven oil reserves, with crude oil production of around 390 000 barrels per day" (EIA 2003), all this production is already presold to foreign governments. Therefore, one of the most challenging matters is to find a different income besides petroleum. It has been discussed in several studies that in order to achieve a more sustainable society, an endogenous thinking would be better than an exogenous one. Endogenous development is to be understood as "local development produced mainly by local impulses and grounded largely on local resources" (Picchi 1994). While exogenous approaches see development as continuous growth which in face of the energy consumption projections, mentioned before, might be disastrous. "Comparing monetary systems, an exogenous approach leads to an accumulation monetary system, where there will be shortage, competition, continued growth, individualism, power concentration and conquest or colonization, in the other hand, endogenous approach leads to no accumulation, sufficiency, cooperation, sustainability, dispersion of power, upkeep and maintenance" (Villacis *et al.* 2017). The difference between acting and thinking local may be the difference when dealing with a global energy crisis.

3 SANTA RITA, PACARI, AND THE ECUADORIAN GOVERNMENT

The Kichwa community of Santa Rita, located in the Ecuadorian Amazon, four hours from Quito the capital city, has a population of 700 inhabitants. It is where one of the finest cocoa from the world comes from. "The community is characterized by the ancestral planting of fine aroma cocoa, typical of the Amazon basin. The families of the community have been engaged in the planting and harvesting of this fruit for decades, which has allowed them to bring sustenance to their homes" (Ecuador Estratégico 2015). The Cocoa Village, highlights the symbolic aspects, history, and traditions and seeks to promote the product to the whole region. Based on this fact there where two different stands within the building process of infrastructure for the specific need of tourism. In one hand the Ecuadorian government agency - Strategic Ecuador, has managed the construction of the "Cocoa Village" (see Figure 1), a fragment of the Bio Route of Cocoa and Chocolate, making the community of Santa Rita the cornerstone of tourism development of Napo Province. The "Cocoa Village" project involves the construction of touristic infrastructure such as communal ranches, plazas, roads, trails, community areas, green areas and spaces for cocoa production" (Ecuador Estratégico 2015).

On the other hand, Pacari Chocolates collaborates for an extended period of time within the Santa Rita community building a relationship that runs from the familiar to the business. "Pacari means "nature" in kichwa language; a name chosen to evoke everything that chocolate represents: the best products from the earth, 100% natural. Pacari's quality reflects the effort with every person involved in the continuous innovation, social responsibility, sustainability and direct business trade with small-scale farmers. This process has one principal player: the Ecuadorian *cacao fino de aroma* known as "Arriba Nacional" (Pacari Premium Organic Chocolate 2015). The collaboration, in this case, came from understanding the way cocoa is produced and commercialized, thus based on these facts find a way for both, the private sector and the community, to improve their productivity. The infrastructure, in this case, comes from the community and its partner´s necessity: a place to show and teach best practices in the cocoa production and commercialization. When the project's base is technology transfer, it has a different means and end. See Figure 2.

We are going to describe three different construction processes, exploring the technologies and knowledge transfer behind the project, making an analysis of product vs process.

The Cocoa Village was first projected and left on standby for several years, in the meantime the Cocoa Cabin was built and when the project was finished the Cocoa Village started to be built. After a year of the completion of the Cocoa Village, Bolivar's Lookout was built.

Figure 1. Cocoa Village. (Government).

Figure 2. Cocoa Cabin, (Pacari, CLQH and Community).

4 COCOA VILLAGE: ECUADORIAN GOVERNMENT AGENCY

- Two years in the planning with almost no communication with the community, the Ecuadorian president suddenly demands the Cocoa Village has to be built and quickly.
- Foundations and slabs were poured concrete and metal plates. All materials and manpower are brought in from distant places. See Figures 3 (a) and (b).
- Wood was used for the structure, it needed to acclimatize and re-treated on site in order to be resistant to local pests and weather.
- The roofs of the Cocoa Village are zinc metal roofing and wood tiles. Zinc a material that has been used in low-cost houses in most areas of Ecuador because it is cheap and placed easily but it has a short lifespan. See Figure 4 (a). The wood tiles a foreign technology that the Builder himself did not guaranty. See Figure 4 (b).
- The contractor needs to assure the performance of workers so the building will be completed on time. The manpower from the community is only used as a complement to the main workers, not allowing a technology exchange but an employee-employer relation. Eighteen months of construction and the only interaction was the meals for the foreign workers.
- This project brings a strong sense of dependence because the know-how was not exchange and maintenance requires materials and assessment from outsiders.
- The Department of Tourism knew the buildings were oversized for this small town and sought for ways to justify the built infrastructure.
- "The Bio Route of Cocoa and Chocolate represents a large investment in terms of financial and human resources. In spite of that, until now it has not meant a significant increase in tourism or income to the community. Currently, Napo has a large offer of community tourism and a limited number of tourists. There is also a lack of links between the Cocoa Route and local and international tourism operators and a lack of information on the cocoa route in Napo (Cummins *et al.* 2016).
- The project is a 1.3 million USD investment, amount that comes from government earning off petroleum extraction. Currently, the government is receiving less income from petroleum. This means that any maintenance of the infrastructure will have to be

done with local funding. No matter how efficient the tourism and entrepreneurships are, the infrastructure is so large and complex that maintenance will be close to nonexistent.

- This space is used more for conferences rather than to attract tourists interested in cacao.

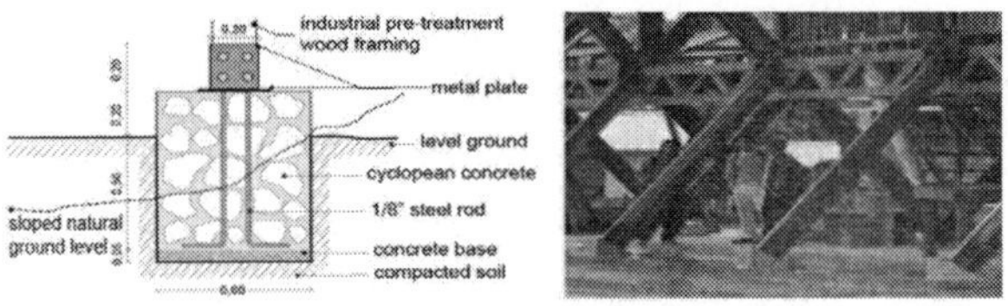

(a) (b)

Figure 3. Foundations with concrete and metal plates.

(a) (b)

Figure 4. Metal roofing and wood tiles.

5 THE COCOA CABIN

- The process began in 2013; Pacari starts organic agriculture, fair trade, and customer service workshops, and an anthropology study. As a result, the possibility for the community to generate agricultural tourism is feasible and Pacari sponsors the building of the Cacao Cabin. Creating a design team with Ensusitio, *Con lo que hay* Workshop *(CLQH)* of the Catholic University, School of Architecture of Quito, and the community.
- The work was with what the site gave us, the existing boulders as the foundation. Both, Cocoa Cabin and bridge were raised on the existing on-site boulders. As a result, the use of the existing rocks reduced the cost and length of the project. See Figure 5.
- The joint between the bamboo made both work as one, by drilling the stone and introducing a steel rod and unite both with an epoxy adhesive, (liquid stone). The bamboo was set over the steel rod and by filling the knot with cement making them bind.
- The structure was built with the wood and *caña guadua* (local bamboo), all bought from the local farmers. This system was developed on site by an engineer, the designers, and the community, taking in consideration what was available and what was required. By using the materials and a shared manpower the appropriate technology transfer is ensured because people from the community shared the conceptualization of this method is now capable of: maintenance of the structure, if needed, replicate or improve the detail.
- The roofing was made with toquilla straw, an endemic leaf used since early settlements in Ecuador. The process starts by knowing the specific moon phase for harvest to the correct placement of the toquilla straw. See Figure 6.
- The sense of independence and empowerment that this process leaves is very strong since the technology was developed on site, based on their ancestral construction knowledge and improvements were shared and mutually developed, then the community by themselves will be able to adapt, built, repair and even develop their own technology. A sense of collaboration and pertinence was created during this process.
- The design and construction team CLQH IV: FADA-PUCE were actually the first clients. They were the practice team on how to present the Cacao Cabin, how to prepare the meals, finding the topics and the essence of the tour. Developing the infrastructure and the tour pan in parallel made the project work as a whole.
- The $21.000 investment by Pacari stayed 99% within the community, purchasing natural materials from the community and hiring local labor. Pacari brings constant international

tourism all year round, the community earns from the meals and guiding and a small percentage is set to pay off the investment, in two years one third is payed off, a win-win situation. Incomes are generated and Pacari promotes the Cabin, and they can show the world where it´s products come from.

- The Cacao Cabin and its process have actually been one way to promote de Cacao Village itself generating additional income for the community.

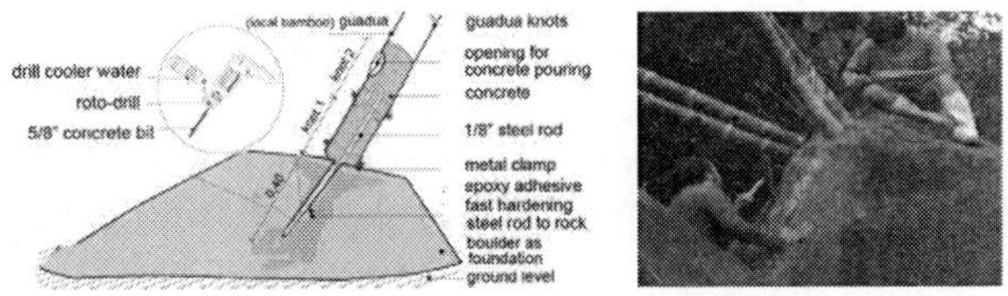

Figure 5. Foundation with local materials.

Figure 6. Toquilla straw roofing.

6 BOLIVAR´S LOOKOUT

Constant contact and visits to the community made it possible for inquiries to be resolved regarding maintenance and expansions, until we receive a phone call from the former community president, Bolivar, "Can you please send us the bits we used to drill the rock? We already have the rods, the epoxy but the bits are of poor quality here". It happens they are building a lookout and they had followed the process from the Cacao Cabin, the difference was that they were not using bamboo, but hardwood. They understood the system and applied it to the materials they had handy. Bolivar Alvarado as president of the community during the Cacao Cabin process actively participated in every step of the construction, learning the foundation system, and teaching us the work in toquilla straw for the roofs and all the bamboo and woodwork, all together with other members of the community. This way their own project was developed.

- In one of the trails to the cacao crops, two more boulders and a stream are passed, and this is a potential for a new lookout and a fishpond. See Figure 7.
- The foundations are big boulders, existing on site, and protect the structure from water.
- The structure is from locally harvested hardwood, resistant to local insects and humidity, anchored to the boulders with steel rods and epoxy as in the Cacao Cabin.
- The roof was made of toquilla straw, having a reinforced detail of the ridge, to protect the wood structure with a long eave. See Figure 7.
- Al materials where harvested from the property. The investment was $100. The process took three months.

Figure 7. Foundation of the bridge, roof structure of Bolivar´s Lookout.

"There is now a community in Napo which makes elevated constructions supported on stones, without concrete. Sometimes it seems that what we do is useful, sometimes it seems like we´re good at what we do. Sometimes we do something," as commented by Gabriela of Pacari´s relations.

By understanding the potential of what is on hand: boulders, wood, toquilla straw, the knowledge learned on the Cacao Cabin, and the view, a complementary project to the Cocoa Cabin was built: Bolivar´s Lookout.

7 CONCLUSION

Four years after, the imposition of the Government of a large-scale infrastructure still overwhelms Santa Rita. The Bio Route of Cocoa and Chocolate, a national project, with a millionaire budget to build but built on a pour study of the scope and upkeep demands. The Cacao Village is struggling to have constant visitors. Roof leaks and needs urgent repair. And all they expect for this infrastructure is what the government brings their way. The government has to do the maintenance, which is logical but unreal and the income from the tours barely cover costs.

With direct knowledge exchange while building the Cacao Cabin, when more tourists keep coming, the surrounding are being rediscovered, new trails, new needs, and opportunities. The community can propose what to do next. They can feel the potential of the project that they built together with Pacari, and CLQH, because it keeps moving and inspiring new work. The Cabin has a new storage space, it has had upkeep, bathroom´s and main roof have been rewoven, the material is handy and the community has the knowledge to work on their own.

- Processes oriented systems are often initiated based on necessity and evolve based on participation, on the other hand product-oriented processes mainly end up creating unnecessary dependence.
- Having a proper understanding and appropriation of the technology leads to the development and evolution of it by understanding the proper use.
- Although the process system can take more time to be implemented rather than the products system, in the long term the technology applied by process will support the development of communities avoiding dependence on maintenance and appliance of it.

Comparing both systems, knowledge transfer product vs. process, with the example of Bolivar´s Lookout, processes is what creates independent and empowered communities in order to ensure their autonomous development, this achievement can be supported by technology.

References

Cummins, I., Alvarado, E., and Andi, B., Actualización del Plan de Manejo de Comunidad Santa Rita 2016-2026, Santa Rita: Runa Foundation, 2016.

Ecuador Estratégico. *Pueblo Del Cacao Fomenta Desarrollo Turístico En Napo*. Retrieved from http://www.ecuadorestrategicoep.gob.ec/15-notefp/923-el-pueblo-del-cacao-fomenta-desarrollo-turistico-en-napo on September 23, 2015.

EIA, U. E., Ecuador: Country Analysis Brief. Ecuador, May 28, 2003.

Gorman, M., Types of Knowledge and Their Roles in Technology Transfer, *The Journal of Technology Transfer*, 219-220, 2002.

Headrick, D. R., *The Tentacles of Progress, Technology Transfer in The Age of Imperialism, 1850-1940.* New York: Oxford University Press, 1988.

Pacari Premium Organic Chocolate, P. Retrieved from *http://www.pacarichocolate.com/*. on January 1, 2016, de http://www.pacarichocolate.com/en/index.php/history, Pacari, 2015.

Perkins, J., *Confessions of an Economic Hit Man*. San Francisco: Berret-Koehler Publishing, Inc, 2004.

Picchi, A., *Born from Within, Practice and Perspective of Endogenous Rural Development.* The Netherlands: Van Grocum Assem, 1994.

Villacis, E., Rodriguez, L., and Ayarza, C., *Resilient Structures and Sust. Constr.*, ISEC Press, 2017.

Responsible Design and Delivery of the Constructed Project
Edited by Abdul-Malak, M., Khoury, H., Singh, A., and Yazdani, S.

INVESTIGATING THE CIRCULAR ECONOMY AS A SOLUTION FOR CONSTRUCTION WASTE: CHALLENGES AND OPPORTUNITIES

LUCIEN EL ASMAR, NAMHO CHO, and MOUNIR EL ASMAR

School of Sustainable Engineering and the Built Environment, Arizona State University, Tempe, USA

The construction industry results in a large amount of material waste, most of which currently ends up in landfills. This study aims to find new ways to reuse waste materials captured as outputs of construction projects. First, the circular economy (CE) concept is explored. Second, the application of CE as a potential solution for minimizing material waste on construction projects is investigated. To achieve a healthy industrial metabolism, the paper illustrates how a complex industrial ecosystem is one that produces little or no waste and whose constituents can be interdependent. Output ratios from the construction industry are compared with previous literature on material waste, as a foundation to present a cycle of material reuse. Reformulating the traditional input-output system into a more circular concept presents a set of challenges; however, the opportunities and significant impacts to the material cycles and landfills have proven beneficial and some major cities are starting to employ this concept with great success.

Keywords: Ecosystem, Waste management, Demolition waste, Industrial metabolism.

1 INTRODUCTION

Although the Circular Economy (CE) concept has gained popularity recently, it actually has been a field of research and documentation for some time. Assets that are at the end of their service lives can be used for other purposes instead of being discarded. Outputs of a production process can be resources for other industries. This idea of finding a use for every waste stream can help close loops in industrial ecosystems. Stahel (2016) states CE has enormous potential in addition to existing waste minimizing concepts such as reduce, reuse, recycle (3R) and industrial parks, presenting a new economic viewpoint centered on sufficiency.

Since year 2000, the U.S. reported generating at least 136 million tons of building construction and demolition (C&D) waste each year. However, only 20 to 30 percent of this waste is recycled (Yuan *et al.* 2010). Some agencies have tried to tackle this issue; for example, the U.S. Environmental Protection Agency (EPA) provides a standard for construction waste management (U.S. EPA 2007). However, C&D waste is still a major contributor to landfills. One possible incentive to help address this problem is focusing on the economic benefits of reducing waste.

Little research focused on examining the feasibility of C&D waste management using a cost-benefit analysis approach. A systematic economically driven approach is needed to reduce construction waste. This paper aims to investigate the state of practice of waste management in construction, and explore the opportunity for CE to impact the waste resulting from the construction industry. To achieve this goal, the authors examine waste statistics from the U.S., South Korea, and the European Union. Then, construction-specific waste ratios are studied and compared, which is one key contribution of this paper. Based on the findings, the paper discusses how CE can help reduce construction waste by examining applications of CE that are economically advantageous and resulting in waste reduction in countries such as the U.K. and the Netherlands.

2 CIRCULAR ECONOMY AS A POTENTIAL SOLUTION TO WASTE

The Ellen MacArthur foundation (2015) defines a Circular Economy as one that keeps products, components, and materials at their highest value and in use. In CE, waste is considered an output and is matched with a process to which it can add value. Figure 1 shows the biological cycles of resources and materials on the left side, and the technical cycles on the right side. This paper focuses on the four technical cycles of share, reuse, refurbish, recycle.

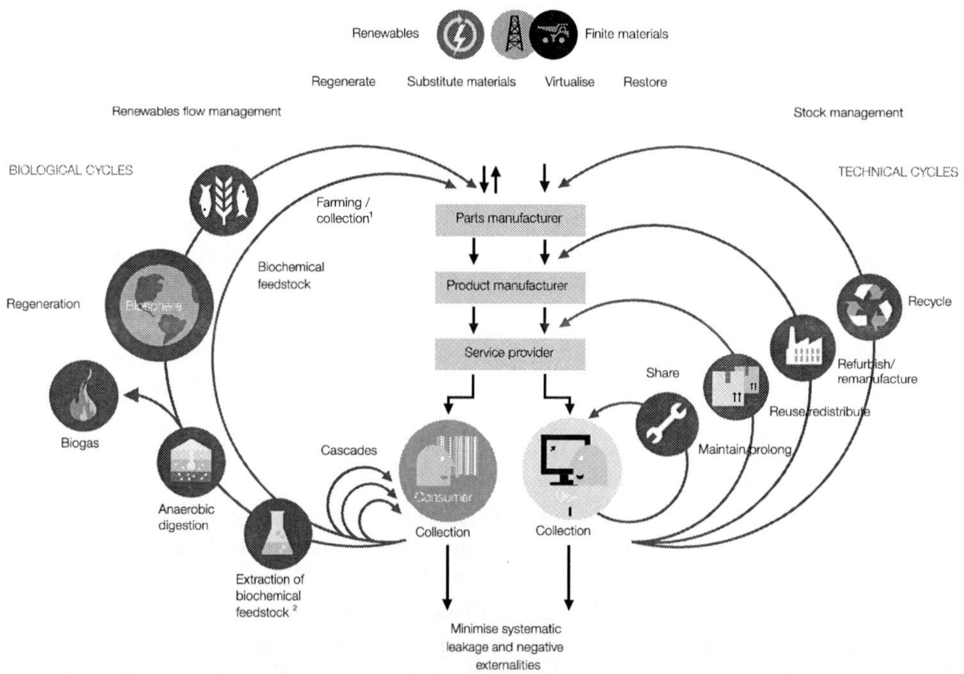

Figure 1. Cycles of the Circular Economy (adapted from Ellen MacArthur Foundation 2015).

ARUP (2016) defines *sharing and maintaining* as maximizing goods utilization and prolonging their life, as well as optimizing an asset's performance. This cycle is preferred when compared with the other cycles due to its least use of new resources. An example in construction is simply reusing existing buildings and spaces as opposed to demolishing and rebuilding a new facility. Next in the figure is *reuse/redistribution* cycle, which means minimizing production and reusing long life components of a system on another structure. One example is modular design in construction. The third cycle is *refurbish and remanufacture*, where components are remanufactured, expending little resources to add considerable value. Finally, recycling is the last and least preferred technical cycle if all others are not possible. One example in construction is crushed concrete, once compliant with earthwork requirements, it can be used as satisfactory soil for fill or sub base. Keeping the resources in closed cycles is key for a successful CE. All technical cycles take part of a model that highlights regenerating and restoring capital, aiming to maintain the resilience of ecosystems.

3 THE CONSTRUCTION INDUSTRY HAS A LOT TO BENEFIT FROM A CIRCULAR ECONOMY

In order to attain and benefit from a CE, the construction industry would require a transformation from its current linear production-consumption pattern, which would first entail a study of how construction material flows are organized. Ness and Atkinson (2001) state that construction material inputs into buildings are growing at an alarming rate. This fact is an indicator of a growing C&D waste problem down the line. Table 1 shows C&D waste ratios of 34.7% and 47.7% for the European Union and South Korea, respectively, based on data from the South Korean Ministry of the Environment (2014) and Eurostat (2017). The table shows that C&D are responsible for a very large portion of total waste. In order to perform a more detailed analysis, Table 2 uses data from the same sources to present C&D waste statistics in these two regions and indicate potential opportunities in C&D waste management. When looking at the 868.5 tons of total C&D waste, the EU recycles 36.20 percent of it while South Korea recycles an enormous 97.88 percent of total construction waste. In the EU, about half of the C&D waste (411.67 million tons per year) still goes to landfills. In a study that investigates the different types of materials existing in demolition debris and construction waste in the United States (U.S. EPA 2007), the numbers put concrete in the first place (by weights) with about 70% of the total C&D waste. Table 3 shows the breakdown of the top C&D waste materials in the U.S.

Table 1. C&D waste ratios in the E.U. and South Korea (2014).

	European Union	**South Korea**
C&D Waste (million tons)	868.5	67.7
Total Waste (million tons)	2,502.9	141.7
Portion of C&D from total (%)	**34.7**	**47.7**

Table 2. Waste management statistics for C&D in E.U. and South Korea (2014).

Waste Management Method	**European Union**		**South Korea**	
	Million tons	**Percentage**	**Million tons**	**Percentage**
Landfill	411.67	47.40	1.08	1.59
Incineration	13.03	1.50	0.36	0.53
Recycle	314.40	36.20	66.23	97.88
ETC (Energy and Backfilling)	129.41	14.90	N/A	
Total C&D Waste amount	**868.50**	**100.00**	**67.67**	**100.00**

Table 3. C&D waste materials in the U.S. in million tons (2014).

Materials	**Demolition debris**	**Construction waste**	**Total**	**Percentage**
Concrete	353.6	21.7	375.3	70.3%
Asphalt concrete	76.6	0	76.6	14.3%
Wood products	35.8	2.9	38.7	7.2%
Asphalt shingles	12.7	0.8	13.5	2.5%
Brick and clay tile	11.8	0.2	12	2.2%
Drywall and plasters	10.3	3.3	13.6	2.5%
Steel	4.3	0	4.3	0.8%
Total	**505.1**	**28.9**	**534**	**100.0%**

In the case of concrete, the U.S. numbers present an excellent opportunity since South Korea showed almost 100% recycling for demolished concrete and asphalt concrete. Traditionally wasted materials can be reused or recycled and used as inputs to the construction industry as well as other industries. The next section illustrates additional opportunities by highlighting current CE applications in the construction industry.

4 CURRENT APPLICATIONS OF CIRCULAR ECONOMY

Circular economy concepts are being used around the world, leading to positive results. For example, a multi-story modular housing block in Stoke Newington, London, UK is adapting the concept of design for disassembly. Modular construction using 3D printing and additive manufacturing could reduce structural waste and construction times, delivering a benefit of over 800 million U.K. pounds per year (ARUP 2016). Another example is Denmark, which is mandating new requirements and presenting new standards for material waste management. As politically encouraged, the total cost of ownership (TCO) of each type of material needs to be comprehensively studied for each product. Waste containers, for instance, are made from one plastic polymer type that is highly recyclable. Reprocessing of worn out plastic containers, by the producer, saves around 80% of the fossil fuel used for manufacturing of virgin plastics. Furthermore, plastic reprocessing saves on average 1.15 tons of CO_2 per ton of plastics in this small example alone (Nielsen 2014). Another example is Japan, which implemented the "reduce, reuse, recycle" (3R) concept in 2005 and presented several regulations including fundamental construction material recycling laws. Incineration in Japan is the highest compared to other countries (Ministerial Conference on the 3R Initiative Website 2005) and data show a growth in incineration from around 20% in 1955 to 75% in 2000. With the introduction of the new regulations, the byproducts of incineration were included in a circular economy cycle instead of being wasted: Gas emissions were controlled through a strict maintenance of incineration temperatures, and the exhaust heat is used for power generation.

In addition to applications in the construction industry, CE concepts are being applied in industries directly related to construction, such as mining. Construction material flows often start with mining. There are opportunities with mining, by capturing byproducts. One example is zinc, used for galvanizing construction materials. Zinc can be captured while processing iron and steel scrap. Another example is Phosphate mining, which allows for a possible recovery of both gypsum and fluorine at the same time. This synergy reduces both the waste disposal from the phosphate mining as well as the mining process of fluorine and gypsum (Ayres and Ayres 1997).

Gypsum is another major construction material. More than 4 million tons of gypsum scrap was recycled in the United States in 2015 alone. Gypsum is primarily used for wallboard, but markets for this particular mineral include athletic fields, cement production, grease absorption, water treatment, and agricultue. The U.S. produced 11.5 million metric tons of gypsum in 2015, making it the largest mine producers of gypsum (U.S. Geological Survey 2016). Recycling C&D gypsum waste can help close the loop and offer a stable supply to the construction market, especially at a time when both U.S. imports of gypsum and mine production are increasing.

5 CHALLENGES AND FINANCIAL OPPORTUNITIES OF CIRCULAR ECONOMY IN THE CONSTRUCTION INDUSTRY

In a survey about waste management, Tam (2007) showed that in construction, cost is still often considered as the major decision factor, while harm to the environment is less important. Financial incentives must be clear in order to attract cross-industrial partnerships within a broader goal of reaching a circular economy. Therefore, the authors analyzed available waste data in an attempt to assign a monetary value to it.

Using the South Korean waste statistics discussed earlier, even though recycling percentages are 98%, additional construction materials can still be included in the waste management plan, such as glass, tiles and ceramics, textile, wallpaper, surface coating materials, and panels. The assumptions for these remaining materials are:

- Construction demolished Soil: $30.62 per 50 lb ($1.35 per Kg)
- Glasses: $26.95 per 2 lb ($24.3 per Kg)
- Tiles & ceramics: $1 per 0.25 kg ($4 per Kg)
- Textile: $350 per ton ($0.35 per Kg)
- Wallpaper: $35 per Kg
- Surface coating board & panel: $200,000 per day
- Sludge, brick & block, wood, resins, and mixed construction materials are negligible

Adding these up for South Korea results in an additional U.S. $1.48 million in benefits per day, or U.S. $532 million per year that can be saved if applying CE succeeds. Recent reports agree that there is a strong economic case for CE. One example is a study on the Indian market, which estimates that applying CE in the construction sector could lead to an annual benefit of 76 US $billion by 2050 (Sukhdev *et al.* 2016). Another example is described by the London Waste and Recycling Board (2015). London consumes more than 20 million tons of construction material per year and produces 10 million tons of construction waste. The latest estimates for the potential benefit from CE opportunities in the built environment in London adds £3 to 5 billion annually to the UK GDP by 2036. Strategies include designing for effective building disassembly, material management and re-use management of building materials.

Moreover, the Ellen MacArthur Foundation (2015) estimates that moving to a circular economy in the EU alone can capture benefits amounting to €1.8 trillion by 2030, with a growth of 3% annually in resource production. Similarly, Japan instituted new regulations for both generators of industrial waste and waste management firms to support the current implementation of the 3Rs concept and ultimately circular economy (Japanese Ministerial Conference 2005).

6 DISCUSSION OF RESULTS

The data presented in this paper showed significant potential of implementing CE in the construction industry. Since monetary benefits constitute key incentives for stakeholders involved in the CE process; this section interprets the results introduced earlier and further discusses their impacts, specifically when it comes to the EU.

In the previous section of this paper, a benefit of $532 Million per year was calculated as the possible result of implementing CE in South Korea, even though its recycling percentages were already at a very high 98% rate. That remaining 2% of C&D waste in South Korea represents about 1.35 million tons of material waste. Dividing $532 Million by 1.35 million tons results in a value of $394 Million for every million ton of material waste. While the EU recycles only 36.20% of their C&D waste, 63.8% (554 million tons) go mostly to landfills, and in smaller shares to incineration and ETC. Assuming the same monetary values of South Korea's materials ($394 Million per million ton of waste) for the sake of computing an estimate for this study, if the EU proceeds to recycle its remaining waste, they could benefit from about $220 Billion per year.

Moreover, ARUP (2016) showed recycling as the least favorite cycle, whereas sharing, maintaining, reusing, and remanufacturing were all preferred venues. Therefore, financial advantages and environmental benefits from preferred cycles can be even greater than the numbers shown here for recycling. In summary, CE can be viewed as a cost-driven concept. The most efficient way of implementing CE is by keeping the products at the end user level and reducing the need for manufacturing new products using new raw materials. Instead, making good use of the large amount of available C&D materials currently being wasted, is a sizeable opportunity for our industry.

7 CONCLUSION

This paper quantified construction and demolition material waste and contrasted it in the EU, South Korea, and the U.S. It proposed circular economy as a solution for construction waste, while presenting a strong case for CE based on current best practices and existing financial opportunities. The opportunities offered by applying a circular economy are contingent on strong partnerships. Moving towards CE requires a cross-industry, cross-performance and multidisciplinary approach in order to find a diverse enough pool of "buyers" so that every product and byproduct finds a home. In the construction industry, there is certainly a need to understand C&D material waste flows in order to identify potential uses and reduce C&D landfill waste. A limitation of this paper is the number of countries for which data was available. Future work will expand the list of countries investigated, and map the material flows out of construction projects to develop a comprehensive set of strategies needed to embed the construction industry in a circular economy.

References

Arup, The Circular Economy in the Built Environment, 2016.

Ayres, R. U. and Ayres L., W., Industrial Ecology: Towards Closing the Materials Cycle, *Journal of Industrial Ecology,* April 1997.

Ellen McArthur Foundation, *Cities in the Circular Economy: An Initial Exploration*, 2015.

Eurostat, Waste Statistics, 2017. Retrieved from http://ec.europa.eu/eurostat/statistics-explained/index.php/Waste_statistics on March 2018.

Japanese Ministry of the Environment, Ministerial Conference on the 3R Initiative, 2005. Retrieved from http://www.env.go.jp/earth/3r/en on March 2018.

London Waste and Recycling Board, London the Circular Economy Capital, 2015.

Nielsen, M., M., Iversen T., L., and Skovgaard M., Green Public Procurement Analyzed in Denmark, Case: City of Copenhagen, Plastic Zero, 2014.

South Korean Ministry of the Environment, *Analysis of National Waste Management Condition* 2014. South Korean Ministry of the Environment, Korea Environment Corp, Report number 11-B552584-000005-10.

Stahel W. R., The Circular Economy, Stiftungsfonds für Umweltökonomie und Nachhaltigkeit, Growth Within: a Circular Economy Vision for a Competitive Europe, McKinsey Center, MacArthur Foundation, 2016. Retrieved from https://www.nature.com/news/the-circular-economy-1.19594#/b2 on March 2018

Sukhdev A., Vol J., Brandt K., and Yeoman R., *Cities in the Circular Economy: The Role of Digital Technology*, Ellen MacArthur Foundation, 2016.

Tam Vivian W. Y., On the Effectiveness in Implementing a Waste-Management-Plan Method in Construction, *Journal of Waste Management,* Elsevier, 2007.

U.S. Environmental Protection Agency (EPA). *Construction Waste Management*. Section 01 74 19, 2007.

U.S. Geological Survey, Mineral Commodity Summaries, January 2016.

Yuan H. P., Shen L. Y., Hao Jane J. L., and Lu W. S., A Model for Cost-Benefit Analysis of Construction and Demolition Waste Management Throughout the Waste Chain, *Journal of Resources Conservation and Recycling,* Elsevier, 55(6), 604-612, 2010.

Responsible Design and Delivery of the Constructed Project
Edited by Abdul-Malak, M., Khoury, H., Singh, A., and Yazdani, S.

THE IMPACT OF POSTS ON INTEREST IN A DEPARTMENT OF CONSTRUCTION MANAGEMENT FACEBOOK PAGE

JOHN SMALLWOOD

Dept of Construction Management, Nelson Mandela University, Port Elizabeth, South Africa

Universities, and units within, use social media to communicate with their respective publics, for varying reasons, namely to promote interest in a page and the related 'owner', to promote events and opportunities, to 'educate', and create awareness. Experience and anecdotal evidence indicate that visitors to Facebook pages find certain posts more interesting than others, which reality administrators of Facebook pages must contend with. The purpose of the study reported on is to determine the impact of a series of posts to determine, which attract the most interest. The research method can best be described as experimental in that posts pertaining to different subjects were undertaken periodically. Posts were constrained to construction, tertiary construction management education, and related issues such as conferences. The salient findings include that graduation publicity in the form of photos, predominates, followed by project related photos, and events in the form of conferences, presentations, and summits in terms of interest. Based upon the findings it can be concluded that certain posts are of greater interest than others. It is recommended that Facebook page statistics be regularly reviewed to determine the impact and relevance of posts. Facebook page administrators must evolve a 'cocktail' of posts to optimize the impact of their pages, which includes promoting interest in a page and the related 'owner'.

Keywords: Social media, Public relations, Communication.

1 INTRODUCTION

Black (1993) defines public relations practice as "The art and science of achieving harmony with the environment through mutual understanding based on truth and full information." He states that there are two distinct branches of practice, namely the reactive and proactive sectors. Reactive includes reacting to problems, dealing with crises, and managing change, and the proactive includes planned programs that serve both the organisation, and the public's interest. Black (1993) describes the role of public relations by presenting the hexagon model, the six sides representing the several factors which influence the role and scope of public relations: the publics of concern; issues of concern; media; the nature of the organisation; situational timing factors, and resources. This paper focuses on the media 'side', and more specifically social media in the form of Facebook.

Then, public relations are frequently 'confused' with marketing. Black (1993) refers to The Chartered Institute of Marketing's definition of marketing: "The management process responsible for identifying, anticipating, and satisfying customers' requirements profitably." Marketing and public relations share many characteristics; however, marketing is an exchange process, which involves four stages, namely: seeking the understanding of the 'other' party; creating value as perceived by the other party; sharing ideas regarding the value of products or services and delivering the goods or services through an actual transaction.

Wright and Hinson (2009) state that social media deliver web-based information created by people with the intention of facilitating communication and now represent one of the world's major sources of social interaction as people share stories and experiences with each other. However, research indicates that public relations practitioners should measure the amount of communication that is being disseminated with respect to their organizations (or client organizations) through blogs and other social media (4.2/5.0), and/or analyze content of what's being communicated with respect to their organizations (or their clients) in blogs and other social media (4.4/5.0). Given the aforementioned, a study was conducted to determine the impact of a series of posts on the Department of Construction Management Facebook page by reviewing issues such as 'reach', 'clicks', 'reactions, comments, and shares'.

2 REVIEW OF THE LITERATURE

2.1 The Advent of Facebook

Deckers and Lacy (2011) relate that in 2004, Mark Zuckerberg, a Harvard sophomore, began developing a new kind of web site for fellow students to track their social lives. It became popular, and membership first expanded to include Stanford, Yale, and Columbia universities, and then most universities across Canada, and the United States of America (USA). By 2006 the site was named Facebook, and any person globally, could become a member of the social networking phenomenon. In July 2010, Facebook announced that it had more than 500 million users (Deckers and Lacy, 2011). Kabani (2012) states "With 750 million users, and growing, this giant can't be ignored." Deckers and Lacy (2011) refer to Facebook as an incredibly powerful tool for personal branding and contend that the main purpose for using Facebook is to create a community and connect people, which helps fuel the influence behind the brand. Furthermore, influence creates viral marketing within a community. Kabani (2012) states that people use facebook to showcase their own identity, which includes both who they are and who they want to be perceived as.

2.2 The Impact of Social Media

Wright and Hinson (2009) conducted their fourth study among public relations practitioners globally. Respondents represented, *inter alia*, a wide variety of segments of the public relations industry, *inter alia*, small agencies or consultancies (25%), corporations (20%), and educational institutions (20%). Respondents were required to indicate the extent to which they agreed that the emergence of social media (including blogs) has changed the way their organisation, or their client organizations: communicate(s) (3.9/5.0); handle(s) external communications (3.7/5.0), and handle(s) internal communications (3.3/5.0). Other findings include: social media have enhanced the practice of public relations (4.0/5.0); and blogs have enhanced the practice of public relations (3.8/5.0), and social media offer organizations a low-cost way to develop relationships with members of various strategic publics (4.0/5.0).

2.2 Current Reality

Evans (2011) states that before the advent of the internet, audiences had to endure listening to or seeing advertisements repeatedly, and that the advertisements were in control of the media that people listened to or watched. Then, in the 1990's the internet evolved, and audiences no longer had to endure commercials, as they could fast forward through commercials, and ignore the 'noise' of marketing. Audiences now turn to communities of likeminded individuals to acquire knowledge of products or services that interest them. Social media provides the forum for people to acquire this information in a natural and more honest way than being subjected to marketing messages. Target audiences can be reached through user-generated content sites, and social media communities. The reality is that the users and community members are now in 'control', and do not want to be 'drilled' with marketing messages. Hajli (2014) cites Senecal and Nantel (2004), who contend that consumers have access to many various sources of information and experiences, which have been facilitated by other customers' information and recommendations courtesy of the internet.

3 RESEARCH METHOD

The research method can best be described as experimental in that posts pertaining to different issues were undertaken periodically by the author, the page administrator. Posts were constrained to construction, particularly construction health and safety (CH&S), and in addition, tertiary construction management education, and related issues such as conferences. In terms of the review of the degree of interest, the review spanned the year 2017, and addressed issues such as 'reach', 'clicks', 'reactions, comments, and shares'. Reach is the number of people who saw any of the page posts. Reach can be broken down into people who saw the posts with or without advertising (paid or organic posts). All posts were organic.

4 RESEARCH FINDINGS

Table 1 presents the Schedule of Department of Construction Management Facebook posts for the year 2017. Reactions, comments, and shares are presented in the extreme right-hand column (R, C, S). A total of 48 posts were made during the period 22 January to 11 December 2017, which resulted in a reach of 14 982, 1 575 clicks, and 444 reactions, comments, and shares. The totals equate to an average reach of 312.1, 32.8 clicks, 9.3 reactions, comments, and shares. The highest reach (2 316) was relative to a BSc (Honours) graduation photo depicting a graduate and the author on 08/04/17, followed by a BSc (Honours) graduation photo depicting two graduates and the author on 08/04/17 with a reach of 1 478. The next highest reach was relative to an ACHASM EC Meeting announcement on 13/03/17 with a reach of 628, and then a PhD graduation photo depicting a PhD graduate and the author on 08/04/17, with a reach of 535. The lowest reach (73) was relative to the 'University www address change' on 18/08/17, which was attributable to the change of the name of the university form Nelson Mandela Metropolitan University to Nelson Mandela University. The highest number of clicks (288) was relative to the BSc (Honours) graduation photo depicting a graduate and the author on 08/04/17, followed by the BSc (Honours) graduation photo depicting two graduates and the author on 08/04/17 (157). The next highest number of clicks was relative to photos depicting Cape Town construction on 24/02/17 (99), and then photos of a mall project on 30/06/17 (85). The lowest number of clicks (1) was relative to the ACHASM Summit reminder on 06/10/17 and the W99 & TG 59 Conference (registration form) on 27/05/17. The highest number of reactions, comments, and shares (141) was relative to the BSc (Honours) graduation photo depicting a graduate and the author on 08/04/17, followed by the BSc (Honours) graduation photo depicting two graduates and the author on 08/04/17 (124). These findings reinforce the contention of Deckers and Lacy (2011) that the main purpose for using Facebook is to create a community and connect people. There were zero reactions, comments, and shares relative to twelve posts.

Table 1. Schedule of Department of Construction Management Facebook posts for the year 2017.

Date	Post	Category	Reach	Clicks	R, C, S
11/12/17	Melbourne buildings and construction H&S (CH&S)	Photos	124	13	3
11/12/17	Saiosh bursary funding	Announcement	105	19	0
26/11/17	New York CH&S	Photos	173	26	4
06/10/17	ACHASM Summit reminder	Announcement	190	1	0
02/10/17	ACHASM Summit final programme	Announcement	143	12	1
24/09/17	ACHASM Summit reminder	Announcement	320	10	5
21/09/17	EPPM Conference banner	Photos	117	3	0
17/09/17	ACHASM Summit flyer	Announcement	150	24	0
12/09/17	Construction & water usage	URL referral	109	5	1
05/09/17	ACHASM Summit alert	Announcement	150	21	3
26/08/17	Publication of book (Property)	Announcement	250	4	2
18/08/17	University www address change	Update	73	3	0
17/08/17	10th cidb PG CID conference	Announcement	234	32	0
16/08/17	Publication of book (Valuing People in Construction)	Announcement	239	8	0
10/08/17	Women in construction	Awareness	276	12	6
09/08/17	Bridge collapse	URL referral	130	6	0
03/08/17	W99 & TG 59 Conference	Feedback article (magazine)	276	13	1
25/07/17	ACHASM Chapter presentation	Announcement	263	8	2
11/07/17	ACHASM Summit alert	Announcement	228	10	0
07/10/17	Melbourne CH&S	Photos	236	67	4
30/06/17	Mall project	Photos	408	85	9
04/06/17	W99 & TG 59 Conference	Announcement	161	43	0
02/06/17	Mall project	Photos	334	32	12
27/05/17	W99 & TG 59 Conference	Registration form	114	1	0
21/05/17	W99 & TG 59 Conference alerts 1 - 4	Announcement	181	23	1
28/04/17	World Day for Safety at Work	Awareness	228	32	2
24/04/17	Hong Kong CH&S	Photos	126	40	2
22/04/17	ICC building, Hong Kong	Video	199	2	1
12/04/17	Honours Team Building Event	Photos	256	17	2
12/04/17	Honours Team Building Event	Photos	250	21	0
12/04/17	Honours Team Building Event	Photos	212	7	1
10/04/17	Honours Team Building Event	Photos	337	34	9
08/04/17	BSc graduation	Photos	369	69	14
08/04/17	BSc graduation	Photos	324	20	5
08/04/17	PhD graduation	Photos	355	37	10
08/04/17	PhD graduation	Photos	535	46	24
08/04/17	BSc (Honours) graduation	Photos	1478	157	124
08/04/17	BSc (Honours) graduation	Photos	2316	288	141
06/04/17	Prize giving	Photos	308	30	6
06/04/17	Port Elizabeth construction	Construction	309	39	3
29/03/17	Port Elizabeth construction	Construction	224	20	3
23/03/17	NHLS tuberculosis (TB) Day	Awareness	263	10	2
14/03/17	W99 & TG59	Announcement	409	13	9
13/03/17	ACHASM EC Meeting	Announcement	628	31	14
13/03/17	ECMBA H&S Conference	Announcement	316	38	8
09/03/17	Johannesburg construction	Construction	135	24	3
24/02/17	Cape Town construction	Construction	264	99	4
22/01/17	Temple at Ta Prohm, Cambodia	Archaeological	157	20	3
Total			14982	1575	444

5 CONCLUSIONS AND RECOMMENDATIONS

Based upon the findings it can be concluded that certain posts are of greater interest than others. The reach relative to the graduation photos clearly indicates the importance of student related posts, 'personalization', 'recognizing of achievement', and 'celebration'. Project related photos and events in the form of conferences, presentations, and summits also realized substantial reaches, and are clearly of interest. The clicks, and reactions, comments, and shares relative to the graduation photos allude to 'communal celebration'.

It is recommended that Facebook page statistics be regularly reviewed to determine the impact and relevance of posts. Facebook page administrators should evolve a 'cocktail' of posts to optimize the impact of their pages, which includes promoting interest in a page and the related 'owner'. Future Department of Construction Management Facebook page posts should focus on various forms of student activity at undergraduate, honors, masters, and doctoral level. Possibilities include first-year orientation, first-year team building, bursary announcements, vacation work opportunities, graduate employment opportunities, and graduate updates.

References

Black, S., *The Essentials of Public Relations*, London, Kogan Page Ltd., 1993.

Deckers, E., and Lacy, K., *Branding Yourself*, Indianapolis, Que Publishing, 2011.

Evans, L., *Social Media Marketing. Indianapolis*, Que Publishing, 2011.

Hajli, M. N., A Study of the Impact of Social Media on Consumers, *International Journal of Market Research*, 56(3), 387-404, 2014.

Kabani, S. H., *The Zen of Social Media Marketing*, Dallas, BenBella Books, Inc., 2012.

Wright, D. K., and Hinson, M. D., An Updated Look at the Impact of Social Media on Public Relations Practice, *Public Relations Journal*, 3(2), 1-27 2009. Retrieved from http://www.prsa.org/Intelligence/PRJournal/Spring_09/ on June 10, 2018.

Responsible Design and Delivery of the Constructed Project
Edited by Abdul-Malak, M., Khoury, H., Singh, A., and Yazdani, S.

THE IMPACT OF THE CONSTRUCTION COMPUTING SOFTWARE (CCS) 'CANDY' COURSE: CONSTRUCTION MANAGEMENT STUDENTS' PERCEPTIONS

JOHN SMALLWOOD and CHRIS ALLEN

Dept of Construction Management, Nelson Mandela University, Port Elizabeth, South Africa

Students' post-course perceptions provide insight relative to their understanding and appreciation of a module, including the impact thereof within the context of the wider course content, hence the study 'Construction management students' post course perceptions regarding a CCS Candy course' presented at a South African university. The salient findings include: 'planning – programming' predominates in terms of the extent Candy training contributed to an increase in knowledge; costing and estimating predominate in terms of an enhancement of skills; coordinating predominates in terms of contributing to an improvement in students' understanding and appreciation of the functions of management work, and cost predominates in terms of an improvement in students' understanding and appreciation of the project parameters. Conclusions include that the CCS Candy course had an impact in terms of: an enhancement of both knowledge and skills aligned with the planning function; a better understanding and appreciation of the practice of construction management, and students understand and appreciate the importance and role of the course in terms of greater knowledge development. It is recommended that the 'add-on' certificated CCS Candy course continue to be included and that the research be conducted annually.

Keywords: Estimating, Planning, Commercial packages, Education, Students.

1 INTRODUCTION

The Department of Construction Management at Nelson Mandela University endeavors to provide relevant and learner centered education in the discipline, develop the necessary competencies, thereby empowering graduates to integrate into the industry as construction managers, and construction project managers. Specialist planning skills education was identified as an area that needed greater focus within the department, and as a result it was decided to align the department with the most widely used planning and estimating package in the South African construction industry, namely CCS Candy. The CCS Candy system is used to control a projects construction process and financial performance from tender stage to the final account stage. Candy has a planning application, but is much more than a planning and estimating package due to its dynamic link between money and time providing a wealth of information for both management and clients (CCS Candy 2013). The subject matter of the course has become highly topical with greater emphasis being placed on the use of software for the planning and estimating of construction projects, with greater interest being shown from commercial entities in students having a practical certification of competency alongside their theoretical knowledge garnered during the degree programme. By engaging with the external training provider of the software supplier, the department has been able to structure the course content to better align students basic understanding of the built environment processes with a more holistic overview of the use of software to deliver projects, simultaneously providing a certificated additional module and a certification that provides those who complete the course a ready entry into the labor market. Given the aforementioned, a study was conducted to determine the impact of the course.

2 REVIEW OF THE LITERATURE

Wickramasinghe and Perera (2010) contend that the possession of subject skills alone is no longer sufficient for a new graduate in meeting employer requirements. Finch *et al.* (2016) state that 40 percent of employers believe that there is a significant skills gap between graduates and requirements, and therefore providing students with the right skills to fill those gaps becomes fundamental for tertiary education as well as construction. Employability of an individual depends upon knowledge, skills, and attitudes, which requires that tertiary education needs to produce quality graduates that employers want to include in their graduate programmes (Wickramasinghe and Perera 2010). Student perceptions of a course's merit can be influenced by many factors, with studies conducted across multiple disciplines showing that "engagement, perceived course value, and the use of deep learning strategies" are all 'integral to a student's positive learning experience' (Floyd *et al.* 2009). Positive learning experiences in turn are likely to lead to positive reinforcement of a course among the student body, relating the perception through deeper learning to the course content becomes a fundamental means to influencing that perception. According to Liu *et al.* (2012), motivation is a key factor in student learning, and Kamardeen (2013) states that self-motivation of students is a prime requirement for their active engagement, which means that the course content needs to stimulate interaction within the study sessions, and into other courses within each discipline's course structure.

3 RESEARCH METHOD

The two-day certificate CCS Candy course is an additional intervention included in the BSc (Construction Studies) programme relative to the subject Construction Management 3. To assess the impact of the CCS Candy course, students were surveyed immediately upon completion thereof to determine the extent to which the CCS Candy training contributed to an: increase in knowledge; enhancement of skills; improvement in understanding and appreciation of the functions in an organisation, the functions and activities of management work, the various project parameters, and built environment processes, and the construction process and activities, and the integration of subjects. The questionnaire consisted of 7 closed-end questions, 97 sub-questions, and 1 open-end question. The closed-end questions entailed a response to a five-point Likert scale preceded by an 'unsure' and 'did not' contribute option. Therefore, respondents effectively responded to a six-point Likert scale. Based upon the number of responses to the six points, a measure of central tendency in the form of a mean score (MS) was computed to enable a relative comparison and rankings. Due to an effective six points on the scale, the MS ranges from 0.00 to 5.00, with a midpoint of 2.50. 15 Students' responses were included in the analysis of the data.

4 RESEARCH FINDINGS

Table 1 indicates the extent to which the Candy training contributed to an increase in knowledge relative to thirty knowledge areas in terms of percentage responses to a scale of 1 (minor) to 5 (major), and MSs between 0.00 and 5.00. It is notable that all the MSs are > 2.50, which indicates that the contribution of the Candy training to an increase in knowledge is major as opposed to minor. Given that effectively a six-point scale ('did not' linked to a five-point) was used, and that the difference between 0 and 5 is five, ranges with an extent of 0.83 (5 / 6) are used to discuss the degree of central tendency. 13 / 30 (43.3%) MSs > 4.17 ≤ 5.00, which indicates that the Candy training made between a near major to major / major contribution to an increase in knowledge. The knowledge areas include 'planning – programming', 'cost control', 'planning – strategic', 'estimating', 'project management', 'measuring (quantities)', 'plant and equipment management', 'materials management', 'cost engineering', 'productivity', 'risk management', and 'procedures', except for 'planning – strategic', these findings are not unexpected. 16 / 30 (53.3%) MSs > 3.33 ≤ 4.17, which indicates that the Candy training made between a contribution to a near major / near major contribution to an increase in knowledge.

Table 1. Extent to which the Candy training contributed to an increase in knowledge relative to thirty knowledge areas.

Knowledge area	Response (%)							MS	Rank
	U	Did not	Minor 1	2	3	4	......Major 5		
Planning - programming	0.0	0.0	0.0	0.0	0.0	26.7	73.3	4.73	1
Cost control	0.0	0.0	0.0	0.0	0.0	30.8	69.2	4.69	2
Planning - strategic	7.1	0.0	0.0	0.0	0.0	42.9	50.0	4.54	3
Estimating	0.0	0.0	0.0	0.0	13.3	20.0	66.7	4.53	4
Project management	0.0	0.0	0.0	0.0	7.1	35.7	57.1	4.50	5
Measuring (quantities)	0.0	0.0	0.0	6.7	6.7	26.7	60.0	4.40	6
Plant and equipment management	0.0	0.0	0.0	0.0	15.4	30.8	53.8	4.38	7
Materials management	0.0	0.0	0.0	0.0	6.7	53.3	40.0	4.33	8
Cost engineering	0.0	0.0	0.0	0.0	6.7	53.3	40.0	4.33	9
Productivity	0.0	0.0	0.0	0.0	6.7	60.0	33.3	4.27	10
Risk management	0.0	0.0	0.0	0.0	13.3	46.7	40.0	4.27	11
Procedures	0.0	0.0	0.0	6.7	0.0	53.3	40.0	4.27	12
Cash flow forecasting	0.0	0.0	6.7	0.0	13.3	26.7	53.3	4.20	13
Training	0.0	0.0	0.0	0.0	26.7	33.3	40.0	4.13	14
Benchmarking	13.3	0.0	0.0	0.0	26.7	26.7	33.3	4.08	15
Final accounts	6.7	0.0	0.0	6.7	6.7	53.3	26.7	4.07	16
Materials	0.0	0.0	0.0	0.0	26.7	40.0	33.3	4.07	17
Specifications	0.0	6.7	0.0	0.0	6.7	53.3	33.3	4.00	18
Methods (construction) - building	0.0	0.0	0.0	0.0	30.8	38.5	30.8	4.00	19
Human resources	0.0	0.0	0.0	7.7	23.1	30.8	38.5	4.00	20
Methods (construction) - civil	6.7	0.0	0.0	0.0	33.3	33.3	26.7	3.93	21
Remuneration	7.1	7.1	0.0	0.0	21.4	21.4	42.9	3.92	22
Financial management	7.1	7.1	0.0	0.0	21.4	21.4	42.9	3.92	23
Subcontractor management	0.0	0.0	0.0	7.7	38.5	30.8	23.1	3.69	24
Work study	7.1	7.1	0.0	0.0	35.7	21.4	28.6	3.62	25
Purchasing	0.0	6.7	0.0	13.3	20.0	26.7	33.3	3.60	26
Information technology	7.1	7.1	0.0	7.1	7.1	35.7	35.7	3.57	27
Design	13.3	6.7	0.0	6.7	13.3	46.7	13.3	3.54	28
Contract documentation	0.0	6.7	0.0	20.0	20.0	26.7	26.7	3.40	29
Management (business)	13.3	6.7	0.0	13.3	0.0	40.0	26.7	3.20	30

Only 1 / 30 (3.3%) MSs > 2.50 ≤ 3.33, which indicates that the Candy training made between a near minor contribution to a contribution / contribution to an increase in knowledge – 'management (business)'.

Table 2 indicates the extent to which the Candy training contributed to the enhancement of twenty-seven skills in terms of percentage responses to a scale of 1 (minor) to 5 (major), and MSs between 0.00 and 5.00. It is notable that all the MSs are > 2.50, which indicates that the contribution of the Candy training was major as opposed to minor. 14 / 27 (51.9%) MSs > 4.17 ≤ 5.00, which indicates that the Candy training made between a near major to major / major contribution – 'costing', 'estimating', 'planning', 'computer', 'organising', 'measuring – productivity', 'financial', 'controlling', 'procedures development', 'coordinating', 'initiating', 'technical', 'systems development', and 'decision making', which are not unexpected as CCS Candy is planning focused and estimating oriented. 9 / 27 (33.3%) MSs > 3.33 ≤ 4.17, which indicates that the Candy training made between a contribution to a near major / near major contribution. 'Supervisory' is followed by 'training', 'measuring – quantities', 'motivating', 'statistical', 'administrative', 'auditing', 'communicating – graphic', and 'negotiating - plant hire'. 4 / 27 (14.8%) MSs > 2.50 ≤ 3.33, which indicates that the Candy training made between a near minor contribution to a contribution / contribution – 'work study', 'communicating – written', 'plan reading', and 'negotiating – subcontractors'.

Table 2. Extent to which the Candy training enhanced twenty-seven skills.

Skill	Response (%)							MS	Rank
	U	Did not	Minor 1	2	3	4	.... Major 5		
Costing	0.0	0.0	0.0	0.0	7.1	7.1	85.7	4.79	1
Estimating	0.0	0.0	0.0	0.0	7.1	14.3	78.6	4.71	2
Planning	0.0	0.0	0.0	0.0	6.7	20.0	73.3	4.67	3
Computer	0.0	0.0	0.0	0.0	0.0	35.7	64.3	4.64	4
Organising	0.0	0.0	0.0	0.0	6.7	33.3	60.0	4.53	5
Measuring - productivity	0.0	0.0	0.0	0.0	0.0	50.0	50.0	4.50	6
Financial	0.0	0.0	0.0	0.0	7.1	35.7	57.1	4.50	7
Controlling	0.0	0.0	0.0	0.0	20.0	13.3	66.7	4.47	8
Procedures development	0.0	0.0	0.0	0.0	21.4	14.3	64.3	4.43	9
Coordinating	6.7	0.0	0.0	0.0	0.0	26.7	66.7	4.40	10
Initiating	0.0	0.0	0.0	0.0	7.1	50.0	42.9	4.36	11
Technical	0.0	0.0	0.0	0.0	13.3	46.7	40.0	4.27	12
Systems development	7.1	0.0	0.0	0.0	0.0	42.9	50.0	4.21	13
Decision making	0.0	0.0	0.0	0.0	6.7	66.7	26.7	4.20	14
Supervisory	0.0	0.0	0.0	0.0	20.0	46.7	33.3	4.13	15
Training	0.0	0.0	0.0	0.0	33.3	26.7	40.0	4.07	16
Measuring - quantities	0.0	6.7	0.0	6.7	6.7	26.7	53.3	4.07	17
Motivating	6.7	0.0	0.0	0.0	26.7	13.3	53.3	4.00	18
Statistical	7.1	0.0	0.0	0.0	14.3	42.9	35.7	3.93	19
Administrative	8.3	0.0	0.0	8.3	0.0	41.7	41.7	3.92	20
Auditing	7.1	0.0	0.0	0.0	21.4	42.9	28.6	3.79	21
Communicating - graphic	0.0	8.3	0.0	8.3	16.7	33.3	33.3	3.67	22
Negotiating - plant hire	8.3	8.3	0.0	0.0	25.0	8.3	50.0	3.58	23
Work study	14.3	7.1	0.0	0.0	14.3	35.7	28.6	3.29	24
Communicating - written	8.3	8.3	0.0	8.3	25.0	25.0	25.0	3.17	25
Plan reading	6.7	20.0	0.0	6.7	13.3	20.0	33.3	3.00	26
Negotiating - subcontractors	7.7	23.1	0.0	15.4	7.7	23.1	23.1	2.62	27

Table 3 indicates the extent to which the Candy training contributed to an improvement in understanding and appreciation of the functions and activities of management work. It is notable that all the activities of the planning, organising, leading, and controlling functions of management work and the coordinating function have MSs > 2.50, which indicates that the

Candy training contributed to more of a major than a minor improvement. In terms of the functions, based upon the mean MSs, coordinating (4.50) achieved the first ranking followed by planning (4.42), controlling (4.09), leading (3.96), and organising (3.72). These findings were expected. It is notable that scheduling (4.79) is ranked first in terms of the function of planning, and overall. Overall, it is followed by programming (4.71), correcting performance (4.50), coordinating (4.50), forecasting (4.43), communicating (4.43), performance measuring (4.43), decision-making (4.36), developing performance standards (4.36), evaluating performance (4.36), budgeting (4.29), developing objectives (4.29), developing procedures (4.21), and developing organization structure (4.21), which all have MSs > 4.17 ≤ 5.00, MSs > 4.17 ≤ 5.00, which indicates that the Candy training made between a near major to major / major contribution to an improvement in understanding and appreciation of functions and activities of management work.

Table 3. Extent to which the Candy training contributed to an improvement in understanding and appreciation of the functions and activities of management work.

Function / Activity	Response (%)							MS	Rank
	U	Did not	Minor ... 1	2	3	4	...Major 5		
Planning:									
Scheduling	0.0	0.0	0.0	0.0	0.0	21.4	78.6	4.79	1
Programming	0.0	0.0	0.0	0.0	7.1	14.3	78.6	4.71	2
Forecasting	0.0	0.0	0.0	0.0	7.1	42.9	50.0	4.43	5
Budgeting	7.1	0.0	0.0	0.0	0.0	35.7	57.1	4.29	11
Developing objectives	0.0	0.0	0.0	0.0	14.3	42.9	42.9	4.29	12
Developing procedures	0.0	0.0	0.0	0.0	21.4	35.7	42.9	4.21	13
Developing organization structure	0.0	0.0	0.0	0.0	28.6	21.4	50.0	4.21	14
Mean								4.42	
Organising:									
Delegating	7.1	0.0	0.0	7.1	21.4	14.3	50.0	3.86	18
Developing organization structure	0.0	7.1	0.0	7.1	21.4	21.4	42.9	3.79	19
Establishing relationships	7.1	7.1	0.0	7.1	14.3	28.6	35.7	3.50	20
Mean								3.72	
Leading:									
Communicating	0.0	0.0	0.0	7.1	7.1	21.4	64.3	4.43	6
Decision-making	0.0	0.0	0.0	7.1	7.1	28.6	57.1	4.36	8
Developing people	0.0	0.0	0.0	7.1	21.4	28.6	42.9	4.07	15
Motivating	0.0	7.1	0.0	7.1	7.1	28.6	50.0	4.00	16
Selecting people	0.0	0.0	0.0	7.1	14.3	57.1	21.4	3.93	17
Mean								3.96	
Controlling:									
Correcting performance	0.0	0.0	0.0	7.1	0.0	28.6	64.3	4.50	3
Performance measuring	0.0	0.0	0.0	7.1	7.1	21.4	64.3	4.43	7
Developing performance standards	0.0	0.0	0.0	7.1	7.1	28.6	57.1	4.36	9
Evaluating performance	0.0	0.0	0.0	7.1	7.1	28.6	57.1	4.36	10
Mean								4.09	
Coordinating	0.0	0.0	0.0	8.3	8.3	8.3	75.0	4.50	4

Table 4 indicates the extent to which the Candy training contributed to an improvement in understanding and appreciation of nine project parameters. It is notable that all the parameters have MSs > 2.50, which indicates that the extent of the contribution is major than minor. However, only 3 / 9 (33.3%) parameters have MSs > 4.17 ≤ 5.00, namely cost, time, and productivity, which indicates that the extent of the contribution is between near major to major / major. These findings were not unexpected. Thereafter, 4 / 9 (44.4%) MSs > 3.33 ≤ 4.17, namely client satisfaction, customer satisfaction, developmental objectives, and quality, which indicates that the extent of the contribution is between a contribution to a near major / near major contribution. Health and Safety, and environment have MSs > 2.50 ≤ 3.33, which indicates that the contribution is between near minor contribution to a contribution / contribution.

Table 4. Extent to which the Candy training contributed to an improvement in understanding and appreciation of the various project parameters.

Parameter	Response (%)							MS	Rank
	U	Did not	Minor 1	2	3	4	.......Major 5		
Cost	0.0	0.0	0.0	0.0	0.0	13.3	86.7	4.87	1
Time	0.0	0.0	0.0	7.1	7.1	21.4	64.3	4.43	2
Productivity	0.0	0.0	0.0	6.7	6.7	46.7	40.0	4.20	3
Client satisfaction	0.0	6.7	0.0	6.7	6.7	40.0	40.0	3.93	4
Customer satisfaction	0.0	6.7	0.0	6.7	13.3	40.0	33.3	3.80	5
Developmental objectives	0.0	6.7	0.0	6.7	13.3	40.0	33.3	3.80	6
Quality	6.7	6.7	0.0	6.7	6.7	46.7	26.7	3.53	7
Health and Safety	6.7	6.7	0.0	13.3	20.0	40.0	13.3	3.13	8
Environment	13.3	6.7	0.0	6.7	26.7	33.3	13.3	2.93	9

5 CONCLUSIONS AND RECOMMENDATIONS

Based upon the findings it can be concluded that the two-day certificate CCS Candy course had an impact in terms of enhancement of both knowledge and skills, and an understanding and appreciation of the practice of construction management in general, and more specifically the functions and activities of management work, and the project parameters, in addition to the planning function and time as a parameter. Although CCS Candy constitutes applications' software, using it requires the use of a range of knowledge areas, and skills, and prompts the users in terms of considering the range of project parameters other than time. Furthermore, the students understand and appreciate the importance and role of the CCS Candy course. Then, it is important to assess the impact of training interventions, regardless of their nature. It is recommended that the two-day certificate CCS Candy course continue to be included as an 'add-on' course, and that the research reported on be conducted on an annual basis.

References

Finch, D., Peacock, M., Levallet, N., and Foster, W., A Dynamic Capabilities View of Employability, *Education + Training*, 58(1), 61-81, 2016.

Floyd, K., Harrington, S., and Santiago, J., The Effect of Engagement and Perceived Course Value on Deep and Surface Learning Strategies, *The International Journal of an Emerging Transdiscipline*, 12, 181-190, 2009.

Kamardeen, I., Motivation-Driven Learning and Teaching Model for Construction Education, *Australian Journal of Construction Economics and Building*, 13(1), 36-49, 2013.

Liu, E., Lin, C., Jian, P., and Liou, P., The Dynamics of Motivation and Learning Strategy in A Creativity-Supporting Learning Environment in Higher Education, *The Turkish Online Journal of Educational Technology,* 11(1), 172-180, 2012.

Wickramasinghe, V., and Perera, L., Graduates', University Lecturers' and Employers' Perceptions Towards Employability Skills, *Education + Training*, 52(3), 226-244, 2010.

Responsible Design and Delivery of the Constructed Project
Edited by Abdul-Malak, M., Khoury, H., Singh, A., and Yazdani, S.

DESIGNING RESILIENT ROOF SYSTEMS

THOMAS W. HUTCHINSON

Hutchinson Design Group, Barrington, USA

Recent powerful climate storms in the USA have the government discussing the need for resilient buildings: Buildings that can withstand severe storms, provide life safety and be quickly repaired. A building's roof is often the first line of defense in the changing climate conditions with its catastrophic natural disasters. The roof, if designed properly, can mitigate the impacts of these emergencies and allow a building to withstand and recover from the disruption caused. Resilience is increasingly being viewed as an important performance objective for governmental, educational, commercial and industrial construction. Interest in resiliency is high and is being actively discussed at all levels of the building industry: Governmental, Codes and Standards and Trade Organizations (ASTM, Resilient). This paper will review key resilient roof system elements that need to be considered by roof system designers: Best design and detail practices for roofing to achieve resiliency. The needed level of design and detailing will be reviewed. Recommendations to achieve resilient roof systems will be provided.

Keywords: Sustainability, Long term service life, Destructive storms, Roof deck, Construction.

1 SUSTAINABILITY TO RESILIENCE: A NATURAL PROGRESSION

For the past two decades the goal of attaining sustainable roof systems was a laudable one. Sustainable low slope roof systems were defined by the CIB W083/RILEM 166-MRS Joint Committee on Roofing Materials and System in there Publication No. 271 as: "A roof system that is designed, constructed, maintained, rehabilitated and demolished with an emphasis throughout its life cycle on using natural resources efficiently and preserving the global environment." In essence the goal was "Long Term Service Life". While new at the time, this concept is now expected by clients and no longer is a request.

The recent rash of violent and destructive storms: Hurricanes, hail, intense rain, wildfires and wind have resulted in a call for improvement in building construction: That improvement is called Resiliency.

Where sustainability calls for a building to minimize the impact of the building (roof) on the environment, resiliency requires a building (roof) to minimize the impact of the environment on the building. Thus, resiliency can be defined as: "The capacity to adapt to changing conditions and to maintain or regain functionality and vitality in the face of stress or disturbance."

Resiliency is the combination of many attributes and can be thought of as: Consisting of Durability, being Sustainability; designed for Resiliency: With the ability to: Adapt – to – Withstand – to – Recover. This concept of Resiliency, designing a roof system to weather intense storms and to be easily repaired (think of Puerto Rico and how you would repair a roof with no power, access to new materials and manpower that cannot get to your site) requires the roof system designer to:

(1) Understand that roofs are systems, only as good as their weakest link;
(2) Remove from their AutoCAD libraries old, out of date and irrelevant details;
(3) Design the roof system integration into associated barrier systems: Wall claddings, air barrier;
(4) Design the roof systems integration into the impinging building elements, such as: Roof curbs, RTU curbs, pipe penetrations, roof to wall, roof edges, roof drains and skylights;
(5) Understand how the roof will be used by the occupants: i.e. Public access, roof decks;
(6) Understand the construction sequencing and its impact on the roof;
(7) Coordinate with other impinging disciplines such as plumbing, mechanical and structural engineers; and,
(8) Develop comprehensive specifications.

1.1 Benefits of Resilient Roof Systems

Loss of property, life, productivity and inability to recover quickly were clearly evident following the severe storms the globe has experienced this year, resulting in both unlivable conditions, loss of jobs and depression. The cost of severe weather disasters in the USA in 2017 is in the billions of US dollars, a troubling trend of billion dollar events.

Resilient roof systems, the first line of a building's defense against severe weather, afford the hope of survival and the ability to be rapidly repaired. Their design doesn't rely on historical climatic events, which are clearly not the norm anymore, but anticipate climate change and repeated increasingly severe climatic events (Baskaran *et al.* 2017) (Baskaran *et al.* 2018). The Multihazard Migation Council (MMC) convened by the National Institute of Building Sciences (NIBS) in their study on mitigation efforts found that for every dollar spent on mitigation against severe weather saved an average of US$4.00. Prevention of destruction is clearly a key element and benefit of resilient roof system design.

2 ROOF SYSTEM COMPONENTS TO CONSIDER IN A RESILIENT ROOF

With resilient Roof Systems needing to be durable, sustainable, and adaptable and have the ability to recover, roof system components need to be consider in that light.

2.1 Roof Deck and Air/Vapor Retarder

The roof deck is the first component to be considered. Often outside the realm of the roof system designer, they must coordinate closely with the structural engineer. The roof deck must be able to support the roof systems anchorage and resist the anticipated forces and more. Thus it is not only the roof deck that needs to be considered, but the structure to which it is attached. Positive attachment and engagement is required. A related structural element to be considered in conjunction with the structural engineer is the roof edge, which must be tied back into the structure. As the first line of defense against wind, the roof edge structure, prior to the application of the roofing and sheet metal, needs to be robust and secure. For reroofing, this will require thorough field investigation. Often remediation of the existing roof edge condition will be required to achieve the desired result.

The air/vapor retarder should be thought of in several ways beyond the traditional value of use over humid interior conditions especially in temperate or cold climates, such as:

(1) Prevention of construction generated moisture movement into the roof system;

(2) The benefit of a temporary roof that will allow for other trades to use the temporary roof, rather than the final roof as a work surface;

(3) As a barrier to the inflow of moisture to the interior if the roof cover is physically damaged in a storm (think windblown debris in a hurricane), partially blown back or otherwise compromised roof cover. A quality vapor retarder is a valued component of most resilient systems.

2.2 Drainage

Draining water off the roof cover is a positive attribute of all roof systems not only resilient roof systems. Removing water from the roof as quickly as possible reduces the potential for that water to move inward should there be a breach in the roof cover. Draining water over a roof edge gutter or over an eave edge has the advantage of always being able to drain and not have the potential to be blocked. Interior roof drains are a predominate method of drainage in North America. For resilient systems the designer should work with the plumbing and structural engineers. Roof drains should be properly located, secured to the structural, so that they cannot be pulled up off the drainpipe if the roof membrane should blow up (see Figure 1). Consideration for when the drain might get clogged in a storm should be given in the form of overflow drainage potential in the form of scuppers or overflow drains that flow free to grade and are not connected to the drain system. Some prudent designers are looking into hidden drains; under the insulation at the vapor retarder as an emergency drain should there be partial roof blow-off that might trap water.

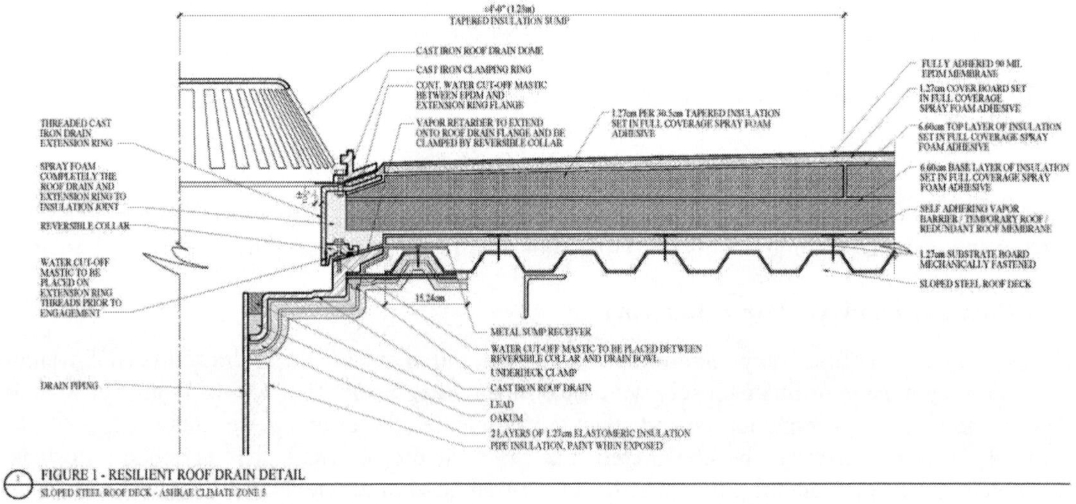

Figure 1. Drains should be properly located, secured to the structure and the vapor retarder integrated.

2.3 Roof Edge

The roof edge needs to be designed, not just be an afterthought. Taking the brunt of the initial wind loading, the roof edge components need to have strength, integrity under load and be integratable with the roof system. The roof edge should always be secured to the structure and all applied components such as wood blocking be robustly secured as well. Parapet walls can provide better wind protection than do gravel stops, but need a higher level of design.

2.4 Insulation and Cover Boards

In addition to its thermal properties, insulation provides the base for the roof cover and often provides the taper for drainage. The insulation layer though can be a weak link: Paper facers are hydroscopic and support mold growth and are susceptible to cohesive failure. Poor installation often leads to separation from the roof deck and/or each other. And low compressive strength can result in crushing of the insulation and resultant roof cover debonding. In North America the primary insulation is polyisocyanurate and this author suggests and specifies 25 psi, coated fiberglass facer (non-hydroscopic and does not support mold growth) polyisocyanurate. For adhered systems the board size should be no larger than 4'-0" x 4'-0" (1.22m x 1.22m) and when set in adhesive, the thickness should not be greater than 2.5" (6.35) so that it can be appropriately set into and bonded in adhesive. The cover board is an important component of a resilient roof system and is not an element of the system that can be value engineered out. The cover board provides a substantial substrate in which to bond the roof cover, protect the insulation from crushing under foot traffic and improved wind uplift resistance. The cover board should be compatible with the adhesives and membrane being used, and when adhered should be 4'-0" x 4'-0" (1.22m x 1.22m in size and a minimum of 1/2" thick (1.27cm). Proven cover board materials include: Cementitious, gypsum-fiber reinforced and glass mat faced gypsum.

2.5 Roof Cover

The roof cover selection should be based on parameters such as:

(1) Availability

(2) Contractor familiarity and quality of installation

(3) Ability to be easily repaired at times of storms, lack of power

(4) Past long-term performance success in the region

(5) Resistance to ultraviolet

(6) Repairable in a crisis and

(7) Redundancy capabilities.

Thicker membrane typically will perform better in harsh and violent climate conditions, especially in regard to puncture resistance.

In the Midwestern United States, EPDM has performed well for over 40 years; contractors are familiar with the product, its installation and nuances and is the only material that has a history of long-term performance. While there are numerous other materials currently available, they do not have a history of performance under the same formulations.

Resilient roof system must be looked at holistically, the sum of all its components. Viewing all the components in a drafted section of the roof system is a suggested way to visually observe how the components complement each other. Alternatively, the section might also show how some components might not work together and thus corrections can be made before moving forward with the design of the various components in the system. An example of a resilient roof system section for a sloped roof deck for drainage either to a roof edge gutter or interior roof drain, for ASHRAE climate zone 5 can be seen in Figure 2.

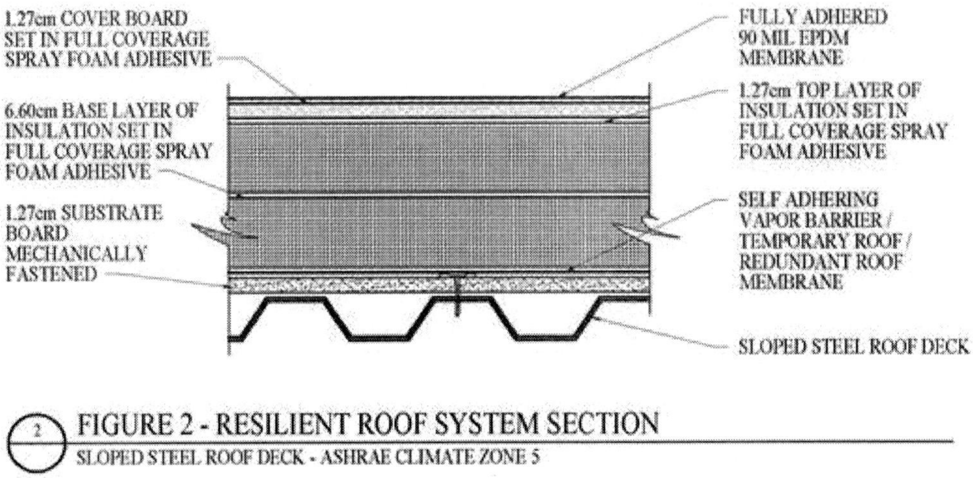

Figure 2. All roof components selected need to work holistically toward a resilient roof system.

By way of comparison, a typical non-resilient roof system is simpler in design with little thought to survivability of severe storms. Figure 3 gives an example of such a system for comparison with the resilient one in Figure 2.

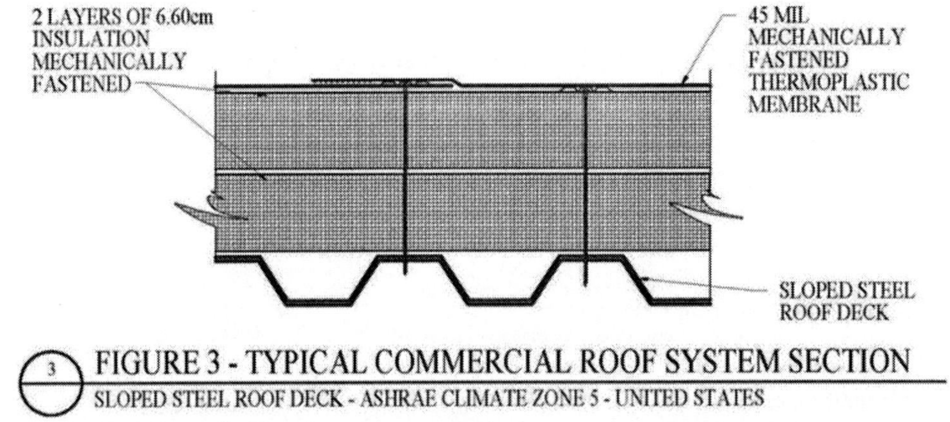

Figure 3: Roof systems that look to achieve only a roof warranty with little thought toward performance in severe are a high risk solution.

3 DESIGN RECOMMENDATIONS FOR ACHIEVING RESILIENT ROOF SYSTEMS

(1) Pre-Design: Gain an understanding of the climatic conditions in which the roof system will be installed: Storm history and 100-year events; New Construction: Work collaboratively with plumbing, mechanical and structural engineers to achieve a holistically design roof system. Reroofing: Evaluate all existing conditions that impinge on the roof and which may be a wind link in your system.

(2) Roof System Design: Think holistically; Think constructability; Select time proven materials in the climatic area of design of which contractors have expertise in installing; Provide redundancies in the roof system design, assuming that in intense climatic events,

a breach in the system will occur; Look at how each component is secured to the structure and each other; Anticipate the unexpected, and review how your design will act: Will it survive or fail?

(3) Construction Documents: Drawings and Specifications: Develop comprehensive drawing and specification that are project specific; and detail, detail, detail all conditions on the roof.

(4) Construction Observations: Be at the project to observe the installation, verify compliance with the drawings and specifications and to answer questions; and, remember you are part of a team working together to achieve a greater good.

4 CONCLUSIONS

This new design consideration, instead of relying on the hundred-year weather history, resilient design has to anticipate the now rapidly changing nature of climate. No longer can engineers and architects rely on what weather events have transpired in the past, where they have occurred or how frequently. Climate scientists and building scientists have to work together with engineers, economists, and manufacturers to determine best products and practices will work.

The concept of resiliency can be challenging when first confronted with altruistic definitions, but if one can think of it as "minimizing the impact of today's ever increasing volatile climate on the roof so that the building can function and the roof be quickly and easily repaired if damaged" the idea becomes clearer. Once crystalized, the design will be natural. Working together with clients, other disciplines and tradesman we can achieve roofs that are resilient to the effects of Mother Nature.

References

Baskaran *et al.*, Climate Change Adaptation Technologies for Roofing and Insulation, *Proceedings of the North American Consultation on Climate Adaptation Technologies, NRC Report #A1-009962.2,* 2017.

Baskaran. A, S., Molleti, D., Lefebvre, and Holcroft, N., Climate Change Adaptation Technologies for Roofing, *Proceedings of the 33rd RCI Annual Convention*, Houston, TX, 2018.

Responsible Design and Delivery of the Constructed Project
Edited by Abdul-Malak, M., Khoury, H., Singh, A., and Yazdani, S.

THE ELEMENTS OF NATURAL CONCEPT IN SUSTAINABLE BUILDING SKINS

TAHER SALAH, OMNIA MOHAMED, and AKRAM AZMY

Dept of Interior Design, Ajman University, Ajman, UAE

Introducing new ideas and solutions that serve environmental functions in sustainable building designs have become an enormous challenge recently, particularly with the current practices of architects and designers in this field. To achieve comfort zone in commercial and residential spaces, the three major aspects that have to be considered are thermal, ventilation and daylight. The proposed project is located in Abu Dhabi, United Arab Emirates, and the main purpose of choosing this location is the hot weather in the U.A.E. Therefore, applying the bio-mimicry design on building skins will react as a solution to improve the indoor thermal comfort in reducing solar heat gain and achieving uniformity in daylight as well as focusing on ventilation in certain seasons, to motivate and enhance the natural ventilation pattern and to reduce the energy consumption as well. This study will mainly focus on developing a building skin that considers locally sourced material such as clay and protect birds' habitat by creating a place for breeding and feeding the birds as well as to integrate plant pots into the skin to reduce pollution. Secondly, combining it with a natural concept strategy that matches the various systems used, to enhance the indoor environment and the indoor comfort needs. The Simulation software, ECOTECT/IES, was used to measure the daylight and thermal performance levels before and after applying the skin concept. The expected findings of this paper may also help in solving part of the existing building envelopes and evaluate their negative impacts in the hot climate region.

Keywords: Thermal comfort, Ventilation, Daylight.

1 INTRODUCTION

Human population is increasing massively; wherever we go, as humans, we completely change the natural habitats of the place we occupy. Studies have shown that the forest has declined by 19% in the last three centuries, the grassland declined by 8%, and cropland increased over 400% (Marzluff 2001). Formerly, buildings were used to perform as a form of housing, making a direct separation between users and their occupied environment. Architecture began to separate from its ordinary characteristics and went outside the box. As much as the designer inspired by nature, and as much as the human needs are reached by doing so, architechs and designers are converting these types of behavioral processes into human-made design solutions (Maglic 2012).Today, the new ideology used is "Biomimicry design" that imitate nature to develop comfortable housings and systems to serve the human needs. The outcome expected from this study is to create a balanced harmony between architecture, human needs, and the environment. This study will mainly focus on the building skin systems that serve the hot climate region. The case study project used in this paper is from Abu Dhabi, United Arab Emirates. The IES and Ecotect software were used for simulation to extract the thermal and daylight figures. The climate – consult was used to extract climatic data information. On the other hand, the proposed bio-mimic concept will have to achieve the optimum desired comfort zone in residential spaces. Therefore, three main features to be focused on are thermal, daylight and improving the natural ventilation pattern in houses.

2 SELECTED CASE STUDY

The Project is located in Al Raha Garden Abu Dhabi, United Arab Emirates. The main reason of choosing this city is the massive construction development taking place. Masdar city is a real project example of the importance of sustainability to this country (Masdar 2018). The construction materials used in the existing villa block walls are plaster concrete and final finish of paint. UAE Climate data indicates the weather seems hot and humid in summer and the temperature rises to about 48 C° and 90% humidity especially in June - August. The average temperature in summer is 32-36 C° (Figure 1). (Casey 2017)

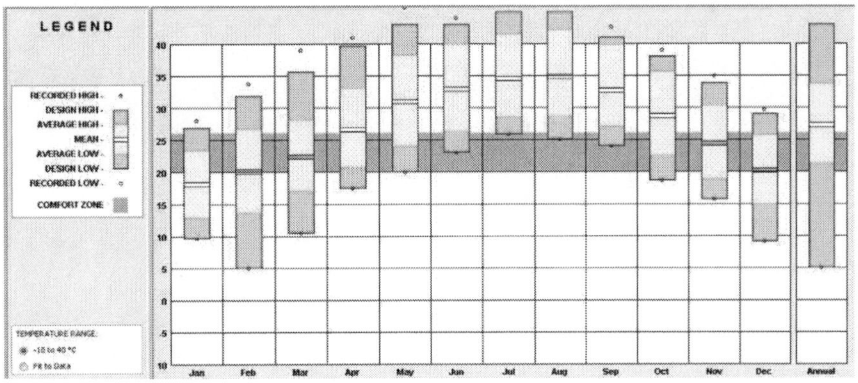

Figure 1. Temperature average. Climate consultant, City of Abu Dhabi, UAE (2016).

3 NATURAL CONCEPT

"It may be true that one has to choose between ethics and aesthetics, but whichever one chooses, one will always find the other at the end of the road" (Hagan 2001). Many habitats have been converted into road and buildings due to human activities. This study focused on developing a building skin inspired by birds in nature. The major idea behind this concept is to develop an eco-friendly building skin including small houses for birds inside the skin that protect these species in UAE (Hockin 2017). Two types of the local birds that live permanently in residential and landscape areas were considered, "House Sparrow" and "Little green Bee-eater" (Kshemkalyani 2009). The birds' wings movement has been studied and matched with the idea of protecting the main building structure by integrating new skin made of solar screen fabric. A modular structure made of clay (terracotta mold) attached to the villa using steel channels and solar screen shades fixed between the clay modular are used to control thermal aspects.

4 EXISTING VILLA ANALYSIS

4.1 Daylight Factor Analysis (Existing Villa)

The percentage of DF was considerably high by 5.5%. This percentage shows the needles of artificial light during daytime, but overheating the space. The current material used is double glass, as to follow the UAE standard requirements. Double-glazing with an absorbent film that transmits 39% of solar heat and reflects 61% U factor is 0.25 and solar heat gain coefficient is 0.39 (Tamco 2017).

4.2 Hourly Heat Gains/Losses – (LV, GF)

This graph (Figure 2) shows the results of material conductivity and HVAC Loads on the existing structure (Plaster Walls and Double Glass). The conductivity of materials increased drastically in noon to 4:00 P.M and reached 3600w. The HVAC increased gradually from early morning of 7:00 A.M. to reach the peak at 12:30 noon by 6800w.

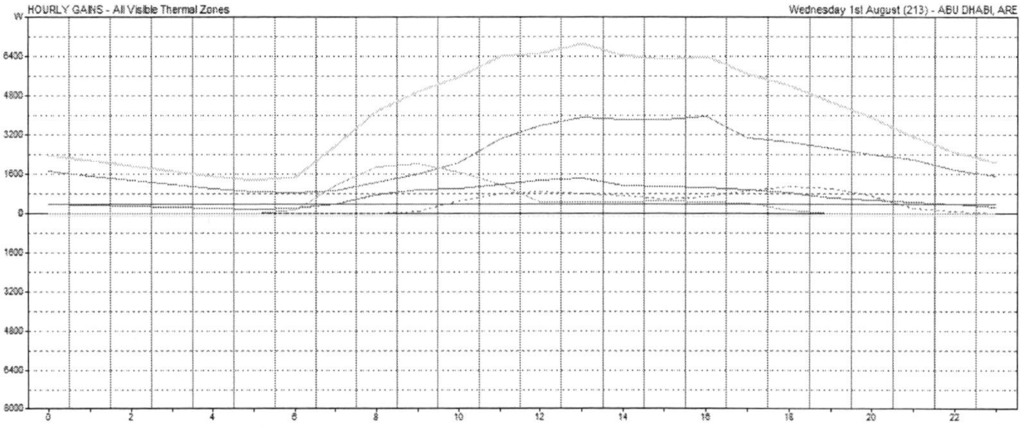

Figure 2. Hourly heat gains/losses (Ecotect 2016).

5 THE PROPOSED MODULAR SKIN DESIGN AND DEVELOPMENT

The developed modular structure is made of clay mold strips divided into pieces, 1m length by 55 cm width (Terracotta Tiles). Each clay tile is erected directly to steel columns that are attached to the building concrete structure vertically with horizontal rails (Figure 3). The structure is creating a double skin façade with the original building structure. In the north façade side, the developed modular (square shape) consists of pot channels in order to grow plants. The roots of the self – climbing vines are placed inside a pot for simple irrigation system by pouring water directly into the pot, and the climbing vines will attach itself to any close rough surface (Figure 3). This will help on feeding the birds as well as a warm place for breeding and creating their nests inside a small opening shaped in the modular. On the other hand, in the south façade, the modular consist of greenery channels that are connected by automated solar fabric louvers to enhance thermal comfort level and control the sunlight and daylight penetration to the original building. (Shen and Tzempelikos 2012). Additionally, this system was made to enhance natural ventilation during cold seasons to improve indoor air quality. This system will give the design flexibility to change and adapt with various weather conditions along the year.

5.1 Exterior Solar Shade

5.1.1 *Solar screen fabric*

Unique PVC-coated solar screen fabrics block up to 97% of the heat without creating a closed in feeling. In addition to superior durability and minimal maintenance, the fabric's 4% openness provides excellent outward visibility. It is a composition of polyester and vinyl polyester (Insolroll 2016).

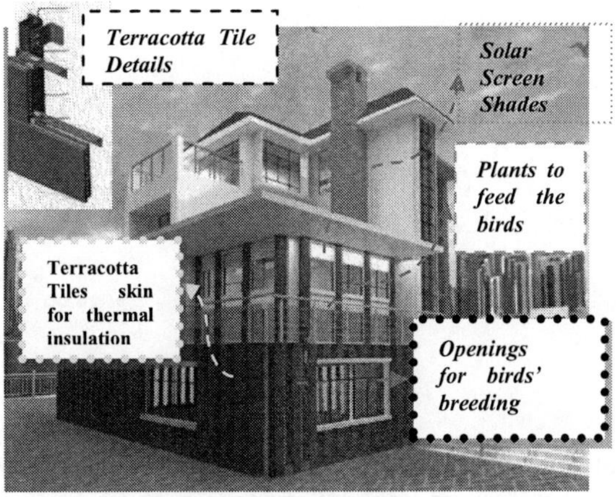

Figure 3. Developed skin.

6 RESULT OF THE APPLIED SOLUTIONS

6.1 Daylight Analysis Pattern (Terracotta Skin & Solar Shades Fabric: Living Room, GF)

The daylight level and daylight factor have been improved significantly after applying the screen shades to the opening in ground floor level.

6.2 Thermal Analysis (Terracotta Tile and Solar Shades Fabric – Living Room, GF)

Figure 4 shows the hourly heat gain/losses of the living room space, while after applying the solar shades screen the conductivity reduced by less than 3200w and the HVAC load decreased to 4900w in August during the peak time 12:00 – 1:00 P.M.

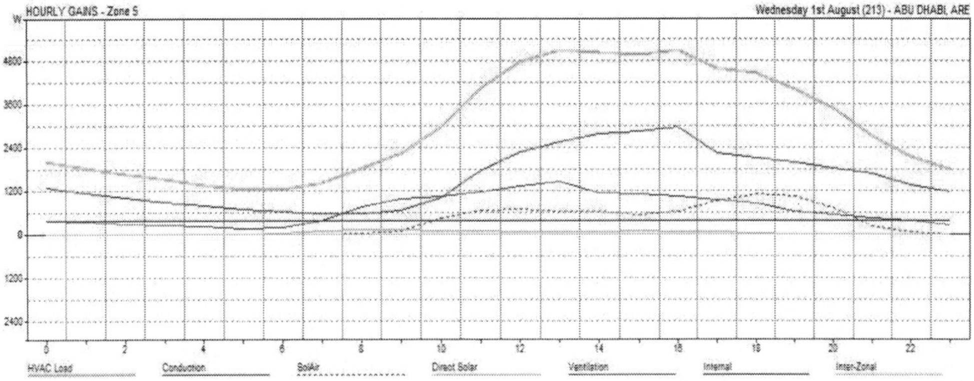

Figure 4. Solar shades screen (hourly heat gains/losses) – Living room, GF (Ecotect 2016).

6.3 Daylight Factor Analysis (Developed Skin – Living Room, GF)

Table 1 demonstrates summary results of daylight factor percentages after adding the proposed skin to the structure. Results have shown improvement and balance in daylight factor with an average of 2.5%. This indicates that thermal level has been improved. However, in a certain time of the day, supplementary artificial light is needed to improve the space lighting (Chou 2004).

Table 1. Daylight factor summary in Living room (Solar Shades and Terracotta Skin) IES 2017.

Surface	Parameter	Values			Uniformity (Min./Ave.)	Diversity (Min./Max.)
		Min.	Ave.	Max.		
Working plane 1 Reflectance=0% Transmittance=100% Grid size=0.30 m Area=13.380 m² Margin-0.30 m	Daylight factor	0.5 %	2.5 %	8.3 %	0.17	0.07
	Daylight illuminance	87.33 lux	577.02 lux	2089.99 lux	0.17	0.07

7 DISCUSSION

Table 2 indicates the final summary results of the studied structure obtained from IES software (The original skin structure vs. the proposed Terracotta modular with solar screen shades). The development of the modular terracotta structure with plant pots and solar screen shades has shown a better improvement to the villa structure in most of the studied factors. Firstly, the daylight factor results in the proposed skin are 2.5% in comparison to the main skin of 5.5%. This percentage of the proposed skin will reduce thermal heat gain and balance DF level. However, artificial light is needed during the day. Secondly, the CO2 concentration in each skin was studied, and results have shown improvement in the clay tile with an average of 360PPM in comparison to the existing skin 455 PPM.

Table 2. Daylight factor and Thermal MWH Analysis comparison (IES 2017).

Residential Villa	Daylight Factor Diffuse Living Room	CO_2 Concentration PPM Living Room	Total Carbon Emission (CE) (Kg O_2)	Cooling load KW Living Room	Total Energy Consumption Yearly
Original Structure Plaster Brick wall + Double Glazing window	5.5 % (Mean Value)	455 PPM (Mean Value)	45,571 $KgCO_2$	0.9 KW (Mean Value)	69.3 MWH
Proposed Skin Terracotta Skin + Solar Screen Shades	2.5 % (Mean Value)	360 PPM (Mean Value)	35,792 $KgCO_2$	0.48 KW (Mean Value)	1.5 MWH

Consequently, the improved space will contribute to reducing the CO2 concentration level and improve the air quality value. Corresponding to the previous result, the total carbon emission value has also been enhanced in the solar screen shades skin. While it is essential to use mineral fibers to have better control in CO2 concentration levels.

Additionally, the usage of plants between the clay structures skins has contributed slightly to the reduction of the total CO2 emission. The third column indicated the cooling loads in each skin. In the proposed skin, the result shows better cooling load with 0.4 KW (mean value) difference than the original skin. Finally, the annual total energy consumption has improved. This result demonstrates a reasonable reduction using the Terracotta tiles. Therefore, if this solution is applied to the full villa's compound, a significant energy efficiency level will be achieved annually. Accordingly, a better sustainable built environment value will be added to the project.

8 CONCLUSION

This study discussed the development of eco-friendly skin that improves building consumption regarding energy loads, daylight factor and natural ventilation. Moreover, create a safe place for birds to regenerate without the need of destroying their natural habitat to build our cities. The proposed skin is made of Terracotta Tiles (steel structure, light in weight) with integrated solar screen shades. A comparative study has been conducted on IES and ECOTECT software to compare between the original structure (Plaster bricks with double glazing – following the UAE standard requirements) of the villa and the proposed modular skin. The developed skin showed a better result in achieving the desired goals of this study that mainly focused on thermal aspect, natural ventilation (to improve IAQ) and daylight control.

REFERENCES

Chou, C., P., The Performance of Daylighting with Shading Device in Architecture Design, *Tamkang Journal of Science and Engineering*, 7(4), 205-212, 2004.

Casey, J., World Climate, *Climate – Charts*, 2017, retrieved from http://www.climate-charts.com on December 10, 2017.

Forys, E. A., and Allen, C. R., The Impacts of Sprawl on Biodiversity: The Ant Fauna of the Lower Florida Keys, *Journal of Ecology and Society*, vol. 10(1), 2005.

Salem, F., A time of simpler houses, and tastes, *Gulf News,* August 2010, retrieved from http://gulfnews.com/uaessentials/a-time-of-simpler-houses-and-tastes-1.674103 on May 12, 2017.

Hagan, S., Taking Shape: A New Contract between Architecture and Nature, *Architectural Press, 2001.*

Hockin, D., Ounsted, M., Gormant, M., Hillt, D., Kellert V., and Barker, M.A., Examination of the Effects of Disturbance on Birds concerning its Importance in Ecological Assessments, *Journal of Environmental Management*, 36, 253-286, 1992. (Online). Accessed on 21-12-2017.

Shen, H. and Tzempelikos, A., 2012. Daylighting and energy analysis of private offices with automated interior roller shades. Solar energy, 86(2), pp.681-704.

Insolroll, Insolroll Solar Shades Control Sunlight, *Insoroll Window Shading Systems*, 2016, retrieved from https://insolroll.com on December 2017.

Maglic, M. J., Biomimicry, Using Nature as a Model for Design, 2012, Master Thesis, University of Massachusetts Amherst, Amherst, Massachusetts.

Marzluff, J. M., Worldwide Urbanization and its Effects on Birds, 2001. Retrieved from https://link.springer.com/chapter/10.1007/978-1-4615-1531-9_2 on December 21, 2017.

Kshemkalyani, A., Guide to Birds of UAE, *Sustainability Tribe*, 2009, retrieved from http://www.sustainabilitytribe.com/guide-to-birds-of-uae/ on May 9, 2017.

Tamco Windows + Doors, 2017, Product Specification, retrieved from http://tamco.ae/ in February, 2018.

Responsible Design and Delivery of the Constructed Project
Edited by Abdul-Malak, M., Khoury, H., Singh, A., and Yazdani, S.

ARCHITECTURAL ENGINEERING IN REHABILITATION AT SOCIAL HOUSING QUARTERS

VITOR ABRANTES[1], NUNO ABRANTES[2], and FERNANDO BRANDÃO ALVES[1]

[1]*Faculty of Engineering, University of Porto, Porto, Portugal*
[2]*CITTA, Faculty of Engineering, University of Porto, Porto, Portugal*

The rehabilitation of social housing quarters is done with the purpose to answer three central questions: (1) to solve the existing anomalies that are often caused by rain water infiltrations through the vertical's building envelope and also through roof, condensation caused by the lack of insulation and poor existent ventilation and existing cracks in building's facades; (2) to improve thermal comfort and increase energy efficiency of the buildings; (3) to increase the satisfaction and self-esteem of the residents promoting a modern and also high quality architectural intervention at social housing quarters. This study will describe the executed solutions for the rehabilitation of this three social housing quarters, three of the biggest social quarters of the Oporto region, each one with a different intervention in the building envelope.

Keywords: Urban requalification, Urban integration, Architecture and urban design, Self-esteem of the residents.

1 INTRODUCTION

This paper attempts to put in evidence the architectural engineering interventions in rehabilitation at social housing quarters implemented to solve the crucial problems, condensation and humidity (Thomson and Sian 2015). Social housing in Portugal represents a very significant portion of housing (50 % of inhabitants of Portugal).

In the 50's, more examples of construction of multifamily housing appeared, primarily through direct state intervention in social housing: "What we see is a spatially monetization of land available and the need for more comprehensive solutions to problems of ordering of metropolitan life" (Baptista 1999).

Around the two important metropolitan areas (Lisbon and Porto) it was verified that over than 20 % of people lives in social housing. A large number of these buildings have an intensive residential use and they need a systematic maintenance that is often ignored for economic reasons and also for insufficient planning.

With the Revolution of April 1974, other initiatives with the same goal of fighting the shortage of housing for people of lower social classes were promoted. Remarkable is the SAAL - Outpatient Services Local Support (1974-1976), created as the Housing Development Fund (FFH) Specialized Technical Corps. This service advocated "a radical model of the city and planning: a city where the poor are also entitled to the historic center" (Costa 1998).

Thus, the rehabilitation of social housing quarters is done with the purpose of answering three central questions:

- To solve the existing anomalies that are often caused by rain water infiltrations through the vertical's building envelope of the building and also through roof, condensation caused by the lack of insulation and poor existent ventilation and existing cracks in building's facades.
- To improve thermal comfort and increase energy efficiency of the buildings.
- To increase the satisfaction and self-esteem of the residents promoting a modern and also high quality architectural at social quarters.

The executed solutions for the rehabilitation of this three social housing quarters, in three of the biggest social quarters of the Oporto region, each one with a different intervention in the building envelope will be described in the present study case. These neighborhoods constitute an experience of the adoption of affordable housing in the modernist principles enunciated in the Charter of Athens (1931) excellent multifamily blocks the object of an extremely well-detailed design, "(...) in intimate relationship with the urban green, (...) with a very strong human scale" (Coelho 2006). The Plan of Improvements (Plano de Melhoramentos) (Câmara Municipal do Porto 1966) is responsible for the emergence of a new urban typology in Porto through new housing estates.

Figure 1. Neighborhood of Contumil①, of Vila D'Este, ② and of Santa Luzia ③.

2 CONDENSATION

Condensation comes from water vapour within the buildings - people, cooking, bathing ... Buildings with poorly insulated walls are very prone to this problem. Due to the lack of thermal resistance, condensation occurs, which leads to water damage in the indoor environment. These results in condensation and risk for mold growth. It will exist condensation if both Eq. (1) and Eq. (2) are true.

$$tsi < tpo \tag{1}$$

$$tsi = ti - \frac{U}{hi}(ti - te) \tag{2}$$

Where, tsi = Interior surface temperature; tpo = Dew point temperature given by the humidity chart. ti = Interior temperature; te = Exterior temperature; U = Thermal coefficient of transmission; and, hi = Thermal interior conductivity.

Figure 2. Examples of observed anomalies (condensation, humidity).

To solve the problem, heat and ventilation should be increased, but in social housing this is not possible. So, it's necessary to improve the insulation of external roofs and walls.

3 ENERGY EFFICIENCY

EU Countries have drawn up strategies to show how they plan to foster investment in the renovation of residential buildings.

One of these strategies is provide an estimate of the expected energy saving that will result from renovations. Most of the social housing has a poor level of efficiency.

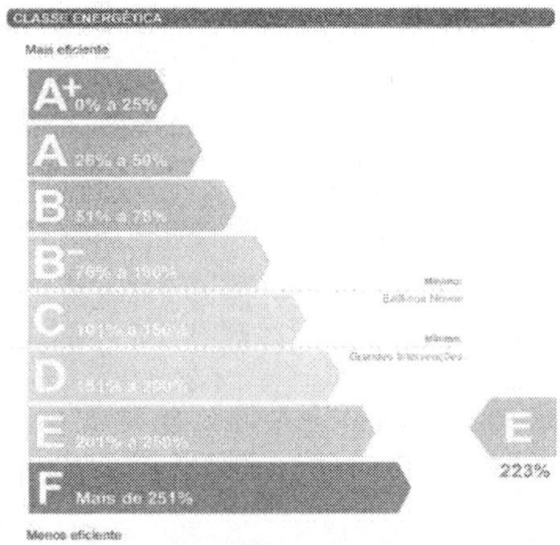

Figure 3. Example of a result of an energy efficiency evaluation (classification E).

The EU program of investment forces that the housing after the rehabilitation reach more two levels.

4 STUDY CASE

4.1 Santa Luzia, Porto

The analysis of the neighborhood of Santa Luzia, at the present time and its redevelopment project, allowed to study the factors that compromise the urban integration and how the requalification met such conditions. The project seeks to understand the relationships between the different spatial components in shaping the urban neighborhood form and the contribution of each of these components to the overall quality of the intervention.

a) Before the architectural engineering in rehabilitation b) After the architectural engineering in rehabilitation

Figure 4. Neighborhood of Santa Luzia, Porto (640 Dwellings).

Each component of the urban space tries to be designed so that the assembly formed by reclassified zones in the shown design, can be understood as a unit - and not the current sum of road sections more or less dispersed buildings and voids between both building an urban fabric.

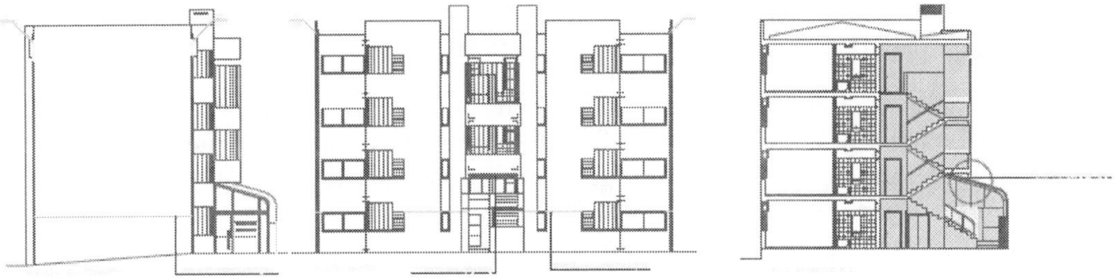

Figure 5. Project of rehabilitation of Santa Luzia, Porto.

4.2 Vila d'Este, Vila Nova de Gaia

The analysis of morphological characteristics of the Vila d'Este urbanization in two stages, before and after the physical rehabilitation - allowed us to study the factors that compromised the urban integration, architectural quality and residential satisfaction of residents.

a) Before the architectural engineering in rehabilitation b) After the architectural engineering in rehabilitation

Figure 6. Neighborhood of Vila d'Este, Vila Nova de Gaia (2085 Dwellings).

The redevelopment meets the full range of objectives, which must meet the physical rehabilitation of public space in affordable housing sets. Like in the previous case, each element of the urban space was designed together with all other elements, so that all the parts reformulated in the rehabilitation work can be seen as a unit. Before upgrading, urban set was only a sum of empty spaces between buildings scattered and fairly disjointed road sections, buildings without quality.

Figure 7. Details of project of rehabilitation of Vila d'Este, Vila Nova de Gaia.

4.3 Contumil, Porto

The intervention in Contumil neighborhood responds to the most critical issues of its urban space and meets the full range of objectives which must meet the physical rehabilitation of public space in the social housing context. Our first goal was to understand the relationships between the different spatial components in the design of the urban form of the whole and the contribution of each part to the overall quality of the intervention.

a) Before the architectural engineering in rehabilitation

b) After the architectural engineering in rehabilitation

Figure 8. Neighborhood of Contumil, Porto (248 Dwellings).

Each part of the urban space is designed so that the whole can be understood as a unit and not as a sum of distinct elements or empty spaces between them; buildings that does not assure continuity if it doesn't benefit from the complementarity between them and the adjacent spaces. Recasting of accessibility plays a very important role.

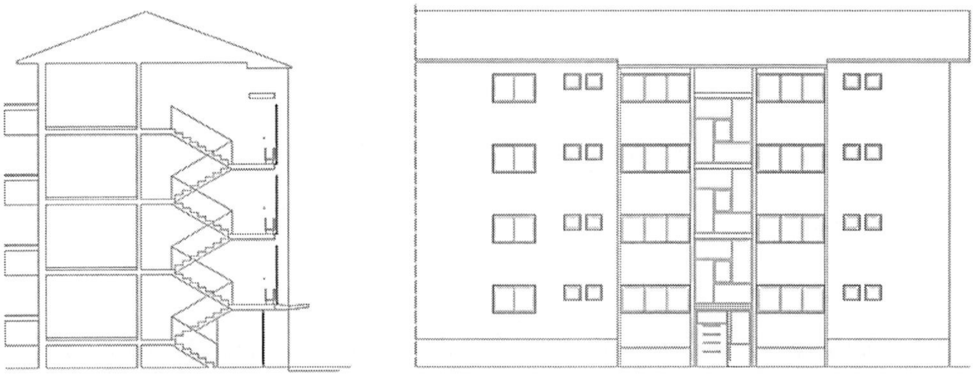

Figure 9. Details. Project of rehabilitation of Contumil, Porto.

The restructuring of the redesigned road network with orthogonal paths and new links that regulate the mesh of the remaining spaces can resolve the gaps in accessibility and proximity of parking.

5 CONCLUSIONS

These three interventions of rehabilitation at social housing quarters show that we can solve the main problems – condensation and lack of energy efficiency - through technical engineering solutions. These solutions combined with architectural interventions and the redesign of the buildings will increase the satisfaction and self-esteem of the residents, qualifying these quarters (ceasing to be considered ghettos) and integrating it in the city.

It is important to refer that the introduction of the architectural component in rehabilitation must be considered as a major and significant value in the image and functionality final result of the rehabilitated buildings (Abrantes 2009).

It is to expect that the architectural valorization work component will be the key for the acceptance of inhabitants and users of the buildings and therefore it induces auto-esteem in people to reinforce their bound with their surroundings.

This point is a value that must be assured for all social habitation.

References

Abrantes V., Conservation Strategy for Large Scale Housing Buildings, *ABACUS - Journal on Architecture, Conservation and Urban Studies*, 4(1), 80-88, May, 2009.

Baptista L. V., *Cidade e Habitação Social*, Celta Editora, Oeiras, 96, 1999.

Câmara Municipal do Porto, *Plano de Melhoramentos 1956-66.* Ed. Direcção dos Serviços do Plano de Melhoramentos, Porto, 1966.

Coelho A. B., *1984-2004 – 20 anos a Promover a Construção de Habitação Social*. Ed. INH, Lisboa, 52, 2006.

Costa A. A.,1974-1975, *o SAAL e os anos da Revolução in "Arquitectura do Século XX – Portugal".* AA.VV. – Deutsches Architektur-Museum, 68, 1998.

Thomson H., and Sian T., Developing Empirically Supported Theories of Change for Housing Investment and Health, *Social Science & Medicine*, 124, 205-214, January, 2015.

Responsible Design and Delivery of the Constructed Project
Edited by Abdul-Malak, M., Khoury, H., Singh, A., and Yazdani, S.

MECHES HOUSE: THE IMPORTANCE OF CHOOSING THE RIGHT BENEFICIARY ON A POST DISASTER ALTERNATIVE CONSTRUCTION

ENRIQUE VILLACIS[1], MARIA LORENA RODRIGUEZ[2], and CYNTHIA AYARZA[2]

[1]*School of Architecture, Design, and Arts, Pontifical Catholic University, Quito, Ecuador*
[2]*Ensusitio Arq., Quito, Ecuador*

More than a year ago, Ecuador suffered a 7.8 M_w Earthquake that was devastating, the automatic response was to rush to the affected areas and rebuild. Meche´s House is an alternative: the crisis as an opportunity to experiment and consolidate popular construction systems based on an endogenous and holistic approach. This study evaluates two different methodologies (*Mutualista Pichincha´s Casa Lista* and Meche´s House) of the selection process for the beneficiaries on a reconstruction post-disaster building program contrasting each case and its needs by analyzing its approaches to the reconstruction procedures: one automatic, non-reflexive and product-based, and one analytic and process-based. Founded on experience and how different processes need different beneficiaries this study will raise some questions about the effectiveness, pros, cons, and perspectives of different approaches to make the correct choice of beneficiary for the reconstruction process after a natural disaster.

Keywords: Ecuador, Earthquake, Resiliency, Endogenous.

1 INTRODUCTION

This study deals with the importance of choosing the adequate beneficiary for a post-disaster construction process. Based on the values of the process and on the social impact in addition to the building itself, we will show the Ecuadorian seismic reality and different approaches for the reconstruction process and how those require different types of beneficiaries and how it affects their lives and their communities.

Ecuador is in constant seismic danger because of its geographical location in the Pacific ring of fire. According to reports; since 1541, there have been 37 major earthquakes, the majority of great magnitude as stated by Sánchez and Limón (2017) On April 16 of 2017, a 7.8 M_w magnitude earthquake hit the coast of Ecuador. The consequences were devastating, 70 % of real estate in the area was destroyed and according to the National Secretariat of Planning and Development (SENPLADES), 13,962 urban and 15,962 rural houses were demolished. Beyond numbers, a lack of technical advice in Ecuador is evident. This fact put in evidence another important issue: over 70% of the buildings that collapsed were built with materials such as concrete, steel, concrete block while most of the buildings that were left standing were made with local construction techniques, such as wood, and bamboo. After his first visits to the affected areas, President Rafael Correa admitted that many buildings collapsed "because of bad construction" Zibell (2016).

Most of the buildings that collapsed were built with no technical support or cutting corners irresponsibly, poor use or misuse of steel, use of aggregates from the beach with high salt content. Several reasons led contractor and construction workers to build carelessly, one is corruption by trying to make a building less costly in order to have more income, and another is ignorance. The bottom line is construction workers adopt these mal-practices and tend to apply them in their own buildings.

An important fact is that "almost 70% of Latin-America is built informally. In Ecuador, 70% of the construction sector is built informally, and three million houses are poorly built" Inmediato (2011). This means that there is no technical support for the construction processes, and building knowledge is empirical, based on malpractice, leaving most of the population at risk.

This evidence puts professionals directly responsible for casualties of the disaster, either through malpractice or because we allow for these mal practices to persist. Evidently, after the disaster, there was going to be a rebuilding process, but as professionals, it would be highly irresponsible to perpetuate those same mistakes: technical malpractices, irresponsible knowledge, and technology transfer.

2 FINANCIAL INSTITUTION "READY MADE HOUSE - *CASA LISTA*"

After the earthquake, well-intentioned support from the government and many institutions was present in the reconstruction process. However, all plans were a massive and immediate answer, and what usually occurs during crisis, "the implicit assumption was that the beneficiaries (...) for the most part, were too backward to initiate development themselves and that local government would be responsible for taking the lead" Finsterbusch and Van Wicklin (1987). After the disaster, most of the reconstruction process was led by government-related organizations. The months after the earthquake hundreds of houses were built but only for those families that follow these organizations own requirements. Families such as Meche's were not able to receive a house. As for most of these cases the project values were: time and cost efficiency and political impact.

A renowned financial institution of Ecuador "*Mutualista Pichincha*" was one of the organizations that through the project *"Échale una mano a Ecuador"* – "Give a hand to Ecuador", built thousands of houses. Similar to many organizations, the requirements to become a beneficiary generally were:

- Ecuadorian citizens or foreigners with permanent residence in the area, of legal age, heads of an organized family nucleus. Single people without dependents and over 50 years.
- Owner receives the poverty stipend.
- The overall monthly income for the family is less than 1087.50 USD.
- Proof of ownership of land.
- Do not own any other house anywhere.
- The owned land must be leveled. *Mutualista Pichincha* will review the land which should be technically leveled and in risk free areas (flooding, landslides, etc.)

Most of these requirements are focused on the person or family itself rather than the impact that the project will have on the community and its construction knowledge. Although it is very important to analyze the circumstances of every specific construction project, the durability, the cost-effective variables or the speed of the construction. Every process needs a specific kind of beneficiary.

3 MECHE´S HOUSE

When the opportunity arrived for Ensusitio Arq[1] to rebuild in the disaster area, values were set to guide the decisions that would affect the development of the project.

- Conceiving the construction not as a product, but as a process for technology and knowledge transfer.
- There must be more people involved in this process than just the beneficiary and family.
- The beneficiary must provide a counterpart or contribution even if it is symbolic: a form of commitment.
- Knowledge transfer needs to be direct and without interference, no volunteers or architecture students. This fact may have led to a slower construction process but an emphasized communication with the beneficiary and its community.
- The beneficiary must be open to experimenting with local materials in order to produce accessible technology that can be replicated.
- The building must help to break the poverty cycle, making the home productive.

Meche, forty-one, is a single mother of five girls and a grandmother of two boys. She is the hairdresser of Pedro Carbo and an active member of the local Cacao Association. She lost her house in the earthquake, she started rebuilding with what was left of the house and lost what little was left in one of the replicas. Although need was evident and dramatic, she was not a candidate for government agencies or organizations as a possible beneficiary because her land was on a slope and all prototypes are for leveled sites.
In addition to the dramatic need, we decided to work with Meche because of:

- Commitment to experiment with local material and technology
- Her house-beauty parlor can be an example of how to break the poverty cycle.
- She was committed to invite interested members of the community in order to build her house in exchange for learning proper practices in construction techniques such as bamboo as structure and bahareque, known as wattle and daub, wall consisting of, bamboo strips coated with a sticky material that combines wet soil, clay, sand, animal dung and straw or fiber.
- Meche had already started her house on her own, by selling the livestock she had, she managed to complete a poured concrete foundation. This was the investment she could make, showing her commitment by selling her valued goods "Commitment can become tangible in the form of financial contribution." Finsterbusch and Van Wicklin (1987).

4 FINDINGS

As mentioned, one of our responsibilities as professionals must be to ensure best practices within the construction processes, in order "to have a proper analysis of each technology and its pertinence for each environment where it would be implemented; this means an appropriate technology transfer process where the product is not only the outcome but also the process itself" (Villacis *et al.* 2017). In other words, the building should not be only a building but a building process that would be a laboratory and workshop in order to ensure an adequate knowledge

[1] Ensusitio Arq: a workshop based in Quito-Ecuador, working with what is available IN ITS PLACE, following participatory and community consensus process for public, private and academic projects.

transfer. The participation of the beneficiary and neighbors in the process "builds a strong sense of community and encompasses benefits that span social, environmental, cultural, economic, and spiritual facets. Furthermore, an appropriate technology best fits with the community it serves because it is created by the people to meet a need. Therefore, the communities are placed at the center of decision making and create technologies that will best serve their communities in the long-term." (Margolus *et al.* 2017). All the reconstruction processes must be thought as a long-term project where the knowledge acquired would transcend the mere house. Knowledge and "development imposed from outside the local setting, no matter how benevolent and well-intentioned, is ultimately counterproductive". Finsterbusch and Van Wicklin (1987), especially after a disaster because external and immediate help in reconstruction process could produce dependence.

By simple observation, the beneficiary selection process in several projects leads sometimes to fractured relationships within the community because of envy or despair, by having a portion of the community sharing the building process and being part of the knowledge transfer the social fabric within the community consolidates rather than dissolves. See Figure 1. This is another reason why choosing the right beneficiary is vital for the entire process.

Even though "projects should involve more participation by beneficiaries, in fact, some would argue that real development, by definition, must involve beneficiaries in their own improvement (e.g. Gran, 1983a, b)", we think not everyone is suited for this kind of process because it requires a great amount of responsibility, but on the other hand this responsibility can build a real impact within the community. Finsterbusch and Van Wicklin (1987) said, "Without participation, the people may benefit but not develop from a project. Thus, participation has intrinsic value".

Figure 1. M. Pichincha house exogenous technologies

Figure 2. Meche's house endogenous technologies

There are several ways to deal with a reconstruction process, but in order to educate responsible local construction workers and at the same time teach people to value their own potentials and local construction materials and technologies an endogenous strategy is needed. Informality in construction will continue once the effects of the earthquake are forgotten. The government should encourage the return to vernacular architecture, with materials from the region, instead of allowing dangerous constructions in concrete of several floors that do not comply with the adequate codes states Zibell (2016). Providing and sharing a construction process with local people and using local materials can ensure best knowledge transfer because "an effective project should transfer essential skills to the beneficiaries themselves where possible so that they can sustain the project after the implementing team has left or have greater capacity for other development tasks." Finsterbusch and Van Wicklin (1987).

Ecuador is not exempt from a future disaster so we considered that one way to deal with the reconstruction was to make the beneficiary an active participant. Thereby their resilience "refers to the ability of human settlements to withstand and to recover quickly from any plausible hazards. Resilience against crises not only refers to reducing risks and damage from disasters (i.e. loss of lives and assets), but also the ability to quickly bounce back to a stable state" UNHabitat (2012), and reaction will be more effective, just because they would not be waiting for external help, charity, but they would be able to rebuild based on their local knowledge and resources: thus the reconstruction process becomes a resilience workshop.

Table 1. Building technologies comparison,

BUILDING EXOGENOUS / ENGOGENOUS TECHNOLOGIES		
	M. PICHINCHA HOUSE	MECHE'S HOUSE
FLOOR	concrete	local wood
STRUCTURE	metal	*guadua* cane (bamboo)
WALLS	prefabricated concrete	*guadua* + soil
ROOF	asbestos cement	metallic

Table 2. Cost and m^2 comparison.

	M. PICHINCHA HOUSE	MECHE'S HOUSE
COST / M2	$ 200	$ 169,50
M2	56	76
N. ROOMS	3	3
TOTAL COST	$11 200	$12 882
CON. TIME	100 units -1 month	1 unit - 9 months

Figure 3. *Mutualista Pichincha* house beneficiaries

Figure 4. Meche's house beneficiaries

5 CONCLUSIONS

- During the selection process for Meche´s House, the decision making was not only the responsibility of the technical team, but also the community, cacao cooperatives, and Pedro Carbo residents had a strong opinion to share. They were conscious of the responsibility and future impact of the project. From the earliest steps of the project this house was not only seen as a house but a living example of good practice for local construction technology catalog and a model of resilience and poverty cycle breaker. Everyone in the process is conscious of the demanding requirements for becoming a beneficiary of this. We consider that providing and sharing a construction process with local people and local material can ensure the best knowledge transfer. When dealing with possible future disasters their resilience capability and reaction will be more effective, because they will not be waiting for external help, charity, but they will be able to rebuild based on their acquired knowledge and resources.

- The beneficiary selection process form the "*Casa Lista*" is far more generic and its needs are based on time efficient products, focused on solving the crisis needs of numbers of people served. These variables are understandable, especially when this process has to face political cycles: cost and time effectiveness.

As a popular Latin American saying goes, *"quien quiere celeste que le cueste"* if you want something priceless much effort has to be put into it. This process has a possible high impact within the community behavior, and that is the reason why a very conscious and careful decision making is important not only to assure commitment but also make a political impact in the community.

References

Finsterbusch, K., and Van Wicklin, W., The Contribution of Beneficiary Participation to Development Project Effectiveness, *Public Administration and Development, Vol. 7,* En J. W. Chichester, 1-2, 1987.

Inmediato, E., *Ecuador Inmediato*, November 30, 2011. Retrieved from http://ecuadorinmediato.com/index.php?module=Noticias&func=news_user_view&id=162607&umt=70_construcciones_en_ecuador_son_informales_advierte_secretaria_nacional_gestion_riesgo on December 2016.

Margolus, J., Nakashima, T., and Orr, C., *Appropriate Technology: Learning from One Another*. Retrieved from http://web.uvic.ca/~essa/wp-content/uploads/2010/03/Reclaiming-Sustainability-Conference-Appropriate-Technology.pdf in 2017.

Sánchez, M. L., and Limón, P., La Construcción De La Noticia Y El Papel De Los Social Media Y Periodismos Ciudadano En La Gestión De Información De Desastres O Catástrofes Naturales, Madrid: Egregius, 2017.

UNHabitat, *unhabitat.org*. Retrieved from http://unhabitat.org/urban-themes/resilience/ in 2012.

Villacis, E., Rodriguez, M. L., and Ayarza, C., Preserving Traditional Construction Techniques and Materials as an Answer to Future Energy-Fuel Crisis, *Resilient Structures and Sustainable Construction*, Pellicer, E., Adam, J. M., Yepes, A. Singh, and. Yazdani, S., (eds.) ISEC Press, 2017.

Zibell, M., April 23, 2016. Retrieved from www.bbc.com/mundo/noticias/2016/04/160422_ecuador_terremoto_problemas_construcciones_arquitectura_ab, on December 15, 2017.

Responsible Design and Delivery of the Constructed Project
Edited by Abdul-Malak, M., Khoury, H., Singh, A., and Yazdani, S.

AN INVESTIGATION INTO CRANE AND SCAFFOLD SAFETY IN CONSTRUCTION INDUSTRY

SWAPAN SAHA, BEN CAUCHI and GRAHAME DOUGLAS

School of Computing Engineering and Mathematics, Western Sydney University, Sydney, Australia

The aim of this study is to determine whether construction sites involving cranes and scaffold in Australian large construction sites have been in compliance with the existing Work Health and Safety (WHS) requirements. The research comprises three case studies which investigated the causes of crane and scaffold safety incidents that occurred in New South Wales. The results of the case study on a number of separate sites in Sydney, NSW have shown that without stringent safety procedures in place to abide by the WHS Act and Regulations, major accidents can occur and can cause serious injury and could cause potential fatalities while working with cranes and scaffold. Results also found that a number of limiting factors including communication skill, regular maintenance, safety checks, safety culture and safety investments could contribute to the severity of the incidents.

Keywords: Scaffolding, Falls from height, Australia.

1 BACKGROUND

According to Safe Work Australia (2012), the Construction industry employed 1.01 million people in 2011–12, or 9% of the Australian workforce. The construction industry has a major impact on Australia's economic growth. Master Builders Australia state that 'the building and construction industry is a key driver of the Australian economy and makes a major contribution to the generation of wealth and the welfare of the community, particularly through the provision of shelter'. To ensure the construction industry stays strong in Australia, it is imperative that strict rules and regulations are set out in regards to workplace health and safety (Safe Work Australia 2012a). Australia's safety standards are among the best in the world, although there are still opportunities for improvement. Over the five-year period from 2007–08 to 2011–12, 211 Construction workers died from work-related injuries and there were 13,735 serious injury claims (Safe Work Australia 2012b). Of these numbers, being hit by moving or falling objects accounted for 29 deaths and nearly 2,200 or 16% of the serious injury claims. This type of incident is often related to the use of cranes and scaffolding in which improper procedures and maintenance can have fatal effects. There is a need to ensure crane and scaffold contractors are following the correct Work Health and Safety procedures that are in place, as well as ensuring plant and equipment is maintained for safe use throughout its operational life. Regulatory issues related to crane and scaffold safety have come under increased scrutiny as a result of recent incidents that occurred in New South Wales, Australia (Safe Work Australia 2013). The aim of this research is to determine whether workers operating cranes and working on scaffolding have the necessary skills to comply with the current WHS laws and regulations in Australia. The objectives of the research are the following:

(i) Compare the procedures set out by the WHS Act and Regulations in regards to cranes and scaffold compared to what is actually happening on construction sites in Australia;

(ii) Investigate the causes of crane and scaffold incidents that have occurred in the New South Wales suburbs of Ultimo, Mascot and Barangaroo; and

(iii) Observe the skills of workers and management operating in and around cranes and scaffolds and their awareness of the required safety standards and the magnitude of the potential consequences.

Previous studies reported the need for safety investments, revised training methods, frequent maintenance, strict certification and the importance of basic communication ensure a safer working environment for workers by enhancing the safety capabilities of the crane and scaffold contractors (Safe Work Australia 2013). Without attention to these key factors, workers on a construction site could be exposed to high risk situations which could result in fatality or in serious injury.

A crane is an item of plant intended for raising or lowering a load and moving it horizontally including the supporting structure of the crane and its foundations (Safe Work Australia 2012b). Cranes are common on construction sites all around the world and are vital to the progression of a project. The main use for cranes on construction sites is to move materials around the site which is much more efficient than manual handling.

The Safe Work Australia's Construction Fact Sheet (2011) identifies that falls from height accounted for 51 fatalities. Of these numbers, 18 involved falls from buildings, 15 involved ladders and 8 involved scaffolding. Statistics on hit by moving or falling objects accounted for 29 deaths (Safe Work Australia 2012b). These statistics are heavily linked to cranes and scaffolding therefore, further research work the safety of cranes and scaffold is important to gauge the status of current safety standards. In Spain, about 40% of serious accidents are caused by falls from height, and of these about 30% involve falls from temporary devices on structures assembled to work at height (Rubio-Romero 2013).

The numbers of persons being hit by moving or falling objects through crane and scaffolding use warrants improve scrutiny by employers and regulators. Workers should be able to go to work without the risk of death or serious injury.

A number of construction contractors own and operate a fleet of tower cranes. Most of these towers are rented and operated by in-house or contracted personnel. Frequent inspections of crane use are essential in order to keep the machinery safe and to ensure that operators of these cranes are obtaining the necessary checks and maintenance procedures, whether on a weekly, monthly or yearly basis. Scaffolding provides access through a construction site as well as being a safety measure. Scaffolding allows workers to work at extreme heights that would otherwise be unsafe to travel without the aid of scaffolding. Scaffolding is designed to prevent falls from heights which can result in either serious injury or death. In the 2009-10, Safe Work Australia, notified fatalities from scaffolding work and stated that two workers died because they fell from scaffolding.

2 RESEARCH SCOPE AND METHODOLGY

In this study three incidents involving crane and scaffold that occurred in and around the Sydney CBD were also investigated to find the causes of these accidents. Data and relevant information on the incidents were sourced from media reports as well as Work Cover NSW incident reports.

Information was gathered under "Freedom of Information" (FOI) laws. Details about the causes of the three incidents on crane and scaffolding failure are presented in the following section.

3 CASE STUDIES ON CRANE AND SCAFFOLD SAFETY

3.1 Ultimo Crane Collapse

On 27th November 2012, news broke that a tower crane had caught on fire at the outskirts of the Sydney CBD at a construction site at the University of Technology, Sydney (UTS) (Figure 1). More than 100 workers and 100 bystanders were in surrounding buildings and were evacuated from the area. The crane operator managed to escape the cabin through the crane ladder. The boom of the crane ultimately collapsed and landed on the building structure below.

Figure 1. The tower crane at Ultimo construction site on fire.

A newspaper article reported that safety maintenance was not up to date and the company had failed to undertake regular safety check even though it was pointed out to them through an earlier safety inspection (Australian Financial Review 2012). After discussion with officers of the WorkCover Authority of NSW and a review of the incident report it was confirmed that the protruding arm of a crane had collapsed on top at the University of Technology, Sydney, construction site after catching fire.

3.2 Mascot Scaffold Collapse

On February 25th, 2014, it was reported in the local media that a scaffolding structure had collapsed on a construction site in Mascot, New South Wales (Yahoo7 2014) (Figure 2). Two men were on a scaffold at the time of collapse and fortunately they escaped with only minor injury. It was also reported that the collapsed scaffold brought down power lines on the street and completely covered two trucks. There was also a concern for leaking fuel from the impacted trucks. As a result of the incident more than 100 people had to be evacuated from the area. The NSW Police and the WorkCover Authority started their investigations immediately after the collapse was reported. After discussion with WorkCover Authority officers and reviewing the information in the 'Factual Inspection Reports', it was found that the scaffold collapse was caused by:

(i) A number of broken cables which were hanging from the services pole (electricity and telecommunication);

(ii) Scaffold and building debris which were resting on the services pole;

(iii) Scaffold and building debris protruded out from the building site onto a Street which blocked one lane of road for traffic;

(iv) Parked trucks were covered with collapsed scaffold, raising concerns of fuel leaks; and

(v) Damage to the exterior wall of the existing building next to the collapsed scaffolding. The workers' union also inspected the site a few months earlier and raised concerns about the safety of the scaffold.

Figure 2. The collapsed scaffold at the Mascot building site.

3.3 Barangaroo Crane Fire

On 12th March 2014, a fire broke out in the basement of a building on the Barangaroo development project in Sydney's CBD (Figure 3). A large blaze resulted in dense smoke being omitted from the construction site raising the alarm to existing surrounding high-rise buildings. Residents from all surrounding buildings were evacuated as a safety precaution (Ralston and Patty 2014). Fire persisted for several hours until the fire was brought under control. There was also a concern that the fire could cause damage to the tower cranes substructure and result in a collapse of the crane. Main roads surrounding the construction site were closed off, causing traffic chaos during the afternoon peak hour commuting period. Fire fighters from NSW Fire and Rescue responded until the blaze was controlled. Fire fighters fought the fire over two days with substantial damage to onsite construction material. The crane did not collapse and after the fire was extinguished, engineers were brought in to assess the structural integrity of the crane. WorkCover mentioned in their report that the fire appears to have been caused by a welding accident which set fire to formwork (Ralston and Patty 2014). Approximately 40 to 60 workers were on the lower levels when the blaze broke out and narrowly escaped death or seriously injured (Ralston and Patty 2014). After discussion with WorkCover officers and review of the Incident Report, it was confirmed that the large fire originated in the formwork at the basement area. The incident report also confirmed that a structural engineer would assess the area and provide a remediation plan to be followed before construction would re-commence.

Figure 3. The crane incident at the Barangaroo development project near CBD.

4 FINDINGS AND DISCUSSION

After considering the news reports on these various incidents and then obtaining incident reports through WorkCover, it is evident that there was a breach of work, health and safety requirements on these construction sites. The first incident in Ultimo could have been fatal had the jib of the crane been pointing out towards the road where hundreds of commuters were passing. It seems that the regular safety and maintenance checks were not carried out on this site. Fuel leaks are reported to have initiated the fire which in turn then caused the crane to collapse. If the appropriate maintenance checks were in place then the incident could have been avoided. It was fortunate that there were no injuries or fatalities although the company in question would still likely receive a heavy fine for an incident of this scale.

The Mascot incident suggests that regular checks of the scaffold were not undertaken and resulted in a collapse in which two workers were seriously injured and narrowly avoided death.

The Barangaroo incident was quite an extraordinary case as the events that unfolded were not directly related to the crane. Although the matter is still being investigated and recently reported that a basement fire was lit in close proximity to the base of the crane and was likely to be the initiating event. It was expected that the crane would have collapsed under the heat of the blazing fire. An interview with the Project Manager and Site Manager identified that the design of crane should also consider fire safety.

5 CONCLUDING REMARKS

Three case studies of incidents at Ultimo, Mascot and Barangaroo have been associated with crane and scaffold safety identified the need for improvements in safety protocols and procedures. Non-compliance with WHS Act and Regulations can lead to large scale incidents especially where cranes and scaffold are involved, resulting in property damage, injury or loss of life. It is evident that crane and scaffold incidents can not only affect site workers, but also impact on surrounding buildings and pedestrians. These large-scale incidents can also lead to tarnished reputation and heavy fines for all companies involved in the events. This study has emphasised the importance of regular safety checks and maintenance of high risk construction activities such as crane operation and use of scaffolding and are important in minimising safety incidents and large-scale damage.

References

Australian Financial Review, Crane Collapses in Inner Sydney, 2012. Retrieved from http://www.afr.com/p/national/crane collapses in inner sydney ZGi5exLXXmzQvc on August 1, 2014.

Ralston, N and Patty, A., Investigators Explore Possible Causes of Barangaroo Blaze, 2014. Retrieved from http://www.smh.com.au/action/printArticle?id=5258622 on October 11, 2014.

Rubio-Romero, J. C., Analysis of the Safety Conditions of Scaffolding on Construction Sites, *Safety Science*. 55(1), 160-164, 2013.

Safe Work Australia, Construction Fact Sheet, 2012a. Retrieved from http://www.safeworkaustralia.gov.au/sites/SWA/about/Publications/Documents/430/Construction-Fact-Sheet-2011-12.pdf on March 27, 2014.

Safe Work Australia, Cranes Code of Practice, 2012b. Retrieved from http://www.safeworkaustralia.gov.au/sites/SWA/model-whs-laws/model-COP/Documents/Draft-COP-December-2012/DRAFT-Cranes-COP.pdf on May 13, 2014.

Safe Work Australia, Scaffolding and Scaffolding Work Code of Practice, 2013. Retrieved from http://www.safeworkaustralia.gov.au/sites/swa/about/publications/pages/scaffolds-scaffolding-work on March 14, 2014.

Yahoo7, Two Men Rode Collapsing Scaffolding, 2014. Retrieved from https://au.news.yahoo.com/nsw/a/21689054/three-injured-in-scaffolding-collapse-at-sy/ on August 1, 2014.

Responsible Design and Delivery of the Constructed Project
Edited by Abdul-Malak, M., Khoury, H., Singh, A., and Yazdani, S.

EVALUATION OF CREDIT ACHIEVEMENTS OF LEED-CERTIFIED BUILDINGS

GUL POLAT[1], HARUN TURKOGLU[1], ASLI PELIN GURGUN[2], and NAZ YIKILMAZ[1]

[1]*Dept of Civil Engineering, Istanbul Technical University, Istanbul, Turkey*
[2]*Dept of Civil Engineering, Yildiz Technical University, Istanbul, Turkey*

Nowadays, the concept of green building is no longer an alternative but a necessity. The acceptance of a building as "green" depends on having a certificate. LEED is one of the widely used green building certification systems in Turkey. The main objective of this study is to find out whether there were statistically significant differences between the credit achievements of 105 newly constructed and certified buildings in Turkey, which were certified according to four different levels, and to determine which of these credit achievements are weak. For this purpose, Kruskal-Wallis and Mann-Whitney U tests were performed and the findings were interpreted. Findings revealed that the highest level of achievement of these buildings was in sustainable sites, water efficiency, innovation in design, and regional priority credits, while the lowest level of achievement was in energy and atmosphere, materials and resources, and indoor environmental quality credits. Analyses revealed that the achievements in the energy and atmosphere, materials and resources, and indoor environmental quality credits of green buildings are relatively low when compared to the ones in other credits. There may be several reasons behind these low levels of achievement such as poor green culture, high investment cost of necessary systems, absence of reusable, recycled, and renewable materials. If these problems, most of which are peculiar to developing countries, can be overcome, higher certification levels can be achieved.

Keywords: Sustainability, Green building, Certification systems, New construction.

1 INTRODUCTION

The growth of the world population over the last 60 years has significantly increased the demand for natural resources such as water, energy, etc. As a result of this increase, environmental problems such as global warming, decrease of water quantity and quality, air pollution, and health problems have emerged. Buildings, which were built by using conventional methods, are responsible for nearly 30% of greenhouse gas production, and consuming approximately 60% of electricity and about 15% of drinking water (Castro-Lacouture *et. al.* 2009, Dwaikat and Ali 2016). Consequently, the concept of green building has emerged as a necessity, not an alternative.

Green building can be defined as eco-friendly buildings using natural and non-waste materials, and consuming resources as much as necessary during their lifecycle (Dwaikat and Ali 2016). Green building certification systems based on specific criteria have been developed to assess whether the constructed structures have the green building feature. Some countries use certification systems that they have developed, while some others adopt standard ones. There are various accepted green building certification systems all around the world. Two of the widely used certification systems are: Leadership in Energy and Environmental Design (LEED) and Building Research Establishment Environmental Assessment Method (BREEAM) (Suzer 2015).

Same as many developing countries, the number of green buildings and the demand for them has considerably increased in Turkey over the last decade. LEED is one of the widely used certification systems in Turkey. U.S. Green Building Council (USGBC) announced the Top 10 Countries for LEED, which is a list that highlights countries outside the United States where LEED originated. In this list, China ranked first on the list as the largest market of LEED, with 34.62 million gross square meters (GSM) of certified LEED space. Canada, India, Brazil and the Republic of Korea, Taiwan, Germany, Turkey, Sweden and United Arab Emirates are the other countries in this list, respectively (Sawit 2016).

There are five rating systems listed in LEED: Building Design and Construction (BD+C), Interior Design and Construction (ID+C), Building Operations and Management (O+M), Neighborhood Development (ND), and Homes. BD+C cover the buildings that are newly constructed or going through a major renovation. BD+C apply to the following project types: new construction and major renovation (NC), core and shell development, schools, retail, healthcare, data centers, hospitality, and warehouses and distribution centers. LEED v4 originated in 2013 and LEED v3 2009 was still in effect till October 2016 (USGBC 2009). Therefore, the number of buildings certified according to LEED v4 is very low when compared to the ones according to LEED v3 2009.

The main objective of this study is to find out whether there were statistically significant differences between the credit achievements of newly constructed buildings in Turkey, which were certified according to four different levels, and to determine which of these credit achievements are weak. For this purpose, Kruskal-Wallis and Mann-Whitney U tests were performed and the findings were interpreted.

2 LEED CERTIFICATION SYSTEM FOR NEW CONSTRUCTION

LEED for Building Design and Construction (LEED BD+C) can be used for the core and shell development of any high-rise building, as well as retail, healthcare or hospitality premises, schools, offices, data centers, warehouses and distribution centers. LEED BD+C are designed for new construction and major renovation projects. At least 60% of the project's gross floor area must be completed by the time of certification (Gurgun *et. al.* 2016).

LEED is a point-based certification system. Any project earns LEED points by meeting certain prerequisites and credits. There are seven credit categories for New Construction with LEED v3 2009, which include: sustainable sites, water efficiency, energy and atmosphere, materials and resources, indoor environmental quality, and regional priority. Credits, their prerequisites and points for each credit are presented in Table 1 (USGBC 2009, Uğur and Leblebici 2017).

Table 1. LEED v3 2009 for new construction.

	Credits and Their Prerequisites	Points
	Sustainable Sites (SS)	**26 Possible Points**
Prerequisite 1	Construction Activity Pollution Prevention	Required
	Water Efficiency (WE)	**10 Possible Points**
Prerequisite 1	Water Use Reduction	Required
	Energy and Atmosphere (EA)	**35 Possible Points**
Prerequisite 1	Fundamental Commissioning of Buildings Energy Systems	Required
Prerequisite 2	Minimum Energy Performance	Required
Prerequisite 3	Fundamental Refrigerant Management	Required

Table 1 (Continued). LEED v3 2009 for new construction.

	Credits and Their Prerequisites	Points
	Materials and Resources (MR)	**14 Possible Points**
Prerequisite 1	Storage and Collection of Recyclables	Required
	Indoor Environmental Quality (IEQ)	**15 Possible Points**
Prerequisite 1	Minimum Indoor Air Quality Performance	Required
Prerequisite 2	Environmental Tobacco Smoke (ETS) Control	Required
	Innovation in Design (ID)	**6 Possible Points**
	Regional Priority (RP)	**4 Possible Points**
	Maximum Total Points	**110 Points**

Rating is based on a 110-point system. Total points and range of credits possible points vary depending on the type of building. A building requires at least 40 points for certification in LEED v3 2009. There are four types of certificates; Certified (40-49 points), Silver (50-59 points), Gold (60-79 points), and Platinum (80 and over) (USGBC 2009).

3 CREDIT ACHIEVEMENTS OF LEED-CERTIFIED BUILDINGS IN TURKEY

Since the LEED for New Construction was first published in 1999, it has been assisting practitioners in improving the quality of buildings and reducing their impacts on the environment. LEED's priority criteria are the physical environment, society, transportation, green energy, heat islands, light pollution, water usage, greenhouse gas emissions, materials, waste, indoor air quality, as well as quantity and user comfort. Energy and atmosphere, and sustainable sites credit categories account for 51% of the maximum achievable points in LEED v3 2009 (USGBC 2009).

According to information obtained from the USGBC project website, there are 425 registered projects in Turkey. 240 of them are categorized as New Construction and 219 of them are evaluated according to the LEED v3 2009. Out of these 219 projects, 105 projects received certificate after the evaluation process. The scorecards of these 105 projects were obtained from the USBGC project website.

Average values, standard deviations, and percentage of credit achievements, and the number of projects for each certification level are presented in Table 2.

Table 2. Credit achievements of LEED-NC 2009 certified buildings in Turkey.

Credits and Max. Achievable Points	**Certification Level and Number of Buildings**											
	Certified (8)			***Silver (17)***			***Gold (71)***			***Platinum (9)***		
	Ave.	Std. Dev.	Ach. %	Ave.	Std. Dev.	Ach. %	Ave.	Std. Dev.	Ach. %	Ave.	Std. Dev.	Ach. %
SS (26)	16.63	2.26	63.94	18.41	2.98	70.81	18.87	3.22	72.59	21.00	3.24	80.77
WE (10)	3.75	1.91	37.50	6.71	2.62	67.06	8.13	2.09	81.27	9.56	0.88	95.56
EA (35)	9.00	2.78	25.71	9.65	4.31	27.56	14.61	4.82	41.73	26.56	5.55	75.87
MR (14)	4.00	1.07	28.57	5.29	1.16	37.82	5.70	1.28	40.74	6.22	1.48	44.44
IEQ (15)	4.88	2.10	32.50	6.76	2.19	45.10	8.39	2.27	55.96	11.44	2.35	76.30
ID (6)	2.88	1.25	47.92	4.12	1.05	68.63	5.00	0.99	83.33	5.56	0.53	92.59
RP (4)	2.63	0.74	65.63	2.88	0.93	72.06	3.65	0.56	91.20	3.89	0.33	97.22

The comparison of the percentages of credit achievements of 105 green buildings with respect to different certification levels is shown in Figure 1.

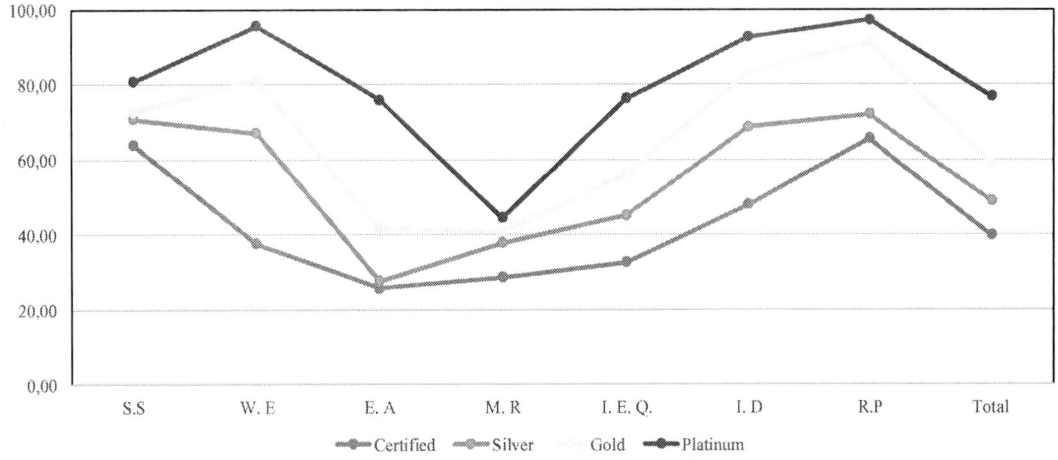

Figure 1. Comparison of percentages of credit achievements of LEED-NC 2009 certified buildings.

According to Figure 1, it can be interpreted that while the percentages of achievements in materials and resource (M.R) credit are very close in all certification levels, the percentages of achievements in energy and atmosphere (E.A) and indoor environmental quality (I.E.Q) credits in platinum certified buildings are very high when compared to the ones in other certification levels.

Another aim of this study is to find out whether there were statistically significant differences between the credit achievements of newly constructed buildings in Turkey, which were certified according to four different levels. For this purpose, Kruskal-Wallis test was conducted via the StatTools v 7.5 software. Kruskal-Wallis is a technique, which is used to test the significance of differences between averages of three or more groups in non-normal distribution groups (Choudhry and Iqbal 2012). The p values obtained from the Kruskal-Wallis test are presented in Table 3. If the p values are less than 0.05, it indicates that there is a statistically significant difference in credit achievements of buildings certified according to different certification levels at 95% significance level.

Table 3. Kruskal-Wallis test p values of credits.

Credits	Kruskal-Wallis Test p Value
Sustainable Sites	**0.0314***
Water Efficiency	**0.0000***
Energy and Atmosphere	**0.0000***
Material and Resources	**0.0036***
Indoor Environmental Quality	**0.0000***
Innovation in Design	**0.0000***
Regional Priority	**0.0000***

* Statistically significant at 95%.

According to Table 3, the Kruskal-Wallis test p values are less than 0.05 for each credit. This means that the credit achievements of buildings certified according to four certification levels are statistically significant different at 95% significance level. Having conducted the Kruskal-Wallis test, the Mann-Whitney U test was performed in order to determine whether there is a statistically significant difference in credit achievements of buildings certified according two different certification levels. The Mann-Whitney U test is a non-parametric technique used to test the significance of differences between two independent group averages in non-normal distribution groups (Chan *et. al.* 2011). If the *p* values are less than 0.05, it indicates that there is a statistically significant difference in credit achievements of buildings certified according to two certification levels at 95% significance level. The *p* values obtained from the Mann-Whitney U test are presented in Table 4.

Table 4. Mann-Whitney U test *p* values of credits.

Credits	Certification Level	Mann-Whitney U Test *p* Values			
		Certified	Silver	Gold	Platinum
Sustainable Sites (SS)	Certified	-	0.2028	**0.0169***	**0.0111***
	Silver		-	0.4823	0.0978
	Gold			-	0.1128
	Platinum				-
Water Efficiency (WE)	Certified	-	**0.0083***	**0.0000***	**0.0000***
	Silver		-	**0.0326***	**0.0047***
	Gold			-	**0.0448***
	Platinum				-
Energy and Atmosphere (EA)	Certified	-	0.9068	**0.0008***	**0.0000***
	Silver		-	**0.0003***	**0.0000***
	Gold			-	**0.0000***
	Platinum				-
Materials and Resources (MR)	Certified	-	**0.0174***	**0.0009***	**0.0055***
	Silver		-	0.2171	0.1471
	Gold			-	0.3684
	Platinum				-
Indoor Environmental Quality (IEQ)	Certified	-	0.0598	**0.0008***	**0.0003***
	Silver			**0.0109***	**0.0006***
	Gold				**0.0015***
	Platinum				-
Innovation in Design (ID)	Certified	-	**0.0308***	**0.0002***	**0.0003***
	Silver		-	**0.0029***	**0.0015***
	Gold			-	0.1254
	Platinum				-
Regional Priority (RP)	Certified	-	0.5526	**0.0000***	**0.0006***
	Silver		-	**0.0004***	**0.0050***
	Gold			-	0.2133
	Platinum				-

* Statistically significant at 95%.

4 FINDINGS AND CONCLUSION

According to the analysis results, the following findings can be highlighted:

- In the sustainable sites credit, the average achievement of the certified buildings remained significantly lower than the gold and platinum buildings, and there is a statistically significant difference between them. However, the average achievement of certified buildings is at the same level of silver buildings, and there is no statistically significant difference between them. Silver, gold and platinum buildings have the same average level of achievement and there is no statistically significant difference among them.
- In the water efficiency credit, the average achievements of all certification levels vary greatly. For this reason, there is a statistically significant difference among all certification levels.

- In energy and atmosphere and indoor environmental quality credits, the average achievement of certified buildings and silver buildings are at the same level, and there is no statistically significant difference between them. There is a considerable difference among the average achievements of the other certification levels, and there is a statistically significant difference among them.
- In materials and resources credit, the average achievement of certified buildings is considerably lower than those of silver, gold and platinum buildings, and there is a statistically significant difference between them. Silver, gold and platinum buildings have the same average level of achievement and there is no statistically significant difference among them.
- In the innovation in design credit, the average achievements of platinum and gold buildings are at the same level, and there is no statistically significant difference between them. There is a considerable difference among the average achievements of the other certification levels, and there is a statistically significant difference among them.
- In the regional priority credit, the average achievements of platinum and gold buildings, and certified and silver buildings are respectively at the same level, and there is no statistically significant difference within each pair.

The analyses revealed that the achievements in the energy and atmosphere, materials and resources, and indoor environmental quality credits of green buildings are relatively low when compared to the ones in other credits. There may be several reasons behind these low levels of achievements. First, the green culture is not very mature. Second, the initial investment cost of necessary systems (e.g., HVAC systems) is very high. Third, renewable energy sources are very limited. And fourth, manufacturing of reusable, recycled, and renewable materials is very limited and most of them are imported, which in turn increases the purchasing cost and brings about logistics problems. In future studies, these reasons can be discussed in more details.

References

Castro-Lacouture, D., Sefair, J. A., Flórez, L., and Medaglia, A. L., Optimization Model for the Selection of Materials Using a LEED-based Green Building Rating System in Colombia, *Building and Environment*, 44(6), 1162-1170, 2009.

Chan, D. W., Chan, A. P., Lam, P. T., Yeung, J. F., and Chan, J. H., Risk Ranking and Analysis in Target Cost Contracts: Empirical Evidence from the Construction Industry, *International Journal of Project Management*, 29(6), 751-763, 2011.

Choudhry, R. M., and Iqbal, K., Identification of Risk Management System in Construction Industry in Pakistan, *Journal of Management in Engineering*, 29(1), 42-49, 2012.

Dwaikat, L. N., and Ali, K. N., Green Buildings Cost Premium: A Review of Empirical Evidence, *Energy and Buildings*, 110, 396-403, 2016.

Gurgun, A. P., Polat, G., Damci, A., and Bayhan, H. G., Performance of LEED Energy Credit Requirements in European Countries, *Procedia Engineering*, 164, 432-438, 2016.

Sawit, A., Infographic: Top 10 Countries for LEED in 2016, usgbc.org, December 14, 2016. Retrieved from www.usgbc.org/articles/infographic-top-10-countries-leed-2016 on January 20, 2018.

Suzer, O., A Comparative Review of Environmental Concern Prioritization: LEED vs other Major Certification Systems, *Journal of Environmental Management*, 154, 266-283, 2015.

Uğur, L. O., and Leblebici, N., An Examination of the LEED Green Building Certification System in Terms of Construction Costs, *Renewable and Sustainable Energy Reviews*, 2017.

USGBC, *For New Construction and Major Renovations*, Washington, DC, US Green Building Council (USGBC), 2009.

BENEFITS OF REMANUFACTURING THE ECONOMY AS A TRANSITION INTO A GREEN ECONOMY

IFIJE OHIOMAH[1], CLINTON AIGBAVBOA[2], and JAN-HARM PRETORIUS[1]

[1]*School of Engineering Management, University of Johannesburg, Johannesburg, South Africa*
[2]*Dept of Construction Management and Quantity Surveying, University of Johannesburg, Johannesburg, South Africa*

Remanufacturing is the restoration of end of life products to almost new. It is a form of sustainability. The research set out to identify the benefits of remanufacturing to the Nigerian economy which could aid with the transition to green economy. A questionnaire survey was used to establish the benefits of remanufacturing to the Nigerian economy. Data analysis involving mean item score revealed development of an Industrial base, industrialisation to rural areas, boosting of export trade, conservation of natural resources are seen as important benefits of remanufacturing to the Nigerian economy. Factor analysis made known four clusters titled in the order of importance as benefits of remanufacturing environmentally, benefits of remanufacturing to trade, benefits of remanufacturing socially, benefits of remanufacturing to entrepreneurs as this clusters signify the benefits of remanufacturing to the Nigerian economy. Further discussed is the implication of the findings, firstly is the development of an industrial base, which will benefit Nigeria, as Nigeria is an under-industrialised nation. Furthermore, the entrenchment of the development of an industrial base would lead to a boosting of export trade in the form of exporting capital equipment to other countries as well as an inflow of foreign exchange. Lastly revealed is the industrialisation to the rural areas, which will stop the social scourge of rural and urban migration.

Keywords: Sustainability, Nigeria

1 INTRODUCTION

Remanufacturing is defined as an industrialized process where parts described as cores, which have reached the end of its lifecycle, are restored to useful life or almost new (Lund 1983). The steps associated with the restoration to useful life are inspection, disassembly, part replacement/refurbishment, cleaning, reassembly and testing to ensure the desired product standard is met (Sundin 2004). Lund (1998) developed seven conditions for remanufacturing. Lund (1983) in his research on remanufacturing identified the following benefits of remanufacturing to countries. Development of an indigenous industrial base through training of a workforce, acquisition of capital goods, and direct technology transfer; Augmentation of foreign exchange by export of remanufactured goods; Development of a service and repair infrastructure to keep industrial and commercial equipment operating; and opportunity for bringing industrialization to rural areas. These benefits enjoyed by countries who develop their remanufacturing industry would also lead to benefits to the remanufacturers, which include the government lowering the entry in terms of capital requirements, leading to more individuals participating in it and increasing the employment rate, labour skills and an increase in the technical expertise and profitability (Ijomah *et al.* 2004). Added benefits to the remanufacturers include broader market appeal. The aim of this study is to identify the benefits of remanufacturing to the Nigeria economy.

2 RESEARCH METHODOLOGY

A quantitative survey was used for this study, whereby Respondents were requested to specify the level of significance of each of the benefits of remanufacturing that play a role in the transition of green economy based on a five-point Likert scale (strongly agree = 5, agree = 4, neutral, = 3, disagree = 2, strongly disagree = 1). Mean item score was used to present the findings for the Likert scale. EFA was performed to gather information about the un-dimensionality of the factors to yield their factor analyzability (Pallant 2010). The EFA was used to confirm the reliability and validity of the factors affecting benefits of remanufacturing the maximum likelihood with an eigenvalue over one, together with Varimax rotation, was specified as the analysis method for this study. The EFA was conducted using SPSS version 21.

2.1 Data Analysis

Two statistical analyses were carried out that is descriptive statistics in the mold of mean item score and factor analysis. The mean item score was used to find the rank of the variables. At the same time as factor analysis was used in instituting which of the variables could be measuring the same underlying effect (Ahadzie *et al.* 2008).

2.2 Mean Item Score

The mean ranking of individual variable was presented to make available a vivid image of the agreement reached by the respondents. A result for the test is revealed in Table 1. The mean for each variable included the standard deviation.

Table 1. Results of mean item score for the benefits of remanufacturing to the Nigerian Economy.

Benefits of remanufacturing to the Nigerian Economy	Mean	Standard deviation	Ranking
Development of an industrial base	4.81	0.397	1
Industrialization to rural areas	4.78	0.438	2
Boosting of export trades	4.65	0.501	3
Conservation of natural resources	4.65	0.500	3
Landfills reduction	4.64	0.598	5
Carbon footprint is reduced	4.64	0.664	5
Inflow of foreign exchange	4.62	0.584	7
Waste reduction	4.62	0.567	7
Less pollution of environment	4.60	0.677	9
Enhance standard of living	4.58	0.745	10
Increased level of technical skills	4.57	0.497	11
Job creation	4.47	0.502	12
Halt urban migration	4.35	0.863	13
Better ways to recovery product	4.32	0.827	14
Better product knowledge	4.18	1.090	15
Manufacturers become responsible for their actions	4.11	1.275	16
Product diversity	3.92	1.181	17
Energy efficiency	3.91	1.026	18

2.3 Factor Analysis

Factor analysis was used to determine which of the variables could be quantifying phases of the same principal dimensions. Factor analysis is suitable for grouping groups of linked variables and thus idyllic for reducing a large sum of variables into a more simply understood framework (Norusis 2006). Tables 1-5 and Fig.1 present the result. The Kaizer-Meyer-Olkin (KMO) measure of sampling adequacy attained a high value of 0,718 (Table 2); the Bartlett test of sphericity was also important which suggests that the population matrix was not an identity matrix (Table 3). Thus, the required tests in respect to adequacy of the sample size were approving for the factor analysis to continue. Cronbach's alpha of 0.714 advocated the reliability of the study instrument used was good.

Table 2: KMO and Bartlett's test.

Kaiser-Meyer-Olkin Measure of Sampling Adequacy.		**0.718**
Bartlett's Test of Sphericity	Approx. Chi -Square	309.999
	Df	91
	Sig.	0.000

The data was subjected to principal component analysis (with varimax rotation). The eigenvalue and factor loading were set at values of 1.0 and 0.5 individually. As shown in table 5, six components with eigenvalues greater than 1.00 were extracted using the factor loading of 0,5 as the cut-off point (see also scree plot in Fig 1). The total variance (see table 6) described by each component extracted is as follows; component 1 (23.13), component 2 (18.72), component 3 (8.67), component 4 (7.24). Thus, the results of the principal component analysis and the components extracted accounted for approximately 57.76% of the total cumulative variance.

Table 3. Rotated component matrix.

Rotated Factor Matrix[a] for benefits of remanufacturing to the Nigerian Economy				
	Factor			
	1	**2**	**3**	**4**
Landfill reduction	.734			.364
Carbon footprint is reduced	.666			
Conservation of natural resources	.536			
Waste reduce	.470			
Development of an industrial base	.376		.305	
Industrialization to rural area	.357			
Product diversity		.799		
Better product knowledge		.688		
Better ways to recovery of products		.447	.363	
Technology transfer	-.261	.350	.666	
Manufacturers become responsible			.606	
Halt urban migration	.321		.473	
Enhance standard of living			.305	
Inflow of foreign exchange	.258			.796

Table 4. Total variance explained.

Total Variance Explained									
Factor	Initial Eigenvalues			Extraction Sums of Squared Loadings			Rotation Sums of Squared Loadings		
	Total	% of Variance	Cumulative %	Total	% of Variance	Cumulative %	Total	% of Variance	Cumulative %
1	3.23	23.13	23.13	2.69	19.22	19.22	2.05	14.65	14.65
2	2.62	18.72	41.85	2.15	15.37	34.59	1.60	11.44	26.09
3	1.21	8.67	50.53	.67	4.80	39.39	1.49	10.64	36.73
4	1.01	7.22	57.75	.54	3.89	43.27	.92	6.54	43.27
5	.968	6.91	64.67						
6	.873	6.23	70.91						
7	.765	5.46	76.37						
8	.683	4.87	81.25						
9	.643	4.59	85.85						
10	.552	3.94	89.79						
11	.447	3.19	92.98						
12	.356	2.545	95.52						
13	.328	2.340	97.863						
14	.299	2.137	100.000						

Based on the examination of the inherent relationships among the variables under each component, the following interpretation was made component one was termed benefits of remanufacturing environmentally; component 2 was termed benefits of remanufacturing to manufacturers; component 3 was termed benefits of remanufacturing socially; and component 4 was termed benefits of remanufacturing to trade. These names were derived from the components by observation of the components. Also, how closely related the variables are using the highest loading factor.

3 DISCUSSION OF RESULTS

3.1 Compared 1: Benefits of Remanufacturing Environmentally

The six extracted variables for the benefits of remanufacturing to the Nigerian economy for component 1 were landfill reduction with (73.4%), carbon footprint is reduced with (66.6), conservation of natural resources with (53.6%), waste reduction with (47.0), development of an industrial base with (37.6%) and industrialization to rural areas with (35.7%). This cluster accounted for 23.13% of the variance. In order *to reduce waste*, the life cycle and efficient use of resources must be extended. Products m be remanufactured to achieve better sustainability. *Reduction of carbon footprint* is another benefit of remanufacturing. In 2005 Caterpillar Global Remanufacturing Operation collected and reused 43 million tons of core material, thereby preventing 52 million tonnes of CO_2 emission into the ecosystem (Caterpillar Sustainability Report 2006). Another benefit for remanufacturing is the *reduction of land fill*. Fuji Xerox is an example of a company which achieved zero landfill after Xerox adopted remanufacturing in the 1980s. Xerox achieved zero landfill in Japan in 2000. By employing a closed loop supply chain whereby, the products are recovered via customer take back policy which is then sent to the remanufacturing plant (Maslennikova and Foley 2000). Lund (1983) in his research on remanufacturing identified the following benefits of remanufacturing to countries identified development of an indigenous industrial base through training of a workforce, acquisition of capital goods, and direct technology transfer and also Lund (1983) added opportunity to bring industrialization to the rural areas as one of the benefits. Conservation of resources is another benefit of remanufacturing to the Nigerian economy, which recaptures value of energy, labour and capital equipment going into the original manufacture of a product and it maintains the product of materials in a high value state (Lund 1983).

3.2 Component 2: Benefits of Remanufacturing to Manufacturers

The three extracted benefits of remanufacturing to the Nigerian economy for component 2 were product diversity (79,9%), better product knowledge (68.8%), better ways to recovery of products with (47.7%). This cluster accounted for (18.72%). Subsequently this component was labelled *benefits of remanufacturing to manufacturers.* In comparison to other forms of product recovery management, remanufacturing of products will retain all the value (Kim *et al.* 2008). That's why it remains the better of other product recovery of products.

3.3 Component 3: Benefits of Remanufacturing Socially

The four extracted benefits of remanufacturing to the Nigerian economy for component 3 were technology transfer (66.6%), conservation of natural resources with (60.6%), halt urban migration (47.3%), enhance standard of living (30.5%). This cluster accounted for (8.68%). Subsequently this component was labelled *benefits of remanufacturing socially.* Lund (1983) in his study on remanufacturing identified the following benefits of remanufacturing; the transfer of technology, opportunity for brings industrialisation to rural area which brings about the halting of urban migration.

3.4 Component 4: Benefits of Remanufacturing to Trade

For benefits of remanufacturing to trade, one variable was extracted that are Inflow of foreign exchange (79.65%) with a cluster which accounted for 7.22%.

4 IMPLICATION OF FINDINGS

From the findings above it was revealed that development of an industrial base is a very important benefit that remanufacturing would bring to Nigeria as opined by the respondents. This is so because Nigeria is an under-industrialized nation. Studies have shown that remanufacturing can bring about industrialization in the form of better product knowledge of products remanufactured, thereby leading to improvement and innovative ideas to better the products remanufactured. Additional benefits are an increased level of technical skills and the creation of jobs. This development of an industrial base once established would give rise to industrialisation to the rural areas, thereby stemming the social scourge of rural-urban migration and leading to an enhanced standard of living. Furthermore, the entrenchment of the development of an industrial base would lead to the boosting of export trade in the form of exporting of capital equipment to other countries as well as an inflow of foreign exchange. This would also lead to the conservation of natural resources as revealed in the above discussion. The carbon footprint is reduced as Nigeria seeks to transition to a green economy which is one of the benefits of remanufacturing activities.

5 CONCLUSIONS

From the primary data, results revealed that the development of an industrial base, Industrialization to rural areas, boosting of export trade, conservation of natural resources, landfills reduction. The development of an industrial base as revealed by the respondents as the main benefits of remanufacturing to the Nigeria economy is so because the Nigeria economy is currently under-industrialised in nature with the development of remanufacturing it would lead to the Nigeria economy a highly-industrialised economy. With the development of an industrialised base this would lead to development of industrial bases in rural areas which could be due to cheap labour or closer to parts to be used. Boosting of exporting trade will arise once the industrial base is fully developed. Presently Nigeria is currently grappling with the issue of landfill reduction as equipment's are abandoned after end of life but with the development of remanufacturing will benefit the economy as landfill will be reduced. The benefits of remanufacturing to the Nigerian economy can be classified into four clusters, namely; benefits of remanufacturing environmentally, benefits of remanufacturing to manufacturers, benefits of remanufacturing socially, benefits of remanufacturing to trade these findings lend support to the benefits of remanufacturing to the Nigeria economy. Lastly remanufacturing would be beneficial to Nigeria, as it is going to bring about the development of technically skilled workers which would be beneficial to the Nigerian economy.

References

Ahadzie D. K., Proverbs D. G., and Olomolaiye P. O., Critical Success Criteria for Mass House Building Projects in Developing Countries, *International Journal of Project Management*, 26(6), 675-87, 2008.

Ijomah, W. L., Childe, S., and McMahon, C., Remanufacturing: A Key Strategy for Sustainable Development, *The Third International Conference on Design and Manufacture for Sustainable Development*, Loughborough, UK, 192-196, 2004.

Kim, H., Severengiz, S., Skerlos, S. J., and Seliger, G., Economic and Environmental Assessment of Remanufacturing in The Automotive Industry, *LCE 2008: 15th CIRP International Conference on Life Cycle Engineering: Conference Proceedings* 2008, 195-196, 2008.

Lund, R., *Remanufacturing: United States Experience and Implications for Developing Nations,* Washington, DC, The World Bank, 1983.

Lund, R. T., Remanufacturing: An American Resource. *Proceedings of the Fifth International Congress for Environmentally Conscious Design and Manufacturing on 16-17, June 1998.* Rochester, RY, Rochester Institute of Technology, 1998.

Maslennikova, I. and Foley, D., Xerox's Approach to Sustainability, *Interfaces,* 30(3), 226-233, 2000.

Norusis, M. J., SPSS 15.0 Guide to Data Analysis, Upper Saddle River, NJ, Prentice Hall, 2006.

Pallant, J., SPSS Survival Manual: A Step-By-Step Guide to Data Analysis Using SPSS (4th edition), Mc-Graw Hill, 2010.

Sundin, E., Product and Process Design for Successful Remanufacturing, *PhD Thesis,* Dept of Mechanical Engineering, Linköpings University, Linköping, 2004.

Responsible Design and Delivery of the Constructed Project
Edited by Abdul-Malak, M., Khoury, H., Singh, A., and Yazdani, S.

IMPACT OF COVERING IRRIGATION CANALS ON EVAPORATION RATES IN ARID AREAS

SHERINE EL BARADEI and MAI ALSADEQ

Construction Engineering Dept, American University in Cairo, Cairo, Egypt

Water is the essence of life; thus preserving enough quantities and good quality of water will elevate the standards of living and health of communities. Egyptian government approved Toshka New Valley project to help deal with the rapidly growing population of the country. The project entails building a series of canals; the main is Sheikh Zayed canal - to irrigate portions of the Western Desert of Egypt. It is expected that the recovered land will become home to three million residents by 2020 and cultivate about 550000 feddans. Toshka is a very arid area with high air temperatures; thus very high evaporation rate; about 3 billion m^3 /year. Researches showed that in the near future Egypt will face scarcity of water. So saving water from Sheikh Zayed canal by reducing evaporation will provide additional source of water to the area. This research aim is to estimate evaporation rate from the canal and its main branches. Variable configurations and spacing of covering canals were chosen to reduce evaporation and at same time allow for re-aeration and other water quality parameters to meet their standards. It was found that full coverage calculated evaporation amount is about 9 million m^3/year which could drip irrigate 35000 feddans of wheat per year which is about 7% of the total area of the land intended to be cultivated in Toshka project.

Keywords: Waterways, Numerical, Modeling, Desert, Water, Sufficiency, Lakes.

1 INTRODUCTION

Water and Energy are the most two vital parameters that would decide on the health and prosperity of a community. If those two parameters could be provided in adequate quantities and managed in an efficient way, then communities are going to flourish. The aim of this research is to reduce water evaporation from irrigation canals, especially in arid areas.

In Egypt, research showed that in the near future we will face scarcity of water. So the amount of water available should be increased by finding new sources of water other than the Nile, at the same time preserving and lessen the waste of water from existing water sources, and hence lessen evaporation from irrigation canals will enhance this idea of water preserving.

The New Valley project or Toshka project -location is shown in Figure 1- is taken as a case study because this area is very arid and it is facing the problem of very high evaporation rate - about 2.5 billion m^3 per year.

In 1997, the Egyptian government approved the Toshka project to help deal with the rapidly growing population of the country. The project entails building a series of canals and a pumping station to carry water from Lake Nasser to irrigate portions of the Western Desert of Egypt. It is expected that the recovered land will become home to over three million residents by 2020. The New Valley will increase Egypt's arable land area by *ten* percent.

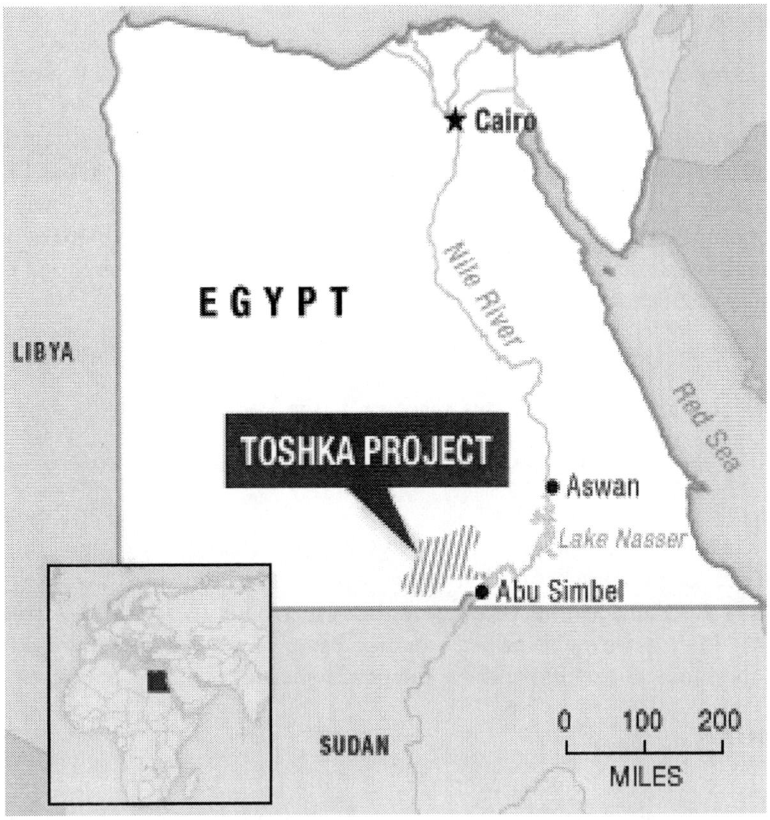

Figure 1. Egypt map showing Toshka area, Sheikh Zayed canal and Lake Nasser.

Toshka Project consists of a system of canals carrying water from Lake Nasser to irrigate the sandy wastes of the Western Desert of Egypt, which is part of Sahara Desert. The canal inlet starts from a site lying eight km to the north of Toshka Bay (Khor) on Lake Nasser. The canal is meant to continue westwards until it reaches Darb el-Arbe'ien route, then northwards along Darb el-Arbe'ien to Baris Oasis, covering a distance of about 310 km. The Mubarak Pumping Station in Toshka was inaugurated in March 2005 with a discharge of $300m^3$/sec. It pumps water from Lake Nasser to be transported by way of a canal through the valley, with the idea of transforming about 550,000 feddans of desert into agricultural land.

There is a very high evaporation rate in Toshka. As for the Toshka lakes that extend on a water surface area of about 1500 Km^2; they lost about 12.69 billion m^3 between years 2002 and 2006 and one of the five lakes completely dried out (this is an indication of how high the evaporation in this area is). This means that for each of those *four* years the estimated loss due to evaporation is about *three* billion m^3/year (El Bastawesy *et al.* 2007).

The actual project is meant to relocate millions of Egyptians from overcrowded cities to the New Valley area. So, this vast amount of people needs to be provided with a sufficient amount of water for drinking and irrigation.

This research is a part of a larger funded research project studying the effect of covering irrigation canals with solar panels.

2 METHODOLOGY

Different coverage scenarios are being studied. As previously mentioned, this is a part of a larger study on covering irrigation canal with solar panels meaning different spacing and hence different covered area strips are assumed and for each case of coverage the evaporation is calculated. The calculations are made for only one km length (960 m exactly) reach of the main Sheikh Zayed canal (total length is 50 km). The coverage pattern could then be repeated for the whole 50 km.

Kindly note that in this research paper covering the canal is taken as a pure action in itself; whereas the material and structure system of the covering will be decided as a step of the larger funded research that was mentioned before.

3 EQUATIONS

The evaporation calculation method for this research is the Penman-Monteith equation, Eq. (1) (Monteith 1965):

$$E = \frac{1}{\lambda} * \left(\frac{\Delta w*(Rn-G)+\gamma*f(u)*(ew-ea)}{\Delta w+\gamma}\right) \quad (1)$$

Where E is the open water evaporation (mm/day), λ is the latent heat of vaporization (MJ/kg), Δ_w is the slope of the temperature saturation water vapor curve (kPa/°C), R_n is the net radiation (MJ m-2day-1), G is the change in heat storage in the water body (MJ/m^2/day), f(u) is the wind function (MJ/m^2/day/kPa), e_w is the saturated vapor pressure at water temperature (kPa), e_a is the vapor pressure at air temperature (kPa) and γ is the psychometric constant (kPa/°C).

4 CALIBRATION

Calibration was done to the Penman-Monteith equation for Lake Nasser real reading of evaporation which is 6.35 mm/day (Elba *et al.* 2014), as shown in figure 2. All parameters of the equation are entered as those of Aswan weather station (where is very close to Lake Nasser).

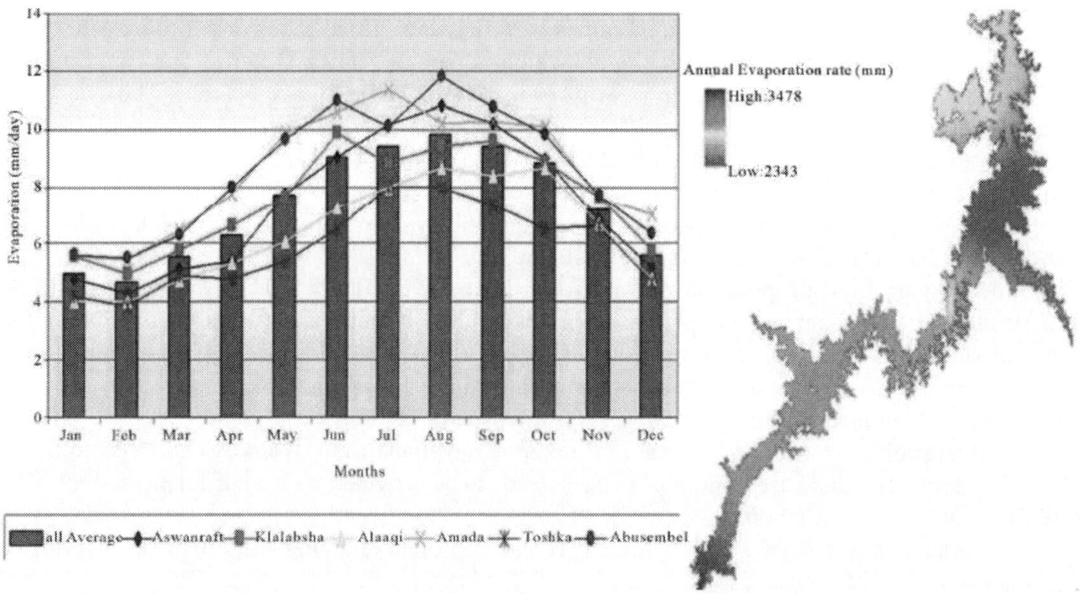

Figure 2. Mean monthly evaporation rate along Lake Nasser (Aswan), including Toshka in mm/day, 2014.

The calibration results are summarized in Table 1, the % error is value is acceptable. The small error could have originated from not very accurate real life readings either of the reference value or of the metrological data such as; water temperature, air temperature, etc. that were used in the Penman-Monteith equation to calculate the evaporation.

Table 1. The percentage error of the calculated average evaporation rate of Lake Nasser.

Calculation method	Avg. annual Evap. (mm/day)	Ref. value (mm/day)	Error %
Penman-Monteith	6.22	6.35	2

5 RESULTS

5.1 Case Study: New Valley Project, Toshka

The case study is taken to be the New Valley project in Toshka in Upper Egypt governorate. This is because this is a very arid area with high evaporation rates as mentioned before. The Sheikh Zayed canal, which will be irrigating the project is about 300 km long containing all structures (culverts, entrance structures…etc). The canal consists of a main part with the dimensions shown in Table 2. There are two sub-branches coming out of this main canal but those are not accounted for in this study.

Table 2. The main Sheikh Zayed canal dimensions.

Parameter	With freeboard	Without freeboard (wetted area)
Length	50 km	50 km
Top width	58 m	54 m
Bottom width	30 m	30 m
Depth	7 m	6 m
Side Slopes	2:1	2:1

5.2 Simulation Results

Table 3. The evaporation rates and volumes for the main canal for different covering spacing.

Spacing length (m)	uncovered area (m^2)	covered area (m^2)	Avg. Evap. rate (mm/day)	Annual Evap. rate (mm/yr)	Evaporation Volume (m3/yr)
960	51840	0	8.77	3202.68	166026.9
48.84	50112	1728	8.78	3205.02	160610.1
22.97	48384	3456	8.79	3207.45	155189.2
14.64	46656	5184	8.79	3209.97	149764.2
10.53	44928	6912	8.8	3212.58	144334.9
8.08	43200	8640	8.81	3215.31	138901.2
4.83	38880	12960	8.83	3222.64	125296.2
3.22	34560	17280	8.85	3230.87	111658.8
2.25	30240	21600	8.88	3240.24	97984.8
1.61	25920	25920	8.91	3251.11	84268.8
1.15	21600	30240	8.94	3264.05	70503.4
0.8	17280	34560	8.99	3279.99	56678.2
0.53	12960	38880	9.04	3300.74	42777.6
0.32	8640	43200	9.12	3330.34	28774.2
0.15	4320	47520	9.27	3381.98	14610.2
0	0	51840	0	0	0

Table 4. Percentage of covered area vs. percentage annual evaporation volume for one km length of Sheikh Zayed main canal for different covering spacing.

Spacing length (m)	% Area Covered	% Annual Evaporation Volume
960	0	100
48.84	3.33	96.74
22.97	6.67	93.47
14.64	10.00	90.20
10.53	13.33	86.93
8.08	16.67	83.66
4.83	25.00	75.47
3.22	33.33	67.25
2.25	41.67	59.02
1.61	50.00	50.76
1.15	58.33	42.47
0.8	66.67	34.14
0.53	75.00	25.77
0.32	83.33	17.33
0.15	91.67	8.80
0	100	0

Mathematical modeling was done to simulate the evaporation rates and volume that would be generated from the canal with varying the coverage area of the canal. An Excel spreadsheet was created to make the simulation. Sample evaporation calculation for *one* kilometer length of the main canal of Sheikh Zayed, the calculations summary is shown in Tables 3, 4 and Figure 3.

It is worth to mention that the calculated total evaporation for the whole uncovered 50 km main reach of the canal is about 9 million cubic meters per year, as shown in table 5.

Table 5. Total annual evaporation volume of the total length of the main canal.

Total Length of the main canal	50	km
Surface Area	2.7	km^2
Total evaporation Volume	8.65E+06	m3/year

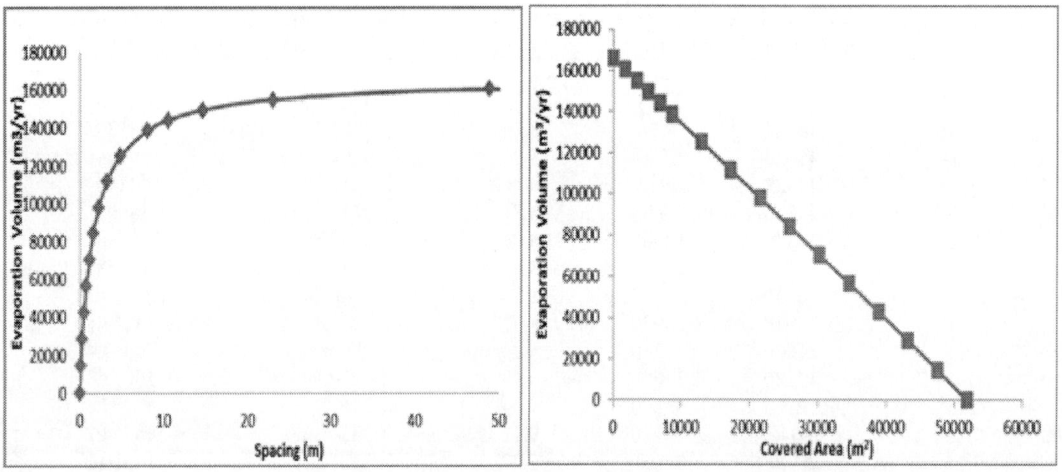

Figure 3. The total evaporation volume vs. different spacing and covered area for the 1 km reach of the main canal.

Because the New Valley project is a project concerned mainly with cultivation and because one of the main crops Egypt cultivates is wheat, the feddans of wheat that could be irrigated with the evaporation amount of water yielded if the total main canal is totally covered are calculated.

Here drip irrigation for the wheat is assumed; this is so because Toshka is a desert area so water is scarce there. As the water requirements for wheat under drip irrigation are 247 cubic meters per cultivation season, once a year (Gamal et al. 2016). The 33,700 feddan of wheat (see Table 6) are equivalent to about 6.5% of the whole 550,000 feddan that are planned to be cultivated in the New Valley project in Toshka.

Table 6. The cultivated area of wheat using the evaporated water of main reach of Sheikh Zayed canal.

Total Evaporation volume (totally uncovered)	8.65E+06	m3/year
# of acre cultivated of wheat	3.50E+04	acre/year
# of feddans cultivated of wheat	3.37E+04	Feddan/year

6 CONCLUSIONS

To summarize, the aim of this research is to study the effect of covering irrigation canals on evaporation rates. The Sheikh Zayed canal in Toshka at Upper Egypt governorate is taken as a case study because Toshka is a very arid area with very high evaporation rates. Also, a new agricultural project is established in this area that would cultivate more than 550,000 feddans that will need lots of water. So, finding ways to preserve water will be of great importance to this project and to all of Egypt as well.

From the evaporation calculation it was found that covering more area of the canal will lessen the evaporation volume. Many possibilities of covering are studied as fully covering will negatively affect the water –as being studied in coming stages of this funded research-.

It was also found that the evaporation volume, which the main canal yields is about 9 million m^3/year which could provide the needed water to cultivate 6.5% of the 550000 feddans by wheat.

To conclude, the evaporation has to be greatly taken into consideration when proceeding with the New Valley project to be able to preserve water that could be otherwise lost.

Acknowledgments

This research is part of a larger research titled "Irrigation canals covered with solar cells". The funding agency is Misr El Kheir. Misr El Kheir Foundation (MEK) is a non-profit development institution established in 2007 with the objective of developing the Egyptian individual in a comprehensive manner. Five key areas have been selected for human development namely: Health, Education, Scientific Research, Social Solidarity and Aspects of Life.

References

Elba E., Farghaly D., and Urban B., Modeling High Aswan Dam Reservoir Morphology Using Remote Sensing to Reduce Evaporation: *International Journal of Geosciences*, 5, 156-169, Feb, 2014.

El Bastawesy, M., Arafat, S., and Khalaf, F., Estimation of Water Loss from Toshka Lakes Using Remote Sensing and GIS, *10th AGILE International Conference on Geographic Information Science Aalborg University,* Denmark, 2007.

Gamal Abdel Rahman, A. M. Talaat and C. Zawe, Water Requirements for Main Crops Grown Under Three Different Agro Ecological Zones, Zimbabwe: *Middle East Journal of Agriculture Research*, 05, 14-28, Jan-Mar, 2016.

Monteith, J. L., *Evaporation and the Environment, in The State and Movement of Water in Living Organisms*, edited by G. E. Fogg, Cambridge Univ. Press, London, 1965.

Responsible Design and Delivery of the Constructed Project
Edited by Abdul-Malak, M., Khoury, H., Singh, A., and Yazdani, S.

YIELD STRESS BASED ON NATURAL STRAIN THEORY UNDER CYCLIC SHEAR LOAD AFTER LARGE SIMPLE SHEAR

YASUYUKI KATO

Dept of Mechanical Engineering, College of Science and Technology, Nihon University, Tokyo, Japan

This paper describes the variation of yield stress under cyclic loads after applying large pre-deformation. In this research, the Natural Strain is used as a strain representation for describing an elastoplastic behavior under large deformation. Using test pieces which is already given a pre-deformation of large simple shear, the yield stress under cyclic shear loads in the forward direction and the reverse direction is examined. As for the method for estimating yield stress in each cycle, the slope of a tangent in the principal deviatoric stress and the principal deviatoric strain curve is adopted. The experiments are conducted with the different sizes of pre-deformation and strain amplitude in order to reveal the effect of pre-deformation on the yield phenomenon. Then, these experimental results are compared with the results of yield stress in the previous study, which is obtained under cyclic load for tension and compression after applying the large uniaxial tension.

Keywords: Elastoplastic behavior, Finite strain, Cyclic loading test, Slope of tangent, Ductile materials, Anisotropy, Strain amplitude, Proof stress.

1 INTRODUCTION

The purpose of this research is to reveal the yield phenomenon under large deformation on the basis of the Natural Strain theory. The Natural Strain is obtained by integrating an infinitesimal strain increment on an identical line element over the whole process of deformation path. Consequently, the shearing strain component is represented by a pure angular strain and it has the merit that can be accurately removed the rigid body rotation from the rotating angle of a line element. Moreover, since the additive law of strain on an identical line element can be satisfied, the strain rate can be clearly decomposed into the elastic component and the plastic component in the same manner as conventional infinitesimal deformation theory (Kato and Nishimura 1996). Therefore, the Natural Strain is effective strain representation which can systematically treat from infinitesimal deformation to the large deformation. In a series of our previous studies (Kato and Moriguchi 2003, Kato and Kazama 2013, Kato 2014, Kato 2016, Kato 2017), the anisotropy of yield surface generated under large deformation has been investigated. In general, when the structure undergoes intense repeated loads due to earthquake etc., a structural plastic failure arises even if the number of cycles is small. Therefore, it is an important issue for structural engineering to clarify the actual mechanism of low-cycle fatigue failure and the yield behavior under cyclic loads. Although the researches on the yield behavior under cyclic loads have been conducted for many years, many of these studies have been done within the range of an infinitesimal deformation. Hence, the detailed studies on the yield behavior under cyclic loads after applying a large pre-deformation have not been fully elucidated. In the case of ductile materials, the yield stress generated after applying large pre-deformation considerably increases as compared with the value of initial yield stress. Moreover, since the anisotropy in material is formed with an increase of deformation, the value of yield stress at pre-deformation side is different from the value of yield stress at opposite side because of the Bauschinger effect. In this research, the variations of yield stress with an increase of cycles when the cyclic load is applied with constant strain amplitude after giving a large pre-deformation is examined. The values of yield stress at pre-deformation side and opposite side gradually decrease and approach the constant value with an increase of cycles because of the effect of material anisotropy reduces with an increase of cycles.

In our recent study (Kato 2016), using the test pieces made from high-purity tough pitch copper, the yield behavior under cyclic loads of tension and compression with the constant strain amplitude after applying a large pre-deformation of uniaxial tension has been examined. As for the estimation method for determining the yield stress, the slope of tangent in the deviatoric stress and strain curves has been used instead of the estimation method by conventional proof stress.

Moreover, in this study, focusing on the pre-deformation of large simple shear that the direction of the principal axis rotate and the line element of it also replace, the yield stress under cyclic shear loads in the forward direction and the reverse direction is examined. Then, in order to reveal the effect of pre-deformation on the yield phenomenon under shear deformation, the experiments are conducted with the different sizes of pre-deformation and strain amplitude.

2 METHOD FOR ESTIMATION OF YIELD STRESS UNDER CYCLIC SHEAR LOAD

Figure 1 (a) shows a deviatoric stress and deviatoric strain diagram when the simple shear in the forward direction and the reverse direction are subjected alternately to the test pieces after applying a pre-deformation of large simple shear. Here, this figure shows the principal deviatoric stress and principal deviatoric strain diagram, however, in the case of simple shear, the principal deviatotic stress S_1 coincide with the shear stress τ_{xy}. Moreover, Figure 1 (b) shows an enlarged view of the first cycle for the principal deviatoric stress and the principal deviatoric strain diagram. In this figure, as shown in point A, the value of yield stress when the shear load is applied again is already determined from the final yield state of pre-deformation of simple shear. In the neighborhood of the yield point A, the principal deviatoric stress and principal deviatoric strain curve becomes a shallow curve, and in this study, it is formulated as follows.

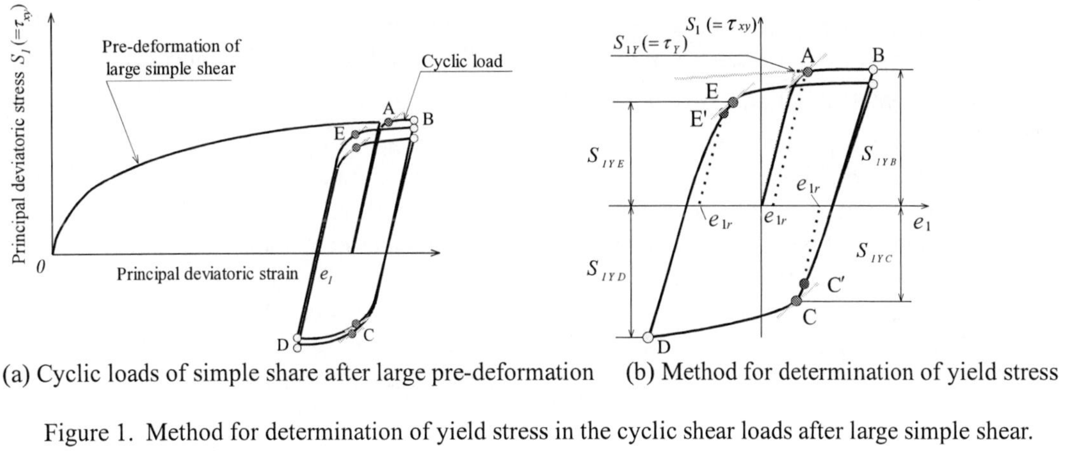

(a) Cyclic loads of simple share after large pre-deformation (b) Method for determination of yield stress

Figure 1. Method for determination of yield stress in the cyclic shear loads after large simple shear.

$$S_1 = a\left(1-exp\left(be_1\right)\right)+ce_1+d \qquad (1)$$

However, a, b, c and d are coefficients determined by using the *Levenberg-Marquardt Method*, which is one of the non-linear least-squares methods. Furthermore, the slope of tangent is also obtained by differentiating Eq. (1), and it is represented by Eq. (2).

$$\frac{dS_1}{de_1} = -ab\,exp\left(be_1\right)+c \qquad (2)$$

Hence, the slope of tangent at yield point A can be derived by Eq. (2). Therefore, the slope of tangent at yielding can be specified in advance and it is indicated by strait blue line in Figure1 (b).

As for the yield stress in a cycle, firstly, the yield stress obtained after applying the shear deformation in the forward direction with specified constant strain amplitude is represented by a point B just before unloading of shear. Secondarily, the yield stress during reverse shear load is obtained by representing the gentle curve of stress by using the experimental equation (1). Then, the stress, which has the same value as the slope of tangent measured at the point A, i.e., the stress at point C, is assumed to be a yield stress. Thirdly, the yield stress obtained after applying the reverse shear strain with specified constant strain amplitude is represented by a point D. Finally, the yield stress at reverse shear side, which is derived by applying a shear load in forward direction again after unloading, is obtained by representing the gentle curve of principal deviatoric-stress by using Eq. (1) once again. And, the stress, which has the same value as the slope of tangent measured at the point A, i.e., the stress at point E, is assumed to be a yield stress.

On the other hand, the yield stress by proof stress is determined by using the value of the residual strain e_{lr} that is obtained by unloading from a point A. Hence, as shown in point C' and E', the proof stress is estimated smaller as compared with the estimated yield stress in this study.

3 EXPERIMENTAL METHOD

In the experiments, the cylindrical specimens, i.e. outer diameter 22[mm], inner diameter 16[mm] and gauge length 30[mm], are used. In order to apply the per-deformation of large simple shear, as for the material quality of specimens, the tough pitch copper, i.e., purity 99.99%, are adopted.

Next, as for the experimental condition, this experiment is composed of two stages. Namely, one is the experiments for applying the pre-deformation of large simple shear. Another one is the cyclic loading tests for the shear loading in the forward direction and the reverse direction.

(1) Experimental condition of pre-deformation of large simple shear

Four different types of shear deformation, namely, values of slip k = 0.37, 0.53, 0.98, 1.37 [-], i.e., values of the principal stretch λ = 1.2, 1.3, 1.6, 1.9 [-], are applied to the test pieces.

(2) Experimental condition of cyclic shear loads

After attaching the strain gauges to the test pieces, the experiments of the cyclic shear loading tests in the forward and the reverse direction are conducted with the constant strain amplitude, i.e., $\Delta e_l=\pm 0.006$ [-], and in these experiments, the numbers of cycles n are all ten times. Then, the values of yield stress at points B, C, D and E in Figure.1 are determined, and the changes of yield stress with an increase of the number of cycle n are investigated. Moreover, as for the values of the principal stretch λ =1.2 and λ =1.3 [-], low-strain amplitude, i.e., $\Delta e_l=\pm 0.003$ [-], is also examined in this research.

4 EXPERIMENTAL RESULTS

Figure 2, 3 and 4 show the experimental results obtained under cyclic shear loading tests after applying the pre-deformation of large simple shear with respect to the constant strain amplitude (the principal strain $\Delta e_I = \pm 0.006$ [-]). In these figures, (a) is the deviatoric stress and deviatoric strain diagrams and (b) shows the variation of yield stress with an increase of the number of cycles, respectively. In these figures, Figure 2 shows the experimental results of the largest principal stretch among them, i.e., λ=1.9[-]. As obviously from this figure, the yield stress at the reverse shear side and the forward shear side, i.e., yield stress at C and D (see S_{IYC} and S_{IYD} in Figure 1 (b)) and yield stress at E and B (see S_{IYE} and S_{IYB} in Figure 1 (b)) have a decreasing tendency as number of cycles increase (see red or black broken lines). It can be seen from this figure that the decreasing tendency of yield stresses at reverse shear side, namely yield stress at point C and D, is small as compared with the forward shear side. Therefore, the back stress also decreases and approaches zero with an increase of number of cycle. On the other hand, results by conventional proof stress, namely, yield stresses at point C' and E', are represented by blue lines in Figure 2 (b). Hence, the estimated yield stresses at point C and E, namely, red curves, become large as compared with the results by proof stress. Furthermore, in these figures, Figure 4 shows the result of the smallest principal stretch, i.e., λ=1.2[-]. It is found from this figure that the decreasing tendency of yield stresses at E and B is small as compared with the results in Figure 2. Next, Figure 5 represents the experimental result of small strain amplitude (the principal stretch λ =1.2 [-], principal strain Δe_I=±0.003 [-]). In the case of small strain amplitude, the decreasing tendency of yield stress at the reverse shear side, namely yield stress at point C and D, slightly increases and approaches to the constant value as the number of cycle increase.

Lastly, Figure 6 shows the experimental result of cyclic load for tension and compression after applying the pre-deformation of uniaxial tension. The yield stress in each position in the cycle gradually decreases with an increase of the number of cycle. However, the decreasing tendency of the yield stresses at tension side, namely yield stress at point B and E, are small as compared with the yield stress in the case of shear deformation.

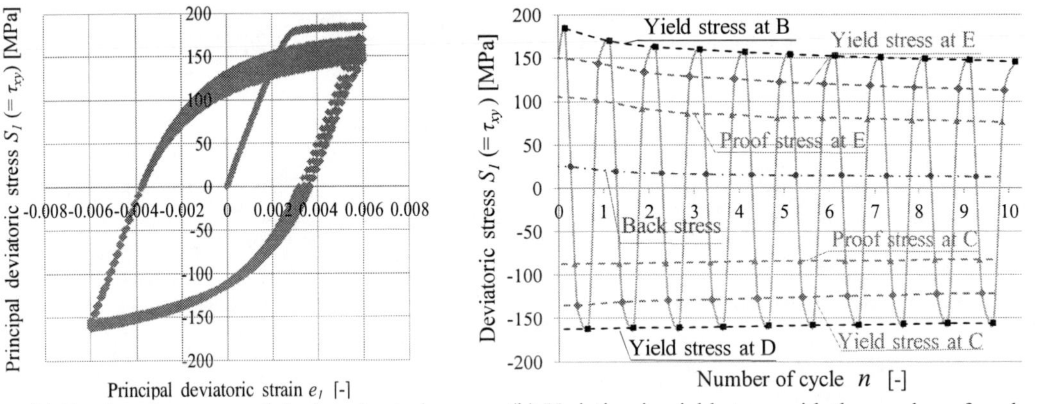

(a) Deviatoric stress and deviatoric strain (b) Variation in yield stress with the number of cycles

Figure 2. Variation of yield stress in cyclic loading for simple shear (λ=1.9 , Δe_I= 0.006).

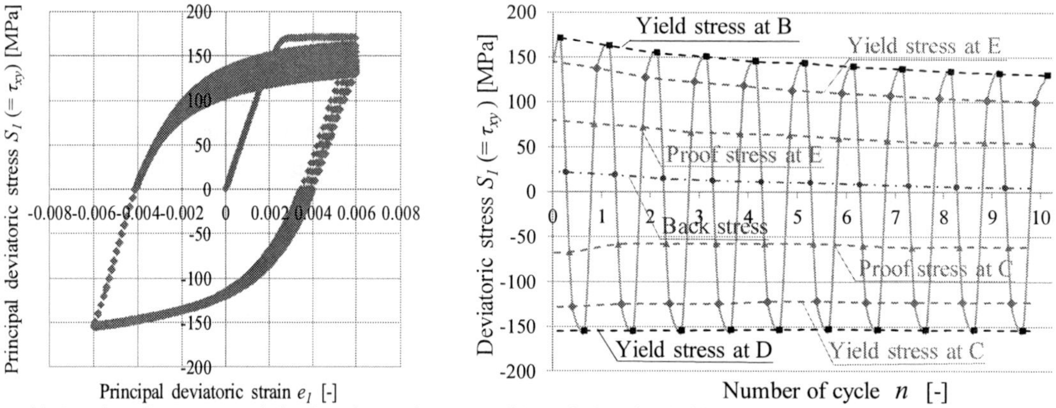

(a) Deviatoric stress and deviatoric strain (b) Variation in yield stress with the number of cycles

Figure 3. Variation of yield stress in cyclic loading for simple shear (λ=1.6 , $\triangle e_1$= 0.006).

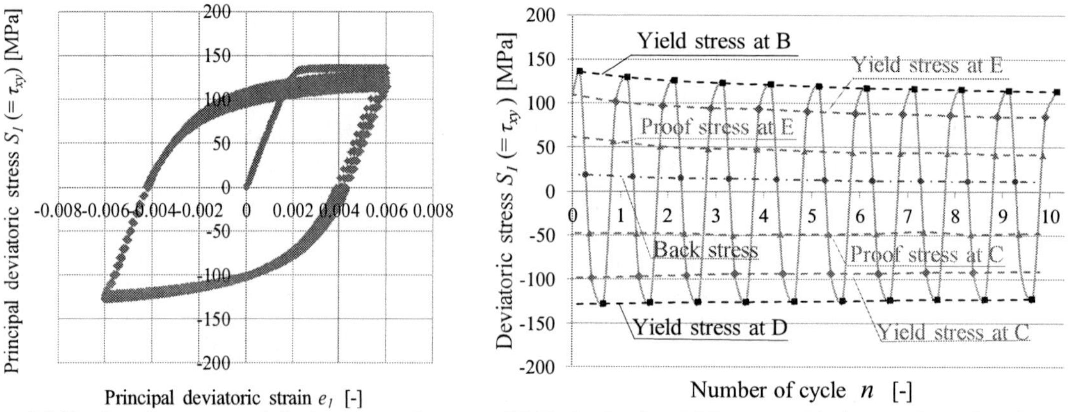

(a) Deviatoric stress and deviatoric strain (b) Varia tion in yield stress with the number of cycles

Figure 4. Variation of yield stress in cyclic loading for simple shear (λ=1.2 , $\triangle e_1$= 0.006).

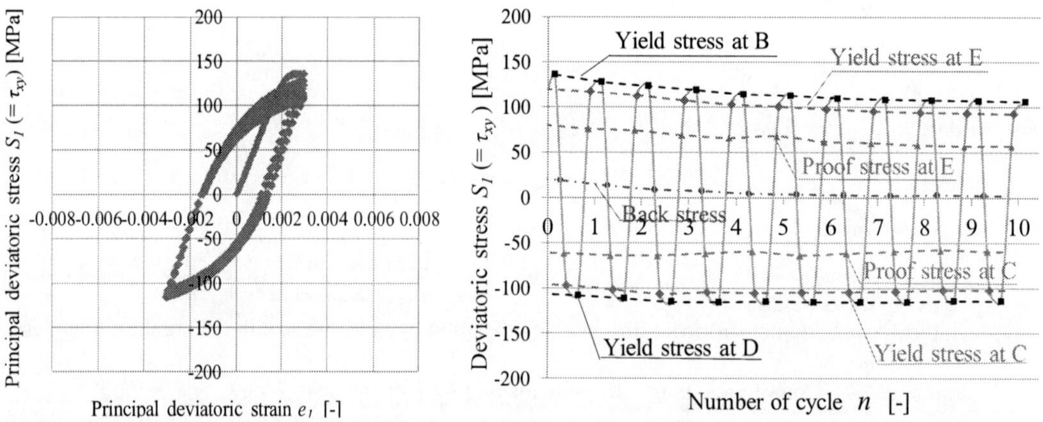

(a) Deviatoric stress and deviatoric strain (b) Variation in yield stress with the number of cycles

Figure 5. Variation of yield stress in cyclic loading for simple shear (λ=1.2 , $\triangle e_1$= 0.003).

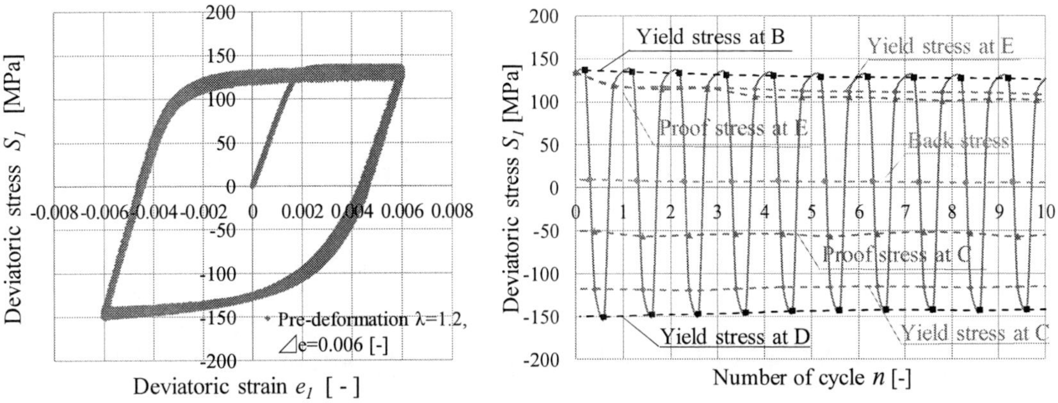

(a) Deviatoric stress and deviatoric strain (b) Variation in yield stress with the number of cycles

Figure 6. Variation of yield stress in cyclic loading for uniaxial tension (λ=1.2 , $\angle e_1$= 0.006).

5 CONCLUSIONS

In this study, using the test pieces, which was already applied the large simple shear, the yield stress under cyclic shear load was estimated. As a result, the following conclusions are obtained:

(1) The yield stresses proposed in this study estimated by using the slope of a tangent have a decreasing tendency with number of cycle increase. However, the decreasing tendency reduces as the number of cycles increases, and they approach to the constant value.

(2) In the case of the large pre-deformation of simple shear, the decrease tendency of the yield stress at the forward shear side becomes larger as compared with the yield stress at the reverse shear side that is the opposite direction to the pre-deformation.

(3) In case of small strain amplitude, the yield stress at the reverse shear side slightly increases and approaches to the constant value as the number of cycle increase.

(4) The yield stress by proof stress is small as compared with the yield stress estimated in this study at both the forward shear side and the reverse shear side.

References

Kato, Y., Natural Strained Yield Surface Shape Estimated Under Pre-deformation of Tension and Torsion, *Sustainable Solutions in Structural Engineering and Construction*, Chantawarangul, K., Suanpaga, W., Yazdani, S., Vimonsatit,V., Singh, A. (eds.), 153-158, ISEC Press, 2014.

Kato,Y., Yield Stress based on Natural Strain Theory under Cyclic Tensile-Compressive Loading After Large Uniaxial Tension, *Resilient Structures and Sustainable Construction*, Pellicer, E., Adam, J. M., Yepes, V., and Singh, A.(eds.), ISEC Press, 2017.

Kato, Y., Yield Surface Shape Estimated by Using Natural Strain under Pre-deformation of Tension After Simple Shear, *Interaction between Theory and Practice in Civil Engineering and Construction*, Komurlu, R., Gurgun, A., Yazdani, S., and Singh, A.(eds.), 247-252, ISEC Press, 2016.

Kato, Y. and Kazama, A., Elasto-Plastic Analysis Using Natural Strain: Shape of Yield Surface Generated after Pre-deformation of Large Simple Shear, *Advanced Materials Research,* 875 and 877, 518-523, 2013.

Kato Y. and Moriguchi Y., Finite Deformation Analysis Using Natural Strain (on Anisotropy in Yield Surface under Large Uni-Axial Tension and Shear). F. Bontempi (ed.), *System-based Vision for Strategic and Creative Design*, 669-675, Rotterdam, Balkema, 2003.

Kato, Y. and Nishimura, T., The Stress Analysis Using the Rate Type Formulation of Natural Strain, *ASME PVP & ICPVT,* 326, 159-166, 1996.

Responsible Design and Delivery of the Constructed Project
Edited by Abdul-Malak, M., Khoury, H., Singh, A., and Yazdani, S.

EFFECT OF CONCRETE CREEP ON THE SHORTENING OF VERTICAL ELEMENTS IN HIGH-RISE BUILDINGS

SANDY CHAABAN, OUSAMA BAALBAKI, and YEHYA TEMSAH

Dept of Civil Engineering, Beirut Arab University, Debbieh, Lebanon

Vertical structural elements are commonly subjected to the time-dependent behavior effects caused by creep which causes shortening in these elements in the direction of the load being applied. Significant amount of creep takes place instantaneously with loading and carries on for the long-term dimensions. In high-rise buildings, the axial shortening in columns is inevitable, so it cannot be ignored. The shortening is differential between vertical structural members due to the difference in axial stiffness and load distribution areas on these elements. Conventional structural analysis assumes that all structural loads are instantaneously applied to the entire completed structure. The construction sequence and loading sequence may be different depending on the construction plan. Therefore, the actual structural behavior can be significantly different from the conventional analytical behavior based on the above assumption. The objective of this study is to highlight the effect of concrete creep behavior on the shortening of vertical elements specifically in high-rise buildings. In this study, we will be considering the ACI 209R-92 model. Based on experimental data, a procedure is developed to compute the elastic shortening due to creep. A 250 meters high-rise is studied taking into consideration the stress and modulus of elasticity depending its construction time and height.

Keywords: Creep shortening, Elastic shortening, Axial shortening, ACI 209R-92 model.

1 INTRODUCTION

The behavior of concrete is affected by its short-term and long-term properties. The short-term properties include strength in compression, tension, bond and modulus of elasticity that occur upon loading. The long-term properties become significant if the applied load is continuously held by the concrete. In buildings, vertical structural elements are subjected to volumetric changes due to axial shortening. Axial shortening is due to elastic shortening, creep and shrinkage of concrete. Most engineers believe that creep and shrinkage are not factors affecting the concrete behavior, so they do not take them into consideration.

Elastic shortening occurs upon hardening of concrete and applying loads. It depends on the applied stress, length and area of element and the time-dependent factor modulus of elasticity. Creep is the deformation of concrete on the long term when subjected to sustained loads. Thus, creep depends on the age of loading of the structure. It originates from calcium silicate hydrates in hardened cement paste (C-S-H) (Gambali and Shanagam 2014). Significant amount of creep takes place with loading and carries out to the long term dimensions causing deformation in the structural elements. The creep strains that occur in high-rise buildings can sometimes be larger than the elastic strains that occur at the age of loading. This makes determining creep properties an essential part before the construction phase. Shrinkage in its simplest definition is the decrease in volume due to the dissipation of water (loss of moisture) in concrete. Both creep and shrinkage are dependent of several factors including: relative humidity, volume-surface ratio (V/S), slump, size of fine aggregates, cement content, and air content and independent of the stress in concrete.

In high-rise buildings, deformations due to creep and shrinkage cannot be ignored (Waghmare *et al.* 2015). There are several prediction models developed to determine the creep and shrinkage. The ACI 209-R92 (2008) model is one of these models which is commonly used due to its practicality. This study highlights the importance of concrete creep and its effect on the axial shortening of vertical elements in high-rise buildings based on the self-weight of the structural elements. In order to achieve this, a 250 meters high-rise building is studied based on the ACI 209-R92 (2008) model.

2 BRIEF REVIEW OF THE LITERATURE

Axial shortening of gravity load bearing structural elements in tall buildings is a phenomenon that was first noticed in 1960's. Ever since, axial shortening has become an interest to many researchers. Axial shortening is cumulative over the height of the structure. Thus, axial deformation that is a combination of elastic, creep and shrinkage shortening effects due to differential shortening becomes critical with the increase in the height of the building. In 1969, Fintel and Khan (1971) developed a method to calculate the total strain in columns. This method is limited because it doesn't take into consideration the relative humidity, water/cement ratio, the percentage of fine aggregates and the percentage of air content. Today, there are four different methods available for predicting the creep and shrinkage behavior of concrete the ACI 209R-92 (2008) model, CEB-FIP model, B3 model and GL2000 model. The commonly used model is ACI 209 (2008). Mola and Pellegrini (2010) discussed the problem of long-term column shortening caused by creep of concrete and derived approximate solutions in reference to the case study of Palazzo Lombardia, the tallest building in Italy by the time the study was conducted. The research done ensures the importance of the topic currently being studied.

3 WORK METHODOLOGY FOR SHORTENING COMPUTATION

3.1 Building Description

A sample 215 meters high-rise (Figure 1) is considered as a case study. The area of slabs and vertical elements were gathered to be used for the calculation of the shortening. The vertical structural elements are distributed symmetrically with respect to the center of gravity of the floor slab. Thus, the axial deformation is assumed to be uniform at the same story level to avoid considering differential shortening which causes additional internal stresses.

3.2 Excel Spreadsheets

An excel workbook is prepared for computing the total axial shortening of vertical elements in the high-rise building. The workbook is divided into several sheets which include the different parameters needed for the calculation as follows: Floor Data, Loads and Stresses, Time Dependent Age of Concrete, Concrete Compressive Strength and Modulus of Elasticity, Creep Coefficient, Creep Strain, Creep Shortening, Elastic Shortening, Shrinkage Strain and Shrinkage Shortening. The objective is to provide a calculation method for the cumulative axial shortening in the studied building taking into consideration the time. The age at loading of concrete is 14 days which is equal to the construction time required for completion of one story.

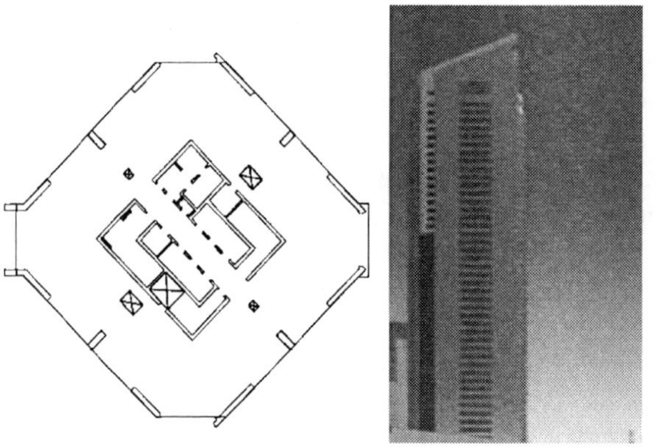

Figure 1. A typical floor plan showing structural symmetry and a side view of the studied tower.

3.2.1 *Elastic shortening*

The elastic shortening is calculated based on the following formula, Eq. (1):

$$\Delta = \frac{PL}{EA} \tag{1}$$

The load induced on each floor due to the self-weight of the structural elements on each floor was computed using the following defined Eq. (2):

$$\text{P} = (\gamma_c \times \text{Area of Vertical Elements} \times \text{Floor Height}) + (\gamma_c \times \text{Area of Slab} \times t_s) \tag{2}$$

where γ_c is the unit weight of concrete (KN/m^3) and E is the modulus of Elasticity (MPa).

The modulus of elasticity is calculated based on the ACI equation, Eq. (3) and (4):

$$E_{cmt} = 0.043{\gamma_c}^{1.5}\sqrt{f_{cmt}}\ (\text{MPa}) \tag{3}$$

$$f_{cmt} = \left[\frac{t}{a+bt}\right] f_{cm28} \tag{4}$$

where f_{cmt} is the compressive strength of concrete at time t (MPa), and γ_c is the unit weight of concrete (kg/m^3). The areas of vertical elements and slabs are in m^2, and the floor heights and slab thicknesses are in (m).

3.2.2 *Creep shortening*

The creep shortening per floor is expressed as per the following Eq. (5),

$$\Delta_{creep}\,(\text{m}) = \text{creep strain} \times \text{floor height} \tag{5}$$

The creep strain was calculated based on the following formula, Eq. (6):

$$\text{Creep Strain}(m/m) = \frac{\sigma}{E} \times \varphi_{(t,t_0)} \tag{6}$$

where σ is the stress on floor, E is the modulus of elasticity at time t and $\varphi_{(t,t_0)}$ is the creep coefficient.

To determine the creep coefficient, the ACI 209R-92 (2008) recommends the following formula:

$$\varphi_{(t,t_0)} = \frac{(t-t_0)^{\psi}}{d+(t-t_0)^{\psi}}\phi_u \tag{7}$$

where, t = age of concrete

t_0 = age at loading of concrete

d = $26.0e^{\{1.42\times10^{-2}(V/S)\}}$ when the shape and size are considered

ψ = psi value = 0.4527; this value was calibrated to fit the high-strength concrete mix

ϕ_u is the ultimate creep coefficient = $2.35\gamma_c$

$$\gamma_c = \gamma_{c,to}\gamma_{c,RH}\gamma_{c,vs}\gamma_{c,s}\gamma_{c,\psi}\gamma_{c,\alpha} \tag{8}$$

$\gamma_{c,to}$, $\gamma_{c,RH}$, $\gamma_{c,vs}$, $\gamma_{c,s}$, $\gamma_{c,\psi}$, and $\gamma_{c,\alpha}$ are factors calculated based on equations proposed by ACI 209R-92 (2008) model and are shown in Eq. (8).

The load induced generates stress on the vertical elements of the floors. This stress is computed based of the following Eq. (9),

$$\sigma\left(KN/m^2\right) = \frac{P\ (in\,KN)}{Area\,of\,Vertical\,Elements\ (m^2)} \tag{9}$$

Table 1. Sample of creep shortening (m) per floor with construction time.

Floor									
7									7
6								6	0.00E+00
5							5	0.00E+00	3.72E-06
4						4	0.00E+00	3.72E-06	4.83E-06
3					3	0.00E+00	5.58E-06	7.25E-06	8.43E-06
2				2	0.00E+00	7.41E-06	7.72E-06	8.98E-06	9.98E-06
1			1st	0.00E+00	4.33E-06	5.15E-06	4.80E-06	5.34E-06	5.79E-06
GF		GF	0.00E+00	3.15E-06	5.31E-06	5.64E-06	5.03E-06	5.46E-06	5.83E-06
Basement 1	B1	0.00E+00	3.57E-06	4.78E-06	7.20E-06	7.32E-06	6.37E-06	6.80E-06	7.19E-06
Basement 2	0.00E+00	3.32E-06	3.98E-06	4.76E-06	6.86E-06	6.80E-06	5.83E-06	6.17E-06	6.47E-06
	Creep Shortening								

After calculating the creep shortening with time, the total creep shortening of each floor was computed which is the summation of the horizontal rows. The total cumulative creep shortening was calculated with the cumulative height as shown in Table 1.

The same procedure was followed to determine the shortening due to shrinkage based on the ACI model.

For shrinkage strain the ACI 209R-92 (2008) model recommends the following Eq. (10):

$$\varepsilon_{sh(t,t_c)} = \frac{(t-t_c)^{\alpha}}{f+(t-t_c)^{\alpha}}\varepsilon_{shu} \tag{10}$$

where $\alpha = 1.0$, $f = 26.0e^{\{1.42\times10^{-2}(V/S)\}}$ and the ultimate shrinkage strain, $\varepsilon_{shu} = 780\gamma_{sh}\times10^{-6}$

$$\gamma_{sh} = \gamma_{sh,tc}\gamma_{sh,RH}\gamma_{sh,vs}\gamma_{sh,s}\gamma_{sh,\psi}\gamma_{sh,\alpha} \tag{11}$$

where, $\gamma_{sh,tc}, \gamma_{sh,RH}, \gamma_{sh,vs}, \gamma_{sh,s}, \gamma_{sh,\psi}$ and $\gamma_{sh,\alpha}$ are factors depending on curing time, relative humidity, volume to surfae ratio, slump, fine aggregates percentage and air content respectively and determined based on the ACI 209R-92 (2008) equations, and shown in Eq. (11).

3.2.3 *Total shortening*

The total shortening of vertical elements is the summation of the elastic, creep and shrinkage shortenings, Eq. (12), (Figure 2.)

$$\Delta_{total} = \Delta_{creep} + \Delta_{shrinkage} + \Delta_{elastic} \tag{12}$$

The total obtained shortening in the 215 meters high-rise building studied is 3.8 centimeters.

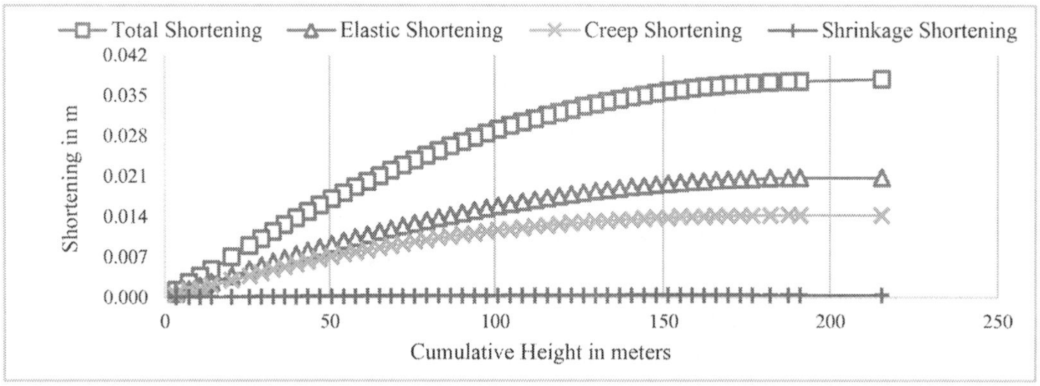

Figure 2. Total, elastic, creep, and shrinkage shortening in tower.

4 RESULTS

The effect of age at loading t_0 on the creep coefficients was studied by considering three different ages at loading 7, 14, and 28 and examining the variation in the creep coefficients. At 7, 14, and 28 days, the age at loading factors ($\gamma_{c,to}$) were 0.994, 0.916, and 0.844 respectively. Thus, the increase in age at loading decreases the value of age at loading factor. At 7, 14, and 28 days the

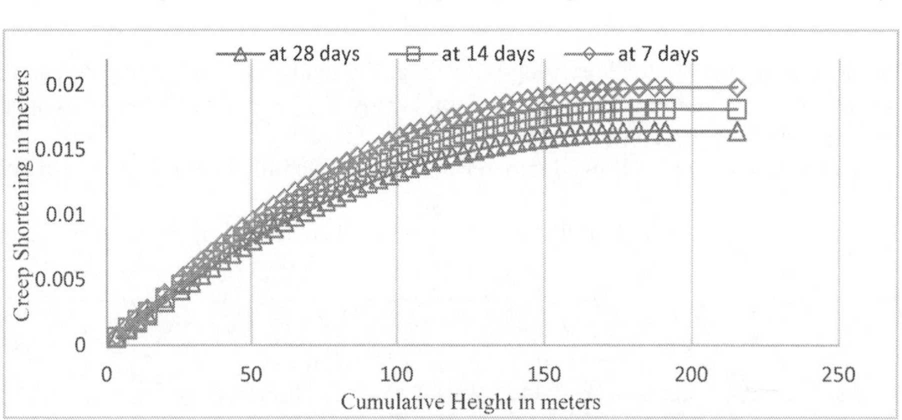

Figure 3. Creep shortening at different ages of loading.

ultimate creep values calculated were 3.45, 3.18 and 2.93 respectively. The obtained values showed that the decrease in age at loading factor decreases the ultimate creep for the same mixture resulting in an increase in the creep coefficients. The earlier the loading the higher the creep. A lower creep coefficient results in lower creep strain hence lower shortening due to creep occurs as shown in Figure 3.

5 CONCLUSIONS

It is of highly importance to study the creep effect in high-rise buildings. The proposed methodology with excel spreadsheets provides a useful tool for calculating axial shortening based on creep, shrinkage and elastic strain. The elastic and creep shortenings obtained are much larger than the shortening due to shrinkage (Figure 2). Though the total shortening in the studied high-rise building is 3.8 cm which is not quite large relative to the total height, the results ensure the importance of taking creep effect into consideration when designing concrete structures. With the increase of height, the importance of time-dependent shortening of vertical elements becomes more critical owing to the cumulative nature of shortening. In this study, the deformation was assumed to be uniform. However, in case the structural system lacked symmetry in stress distribution and design differential shortening occurs which affects the stability of the structure. In addition, the age at loading affects the shortening due to creep (Figure 3). This is due to the fact that the concrete did not reach it is ultimate strength before the load is induced from the new constructed floor.

References

ACI 209.2R-08, Guide for Modeling and Calculating Shrinkage and Creep in Hardened Concrete, *ACI Manual of Concrete Practice*, 2008.

Gambali, A. K. and Shanagam, N. K., Creep of Concrete, *International Journal of Engineering Development and Research (IJEDR)*, 2(4), 3800-3802, 2014.

Fintel, M. and Khan, R., Effect of Column Creep and Shrinkage in Tall Structures-Prediction of Inelastic Column Shortening, *ACI Journal Proceedings, 66*, 1969.

Mola, F. and Pellegrini L. M., Effects of Column Shortening In Tall Structures, *35th Conference on OUR WORLD IN CONCRETE AND STRUCTURES.* Singapore: CI-Premier PTE LTD, 2010.

Waghmare, A. P., Dode P. A., and Chore, H. S., Axial Shortening Effect on Vertical Element Using Elastic and Inelastic Approach by ACI Code, *International Journal for Computer Applications*, 5-11, 2015.

Responsible Design and Delivery of the Constructed Project
Edited by Abdul-Malak, M., Khoury, H., Singh, A., and Yazdani, S.

BEHAVIOR OF FULLY ENCASED COLD FORMED STEEL JOISTS WITH/WITHOUT WEB OPENINGS: EXPERIMENTAL INVESTIGATION

HANADI ELKHANSA, ADNAN MASRI, and ZAHER ABOU SALEH

Dept of Civil Engineering, Beirut Arab University, Beirut, Lebanon

This paper presents an experimental investigation on the behavior of encased light cold-formed steel joists with/without web openings. The study is carried on two sets of composite beams constructed of cold-formed steel joists using I-beam built-up sections. The first set has no web openings while the second set has circular web openings. Each set consists of three beams with various steel section-concrete reinforcing ratio. The constant parameters included in this study are the cross sectional area of the specimen, distribution of web openings in the encased steel I- joist, and the compressive strength of concrete. The tests have been conducted by applying two concentrated identical loads. Consequently, a comparative study is carried out to compare between the behavior of specimens within one set, and a comparison between the specimens of the two sets. This comparison is related to strength, stiffness, ductility, energy absorption capacity and failure modes. Based on the testing results, it is concluded that, the load capacity of the composite beams without web openings is greater than that with web openings for the same steel ratio. But the existence of web openings in the encased steel joists enhances the ductility and the energy absorption of the composite beams. Also using cold formed steel I-joists in the composite beams allow the failure mode to be ductile flexural failure.

Keywords: Composite beam, Stiffness, Energy absorption capacity, Failure mode, Web openings, Experimental study.

1 INTRODUCTION

Cold-formed steel (CFS) is the common term for products made by rolling or pressing thin gauges of steel sheets into goods. There are many ways to create cold-formed steel section. One way to deform the sheet into a usable product includes working sheet steel using folding, rolling, or presses (Yu and Wiley 2000).

Some of the main properties that make cold-formed steel more preferable are (Hancock *et al.* 2001, Dandens *et al.* 2011) lightness in weight, ease of prefabrication and mass production, fast and easy erection and installation, economic transportation and handling, and recyclable material.

Composite construction achieves important benefits by making steel and concrete work together. These advantages may be improved by using light cold formed steel sections. Light sections have been introduced in the composite construction and the general principles of composite design using light steel sections are the following:

- During construction, light steel beams are designed elastically to support the construction loads.

- Once the steel and concrete are acting compositely, composite light steel beams are designed plastically to support the loads acting at the ultimate limit state (Hamidi 2013).

Wehbe *et al.* (2013) carried out a series of experimental studies on full-scale test specimens representing concrete/CFS flexural elements under gravity loads. These studies were designed to investigate the structural performance of concrete/CFS simple beams.

2 AIM AND OBJECTIVES

The primary goal of the research was to conduct an experimental investigation of the encased composite beams with and without web openings and compare them to each other. The reason of using web openings is to replace the stand-off screw done by the previous investigation. The tests were conducted by applying two concentrated identical loads, each load is acting at a distance 250 mm from the mid-span on each composite beam. Specific objectives of the test program to this report are to:

(i) Investigate the effectiveness of using the web openings.

(ii) Observe the mode of flexure behavior.

(iii) Figure out if the composite beam is behaving as one unit.

3 EXPERIMENTAL INVESTIGATION

3.1 Test Specimens

Six full-scale rectangular composite beams were constructed and tested in this study. All the beams have a 3000 mm clear span and a rectangular section of 300 x 200 mm, reinforced by cold formed steel plates (Top flange, Bottom flange, Web) welded together to form the steel I-joist (built-up section) encased in the concrete to form a new concept of composite beam.

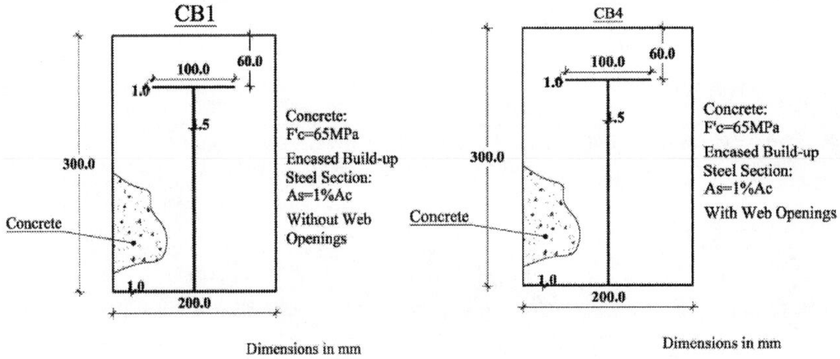

Figure 1. Typical cross-sections of the composite beams with/without web openings.

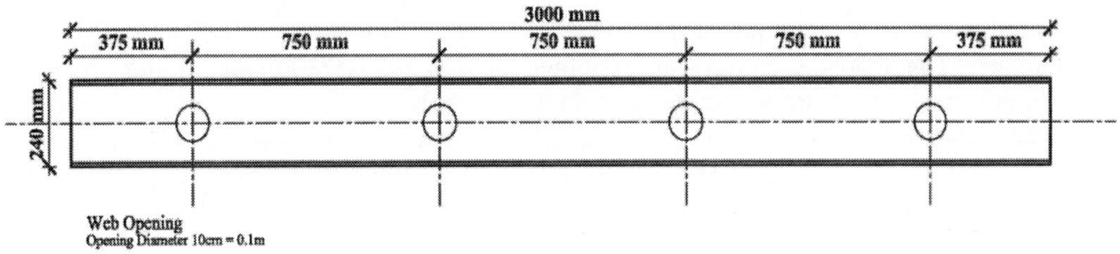

Figure 2. Side view for the steel I-joist with web openings (set 2).

Table 1. Characteristics of the experimental test specimens.

Sample	CB1	CB2	CB3	CB4	CB5	CB6
Percentage of steel (%)	1%	2%	3%	1%	2%	3%
Width of beam (b) (mm)	200	200	200	200	200	200
Depth of beam (d) (mm)	300	300	300	300	300	300
Width of top flange (mm)	100	100	100	100	100	100
Thickness of top flange (mm)	1	2.5	4	1	2.5	4
Width of bottom flange (mm)	200	200	200	200	200	200
Thickness of bottom flange (mm)	1	2.5	4	1	2.5	4
Height of the web (mm)	238	235	232	238	235	232
Thickness of the web (mm)	1.5	2	3	1.5	2	3
Number of web openings	NA	NA	NA	4	4	4
Diameter of web openings (mm)	NA	NA	NA	100	100	100

Composite beams were divided into two sets (Table 1, Figures 1 and 2). Set 1 contains beams CB1, CB2 and CB3 without web openings. Set 2 contains beams CB4, CB5 and CB6 with web openings, see Figure 2 for locations of web openings (AISI Standard S201-07 2007).

3.2 Materials

Eleven standard cylinders of 152.5mm (6in) diameter x 305mm (12in) height were cast from each batch and kept in the same environment as the composite beams. The average compressive strength f'_c at the time of testing (after 28 days) was 65MPa (ASTM C39 (2000), Carreira and Chu 1985).

The steel plates used in the composite beam samples were also tested and showed a yield strength of 161.91 MPa and an average modulus of elasticity of 211048 MPa.

3.3 Test Setup and Instrumentation

The tests were carried out by increasing the pressure in the hydraulic jack and recording the central deflection of the composite beam at each increment of the load as shown in Figure 3.

Figure 3. Installation of the samples with the testing machine.

4 EXPERIMENTAL BEHAVIOR

All the specimens were tested after curing. The beams were subjected to flexural loads to study their behavior. The load deflection curve is then plotted online allowing monitoring of the behavior in the zone within elastic, plastic then damaged phase.

The load-deflection curves for composite beams CB1 and CB4 are shown in Figure 4.

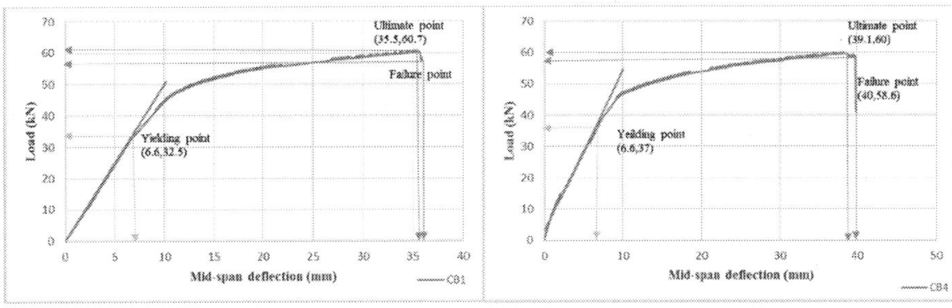

Figure 4. Typical Load-Deflection curves for the composite beams with/without web openings.

Table 2. The experimental results of the composite beams.

Composite Beam	% of As	P_y (kN)	Δ_y (mm)	P_u (kN)	Δ_u (mm)	P_f (kN)	Δ_f (mm)	Δ_f/Δ_y
CB1	1%	32.5	6.6	60.7	35.5	57.4	36	5.45
CB2	2%	87	7.35	127	24	113	26.4	3.5
CB3	3%	98	9	137	16.5	133	16.5	1.83
CB4	1%	37	6.6	60	39.1	58.6	40.5	6
CB5	2%	60	9	95.6	31.6	71.2	32.6	3.56
CB6	3%	82	8.87	97.9	17.8	86.7	23.5	2.72

5 DISCUSSION AND COMPARISON OF THE EXPERIMENTAL RESULTS

5.1 Load-Deflection Behavior

The beams will be compared according to stiffness, load-deflection, ductility, energy dissipation and failure modes. The load-deflection curves of the composite beams with and without web openings for different steel ratios are shown in Figure 5.

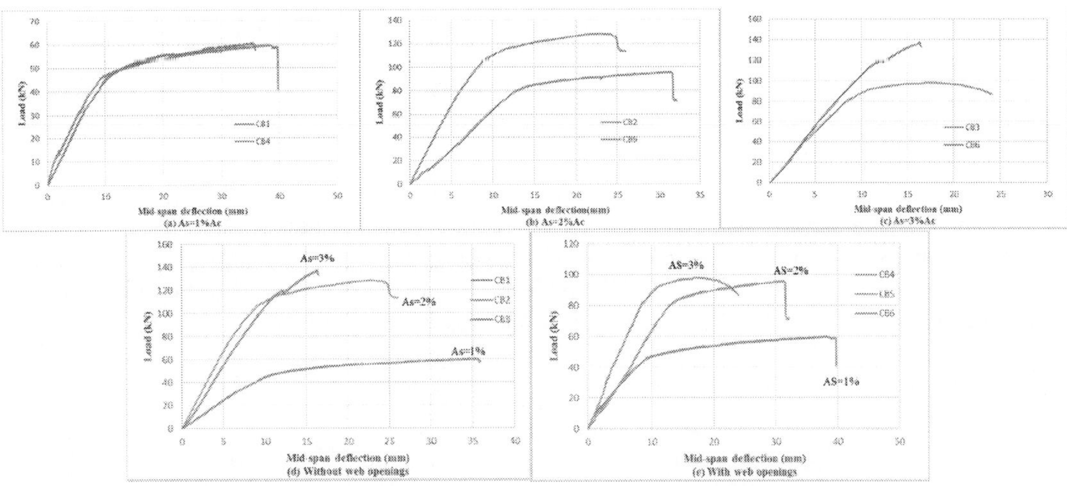

Figure 5. Load-deflection curves for the tested composite beams.

5.2 Stiffness

Stiffness is defined as the load required causing unit deflection. The stiffness values of the specimens at ultimate loads are shown in Figure 6.

5.3 Ductility

The ratio of the failure deflection to the deflection at first yield is known as ductility factor (μ). Figure 7 represents the ductility factor of the tested composite beams.

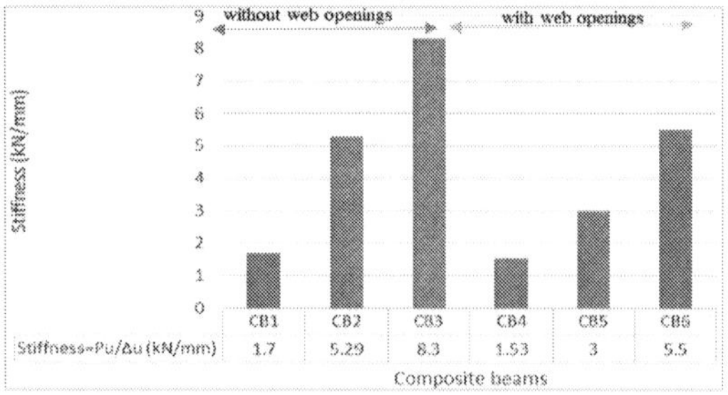

Figure 6. Stiffness values.

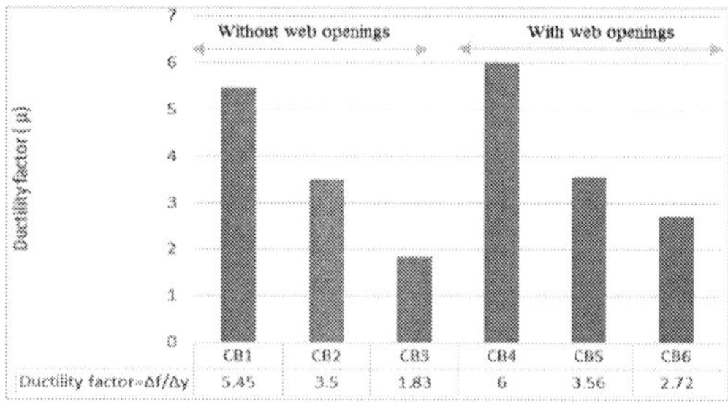

Figure 7. Ductility factor.

5.4 Energy Absorption Capacity

Energy absorption capacity of the beam members can be approximated as the area under the load-deflection curve. Figure 8 shows the energy absorption capacities of the tested composite beams.

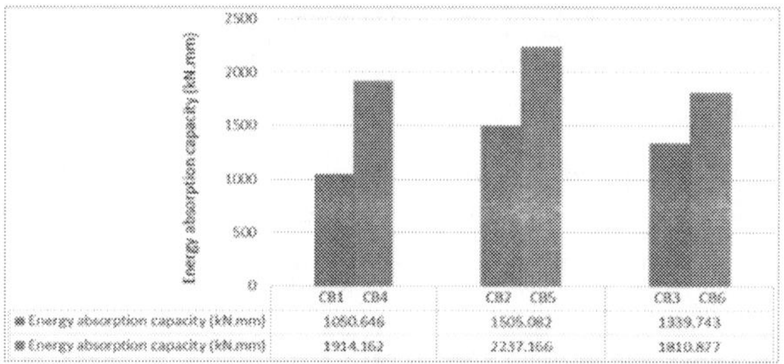

Figure 8. Energy absorption capacity.

5.5 Failure Modes

The cracks were formed in the center of the samples and originated from the top (tension zone) of the specimen and extended towards the bottom (compression zone) of the specimen. The majority of cracks are formed near the zone of the points loading. As the load increases, the initial cracks have widened till the ductile flexural failure occurred (plastic zone) in all the composite beams with or without web openings in the encased steel beam.

6 CONCLUSIONS

Six specimens were designed, fabricated and tested to investigate the flexural failure and the behavior of the composite beams with cold formed steel I-joist encased in concrete, with different percentages of steel-concrete reinforcing ratio, with or without web openings.

The ultimate strength, load-deflection curve, ductility, stiffness, energy absorption capacity and the failure mode of each specimen were recorded and studied carefully.

The major results obtained from this study are summarized as follows:

(i) The ultimate load capacity of composite beam without web openings and with steel ratio 3% (CB3) increases by 125.7% and 7.9% more than CB1 having steel ratio 1% and CB2 having steel ratio 2% respectively and the ultimate load capacity of CB2 increases by 109.2% more than CB1. While the ultimate load capacity of composite beam with web openings and with steel ratio 3% (CB6) increases by 63.2% and 2.4% more than CB4 having steel ratio 1% and CB5 having steel ratio 2% respectively, and the ultimate load capacity of CB5 increases by 59.3% more than CB4. Thus, the ultimate load capacity of the composite beams increases as the percentage of steel increases.

(ii) The ultimate load capacity of composite beams without web openings with steel ratio 1% (CB1), 2% (CB2) and 3% (CB6) increases by 1.2%, 32.8% and 40% more than composite beams with web openings with steel ratio 1% (CB4), 2% (CB5) and 3% (CB6) respectively. It can be concluded that the load capacity of the composite beams without web openings is greater than that with web openings for the same steel ratio.

(iii) Stiffness values of the specimens without web openings are higher than that with web openings approximately by 11.11%, 76.33%, and 50.9% for steel ratios of (1%, 2%, and 3%) respectively, means that the stiffness increases as the percentage of steel increases.

(iv) Web openings improve the ductility of the composite beams of the same steel ratio. Also, it shows that the ductility of the samples with and without web openings decreases as the steel ratio of the encased steel beam increases.

(v) Specimen with/without web openings have the same capacity at 1% of steel (lower ratio).

(vi) The energy absorption capacity for the composite beams with web openings is higher than that for the composite beams without web openings for all steel ratios 1%, 2% and 3%, means that the introducing of web openings in the composite beam enhances its energy absorption capacity.

(vii) Using Cold formed steel I-joist (built-up section) in the composite beams allow the failure mode to be ductile flexural failure.

Acknowledgments

I would like to express my sincere appreciation to my professors at BAU for providing enormous support and encouragement, in addition to their valuable assistance in many ways over the past years. Many thanks to the Structural Laboratory staff members for their assistance with fabricating, preparing and testing the composite beams with cold-formed steel I-joists. I would also like to thank my family and friends for their support and contribution towards this research.

References

AISI Standard S201-07, American Iron and Steel Institute, 2007 Edition, *North American Standard for Cold-Formed Steel Framing-Product Data*, June 2007.

ASTM C39, *Manual of Aggregate and Concrete Testing,* American Society for Testing and Materials, West Conshohocken, Pennsylvania, 2000.

Carreira, D.J. and Chu K.H, Stress-Strain Relationship for Plain Concrete In Compression, *ACI Journal*, 82(6), 797–804, 1985.

Dandens, A., Kreilis, J., and Andersons, G., Properties of Cold-Formed Steel Sections, Civil Engineering '11-3rd International scientific conference', 166-170, January 2011.

Hamidi, B., *Design Of Composite Beams Using Light Steel Constructions*, 2013.

Hancock, G., Murray, T., Ellifritt, D., and Dekker, M., *Cold-Formed Steel Structures to the AISI Specification*, 2001.

Wehbe, N., Bahmani, P., and Wehbe, A., Behavior of Concrete/Cold Formed Steel Composite Beams, Experimental Development of a Novel Structural System. *International Journal of Concrete Structures and Materials*, 7(1), 51-59, March 2013.

Yu, W., *Cold-Formed Steel Design*, Wiley, J., and Sons Inc., New York, 2000.

Responsible Design and Delivery of the Constructed Project
Edited by Abdul-Malak, M., Khoury, H., Singh, A., and Yazdani, S.

INFLUENCE OF BOUNDARY CONDITIONS VARIATION ON TOPOLOGY OPTIMIZATION OF LOAD-CARRYING PARTS

BOŠTJAN HARL[1], MARKO KEGL[1], JOŽEF PREDAN[1], and DEJAN DINEVSKI[2]

[1]*Faculty of Mechanical Engineering, University of Maribor, Maribor, Slovenia*
[2]*Faculty of Medicine, University of Maribor, Maribor, Slovenia*

The paper discusses the problem of boundary conditions formulation in FEA-based design procedures of load-carrying parts driven by topology optimization. The emphasis is on the correct formulation of boundary conditions and the definition of corresponding load cases so that topology optimization will deliver a usable design. It demonstrates that inadequate preparation of load cases may lead to a result that seems to be reasonable, but may behave badly in practical application due to possible lack of robustness and reliability. To illustrate this, an example study is performed for a load-carrying bracket that is fastened by four screws. Namely, in such situations it may quickly happen that one or more screws become at least slightly loose. This changes the stress fields in the part dramatically. If this is not captured correctly in the applied load cases, this means that the topology optimizer will deliver some design that might be extremely sensitive to a loose-screw situation. Such a design may fail very quickly during normal operation.

Keywords: FEA, Lightweight design, Optimal design, Support conditions.

1 INTRODUCTION

Design of load-carrying parts is becoming a more and more challenging task for an engineer. Reasons for that are higher requirements related to longer lifespan, reduced material consumption, and specific demands of manufacturing technologies. Fortunately, the development of structural design and optimization procedures is able to follow those increasing demands. The most vital tools in this context are the procedures for numerical analysis of structural response and the corresponding numerical optimization procedures. Unfortunately, proper engagement of all these tools may be quite tricky in real-life applications, which may lead to disappointing optimal designs. In many cases, the reason for a bad result is insufficient knowledge in the modeling of boundary conditions and poor understanding of their actual role within the optimization process.

In the field of development of structural optimization methods, one of the most important challenges is topology optimization of load-carrying lightweight parts (Rozvany 2009). Topology optimization can result in substantial benefits, compared to other optimization procedures (Bendsøe and Kikuchi 1988, Huang and Xie 2010). Nevertheless, its development was rather slow in the past, because topology optimization often generates shapes, which are practically impossible to produce with conventional technologies. However, these circumstances are currently changing practically daily due to the impressive development of modern multi-axis

CNC machines and additive manufacturing (AM) technologies. The latter ones seem to be especially well suited for topology optimized parts.

By engaging AM processes, almost every custom shape can be produced and excellent new materials have been emerging regularly in the past few years. This specific circumstance created new potentials for employing the topology optimization procedures in the design of load-carrying parts. In fact, those procedures have become vitally important since AM produced parts are typically sensitive to crack initiation, which can cause early and unexpected structural failure of a mechanical part. In order to minimize the risk of potential structural failure of such parts, probably the most important requirement is to remove stress concentrations.

Design of a load-carrying part with prescribed material volume and without stress concentrations (and with relative low stress levels) is an extremely demanding task, which requires the employment of adequate numerical procedures, typically FEA combined with topology optimization (e.g., Harl *et al.* 2017). What is often underestimated in these procedures is the importance of accurate modeling of boundary conditions and the need to understand fully what the optimizer is actually doing. Consequently, maybe two of the most exposed reasons for bad topology optimization results can be summarized as follows:

- Misuse of standard FEA modelling practices. Namely, in standard FEA procedures it is quite reasonable to engage many simplifications or procedures that may reduce the modeling and computation time, while preserving reasonable accuracy. For example, variable size finite elements are used in mesh generation, or, in a bolt fastening situation the nodes being in contact with the bolt are simply fully supported. This introduces errors that may often be neglected in standard FEA. In topology optimization, however, these errors may quickly cause the optimizer to generate a bad design.
- Misunderstanding of the considered mechanical problem and the optimization process. What is often forgotten is that a topology optimizer will generate a design that is totally and exclusively adapted to the underlying boundary conditions. This requires an extremely careful analysis of all possible supporting and loading situations that might eventually (maybe accidentally) appear during the life time of a part. All such situations have to be added to the originally prescribed set of loading conditions.

The paper is intended to improve the understanding of the topology optimization process of load-carrying parts. More specifically, it addresses the problem of identifying properly all possible loading conditions in case of a screw fastening of a part. The outline of the paper is as follows. Section 2 describes briefly the considered example and the optimization task. In Section 3 all applicable load cases are formulated. In Section 4 the results of running various optimization tasks are presented and analysed.

2 DESCRIPTION OF THE EXAMPLE BRACKET AND OPTIMIZATION TASK

To illustrate the influence of boundary conditions on the computed optimal design, an example bracket, fastened with four screws, is selected. The bracket is mounted to a frame of a construction machine vehicle and carries a hydraulic motor and a wheel of the vehicle. This means that is has to carry all loads transmitted from the wheel to the frame. In this example, special attention will be given to consequences of changed boundary conditions caused by a screw becoming loose.

For the purpose of optimization, the wheel bracket is partitioned into two domains as follows: the subdomain that is free for optimization (Figure 1; light blue color), and the rest of the total domain that has to remain fixed and is thus not optimized (Figure 1; gray color).

The bracket is fastened by four screws and at the top additionally supported by the frame along a banded contact area. Thus, the supported surfaces are the upper frame contact area and the screw bores (Figure 1; dark blue color).

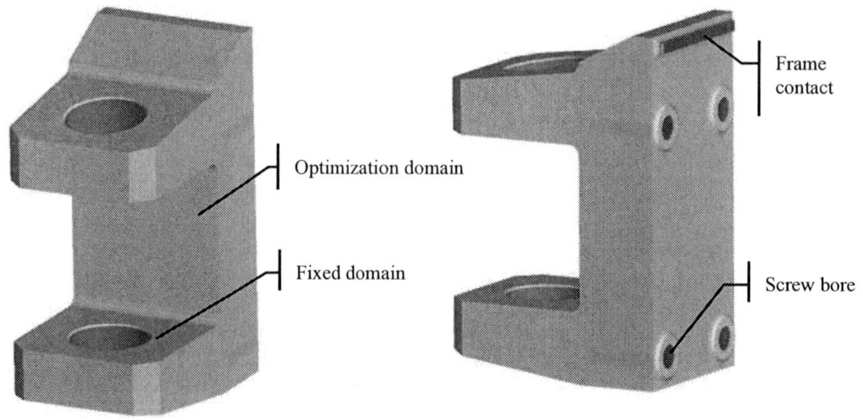

Figure 1. Domains and supported surfaces of the example wheel bracket.

The bracket is loaded by 5 basic load sets A-E (Figure 2), as follows:

(a) vertical load (A; 7.5 kN),

(b) loads due to vehicle acceleration (B; 2x40 kN) or deceleration (C; 2x40 kN),

(c) loads due to vehicle driving through the left (D; 2x40 kN) or right (E; 2x40 kN) bend.

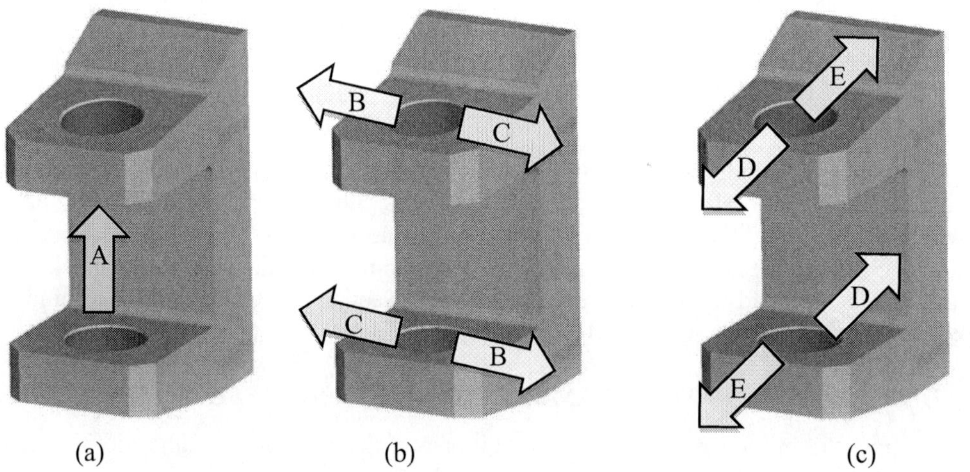

Figure 2. Basic load sets A-E acting on the wheel bracket.

The optimization task is approximately defined as follows: find the optimal topology of the bracket if the volume of the free domain has to be reduced by 50%. By the term optimal, we mean maximal structural stiffness and minimal attainable stress levels. This is a classical topology optimization problem that can be solved by minimizing the structural strain energy.

3 DEFINITION OF ALL POSSIBLE LOAD CASES

An important (desired, but potentially dangerous) effect of topology optimization is that the final design is fully and exclusively adapted to the specified boundary conditions. Therefore, it is of extreme importance that the set of all applicable boundary conditions is carefully analyzed, defined, and included into the definition of the optimization task. Any failure in capturing all possible situations may result in designs that will not fulfill the expectations.

In the context of optimization, a particular support/loading situation of a part is termed a load case (LC). In a typical optimization task, several different load cases are needed in order to describe all applicable situations. It should be noted that, in general, all load cases may differ in both, supports and applied loads.

The applied loads are described by 6 different situations; each of these situations is defined as a combination of basic load sets A-E as given in Table 1.

Table 1. Loading situations – combinations of basic load sets A-E.

Loading situation	Set A	Set B	Set C	Set D	Set E
1	x	x			
2	x		x		
3	x	x		x	
4	x	x			x
5	x		x	x	
6	x		x		x

Regarding the supports: the part is theoretically supported at four screw bore surfaces and at the frame contact area. Theoretically, these support conditions are the same for all loading situations, meaning that we basically have 6 load cases denoted as LC1 through LC6. Let us denote the corresponding optimization task as T1.

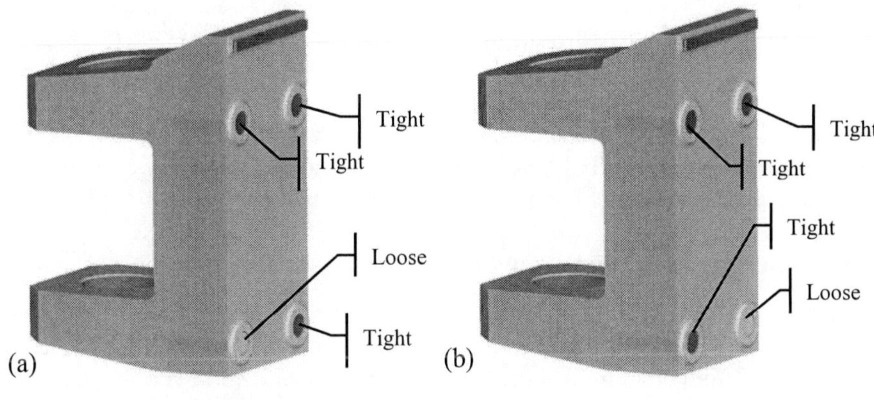

Figure 3. Boundary conditions of the hydraulic engine beam – one of the screws is loosened.

Let us now assume that one of the lower screws becomes loose (Figure 3). This changes the support conditions and to include this possibility into the optimization process, 6 additional load cases have to be defined: LC1A through LC6A (6 loading situations at changed support conditions). In a similar way, other screws may also be become loose. In this work, we will just add another set of load cases LC1B through LC6B covering the situation of another lower screw becoming loose (Figure 3b). So, we now have a total of 18 load cases LC1-LC6B which define another optimization task, here denoted as T2.

4 OPTIMIZATION RESULTS AND ANALYSIS

Both topology optimization tasks (T1 and T2) were solved by CAESS ProTOp. The results are analyzed briefly in the following.

4.1 Optimal Design T1 (All Screws Tight; 6 Load Cases)

Figure 4 illustrates the stress fields within the optimal design T1. Figure 4a shows the stresses computed within the optimization process, where all screws were tight. One can see that maximal stresses (excluding numerical peaks) are mainly around 200 MPa.

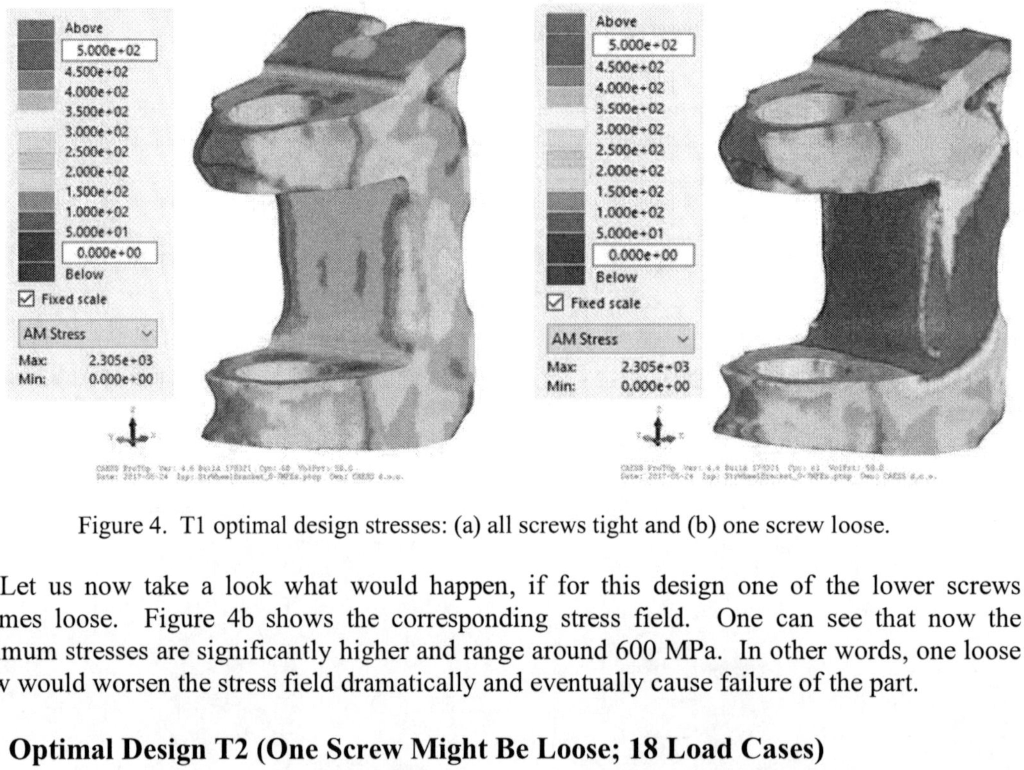

Figure 4. T1 optimal design stresses: (a) all screws tight and (b) one screw loose.

Let us now take a look what would happen, if for this design one of the lower screws becomes loose. Figure 4b shows the corresponding stress field. One can see that now the maximum stresses are significantly higher and range around 600 MPa. In other words, one loose screw would worsen the stress field dramatically and eventually cause failure of the part.

4.2 Optimal Design T2 (One Screw Might Be Loose; 18 Load Cases)

Figure 4 illustrates the stress fields within the optimal design T2. Figure 5a shows the stresses computed within the optimization process, where one screws might be loose. One can see that maximal stresses (excluding numerical peaks) are mainly around 350 MPa.

Let us now take a look how this part would perform under normal conditions (all screws tight). Figure 5b shows the corresponding stress field. One can see that now the maximum stresses are somewhat above 200 MPa.

By comparing these stress fields with the stress fields obtained with optimal design T1, one can see the following: (i) under normal conditions, design T2 performs only slightly worse than design T1 and (ii) under the loose-screw conditions, design T2 performs ways better than design

T1. Obviously, design T2 can be seen as a much better alternative because of its much lower failure probability under non-ideal support conditions.

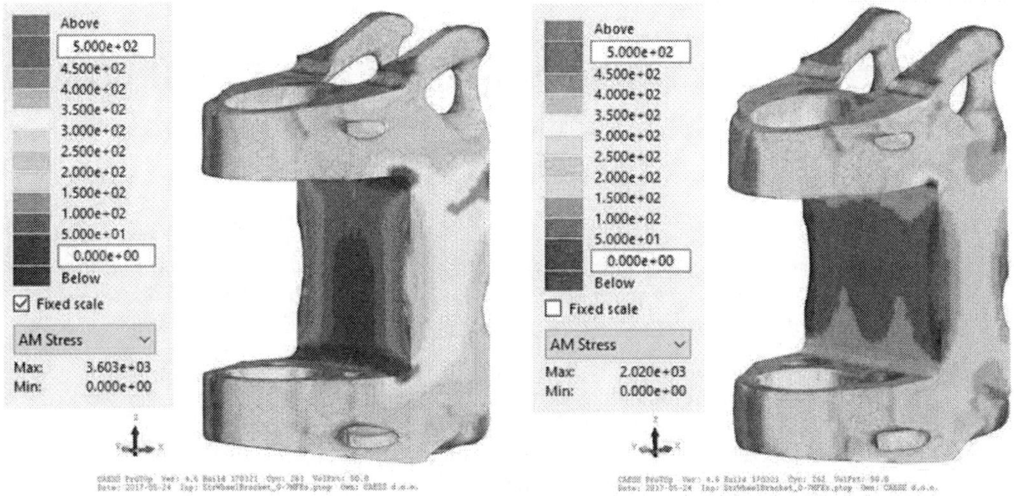

Figure 5. T2 optimal design stresses: (a) one screw loose and (b) all screws tight.

5 CONCLUSION

Topology optimization procedures are becoming indispensable in the design of light-weight and efficient load-carrying structural parts. Unfortunately, their engagement is more sophisticated as this might seem to be at a first glance. Perhaps the most exposed problem is the difficulty to describe and capture accurately enough all possible support/loading situations. Actually, one missed load case may deliver a design that may fail rather quickly under real-life operating conditions. Therefore, for any topology optimization task, careful analysis, formulation, and verification of all possible support /loading situations should be done with greatest care possible.

References

Bendsøe, M. P. and Kikuchi, N., Generating Optimal Topologies In Structural Design Using a Homogenization Method, *Computer Methods in Applied Mechanics and Engineering*, 71(2), 197-224, doi:10.1016/0045-7825(88)90086-2, 1988.

Harl, B., Predan, J., Gubeljak, N., and Kegl, M. On Configuration-Based Optimal Design of Load-Carrying Lightweight Part, *International Journal of Simulation Modelling*, ISSN 1726-4529, 16(2), 219-228, doi: 10.2507/IJSIMM16(2)3.369, June 2017.

Huang, D. W. and Xie, Y. M., Evolutionary Topology Optimization of Continuum Structures, *John Wiley & Sons*, 2010.

Rozvany, G. I. N., A Critical Review of Established Methods of Structural Topology Optimization, *Structural and Multidisciplinary Optimization,* 37(3), 217-237, doi:10.1007/s00158-007-0217-0, 2009.

Responsible Design and Delivery of the Constructed Project
Edited by Abdul-Malak, M., Khoury, H., Singh, A., and Yazdani, S.

NONLINEAR FINITE ELEMENT ANALYSIS OF TRANSVERSELY PRESTRESSED CONCRETE DECK SLABS

SANA AMIR[1], COR VAN DER VEEN[2], and ANE DE BOER[3]

[1]*Civil Engineering Dept, American University in Dubai, Dubai, United Arab Emirates*
[2]*Dept of Design & Construction, Delft University of Technology, Delft, Netherlands*
[3]*Ministry of Infrastructure and the Environment (Rijkswaterstaat), Utrecht, Netherlands*

This paper describes the modeling and analysis procedure of a 3D, solid, nonlinear finite element (FE) model of a bridge developed in the finite element analysis software package TNO DIANA to study the structural behavior in punching shear of transversely prestressed concrete deck slabs cast between flanges of long, pretensioned girders, and compressive membrane action. The numerical research was part of a broad project involving laboratory experiments carried out on a 1:2 scale model of such a bridge in Delft University of Technology. Both the experimental and numerical results showed much higher capacities than expected and this was attributed to the development of compressive membrane action in the plane of the slab. The numerical results were then compared with the experimentally found ultimate loads of eight basic test cases and it was discovered that the nonlinear FE models can predict the load carrying capacity quite accurately with a coefficient of variation of only 11%. It was concluded that punching shear failures can be reasonably modeled with non-linear finite element analysis of 3D solid models. Furthermore, using composed elements can lead to the determination of compressive membrane forces developed in a laterally restrained slab, which was previously difficult to determine using analytical techniques.

Keywords: Bridge, Bearing capacity, Punching shear, Compressive membrane action, Numerical modeling, DIANA, Composed elements.

1 INTRODUCTION

1.1 Background

Safety of existing structures is a pressing matter being investigated by designers all over the world. A lot of research has been going on in the Netherlands on this subject as there are a large number of bridges built back in the 60s and 70s of the last century. One particular type of bridge, whose safety was questionable, consists of a thin, transversely prestressed bridge deck slab cast between flanges of long, pretensioned girders. There are 69 bridges in the Netherlands of this type and it was found out that the shear capacity of the deck slabs as prescribed by the codes is more conservative in the recently implemented EN 1992-1-1:2005 (CEN 2005) than in the former Dutch NEN 6720:1995 (1995). Also, modern traffic loads are higher than assumed in the original design. Therefore, it was decided by Rijkswaterstaat (Ministry of Infrastructure and the Environment, Netherlands) to carry out a comprehensive research program to investigate the bearing capacity of these bridge decks. The first part of the research involved nineteen experiments being carried out in the Stevin II laboratory, Faculty of Civil Engineering and Geosciences, Delft University of Technology on the scaled model of such a bridge. A 1:2 scaled model of the bridge was constructed consisting of a thin transversely prestressed concrete deck slab cast between precast concrete girders. The slab panels were subjected to concentrated loads simulating Eurocode Load Model 1 (CEN 2005) and the bearing (punching shear) capacity was observed. This paper describes the numerical research that followed the experimental program to investigate the bearing capacity of the model bridge deck. A comparison of the nonlinear finite element analysis and the experimental analysis of eight typical test cases is made in this paper.

1.2 Compressive Membrane Action

The conventional methods of bridge design were based on conservative flexural theories but it has been discovered that under concentrated wheel loads, the laterally restrained deck slabs mostly fail in punching shear rather than in flexure (Batchelor 1990, Fang *et al.* 1994). The reason behind this is the development of compressive membrane forces in the deck slab. When a load is applied on a laterally restrained slab, its edges tend to move outside and the restraint of the boundary elements produce a compressive membrane force in the plane of the slab enhancing the bearing capacity in both flexure and punching shear. This phenomenon is called compressive membrane action (CMA).

Currently, codes like Eurocode 2 (CEN 2005) and ACI 318 (2005) do not consider CMA in their capacity formulae. However, there are some codes that do consider CMA like CSA: CHBDC (2005), the Transit New Zealand (2003) code and UK HA, BD81/02 (2002), but these are only applied for reinforced concrete slabs. It is hypothesized, in this research, that the in-plane forces arising from the combined action of prestressing and membrane forces will increase the bearing capacity to a large extent allowing thinner deck slabs to be applied with no problems of serviceability and structural safety.

2 EXPERIMENTAL PROGRAM

The 1:2 scale model bridge deck consisted of three transversely posttensioned deck slab panels cast in situ between precast, pretensioned girder flanges. The interface between the deck slab and the girder flanges was indented to generate sufficient shear capacity and had an inclination of 1:20. The transverse prestressing bars (15 mm ϕ bars in 400 mm c/c ducts) in the deck slab were unbonded so that the prestressing level could be varied during the experiments. Two transverse cross-beams were also provided and were posttensioned to the same level as the deck slab. The test setup is shown in Figure 1.

For the deck slab and the transverse cross-beams, the concrete compressive cylinder strength was 65 MPa, the tensile strength was 5.41 MPa and the modulus of elasticity was calculated as 39 GPa (as per Eurocode 2). For the girders, the concrete compressive cylinder strength was 75 MPa, the tensile strength was 6.30 MPa and the modulus of elasticity was 41 GPa. The steel reinforcement had a yield strength of 525 MPa and the prestressing steel bars had a characteristic tensile strength of 1100 MPa.

Figure 2 shows the model of the bridge deck in the laboratory and the test loading positions in the plan view of the deck slab. In all the tests, a concentrated load (wheel print load) was applied through a 200×200 mm, 8 mm thick rubber bonded to two 200×200×20 mm steel plates. The concentrated load was according to Eurocode 1 Load model 1, NEN-EN 1991-2:2003 (CEN 2005) scaled down according to 1:2.

Two levels of transverse prestressing were mainly investigated: 1.25 MPa and 2.5 MPa. Four types of tests were performed: a) Single point load acting at mid span of deck slab panel (P1M);

b) Single point load acting close to the girder flange-deck slab interface/joint (P1J); c) Double point loads at 600 mm c/c acting at mid span of deck slab panel (P2M); d) Double point loads at 600 mm c/c acting close to the girder flange-deck slab interface/joint (P2J). In the tests performed close to the girder flange-deck slab interface, load was placed at 200 mm c/c from the joint except in two tests, BB3 and BB4 where it was placed at 110 mm c/c. For details of the real bridge, the prototype and the experimental setup, reference is made to Amir (2014) and Amir *et al.* (2016).

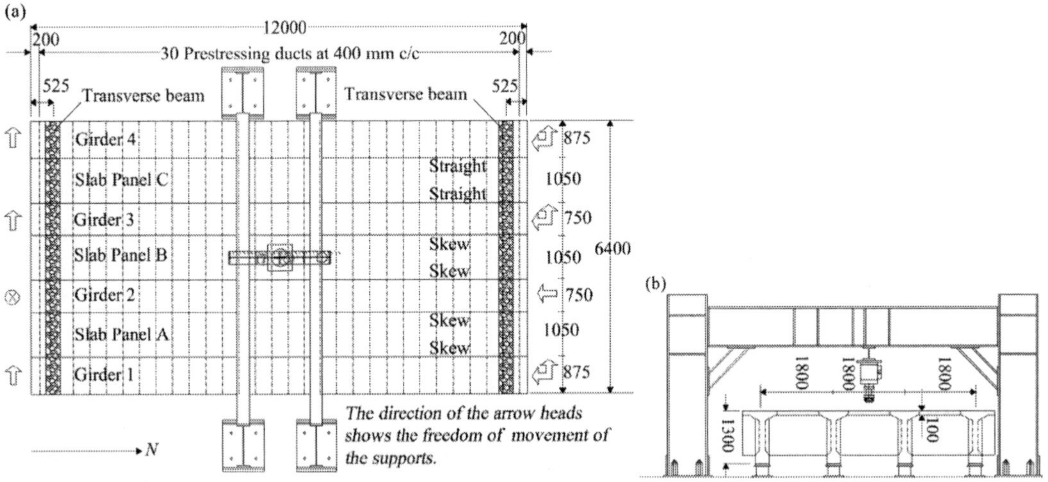

Figure 1. Test setup: a) Plan view b) Side view (transverse direction). All dimensions are in mm.

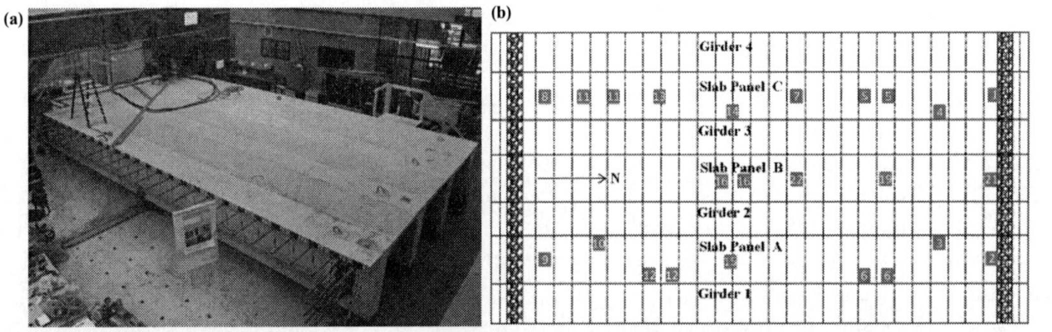

Figure 2. a) Bridge model; b) Deck slab test positions (BB1-BB22). Duct positions are also labeled.

3 NUMERICAL ANALYSIS

For the numerical analysis, a 3D solid finite element model of the prototype bridge deck (Figure 3) was made in the FEA software package DIANA (FX+ 9.4.4) (2012). The model consisted of 3D solid elements (CHX60 and CTP45) with a fine mesh around the loading area and a course mesh away from the loading to reduce the time for computation. A layer of composed elements (CQ8CM) was provided in the fine mesh area to calculate compressive membrane forces. Ducts at 400 mm c/c were provided only in the fine mesh area around the loading. Prestressing pressure was applied according to the required level of transverse prestressing in the deck slab and the transverse cross-beams.

For most cases the deck slab was analyzed nonlinearly while the girders and the transverse cross-beams remained in the linear range. The only exceptions to this were the tests BB3 & 4. The flange of the adjoining girder was analyzed as nonlinear since the experimental load was too close to the interface (110 mm c/c) and linearity of the flange would have induced a much higher capacity than in reality. An embedded reinforcement grid based on the actual steel reinforcement ratio was provided in the deck slab panels at the top and bottom in the horizontal as well as the vertical direction.

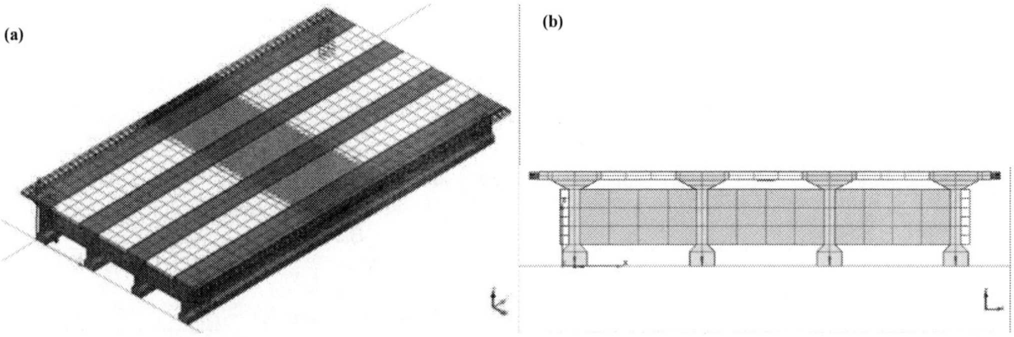

Figure 3. a) 3D solid finite element model developed in DIANA (2012); b) Cross-section of the model.

3.1 Material Models and Additional FEA Nonlinear Material Properties

For the nonlinear analysis of the deck slab, a smeared cracking "Total strain crack rotating model" was selected. An elastic-perfectly plastic model, CONSTA, was used for the concrete behavior in compression, whereas, an exponential softening curve, HORDIJK, (Hordijk 1991) was used for the concrete behavior in tension. A fracture energy (G_f) of 0.15 N/mm was assumed for the deck slab concrete (for a maximum aggregate size of 20 mm, MC90 (1990) gives a value of 0.135 N/mm for the fracture energy, whereas MC2010 (2012) gives a value of 0.21 N/mm). The Poisson ratio, ν, for all the concrete components, was taken as 0.2. For the embedded grid reinforcement, the von Mises plasticity criterion was used with a Poisson ratio of 0.3.

3.2 Iteration Method and Convergence Criteria

Both physical and geometrical nonlinearities were applied to the system. Composed elements were generated while giving the analysis commands. An incremental-iterative procedure was used for the nonlinear analysis and modified Newton Raphson method was used for the solution. The prestressing load was applied to the bridge deck in a single step. After that a displacement-controlled load was applied with a step size of 0.1 mm unless the solution diverged, in which case the displacement increment was reduced to 0.05 mm. Since the applied load was displacement-controlled, the default force and energy-based convergence criterion was employed.

4 COMPARISON OF EXPERIMENTAL AND NUMERICAL RESULTS

Results for ultimate load and failure mode of the typical test cases are described in Table 1. It was observed that for all test cases, failure always occurred by punching shear. Flexural punching (large rotations but final failure was in punching shear) was observed for double loads applied in the midspan which is consistent with findings of Zheng *et al.* (2010), and brittle punching was observed for the rest. It was also observed that the ultimate bearing capacity increased with the increasing prestressing level. Both the experimental and numerical results show comparable ultimate loads and similar failure modes, see Table 1.

Table 1. Comparison of finite element analyses and experimental ultimate loads.

Test BB.	TPL (MPa)	Designation	P_T (kN)	P_{FEA} (mm)	Test and FEA Failure Mode	P_T/P_{FEA}
1.	2.5	C-P1M	348.7	302.3	Brittle punching	1.15
2.	2.5	A-P1M	321.4	302.3	Brittle punching	1.06
3.	2.5	A-P1J	441.6	429.9	Brittle punching	1.03
4.	2.5	C-P1J	472.3	429.9	Brittle punching	1.10
5.	2.5	C-P2M	490.4	529.9	Flexural punching	0.93
6.	2.5	A-P2J	576.8	537.0	Brittle punching	1.07
7.	2.5	C-P1M	345.9	302.3	Brittle punching	1.14
8.	1.25	C-P1M	284.5	271.4	Brittle punching	1.05
9	1.25	A-P1M	258.2	271.4	Brittle punching	0.95
10.	1.25	A-P1J	340.3	300.7	Brittle punching	1.13
11.	1.25	C-P2M	377.9	453.4	Flexural punching	0.83
12.	1.25	A-P2J	373.7	454.9	Brittle punching	0.82
					Mean	1.02
					Standard deviation	0.11
					Coefficient of variation	0.11

Note: TPL: Transverse prestressing level, P_T: Test ultimate load, P_{FEA}: Finite element analysis ultimate load.

5 RELATIONSHIP BETWEEN THE IN-PLANE FORCES AND FAILURE LOADS

It is evident from the experimental and finite element analysis results that sufficient membrane action had developed in the plane of the deck slab and combined with the transverse prestressing, it positively enhanced the ultimate bearing capacity of the deck slab. It was observed both experimentally and numerically that for each type of the load, the deck slab showed horizontal or lateral displacements only after the initial cracking (Amir 2014) which correlates well with the findings of Liebenberg (1966), He (1992) and Fang (1985).

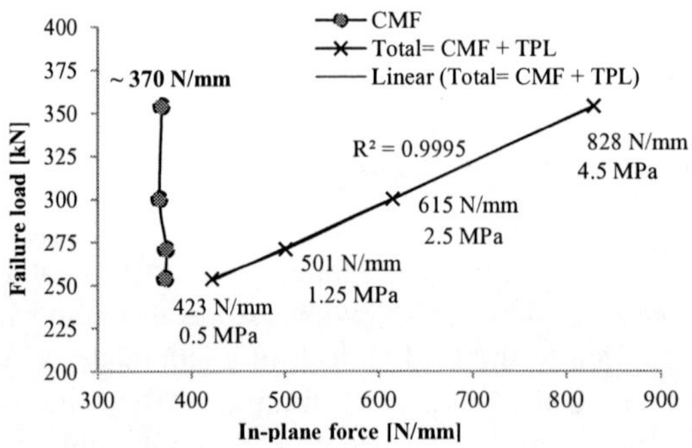

Figure 4. Relationship between the failure load and the in-plane forces for various TPLs.

Figure 4 shows the relationship between the distributed in-plane force developed (N_{xx} obtained from composed elements in DIANA (2012)) and the failure load for various levels of transverse prestressing level (0.5, 1.25, 2.5 and 4.5 MPa). It can be observed that for all cases, the overall in-plane force increases with the increasing prestress force and the relationship is almost linear. However, subtracting the initial prestressing from the overall in-plane force corresponding to that particular TPL gives a constant value of the compressive membrane force (CMF ~ 370 N/mm) for the deck slab showing that CMA is independent of the transverse prestress level. This would mean that for a particular deck slab having a certain lateral stiffness, the membrane action developed remains constant if all other parameters remain the same.

6 SUMMARY AND CONCLUSIONS

A 3D, solid, 1:2 scaled model of a real bridge was developed in DIANA (2012) and non-linear analyses were performed to simulate the experiments done in the laboratory on the same prototype. A basic analysis comprising of eight test cases was performed to study the ultimate bearing capacity of the bridge deck and compressive membrane action. It was observed that substantial CMA develops in the deck slab and transverse prestressing affects the bearing capacity positively. Furthermore, failure always occurred by punching in the deck slab span, regardless of the position of the load. The interface proved to have sufficient strength and was never governing. It can be concluded that punching shear failures can be reasonably modeled with nonlinear finite element analysis of 3D solid models. Sufficient cost savings can be made if numerical studies are performed over expensive experiments, but it is recommended that the results of applied computer models are first calibrated with experimental results.

References

ACI 318-05, *Building Code Requirements for Structural Concrete and Commentary (318R-05)*. Farmington Hills, Michigan: American Concrete Institute, 2005.

Amir, S., *Compressive Membrane Action in Prestressed Concrete Deck Slabs*, PhD thesis, Delft University of Technology, Delft, the Netherlands, 2014.

Amir S., van der Veen, C., de Boer, A., and Walraven, J. C., Experiments on Punching Shear Behavior of Prestressed Concrete Bridge Decks, *ACI Structural Journal*, 113, 627-636, 2016.

Batchelor, B. de V., Membrane Enhancement in Top Slabs of Concrete Bridges, *Concrete Bridge Engineering, Performance and Advances*, 189-213, London: Routledge, 1990.

BD 81/02., Use of Compressive Membrane Action in Bridge Decks, *Design Manual for Roads and Bridges, UK Highways Agency,* 3, 4(20), 2002.

CEN 2005, Eurocode 2: *Design of Concrete Structures - Part 1-1 General Rules and Rules for Buildings. NEN-EN 1992-1-1:2005,* Brussels, Belgium: Comité Européen de Normalisation, 2005.

CHBDC, Canadian Standard Association: Canadian Highway Bridge Design Code, CAN/CSA-S6-00 (R2005), Canada, 2005.

DIANA, *User's Manual - Release 9.4.4*. Delft: TNO Building and Construction Research, 2012.

Fang, I. K., *Behavior of Ontario-Type Bridge Deck on Steel Girders*. PhD. Dissertation. University of Texas, Austin, USA, 1985.

Fang, K., Lee, J. H., and Chen, C. R., Behavior of Partially Restrained Slabs Under Concentrated Load. *ACI Structural Journal,* 91(2), 133-139, 1994.

He, W., *Punching Behavior of Composite Bridge Decks with Transverse Prestressing*. PhD Dissertation, Queen's University, Ontario, Canada, 1992.

Hordijk, D. A., *Local Approach to Fatigue of Concrete*, PhD Dissertation, Delft University of Technology, Delft, the Netherlands, 1991.

Liebenberg, A. C., *Arch Action in Concrete Slabs*. CSIR Research report (234). Bulletin 40. Pretoria: National Building Research Institute, 1966.

MC90, CEB-FIP Model Code 1990, London, Thomas Telford Ltd, 1993.

MC2010, Fédération Internationale du Béton (fib), Final draft (1), Bulletin 65. Lausanne, 2012.

NEN 6720:1995. *Regulations for Concrete: Structural Requirements and Calculation Methods (in Dutch),* Dutch Normalisation Institute (NEN), 1995.

Transit New Zealand Ararau Aotearoa, *New Zealand Bridge Manual* (2nd Ed), New Zealand, 2003.

Zheng, Y., Taylor, S., Robinson, D., and Cleland, D., Investigation of Ultimate Strength of Deck Slabs in Steel-Concrete Bridges. *ACI Structural Journal,* 107(1), 82-91, 2010.

Responsible Design and Delivery of the Constructed Project
Edited by Abdul-Malak, M., Khoury, H., Singh, A., and Yazdani, S.

CASE STUDY OF SEISMIC EVALUATION AND STRENGTHENING OF EXISTING LOW-RISE REINFORCED CONCRETE BUILDING

AMTHAL HAKIM and ADNAN MASRI

Dept of Civil Engineering, Beirut Arab University, Beirut, Lebanon

All new buildings nowadays have to be designed and executed to overcome any imposed type of loading (lateral/vertical). On a universal scale, the stock of buildings built before 1980's is believed to be many times more than the number of newer buildings in most urban cities. In Beirut, as an example, a large proportion of Reinforced Concrete (RC) structures were constructed in the absence of mandatory earthquake design requirements, and unquestionably recognized as the type of construction most vulnerable to earthquakes. The performed research focused on how to evaluate the status of old building and how to design and execute the convenient seismic strengthening schemes. A case study has been selected to implement the evaluation process and design proposals. Conventional seismic upgrading technique has been assessed like the addition of shear walls in addition to more innovative approach which is the installation of steel bracing system. The strengthening schemes proposed aimed to create an ideal harmonization of the technical, economic and social aspects of the issue in hand. Analysis of the three structural systems (existing, modified with shear walls and with bracing systems) has been performed using the ETABS software including static equivalent, dynamic and pushover analyses. The research sorted out with a comparison between the systems based on different structural criteria followed by general recommendations and suggestions.

Keywords: Earthquake, Retrofitting, Rehabilitation, Shear walls, Steel braces, Lateral displacement.

1 INTRODUCTION

A remarkable seismic activity is carefully observed in Lebanon and the Region revealing the need for seismic assessment and strengthening for all new and some existing construction projects. Specialists keep sending alerts for many old low-rise buildings in Lebanon. Those buildings, more than 50% of Beirut Reinforced Concrete structures, are constructed in the absence of mandatory earthquake design requirements and conclusively vulnerable to earthquakes. The Lebanese Government imposed many regulations concerning the seismic activity and the needed considerations in the structural design, execution or modification of new or existing buildings. It refers to the seismic hazard level specified in the Decree 7964 dated April 7, 2012 which considers the national territory as one seismic zone. The requirements of the present Standard apply equally to new and some existing structures such as power plants, electrical substations, pumping and water treatment stations, water tanks of a capacity greater than $1000m^3$, educational occupancies, hospitals and health centers, fire stations public buildings and civil works and structures including large cantilevers or deep foundations (NL-135:2012).

The main challenge remains to assess and choose the best seismic strengthening techniques that are technically, economically and socially acceptable. The most common process to start is to set a screening criteria, then proceeding to the evaluation and ending up with applying the convenient seismic strengthening design (Cheung *et al.* 2004). In screening, buildings are classified as high, medium or low need for rehabilitation. The evaluation process takes place accordingly for the "high need" classified structures at first; it identifies the susceptibility of the structural components to seismic loads where the building performance is described in terms of safety during and after the event. Strengthening becomes urgent if the evaluation shows that the building does not meet the minimum requirements up to the existing code. This means that it could be subject to local or even full collapse during any seismic event. (Hakim 2016).

2 CASE STUDY CONFIGURATION

In this paper, the three steps process are applied to a representative case study in Beirut: "The Beirut Arab University (BAU), Block C", (Figure 1), denoting a suitable application for this process.

It is a university complex located in the heart of Beirut, built five decades ago, and serving more than 1000 students and staff members. It consists of a five-story Reinforced Concrete structure with inter-story heights ranging from 3.5m to 4.0m and a total height of 23.5m at the top roof. The BAU main building was structurally designed in the late 1950s. The construction works were done in 1958 with the absence of any seismic codes or requirements. The structural system is a "simple beam-column" system known also as Gravity Load Design (GLD) framing system. The load's path in GLD is transmitted from slabs to beams then to columns. Gravity Loads applied to this kind of structural systems will not transmit considerable moments to the columns.

A full structural investigation of the building confirms the efficiency of the as-built structure in terms of material type/performance and the soil/foundation types under gravity loads. Yet, the 3D numerical modeling evaluation endorses the vulnerability of this structure to seismic loads. The Lebanese Decree 7964 automatically screens this building as "High Need" for seismic strengthening. Therefore, two strengthening schemes are proposed, evaluated, designed and compared which are the addition of shear walls and the addition of bracing system.

Figure 1. Beirut Arab University Block C eastern facade.

3 ANALYSIS

ETABS 15.2.0 Modeling software is used to perform the Static Equivalent, dynamic and pushover analyses according to the below design criteria. Three prototypes, as shown in Figure 2, are considered:

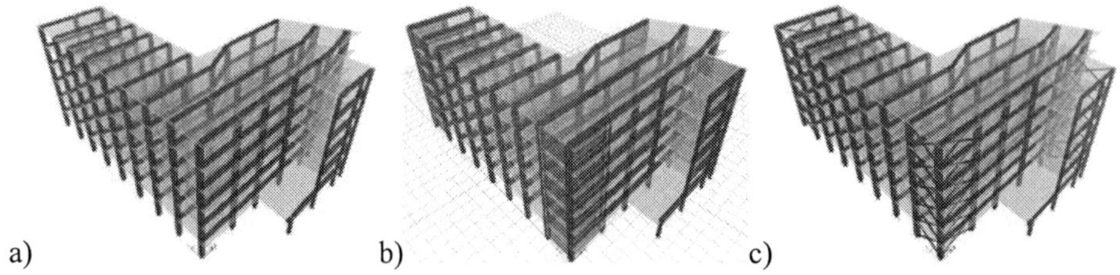

Figure 2. ETABS 3D Views of a) Existing structure, b) Existing structure with addition of edge shear walls, c) Existing structure with addition of edge steel braces.

- The material input for the model of Figure 2.a. is based on set of standard tests performed on existing structural elements; ASTM C39 for the evaluation of concrete compressive strength (f'_c) and ASTM A-370 for evaluation of the steel bars yielding stress. (f_y)
- The concrete compressive strength (f'_c) is at minimum 20 MPa for most of the cored samples.
- The Mild Steel Reinforcement found with 14mm and 16mm plain diameter has a yielding stress (f_y) near 260 MPa.
- The added shear walls and steel braces (Figure 2.b and 2.c) are designed with the following material parameters; (f'_c=30 MPa) and grade 40/60 high strength steel with (f_y = 420MPa).
- The Gravity Loads are assigned according to ASCE/SEI 7-05 (ASCE 2006) (Self-weight, Super Imposed Dead Loads = 5.0 KN/m^2 and Live Loads = 5.0 KN/m^2).
- Seismic Loads are assigned according to UBC 97 (ICBO 1997) (Seismic Zone: 3, Soil Type: C).
- The design of 25cm thickness shear walls considers the existing columns as boundary members. They are vertically continious, joined to the existing frames and fixed to the foundation. The implantation of these walls causes closure of some exterior bays as shown in Figure 3. (Grids: A-2-3, 2-A-B, F-1-2, 5-E-G, C-8-9, A-10-11, 11-A-B).
- The added braces have compact and non-slender steel sections. They are concentric X-Braces (25cm diamater and 1cm thickness) placed in the same location of the added shear walls (Figure 3). The newly introduced system utilizes the strength of existing column-beam connections. Noting that, braces are pin connected to the column-beam joint and require simple and clean connections.

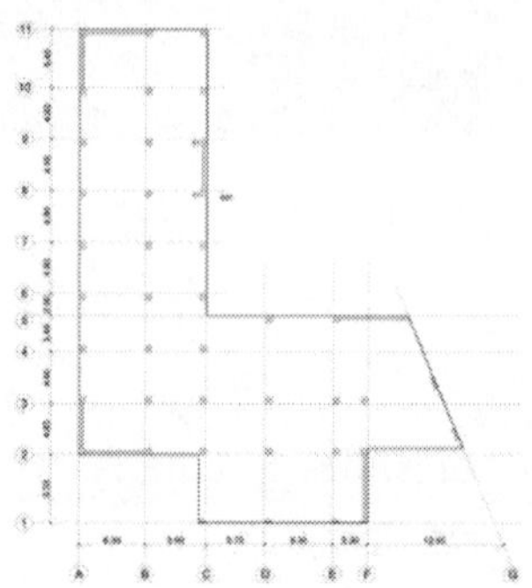

Figure 3. Plan view of BAU typical floor modified with shear walls.

4 RESULTS

4.1 Drift

Drift values are checked for the three models of Figure 3. The proposed check by UBC 97 (ICBO 1997) code evaluates the maximum inelastic response displacement, ΔM, of the structure caused by the Design Basis Ground Motion (see Eq.1). Table 1 includes results on roof floor where the modified structure with shear walls and steel braces show satisfactory results in compare to those of the existing structural system.

$$\Delta M = 0.7\,R\,\Delta S \leq 0.02 \times H_s \tag{1}$$

Where **R**, Δ**S** and $\mathbf{H_S}$ denote respectively the over strength factor, the total story drift, and the total story height. The overstrength factor is defined as numerical coefficient representative of the inherent overstrength and global ductility capacity of lateral force resisting systems. In the three cases, R is set to 5.5. The assumption considers the basic structural system as Building Frame System where lateral forces are resisted by shear walls or steel braces. (ICBO 1997).

Table 1. Interstory drift results – roof floor.

	Story	**Direction**	$\Delta S/H_S$	$0.7*R*(\Delta S/H_S)$	**Status**
Existing Structure	Roof	X	0.007202	0.0277277	NOT Adequate
	Roof	Y	0.019853	0.0764341	NOT Adequate
	Roof	X	0.004735	0.0182298	Adequate
	Roof	Y	0.018177	0.0699815	NOT Adequate
	Roof	X	0.062028	0.2388078	NOT Adequate
	Roof	Y	0.060738	0.2338413	NOT Adequate
Existing structure with Shear Walls	Roof	X	0.000541	0.0020829	Adequate
	Roof	Y	0.000192	0.0007392	Adequate
	Roof	X	0.000392	0.0015092	Adequate
	Roof	Y	0.000136	0.0005236	Adequate
	Roof	X	0.000296	0.0011396	Adequate
	Roof	Y	0.001008	0.0038808	Adequate
Existing structure with Steel Braces	Roof	X	0.000541	0.0020829	Adequate
	Roof	Y	0.000192	0.0007392	Adequate
	Roof	X	0.000392	0.0015092	Adequate
	Roof	Y	0.000136	0.0005236	Adequate
	Roof	X	0.000296	0.0011396	Adequate
	Roof	Y	0.001008	0.0038808	Adequate

4.2 Lateral Displacement

Lateral displacement of the structure cannot exceed the $\frac{H}{500} = 47mm$ conservative limiting value. As such, H is the total height of the building (23.5 m). Graphs in Figure 4 show that the existing structure's lateral displacement values exceeds the allowable. Whereas, values of lateral displacement in the two modified structural system remains less than 47mm.

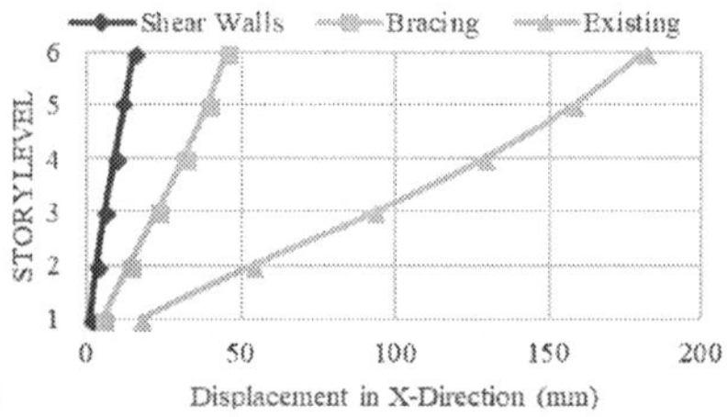

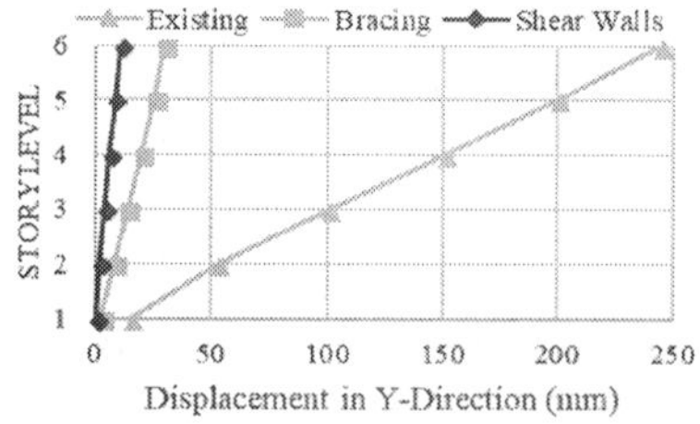

Figure 4. Horizontal displacement of the 3 models at different stories for both earthquake forces directions a) X-Direction b) Y-Direction.

4.3 Performance Based Check

Performance Based Check is done through the pushover analysis where beams and columns are modeled with concentrated plastic hinges at the column and beam faces, respectively. The beams have only main moment hinges whereas columns have axial load and biaxial moment (PMM) hinges. It is performed in presence of gravity loads, with monotonically increasing lateral loads until reaching maximum force value. These loads are distributed according to the ASCE requirements independently in the X and Y directions. The Performance Based curve deduced form the analysis is shown in figure 5 and denotes the following results.

- The as-built structure illustrates the lower stiffness and strength values between the three models in the elastic phase (54735 kN/m and 5000 kN).
- The bracing system has higher stiffness and strength (140882KN/m and 5500 kN).
- The shear walls system offers the largest stiffness, 4.5 times higher than the bracing system, and largest strength values with (511339 KN/m and 11000 kN)
- The ductility of the system is undesirably affected when shear walls are applied. Yet, it is highly enhanced in the bracing system alternative (the curve extends largely).
- To be noted that the Y-direction forces curves show similar behavior for the three models.

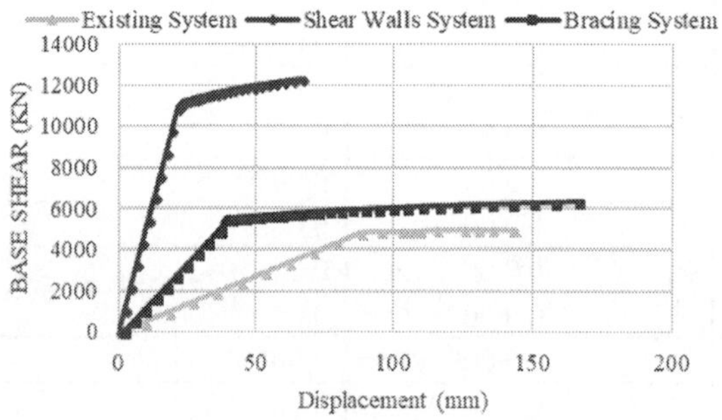

Figure 5. Base Shear versus Displacement for X-Direction earthquake forces.

4.4 Foundation

A test pit under the column on grid A-11 (Figure 3) shows a 230x230x70 cm spread footing.

- Using the Autodesk Robot software, the status of this foundation is evaluated.

- It shows that a failure is expected in both strengthening schemes: soil pressure is greater than the allowable soil bearing capacity (360 kPa).
- In the case of bracing system, a simple footing modification is required (e.g. increase of footing's area by 20%).
- In the case of shear walls system, strip footing should be implanted.

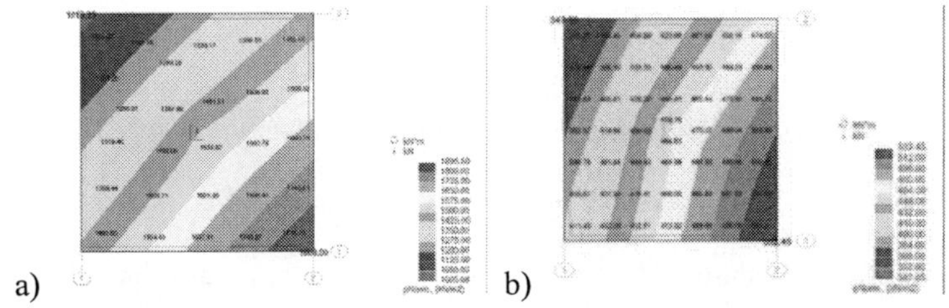

Figure 6. Soil Pressure below footing of Column (A-11) a) Shear Walls system, b) Bracing system.

5 CONCLUSION

The comparison between the shear walls system and steel braces system shows that both of them provide the desired seismic performance. For any earthquake event, similar to that simulated according to the Lebanese Norms and UBC 97 (ICBO 1997) code, the considered case study should show acceptable drift, stiffness and strength values.

However, the addition of steel braces to the existing structure provides a better ductility than the shear walls addition system and requires a minor modification for the foundation scheme. The bracing system provides an easier construction process and it is aesthetically acceptable if matched with the general architecture of the building.

In conclusion, the seismic evaluation of the Beirut Arab University main building confirms that an implementation of steel braces on the edge bays of the building could achieve an adequate seismic strength. Hence, it should prevent any collapse or significant damage during earthquakes.

6 GENERAL RECOMMENDATIONS

(i) The seismic evaluation of every old building is an urgent need in Beirut. The case study evaluated in this paper is a prototype of the majority of Beirut old buildings.

(ii) A national strategy should be adopted to find solutions against the high threats of earthquakes in Lebanon.

(iii) Strong, fair and sustained strategy could be summarized in the 3 steps process adopted in this research: Screening, Evaluation and the strengthening application.

References

American Society of Civil Engineers (ASCE), ASCE/SEI 7-05, *Minimum Design Loads for Buildings and Other Structures*, USA, 2006.

Cheung, M., Foo, S., and Granadino, J., *Seismic Retrofit of Existing Buildings: Innovative Alternatives*, Public Works & Government Services Canada, 2004

Hakim, A., *Seismic Assessment and Strengthening of Existing Low - Rise Reinforced Concrete Building in Beirut*, Beirut Arab University, Lebanon, 2016.

International Conf. of Building Officials (ICBO), *UBC 97, Uniform Building Code*, Vol 2, USA, 1997.

Lebanese Public Safety Law, Act of 2012, Decree 7964.

Responsible Design and Delivery of the Constructed Project
Edited by Abdul-Malak, M., Khoury, H., Singh, A., and Yazdani, S.

THE EFFECT OF IMPACT LOADS ON PRESTRESSED CONCRETE SLABS

YOUMN AL RAWI, YEHYA TEMSAH, HASSAN GHANEM, ALI JAHAMI, and MOHAMAD ELANI

Dept of Civil Engineering, Beirut Arab University, Beirut, Lebanon

Many research studies have been conducted on the effect of impact loading on structures, and design procedures were proposed for reinforced concrete (RC) slabs; however the availability of these studies and procedures are limited for prestressed slabs. The proposed research will examine, using numerical analysis, the impact of rock fall on prestressed concrete slabs with equivalent moment capacity reinforced concrete slabs. It is expected that prestressed concrete slabs will have different behavior to resist impact loading compared with traditional reinforced concrete slabs. The thickness of the prestressed concrete slab will be 25cm whereas that of the reinforced concrete slab will be 30cm. The impact loading consists of 500Kg drop weight. The drop height will be 10m, 15m and 20m.The structural analysis is performed using a Finite Element program "ABAQUS". A comparison will be done between both slab types in terms of failure mode, damage, and deflection. It has been found that both slabs failed in punching. However, the RC slab performed better than the prestressed concrete slab with respect to the value of the deflection at mid-span, while both showed punching shear mode of failure.

Keywords: Prestressing, Punching, Deflection, Capacity, Drop weight, Dynamic behavior.

1 INTRODUCTION

Prestressed concrete slabs are widely used in various floor types such as residential, industrial, and commercial buildings as well as in special constructions like bridges and complex structures. Usually, the floor system forms the major part of the cost of a building. The use of prestressed slabs reduces the cost of the construction and the time of execution. Also, in some special structural cases such as large spans, prestressed slabs are more economical and less complicated to implement. However, their resistance to impact damage caused by threats such as rock fall or other impact loads is currently under high interest.

The risk of rock fall accidents is increasing in habited mountain zones due to environmental changes, thawing of the forests, and manmade modifications on the natural landscape. Experts need to expand their knowledge and safety against rock fall threats especially with the growth of population and the high mobility, such as reported into and through Alpine areas by Schellenberg *et al.* (2016).

There is growing recognition of the importance for resistant structures and, in consequence, a need for tolerable, related evaluation and design processes. As to impact resistant structures, many studies have accounted an assessment of the behavior of reinforced concrete slabs under impact load such as Berthet-Rambaud *et al* (2002), Hrynyk and Vecchio (2014), and Sagals *et al.* (2015).

The majority of impact-related research has been concentrated on developing empirical formulae to estimate the element damage and capacity under precise types of impact-loading circumstances. However, the success in developing these formulae was limited. Moreover, with respect to prestressed slabs, much of the work has been exclusively focused on localized damage mechanisms such as concrete scabbing and missile penetration, (Thai and Kim 2017). Other aspects such as quantification of the damage in terms of width and depth of the scabbed area, or in terms of the failure mechanism, are rarely mentioned (Orbovic *et al* 2009).

2 NUMERICAL MODELING

2.1 Size and Geometry

Six numerical slab models have been studied using finite element software "ABAQUS". The aim is to evaluate the behavior of prestressed slabs compared to an equivalent in moment capacity reinforced concrete slab under impact load. This load is a falling weight from three different heights: 10m, 15m and 20m respectively as shown in Table 1. The drop weight consists of 500kg reinforced concrete block as shown in Figure 1. As for slabs' dimensions, prestressed slabs (PT Slabs) and reinforced concrete slabs (RC slabs) have the dimensions of 3m*6m with different thickness of 0.25m and 0.30m, respectively. Both slabs have longitudinal reinforcement of T16 at 150mm in both directions. The PT slabs have, also, 6 tendons made up of 2 strands (area of $99mm^2$ each). All slabs are simply supported in the long direction. The supports are steel plates resting on neoprene pads as shown in Figure 2.

2.2 Finite element Modeling

For modeling procedure in ABAQUS, concrete elements and supports were modeled using an eight-node linear brick element (C3D8R) as shown in Figure 3. Each node has three degrees of freedom, (Ellobody and Bailey 2009). Steel reinforcements and tendons were modeled as two-node linear 3-D truss elements. All steel reinforcement was assumed to be fully embedded inside the concrete body.

The "Concrete Plasticity Damage (CDP)" model (built-in ABAQUS model) was used to represent the concrete material stress-strain non-linear relationship. This model has high performance in dynamic and impact load analysis. The compressive strength considered in this study was 30 MPa. For steel reinforcement an elastic-perfectly plastic behavior was considered with yield strength equal to 420 MPa. On the other hand, steel tendons were defined using bi-linear behavior with yield strength of 1860 MPa.

Table 1. Study cases.

Drop Height(m)	Type of Slab	Model Name
10	PT	M1
	RC	M4
15	PT	M2
	RC	M5
20	PT	M3
	RC	M6

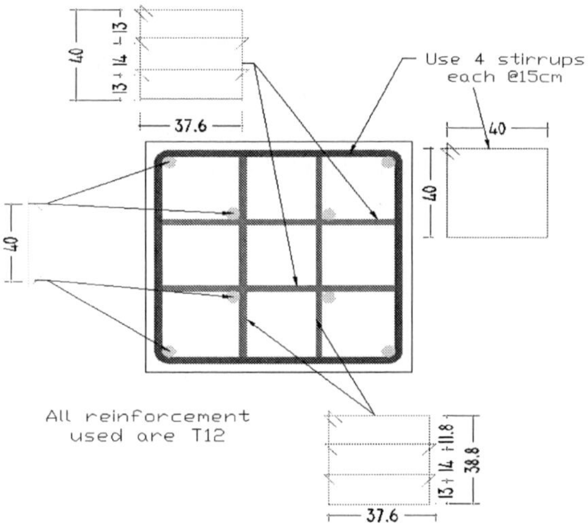

Figure 1. Drop weight reinforcement (cm).

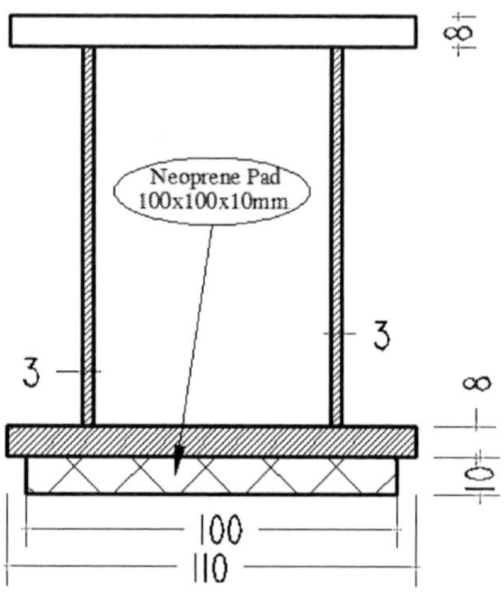

Figure 2. Steel cylinders (mm).

Regarding the prestressing force, a predefined field of type "initial stress" in the axial direction was defined, and a value of 1440 MPa was considered as a jacking stress. An initial speed was defined for the falling object equivalent to its real speed when reaching the top surface of slab. Block speed can be determined from Eq. (1), where "g" is the gravity acceleration and "h" is the height of falling object.

$$V = \sqrt{2gh} \tag{1}$$

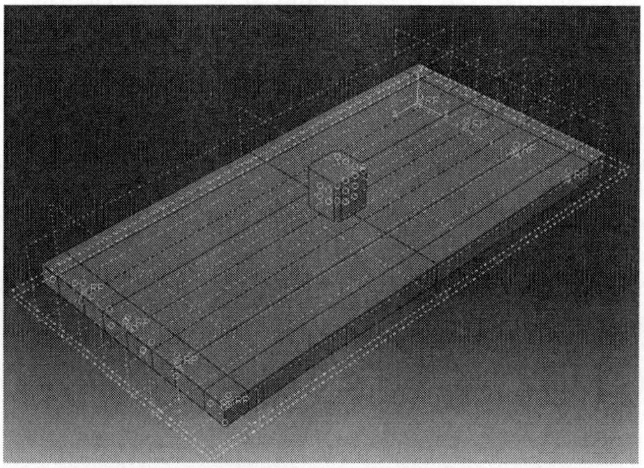

Figure 3. Finite element model.

2.3 Material properties under impact loads

Concrete behaves in a different way when subjected to impact loading. Its strength and stiffness under impact will be higher when subjected to static loading (Temsah *et al.* 2017). Many researches were conducted to study the tensile and compressive strength of concrete at different applied load rates. Some of the published results are presented in Malvar and Crawford (1998). As for reinforcing steel, the behavior is also affected when applying impact load. The yield strength will be increased as well as the ultimate strength, but the modulus of elasticity will not be affected. Figure 4 shows the stress strain curve of the concrete under different load rates.

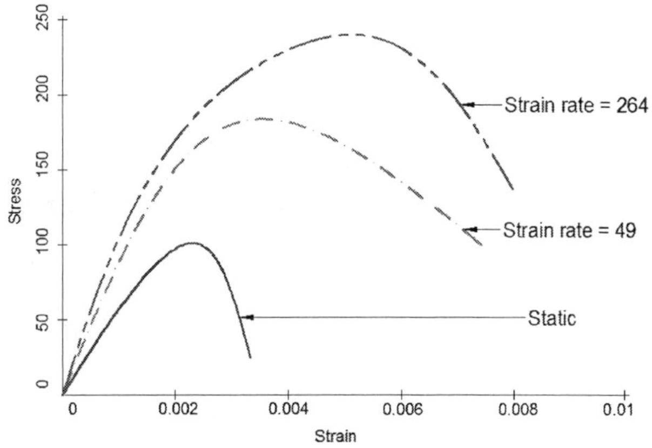

Figure 4. Concrete stress strain curve.

3 RESULTS AND DISSCUSION

3.1 Failure mode

The most critical impact effect was on the slabs subjected to drop weight from 20m. The governing mode of failure was punching shear failure as can be seen in Figures 5 and 6.

Figure 5. Punching failure in RC Slab due to 20 meters drop.

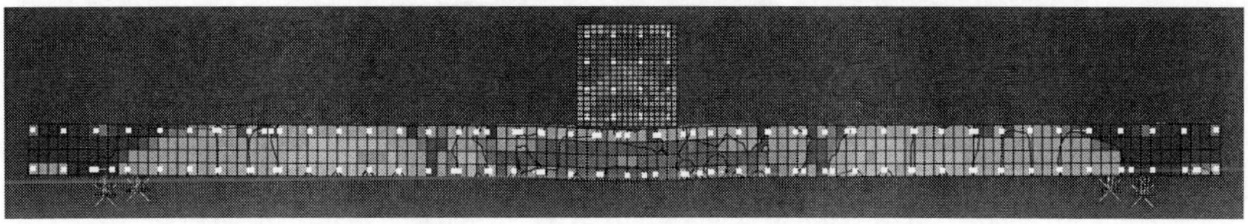

Figure 6. Punching failure in PT Slab due to 20 meters drop.

The punching shear mode of failure under impact load was shown through the experimental work of many researchers such as Orbovic *et al.* (2009), Hrynyk and Vecchio (2014), and Sagals *et al.* (2015). Then, the numerical analysis complies with the previous research studies as for the mode of failure.

The extent of the tensile damage zone was about 1.6m in PT slab and 1.5m in RC slab, whereas in the compression zone the damage was limited due to the top steel reinforcement used.

3.2 Mid-Span Deflection

The mid span deflection results are summarized in Table 2 and shown in Figure 7 for all slabs. The difference in deflection between the prestressed slab and reinforced concrete slab was about 10% when subjected to the impact load from both heights 10m and 15m. This percentage was dramatically raised to about 35% when the height was changed to 20m. It can be realized that the RC slab is better performing than the equivalent PT slab in terms of deflection in all cases. Again that can be justified by the inertia of the RC slab due to the increased thickness. This is different than what previous studies have shown since none of them compared PT slab to an equivalent moment capacity RC slab.

Table 2. Mid span deflection for all samples.

Drop Height(m)	Type of Slab	Model Name	Velocity at impact(m/s)	Energy at impact(KJ)	Mid-span Deflection (mm)
10	PT	M1	14	49	17.324
	RC	M4			15.678
15	PT	M2	17.5	73.5	23.639
	RC	M4			21.277
20	PT	M3	19.8	98	34.377
	RC	M6			26.111

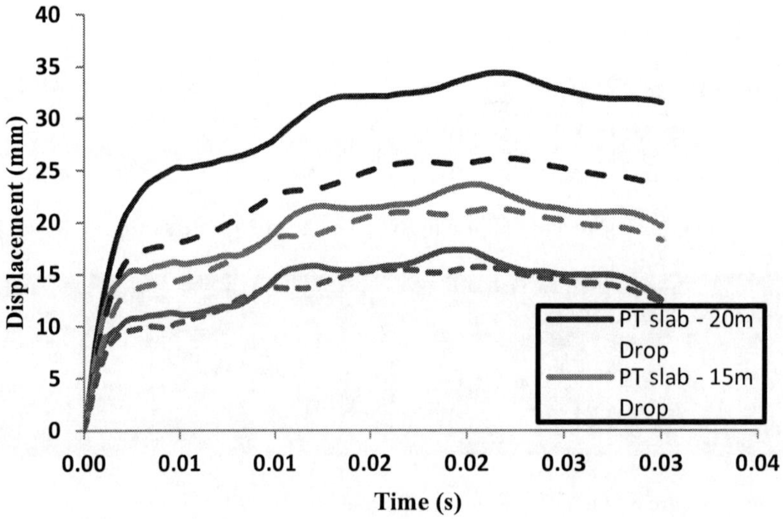

Figure 7. Deflection time history for all slabs.

4 CONCLUSION

The following conclusions can be drawn from the current investigation:

(i) The behavior of prestressed concrete slabs and its equivalent in moment capacity reinforced concrete slabs is different under impact of falling load.

(ii) The analysis showed that both types of slabs subjected to impact load due to drop weight are likely to fail due to local punching shear. Therefore, this mode of failure should be considered in any design procedure in the future.

(iii) The area of damage zone in prestressed concrete slabs was greater than those of reinforced concrete slabs, due to higher damage level of prestressed concrete slabs.

(iv) The inertia has a great effect on the behavior of slabs when subjected to impact loading, especially when the mode of failure is Punching shear failure. That was concluded from the difference in deflection between RC and PT slabs.

References

Berthet-Rambaud, P., Mazars, J., and Daudeville, L., Numerical Modeling of Rockfall Impacts on Reinforced Concrete Slabs for the Design of New Rock Sheds, FraMCoS-5, Vail, 2004.

Ellobody, E., and Bailey, C., Modeling of Unbonded Post-Tensioned Concrete Slabs under Fire Conditions, *Fire Safety Journal,* Elsevier, 44(2), 159-167, February, 2009.

Hrynyk, T., and Vecchio, F., Behavior of Steel Fiber-Reinforced Concrete Slabs under Impact Load, *ACI Structural Journal,* 111(5), 1213-1224, September-October, 2014.

Malvar, L. J., and Crawford, J. E., Dynamic Increase Factors for Concrete, *28thDDESBSeminar,* Orlando USA, August, 1998.

Orbovic, N., Lee, N., Elgohary, M., and Blahoianu, A., Tests on Reinforced Concrete Slabs with Pre-stressing and with Transverse Reinforcement under Impact Loading, *20th International Conference on Structural Mechanics in Reactor Technology (SMiRT 20)*, Espoo, Finland, 2009.

Sagals, G., Orbovic, N., and Blahoianu, A., Numerical Simulation of Missile Impact on Reinforced Concrete Slabs: Effect of Concrete Pre-stressing, *Transactions, 23rd Conference on Structural Mechanics in Reactor Technology, SMiRT-23 Manchester, United Kingdom,* August, 2015.

Schellenberg, K., Khasraghy, S. G., and Vogel, T., Impact Behavior of Reinforced Concrete Slabs Subjected to Rock Fall Loading, *WIT Transactions on Built Environment*, 98, 25-24, 2008.

Temsah, Y., Jahami, A., Khatib, J., and Sonebi, M., Numerical Derivation of Iso-Damaged Curve for a Reinforced Concrete Beam Subjected to Blast Loading, *2nd International Congress on Materials & Structural Stability*, Diouri, A., Boukhari, A., Ait Brahim, L., Bahi, L., Khachani, N., Saadi, M., Aride, J., and Nounah, A. (eds.), Rabat, Morocco, 22-25 November 2017.

Thai, K., and Kim, S., Numerical Simulation of Pre-stressed Concrete Slab Subjected to Moderate Velocity Impact Loading, *Engineering Failure Analysis,* Elsevier, 79, 820-835, September, 2017.

Responsible Design and Delivery of the Constructed Project
Edited by Abdul-Malak, M., Khoury, H., Singh, A., and Yazdani, S.

THE EFFECT OF SHEAR REINFORCEMENT RATIO ON PRESTRESSED CONCRETE BEAMS SUBJECTED TO IMPACT LOAD

MOHAMAD ELANI, YEHYA TEMSAH, HASSAN GHANEM, ALI JAHAMI, and YOUMN AL RAWI

Dept of Civil Engineering, Beirut Arab University, Beirut, Lebanon

Structural elements subjected to impact loads have a different response than those subjected to static loads. This research studied the effect of using shear reinforcement to reduce the local damage occurred when an impact load applied on a prestressed concrete beam. An accurate finite element model was provided for the analysis using the advanced volumetric finite element modeling program (ABAQUS). The concrete material was defined using the built in concrete damage plasticity model (CDP), that considers the nonlinear behavior of concrete when subjected to dynamic loading. All material properties were modified using the dynamic increase factor (DIF) to consider the effect of impact loading. It was realized that the failure was concentrated in the impact zone. However, using shear reinforcement reduced the permanent damage occurred due to impact.

Keywords: Numerical, Damage, Camber.

1 INTRODUCTION

Buildings constructed on steep slope, are vulnerable to rock falling. Rock falling causes impact loading which leads destruction in property. Many researchers have been conducted to study the effect of impact load on Reinforced concrete structures (Chen and May 2009), but very few studies were done on their effect on prestressed concrete beams. Use of prestressed concrete structures is increasingly popular (Kang and Huang 2012), whereas lack basic knowledge of designing prestressed beams subjected to impact loads in international codes exists. Using finite element model give us real results (Joshuva *et al.* 2014). They used 3D finite element model using ANSYS to compare RC beams and prestressed concrete beam. Applying the same load, the prestressed concrete beam failed at 143.6 KN, which is about 22% greater than RC beam of same dimension. Prestressed concrete beam of same dimensions, with approximate same area of reinforcement gave higher flexural capacity and better efficiency. PT increases stiffness of concrete and flexural capacity (Kumar *et al.* 2016). Moreover, researches have worked on modeling of post tension slabs using Abaqus (Kung 2012). The research discussed possible methods of modeling bonded and unbounded post tension beams, and recommended usage of tangential and normal contact conditions for modeling bonded and unbounded cases.

2 AIM AND SCOPE

The aim of this research, is to model prestressed concrete beam subjected to direct impact from a falling object using the commercial finite element program ABAQUS. Then a parametric study will be insured to investigate the effect of shear reinforcement ratio on decreasing damage and deflection. Final conclusions will be drawn at the end of the research.

3 NUMERICAL MODELING

A 300x600mm prestressed concrete beam was modeled with span of five meters, to benefit from cambering effect. The element used was C3D8R "An eight-node linear brick". 3φ14mm were added as minimum reinforcement at TOP and BOTTOM of the beam, and 4φ14mm as shrinkage reinforcement was used at each side, Figure 1. As for stirrups, a ten mm bar size was used. Three different spacings (100mm, 200mm, and 300mm) were used in this study. All these ordinary reinforcements were modeled as rebar elements "Two nodded linear elements". Eight tendons of diameter 12.7mm were added at 100mm height from the bottom face of the beam using the same element used in modeling the concrete body (C3D8R). The falling object in this study is a cubic reinforced concrete block with a dimension of 300x500x500mm with a weight of 187.5kg, and was reinforced by four 10mm stirrups in all directions (Figures 2 and 3).

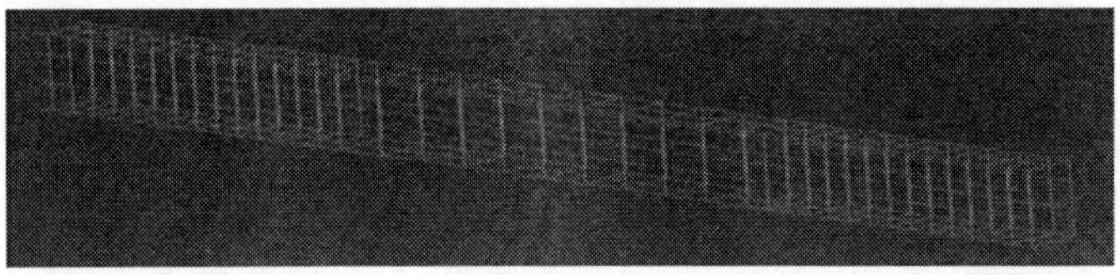

Figure 1. 3D view for all ordinary beam reinforcements.

Figure 2. 3D view for falling object.

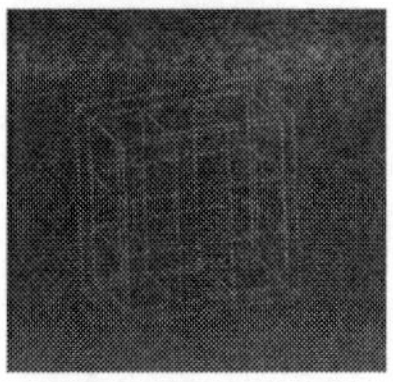

Figure 3. 3D view for Block reinforcements.

The Concrete Damage Plasticity Model will be used to define the behaviour of concrete. This model works with static and dynamic load conditions. It was derived by Lubliner (1989) and developed by Lee and Fenves (1998). It represents the nonlinear behaviour of concrete using different input parameters such as: inelastic strain, cracking strain, stiffness degradation and recovery and other parameters. (Figures 4 and 5). As for steel reinforcement and tendons, the plastic behaviour was considered in material definition. Moreover, the effect of impact conditions on all material used in this analysis was considered by using the dynamic increase factor (DIF); since both steel reinforcement and concrete will behave stiffer when subjecting to an impact load conditions (Temsah *et al.* 2017a, 2017b). A trial run was done to find the DIF when subjecting the beam to the falling object. The strain rate was found to 4.45 s^{-1}, and the corresponding DIF will be equal to 1.46.

Reinforcing steel can be modelled using several methods. For this simulation we will consider the elasto-plastic behaviour of reinforcing steel, and we will assume a perfect bond between concrete and steel. Whereas for the concrete - tendon interaction there were some differences (Kung 2012). A tangential and normal behaviour has been added allowing bit of slip. A predefined field condition was used to define the initial stresses in tendons, and gravity was defined for the block so it fell by the earth gravity.

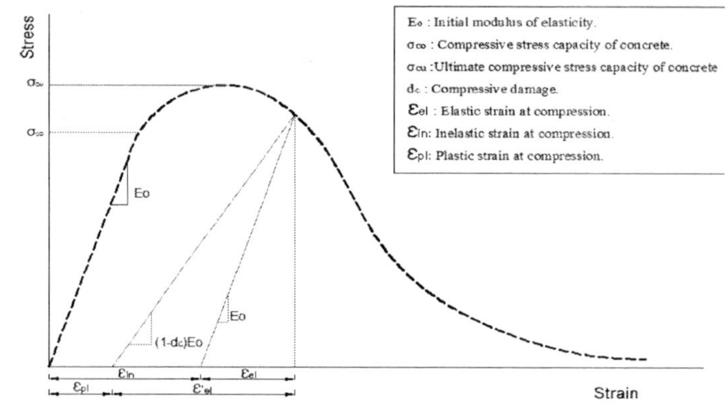

Figure 4. Compressive behaviour of concrete (CDP).

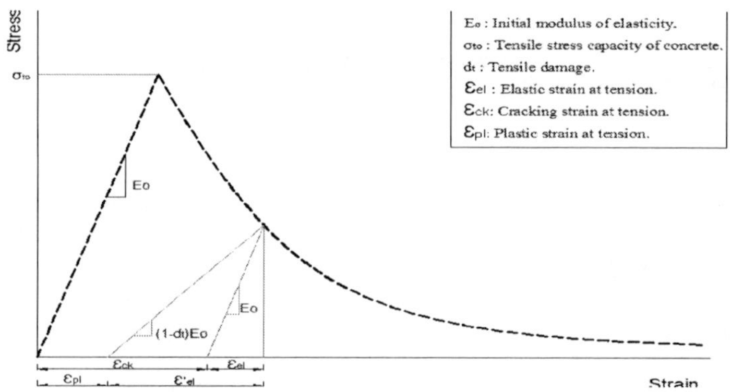

Figure 5. Tensile behaviour of concrete (CDP).

4 ANALYSIS RESULTS

4.1 Beam Camber

Figure 6. The effect of cambering on beam.

The prestressing effect was successfully represented. Referring to Figure 6, the camber of the beam was about 4.5 mm. To check this number, a comparison done between the equation of camber calculation and the result from ABAQUS. The following equation was used to calculate the camber:

$$\delta = \frac{P.e}{E.I} \times \frac{L^2}{8} \tag{1}$$

Where "P" is the prestressing force, "e" is the eccentricity, "E" is the modulus of elasticity, "I" is the moment of inertia of the beam, and "L" is the span length of the beam. Applying this equation, the value of cambering was 4.32 mm which is close to the one calculated by finite element modeling.

4.2 Beam Stresses

Top and bottom stresses were checked manually due to own weight and prestressing effect using the following equations (Nawy 2012):

$$F_b = -\frac{P}{Ac} + \frac{P.e.c}{Ig} - \frac{M.c}{Ig} \tag{2}$$

$$F_t = -\frac{P}{Ac} - \frac{P.e.c}{Ig} + \frac{M.c}{Ig} \tag{3}$$

Where "P" is prestressing force, "Ac" is the cross section area of beam, "e" is the distance between c.g of tendons and c.g of beam cross section, "c" is the distance from c.g of beam and the top/bottom face, and "Ig" is the gross moment of inertia.

Table1. Numerical analysis vs. theoretical equations (Nawy 2012).

	Finite Element	(Nawy 2012)	Error
TOP STRESSES	1.52	1.74	12.6%
BOTTOM STRESSES	-7.8	-9.33	16.4%

4.3 Impact Load and Mid Span Deflection

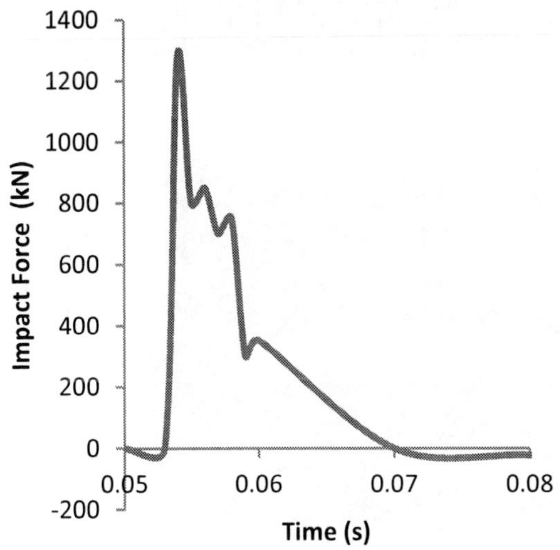

Figure 7. Equivalent impact load.

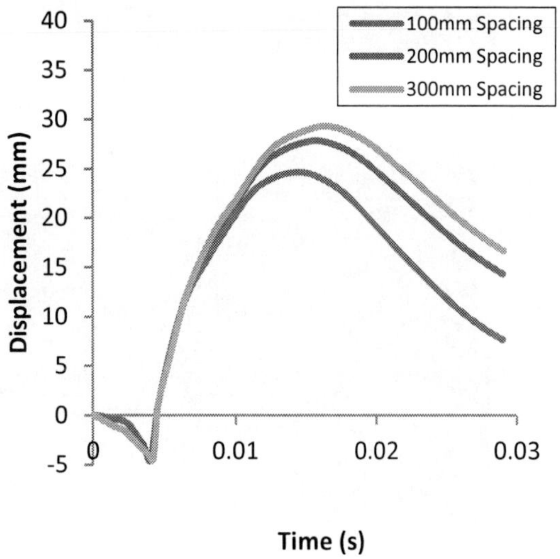

Figure 8. Displacement curve for all cases.

The equivalent peak load of the impact was 1278 KN as shown in Figure 7. As for mid span deflection curves (Figure 8), it can be divided into two zones: the first one is the cambering zone (with negative values) where cables were prestressed and the beam was deflected upwards. Then the test started and the drop weight hit the beam causing the deflection to go to the opposite direction. As can be seen the deflection was improved when using less shear reinforcement spacing (100mm).

4.4 Beam Damage

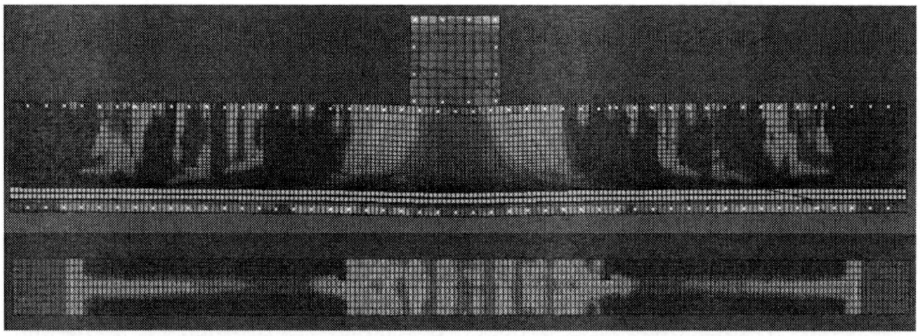

Figure 9. Tensile damage for 100mm spacing.

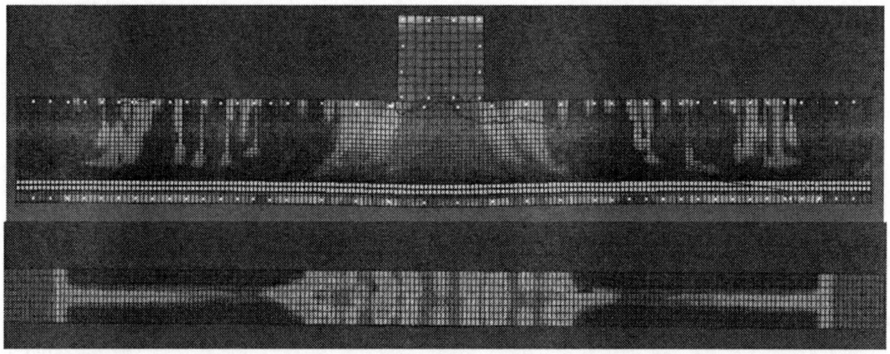

Figure 10. Tensile damage for 200mm spacing.

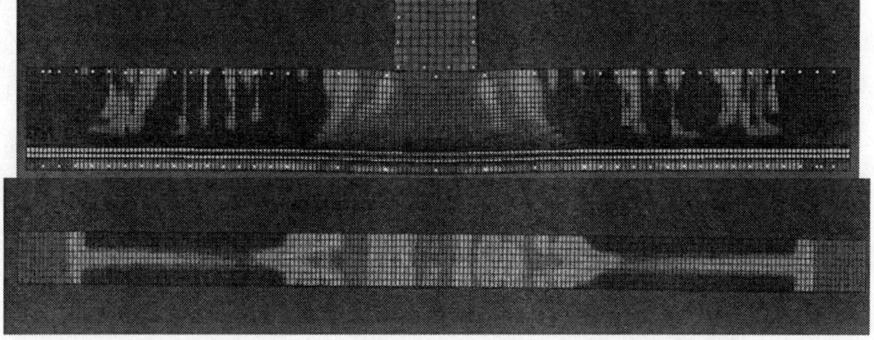

Figure 11. Tensile damage for 300mm spacing.

Table 2. Tensile damage zone length.

	Tensile Damage Zone		
Stirrups spacing (mm)	100	200	300
Zone length (m)	1.35	1.6	1.9

The Damage mode was local and concentrated at the impact point as can be seen from Figures 9 to 11, and a local punching occurred. As shown, using stirrups reduced the tensile damage in prestressed concrete beam which indicates that these reinforcements have a great role in resisting local punching.

5 CONCLUSION

The following conclusions can be drawn from the current investigation:

(i) Finite element modelling can successfully simulate prestressed concrete structures with all its stages starting from prestressing stage till load application stage. This type of simulations must consider the nonlinearity of materials in order to reach acceptable results, and to take into consideration the effect of impact loads on the mechanical properties of materials.

(ii) The analysis showed that prestressed concrete beams subjected to impact loads due to drop weight are likely to be damaged locally at the impact point in both tension and compression zones. However, by using shear reinforcement the local damage decreased, and that was shown in terms of tensile damage and mid-span deflections.

References

Chen, Y., and May, I. M., Reinforced Concrete Members under Drop Weight Impacts, *Structures and Buildings,* 162, 45-56, 2009.

Joshuva, N., Saibabu, S., Sakaria, P., Lakshmikandhan, K. N., and Sivakumar, P., Finite Element Analysis of Reinforced and Pre-Tensioned Concrete Beams, *International Journal of Emerging Technology and Advanced Engineering,* 4(10), 444-457, October, 2014.

Kang, T., and Huang, Y., 4th International Conference on Computer Modelling and Simulation, 41, 45, 2012.

Kumar, V., Iqbal, M., and Mittal, A., Behaviour of Prestressed Concrete Under Drop Impact Load, *11th International Symposium on Plasticity and Impact Mechanics, Implast 2016,* Elsevier, Procedia Engineering, 173, 403-408, 2016.

Temsah, Y., Jahami, A., Khatib, J., and Sonebi, M., Numerical Derivation of Iso-Damaged Curve for a Reinforced Concrete Beam Subjected to Blast Loading, *2nd International Congress on Materials & Structural Stability,* Rabat, Morocco, DOI:10.1051/matecconf/201714902016, November, 2017a.

Temsah, Y., Jahami, A., Khatib, J., and Sonebi, M., Numerical Analysis of a Reinforced Concrete Beam under Blast Loading, *2nd International Congress on Materials & Structural Stability,* Rabat, Morocco, DOI:10.1051/matecconf/201714902063, November, 2017b.

Responsible Design and Delivery of the Constructed Project
Edited by Abdul-Malak, M., Khoury, H., Singh, A., and Yazdani, S.

RAPID LOAD TESTING OF STONE COLUMNS TO AID DESIGN OF FOUNDATIONS ON IMPROVED GROUND

JOHN FAHD TOUMA, SALAH SADEK, and SHADI NAJJAR

Dept of Civil and Environmental Engineering, American University of Beirut, Beirut, Lebanon

Building Codes and best practice require load testing of embedded structural foundation elements to validate design and construction execution quality. This requirement is particularly challenging when associated with ground improvement schemes relying on granular reinforcing/stone columns. Stone columns present an economic solution for improving the bearing capacity of shallow foundations on soft soils. A novel impulse load test was developed and used to quantify the load-displacement response of shallow foundations supported on stone columns at a clay site. The device is referred to as the Rapid Plate Load Tester (RPLT) and is a modified version of the Axial Compressive Force Pulse test for deep foundations. In this paper, the comprehensive site investigation, stone columns construction, load testing procedure, and data analysis are described. Static and dynamic field tests were performed to target loads of 2000 kN and equivalent bearing pressures of 500kPa. The results obtained from the RPLT tests were used to derive equivalent static load settlement curves for footings on both the natural clay ground and improved ground and compared with the results obtained from the full-scale static load tests.

Keywords: Raft, Clay, Unloading point method, Shallow foundations, Static load test.

1 INTRODUCTION

Economic activity has been on a consistent upward trajectory in the Gulf Cooperation Council (GCC) states since 2004. The region has witnessed the construction of numerous new airports, large industrial zones with downstream processing plants, and waterfront reclamation for both low and high rise residential projects. Many of these large projects were undertaken in marginal coastal areas with poor ground conditions unsuitable for sustaining high structural loads without large associated settlements. Large scale ground improvement schemes have repeatedly proven to be a cost effective alternative to pile foundations for structures with working bearing pressures on the order of 200 kPa such as pipe racks, warehouses, and low-rise buildings. Ground improvement is typically executed by specialist contractors on a design build contract basis. Such contractors will base their design on a combination of recommendations available in the literature and their observations on the performance of past designs in local geologies. Acceptance criteria are generally specified in terms of a full scale static load test on a representative foundation, such as a two meter wide square footing, using a test method such as the American Standard for Testing & Materials (ASTM) D1194 (1994) for the Plate Load Test (PLT).

The high cost, technical and logistical complexity, and market availability of suppliers providing full scale load testing has led engineers to alternative test methods. Touma *et al.* (2016) described the development of the Rapid Plate Load Test (RPLT), a device to perform rapid load tests on footings, inspired by similar tests on piles as described in ASTM D7383 (2010). A schematic of the proposed RPLT is presented in Figure 1. The primary components are the steel footing, instrumentation anvil, guide frame, and an eight ton falling weight. The secondary components consist of acquisition devices and electronic instrumentation. The instrumentation consists of a photocell, reference laser, load cell, and accelerometers. The simplified procedure consisted of raising the falling weight within the confines of the guide frame to a designated height of up to 2m. The load pulse generated by the falling weight striking the anvil is measured by a load cell embedded within the anvil, and the vertical displacement of the steel footing is measured by a photocell and an accelerometer. Two additional accelerometers are fastened to the footing to jointly acquire acceleration and velocity. The acquired signals from the instrumentation are used to generate time history plots, from which a relationship between applied load and vertical settlement can be determined. A research program undertook several RPLT tests at a clay site in Lebanon. One of the outcomes of the research program was a comparison of load settlement curves obtained by static load test and RPLT.

The objective of this paper is to investigate the RPLT performance on ground improved with stone columns. For this purpose, stone column groups with different configurations were installed in the same test zone in which the tests reported in Touma *et al.* (2016) were conducted. One static load test and three dynamic RPLT tests were conducted using a steel grillage footing that was supported on clay reinforced with stone columns in three configurations providing an area replacement ratio ranging between 16.5 and 29%. In addition, a comprehensive field and laboratory investigation campaign that consisted of standard penetration tests (SPT, ASTM D1586-11), pressuremeter tests (PMT, ASTM D4719-07), and CU triaxial tests with pore pressure measurement was performed (ASTM D4767-11) to characterize the soil profile at the site. The results of the site investigation in addition to the results of the static and RPLT tests are presented in this paper.

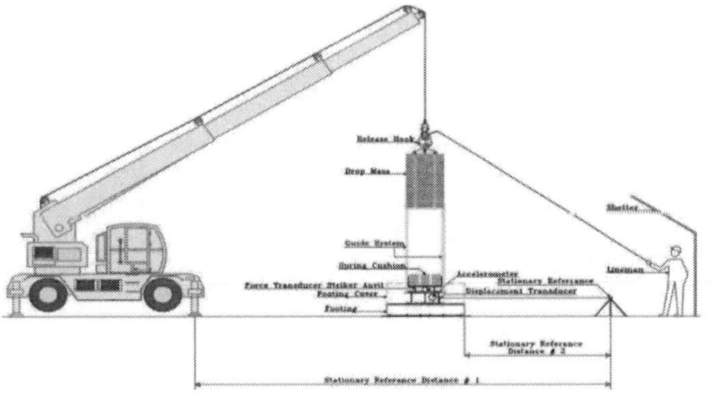

Figure 1. Schematic of the RPLT test device.

2 SITE INVESTIGATION

The investigation consisted of two boreholes and three test pits in the immediate vicinity of the test footing locations. The first borehole was used to conduct SPT tests and the second borehole was used to conduct pressure meter tests. Both boreholes were also used to retrieve undisturbed samples using thin-walled Shelby tubes. Data from two additional boreholes was also available in the site from previous testing. Test pits were excavated up to a depth of 3 m below NSL with the aid of an excavator so as to visually inspect the ground conditions up to a shallow depth. The sampling, preservation and transportation of the samples were carried out as per ASTM D4220-14. The disturbed overburden samples and the relatively undisturbed soil samples that were collected from the boreholes were labeled and preserved properly before transportation to the laboratory for testing. Undisturbed specimens were used as a basis for determining the shear strength parameters of the soil using triaxial testing. Field borehole logs were developed on the basis of the material encountered at the site, which was then confirmed with laboratory test results.

Groundwater was encountered at a depth of 2.5m below NGL at the time of the investigation. The investigation revealed generally uniform soil conditions across the site. Density and consistency were assessed using SPT tests and Atterberg limits and used in conjunction with the boring logs to determine the generalized soil profile. The top overburden soil comprises sandy Lean Clay/Fat Clay up to a depth of 7.5 m below the NSL. The top layer is underlain by Clayey Gravel with sand present up to a depth of 18 m below NSL. Below the gravel layer, a sandy Lean Clay layer is present up to the maximum investigated depth of 22 m below NSL. The results of the pressuremeter tests that were conducted at depths of 2.5m, 5.75m, and 7.75m are presented in Figure 2a. The pressuremeter results suggest an undrained shear strength ranging between 100 and 130 kPa, with an associated pressuremeter modulus ranging between 8 to 12 MPa. Similarly, sample results from CU triaxial tests that were conducted on specimens obtained from a depth of 3.0 to 3.8m are presented in Figure 2b. Other triaxial tests were conducted on specimens obtained from depths ranging from 1.5m to 5.0m, but their results are not shown for length limitations. The results of the triaxial tests indicated a response that is typical of a slightly overconsolidated to overconsolidated clay, with most of the tested specimens showing a strain-hardening behavior at large strains. This behavior was associated with the generation of negative pore water pressure as a result of the dilative tendency at larger strains. Positive pore water pressures were initially recorded at very small strains and were followed by a tendency for dilation. This dilative tendency was stronger in specimens obtained from depths ranging from (3.0 to 5.0m) compared to specimens obtained from shallower depths (1.5m to 3.0m). These results indicate that the clay at shallower depths may be less overconsolidated than the deeper clays which show larger undrained shear strength and stiffness values.

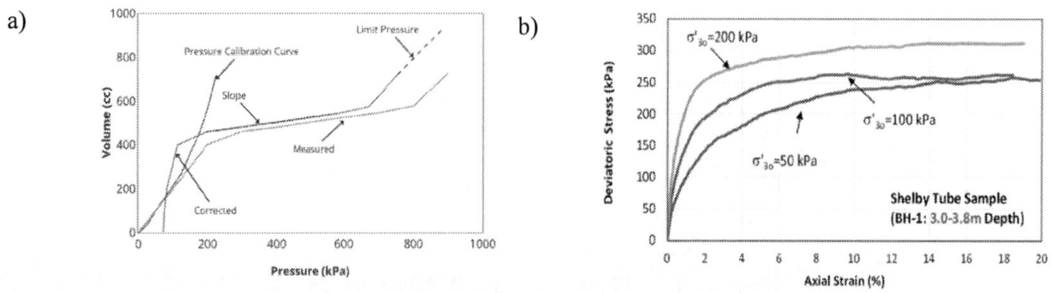

Figure 2. Representative results from a) Pressuremeter and b) CU triaxial test.

3 SITE PREPARATION AND STONE COLUMNS CONSTRUCTION

The research plot boundary was 18m x 18m and consisted of a grid of twelve locations intended for load tests. The site was cleared of vegetation, grubbed, and leveled. Stone columns installation used techniques similar to those used in installing Rammed Stone Columns, as described by Barksdale *et al.* (1983). A percussion drilling rig was used to excavate the 50-cm and 60-cm diameter columns. The excavated bore remained stable. After removal of the drill string, the depth of the excavation bottom was checked. The column was constructed in four to five lifts, each of approximately 0.25 m^3. A hopper with the measured lift volume was used in discharging gravel through a five inch diameter pipe to the bottom of the excavated hole. A fabricated steel mandrel was fastened to the rig chisel and repeated blows were used to compact the lift. Sample configurations are illustrated in Figure 3 showing the the two five column configurations and the three column configuration. The unit cell has been delineated in each configuration and the area replacement ratio is calculated as the ratio of the column area to the unit cell area. The baseline reinforcement ratio of 17.4% was tested by static load test, and using multiple and single cycle rapid load tests at an adjacent location. An over-reinforced location and an under-reinforced location, with area replacement ratios of 28.9% and 16.6%, respectively, were tested with multiple cycle rapid load tests. In total, eighteen stone columns were installed across four locations. The remaining locations within the testing grid were tested in an unreinforced state and were subjected to a selection of multiple and single cycle rapid load tests.

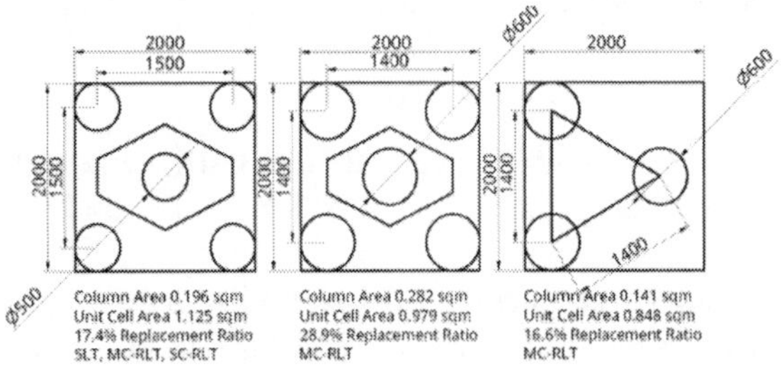

Figure 3. Column configuration at a) Location 2, b) Location 3, and c) Location 9 with dimensions in millimeters.

4 ASSESSMENT OF STATIC RESPONSE

Static load tests were performed on unimproved ground and on the improved footing locations described in Figure 3 in general accordance with ASTM D1194 (1994). A defining characteristic of the RPLT setup is that the grillage foundation and bottom plate can be fastened to either a reaction assembly for a static load test or to a drop hammer for a dynamic test, and the bottom plate in this instance had dimensions of 2m x 2m. Load was increased at a nominal increment of less than 200 kN and displacements were measured at the end of fifteen minute intervals. In the case of the test on stone columns, the footing continued to creep beyond the measurement time interval of fifteen minutes, and after one hour of maintaining the load, the footing was deemed to have failed as shown in Figure 4a. In contrast, the footing on unimproved ground settled beyond the limit of the load jack at approximately 1600 kN. A second cycle was performed which also displayed settlement beyond the fifteen minute measurement interval and the test was hastily terminated.

The load settlement curves for footings supported on unimproved and improved ground are presented in Figure 4b. Ultimate loads of 1610 kN and 1980 kN (pressures of 402.5 kPa and 495 kPa, respectively) were applied to the unimproved and improved footing locations. The curves exhibited an initial linear region that is characterized by a linearly proportional relationship between load and settlement. It is clearly evident that the load test pertaining to a footing on improved ground exhibits an initial stiffness that is improved in comparison to that of the unimproved ground. The linear region is followed by a nonlinear transition region and a final linear region, evidencing behavior described by Akbas and Kulhawy (2009). The tests were terminated due to continuing creep beyond the fifteen minute measurement interval.

a) b)

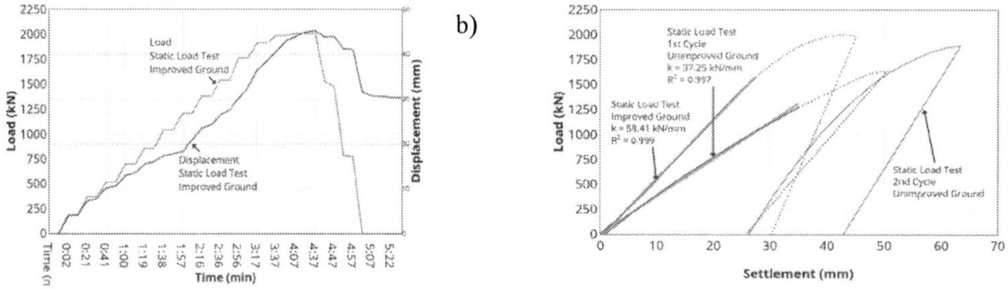

Figure 4. Static load tests results showing a) Variation of load and displacement with time and b) Load-settlement curves on unimproved & improved ground.

5 COMPARISON OF STATIC AND RPLT RESPONSE

RPLT tests were performed at locations adjacent to the static load test locations. A cushion comprised of alternating 25mm thick layers of Styrofoam and plywood was used at varying total thicknesses. The load-time history for any given blow was measured using a single load cell with a capacity of 5000 kN located in the center of the anvil. The acquired time histories are preprocessed by application of calibration constants, signal isolation, and noise filtering. Sample force and displacement time histories are shown in Figure 5a. ASTM D4945-00 mentions the need to monitor proportionality of force and displacement in the case of dynamic tests on piles. Dynamic load tests typically apply a force pulse duration of 10 to 30ms, whereas rapid load tests are performed over a longer period on the order of 50ms to 500ms. At these long durations, it is impossible to maintain proportionality indefinitely beyond the point of maximum displacement, and the loss of proportionality can be attributed to arrival of high frequency vibrations at the laser reference, and the tendency of accelerometers to drift when near stationary.

The Unloading Point Method (UPM), as described by Middendorp *et al.* (1992), was used in processing time histories and deriving an equivalent static load settlement curve from the RLT results. Multiple and single cycle RLT were performed. It was observed that the single cycle test methodology ensures that the result of the RLT corresponds to virgin soil that has not been disturbed by previous blows or loading cycles. Results on Figure 5b compare acquired load settlement curves (measured force versus settlement) with those obtained following application of the UPM method, and indicates an effective correlation level between the load settlement curve obtained by the static load test and the single cycle RLT curves (UPM).

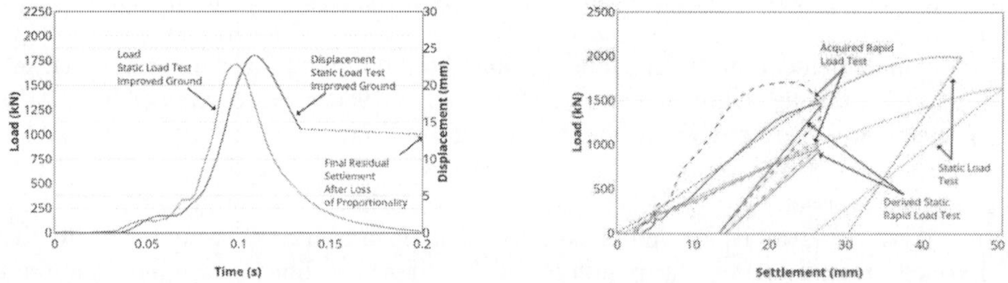

(a) (b)

Figure 5. Rapid load test results showing a) Load and displacement variation with time and b) Comparison of static and rapid load tests results before and after improvement with stone columns.

The fact that the RLT-derived curves follow the static curve is very promising, despite the fact that the RLT curves fell short of capturing the full spectrum of load. Future tests at the site will aim at increasing the loading energy by increasing the drop height to try and achieve higher derived static loads on the footing.

6 SUMMARY AND CONCLUSIONS

The results presented in this paper demonstrate the promise of rapid load testing as a technique to evaluate compliance of shallow foundations with settlement specifications of ground improvement programs. One to one comparisons between results from static load tests and rapid tests showed that the RPLT constitutes a feasible testing method that could supplement quality control measures for foundations on improved or unimproved ground.

Acknowledgements

This study was made possible by the collaboration and contribution of Engineering & Research International (ERI). The authors thank Dr. Fadlo Touma for his support and thoughts on the subject.

References

Akbas, S. O., and Kulhawy, F. H., Axial Compression of Footings in Cohesionless Soils. I: Load-Settlement Behavior, *Journal of Geotechnical and Geoenvironmental Engineering*, 135(11), 1562-1574, 2009.

ASTM D1194, Standard Test Methods for Bearing Capacity of Soil for Static Load and Spread Footings, *Annual Book of ASTM Standards*, ASTM International, Pennsylvania, 1994.

ASTM D7383, Standard Test Methods Piles Under Rapid Axial Compressive Load, *Annual Book of ASTM Standards*, ASTM International, West Conshohocken, Pennsylvania, 2010.

Barksdale, R. D., and Bachus, R. C., *Design and Construction of Stone Columns Volume I*, Federal Highway Administration, Washington, DC, 1983.

Middendorp, P., Bermingham, P., and Kuiper, B., Statnamic Load Testing of Foundation Piles, In *Proc. 4th Inter. Conference on the Application of Stress Wave Theory to Piles,* 1992.

Najjar, S. S., A State-of-the-Art Review of Stone/Sand-Column Reinforced Clay Systems, *Geotechnical and Geological Engineering*, 31(2), 355-386, 2013.

Touma, J., Sadek, S., and Najjar, S., Development of a Rapid Load Test for Shallow Foundations, In *Proc. 4th International Conference on New Developments in Soil Mechanics and Geotechnical Engineering*, Nicosia, North Cyprus, 2016.

Responsible Design and Delivery of the Constructed Project
Edited by Abdul-Malak, M., Khoury, H., Singh, A., and Yazdani, S.

NUMERICAL ANALYSIS OF COMBINED-SECTION STEEL COLUMNS

GEORGE ISKANDER[1], EZZELDIN SAYED-AHMED[2], and EMAM SOLIMAN[1]

[1]*Dept of Structural Engineering, Ain Shams University, Cairo, Egypt*
[2]*Dept of Construction Engineering, American University in Cairo, Cairo, Egypt*

Combined-section columns composed of two separate steel sections present an ideal and economic design for long columns subject to high values of bending moments and axial forces. Research dealing with design of these columns is currently insufficient. At present, most codes of practice consider each column's component to behave either separately or rigidly connected to the other components. As such, the main objective of this research is to scrutinize the behavior of laced-section columns, which are subjected to eccentric loading and propose design criteria for them. In this paper, a non-linear numerical model for these columns is developed based on the finite element method. The results of the numerical model are first verified against the outcomes of experimental investigations available in literature. Then, the model is adopted to simulate the behavior and the capacity of the combined laced columns. The numerical model includes both the geometric and materials nonlinearities along with the effect of initial imperfections.

Keywords: Built-up columns, Laced columns, Combined steel column, Numerical modelling.

1 INTRODUCTION

Combined columns (also referred as built-up columns) are columns, which consist of two or more main chords connected by lacing bars or batten plates (Figure 1(a)). Main chord sections may be UPN sections, IPE sections, angles, ...etc. There are many configurations for the main chord including the used sections type and the way of their arrangement as shown in Figure 1(b).

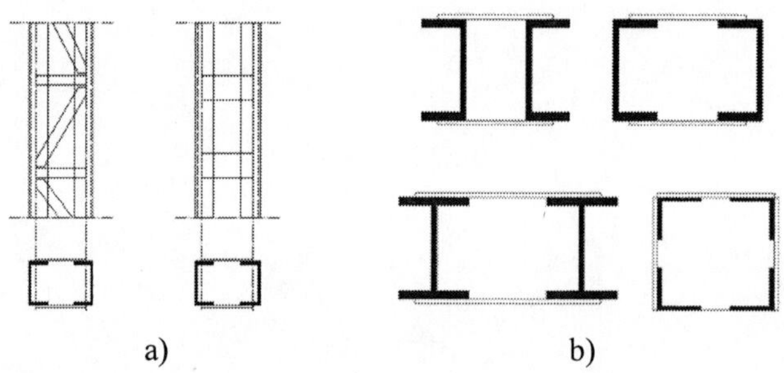

Figure 1. Built up columns (a) and sections (b).

The main advantage of using such column is the large moment of inertia resulting from the distance separating between the centroids of the chords, which decreases the slenderness ratio and increases both axial force and bending moment capacity. Usage of combined sections steel columns presents an ideal and economic design for long columns subject to high values of bending moments and axial forces. Research dealing with design of these columns is currently insufficient with the currently available ones reported elsewhere (Iskander 2018).

2 CURRENT DESIGN APPROACH FOR COMBINED COLUMNS

2.1 Eurocode, EN 1993-1-1 (2005)

EN 1993-1-1 (2005) proposes Eq. (1) and (2) for designing combined columns subject to axial compressive forces as well as bending moments in order to check the capacity of the critical chord subject to compressive force due to the effect of the axial force and the bending moment;

$$N_{ch,\,Ed} = 0.5\,N_{Ed} + \frac{M_{Ed}\,h_o\,A_{ch}}{2I_{eff}} \tag{1}$$

Where,

$$M_{Ed} = \frac{N_{Ed}\,e_o + M^I_{Ed}}{1 - \frac{N_{Ed}}{N_{cr}} - \frac{N_{Ed}}{S_v}} \tag{2}$$

The verification check for buckling should be performed for the chord with critical compression force using Eq. (3) which is commonly used for any compression member.

$$\frac{N_{ch,Ed}}{N_{b,Ed}} \leq 1 \tag{3}$$

where, N_{ed} and M_{ed} are the external axial force and bending moment, h_o is the distance between chords centroids, A_{ch} is the chord area, I_{eff} is the effective moment of inertia, e_o is the initial imperfection, N_{cr} is the Euler buckling load, $N_{b,Ed}$ is the chord axial capacity and S_v is the shear stiffness of the column. This method is discussed in detail elsewhere (Sayed-Ahmed and ElSerwi 2017).

2.2 AISC 360-10 (2010)

AISC 360-10 (2010) recommends that the slenderness ratio of single chord between the lacing bars about its least radius of gyration should not exceed 75% of the slenderness ratio of the column as a whole. To take account for shear deformation Eq. (4) is used to calculate modified slenderness ratio;

$$\left(\frac{KL}{r}\right)_m = \sqrt{\left(\frac{KL}{r}\right)_0^2 + \left(\frac{a}{r_i}\right)^2} \tag{4}$$

where, $(KL/r)_0$ is the slenderness ratio of the column as a single unit, a is the distance between laced points and r_i is the radius of gyration of the chord about its minor axis.

2.3 CSA-S16-14 (2014)

CSA-S16-14 (2014) recommends that the slenderness ratio of each combined column's chord about its least radius of gyration should not exceed that of the built-up member. The compressive resistance of the column should be based on:

1. The slenderness ratio of the built-up column about appropriate axis when buckling mode does not involve relative deformations to produce shear force in lacings.
2. The equivalent slenderness ratio about axis orthogonal to that of item (1) when buckling mode involves relative deformations that produce shear force in lacings as shown in Eq. (5):

$$\rho_e=\sqrt{\rho_o^2+\rho_i^2} \tag{5}$$

where ρ_o is the slenderness ratio of the column as a single unit and ρ_i is the local slenderness ratio of the chord about its minor axis.

3 NUMERICAL MODELLING OF COMBINED STEEL COLUMNS

To predict the failure load of combined (built-up) columns, a non-liner finite element model is developed. The commercial program (ABAQUS) is used as a platform for processing the model solution of the FE equations and post processing the model results. The model assembly, element type, material model, boundary conditions, load assignment, constrains and the mesh size are used as follows:

3.1 Model Assembly

Laced columns with different configuration for lacing members are numerically modelled using the proposed FE model. The lacing members are configured as shown in Figure 2 as W-, WH- and X-type. 8-node shell element is used to model both the chords and the laced members.

3.2 Material Model

Steel S235 of 235 MPa engineering yield stress and 360 MPa ultimate engineering stress is adopted for the column chords and the lacing members. The steel material is modeled as an elasto-plastic material with true stress-strain curve following the DNVGL-RP-C208 (2013). The modulus of elasticity used is $2.1\text{x}10^5$ MPa and Poisson's ratio is 0.3 for the elastic part. For the plastic part, the provided data from the DNVGL is adopted based on the element thickness. Figure 3 schematically shows the adopted stress-strain curve for the steel.

3.3 Boundary Conditions

To model the hinged-hinged W- and X-type columns with uniform bending moment over its length, the following boundary conditions are used:

- For the upper end of the column, all nodes of the chords are connected to a reference point with a rigid body constraint of tie type, the reference point is located at a certain distance from the geometric center to model the eccentric load (Figure 4). Then, a displacement restraint is assigned to the reference point to model the hinge; it sets the displacement in X, Z direction and rotation about Y-axis to zero.

- Only half of the column length is modelled to make use of symmetry as such plane of symmetry is used at the column mid-height as the lower boundary condition. For WH-type columns, both the upper and lower boundary conditions are modeled as the upper boundary condition of W and X type columns, since full column must be modeled as there is no plane of symmetry at column's mid-height.

3.4 Load Assignment

The load is assigned to the eccentric reference point with different eccentricities to model different M/N ratio as shown is (Figure 5). The load is given a value of unity and RIKS algorithm is adopted to increase it step by step: RIKS algorithm is used for the iterative process of solving the equation of the non-linear FE model.

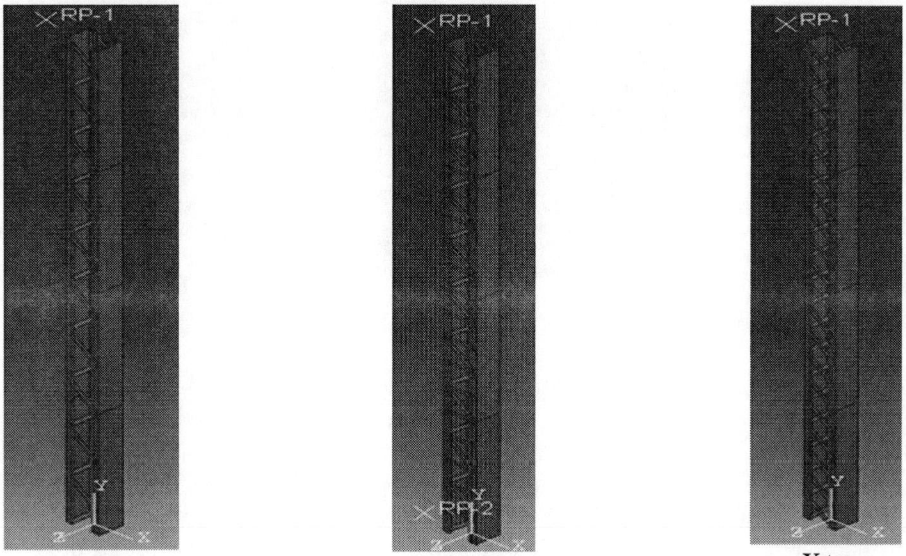

W type | WH type | X type

Figure 2. Model assemblies.

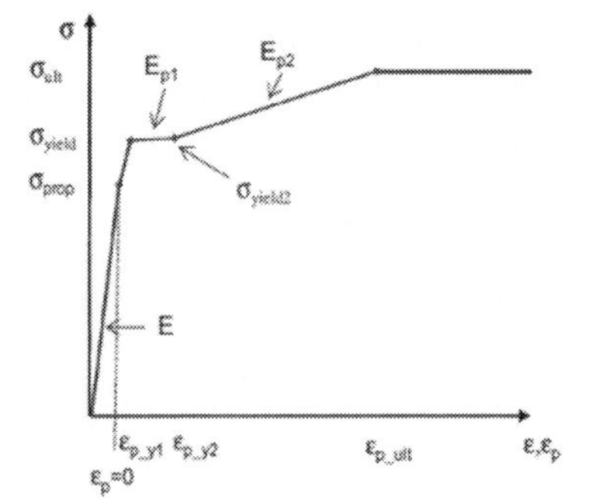

Figure 3. Stress strain curve.

Figure 4. Upper nodes constraint.

3.5 Lacing Bars- Chords Connection

To model the hinged connection between the lacing bars and the chord, coupling constrain is used with the degrees of freedom to be constrained: U1, U2, U3, UR1, UR2, the chord and lacing geometries are sketched to create a point of attachment to be used by the constraint (Figure 6).

4 MODEL VERIFICATION

The numerical model is verified with the results of experimental investigation program obtained by Kalochairetis *et al.* (2014). Their research program consisted of an experimental investigation and numerical analysis performed on 5 groups of laced built-up beam-columns. The current verification is done on three of these groups, which are close to the model used in this research.

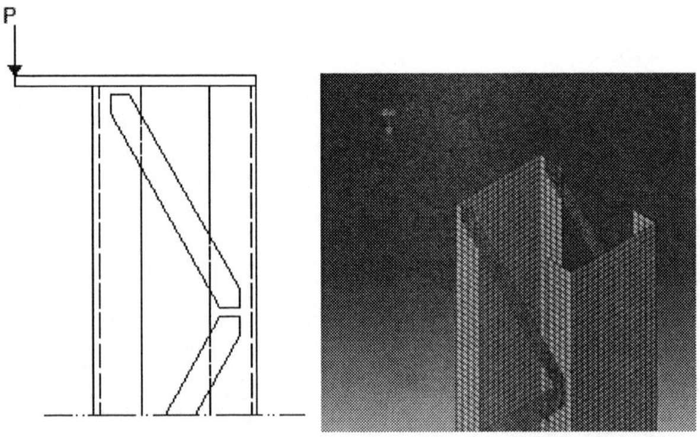

Figure 5. Load assignment.

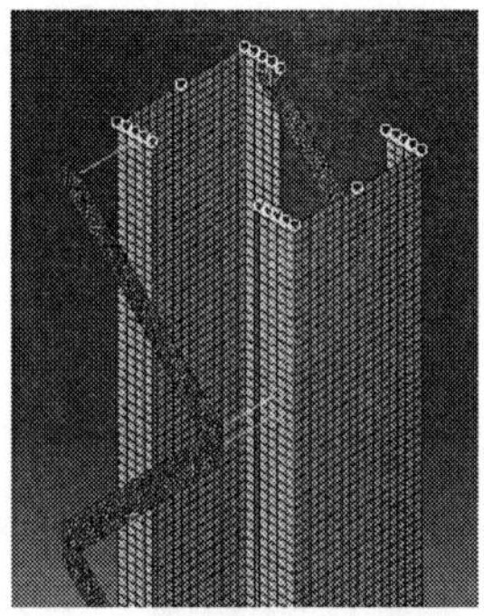

Figure 6. Lacing-chord constraint.

The considered groups are 2000 mm long built-up beam-columns with two UPN 60 chords and lacing members composed of 25×25×3 angles, the columns are divided into five parts with panel length of 400 mm. The total length of the column including the upper and lower pin supports is 2300 mm. The top/bottom eccentricities of the pin supports are 100/100 (same side), 100/80 (different sides) and 50/50 (same side) for groups 1, 4 and 5, respectively.

Load-displacement curves for mid height of column (Groups 1 and 5) resulting from both the experimental investigation and numerical model are plotted in Figures 7 and 8. Further, the load-displacement curves for the first and last panel for Group 4 are plotted in Figures 9 and 10.

Table 1 summarize the results of the considered experimental data and compare it to the outcomes of the numerical model. Both the stiffness and the capacity of the columns are listed in this table with the ratio between the experimental program and the numerical analysis results.

The verification of the three groups shows that the developed model predicts the load, displacements and failure mode in an acceptable way and the modeling technique could be reliable to perform the intended parametric study using it.

Table 1. Results of the numerical model verification analysis.

Group number	P_{FEA} (kN)	P_{exp}(kN)	K_{FEA} (kN/m)		K_{exp} (kN/m)	P_{FEA}/P_{exp}	K_{FEA}/K_{exp}
1	196	200	50.8		38	0.98	1.34
4	255	220	261(upper)		240	1.16	1.09
5	255.5	241.5	100.4		93	1.06	1.08

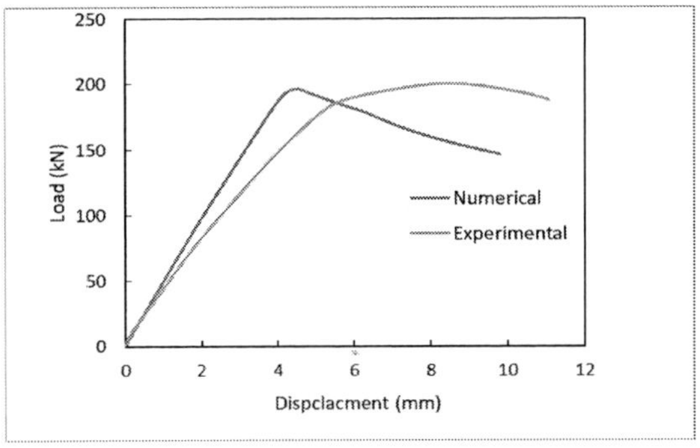

Figure 7. Load-displacement curve for Group 1.

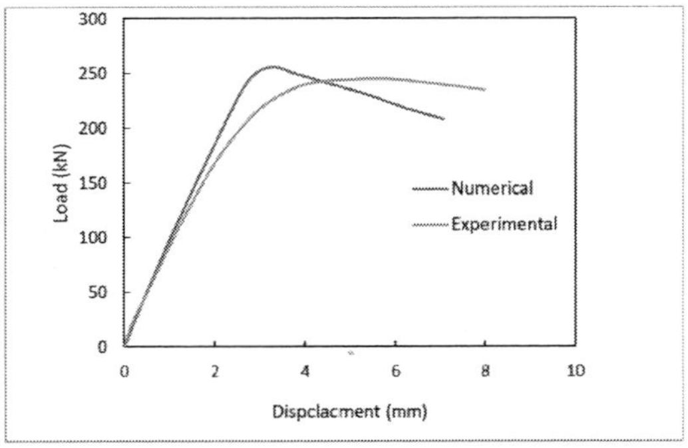

Figure 8. Load-displacement curve for Group 5.

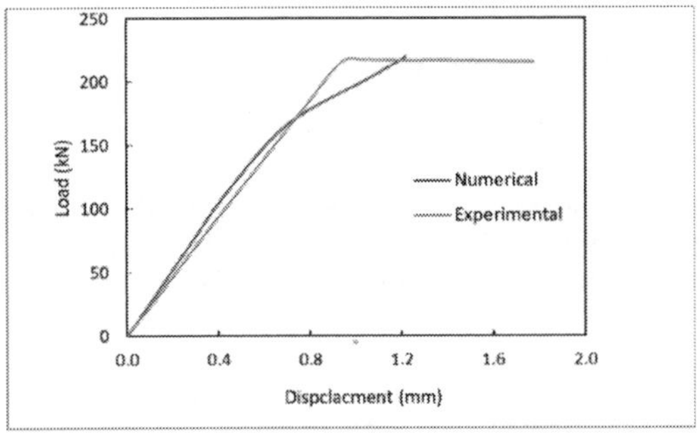

Figure 9. Load-displacement the upper panel for Group 4.

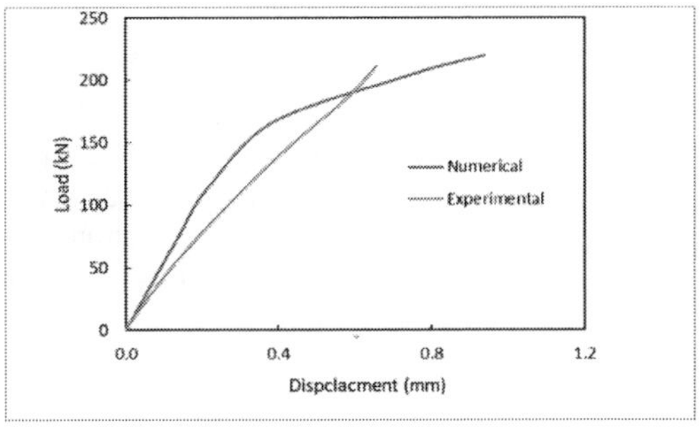

Figure 10. Load-displacement the lower panel for Group 4.

5 CONCLUSIONS

A nonlinear numerical model for combined-section steel columns is built-up and verified against experimental results available from literature. The model can successfully predict the failure load of such columns when subject to eccentric loading. The model will be used to perform a further parametric study and develop procedures to design these columns.

References

AISC 360-10, American Standards for Designing Steel Structures, 2010.

CSA S16-14, Canadian Standards for Designing Steel Structures, 2014.

DNVGL-RP-C208, Determination of Structural Capacity by Non-linear FE Analysis Methods, 2016.

EN 1993-1-1, Eurocode 3: Design of Steel Structures-Part 1-1: General Rules and Rules for Buildings, European Committee for Standardization, 2005.

Kalochairetis, K. E., Gantes, C. J., and Lignos, X. A., Experimental and Numerical Investigation of Eccentrically Loaded Laced Built-Up Steel Columns, *Journal of Constructional Steel Research,* 101 (1), 66-81, April, 2014.

Iskander, G., *Behaviour and Design of Combined Section Steel Beam Columns*, MSc Thesis, Faculty of Engineering, Ain Shams University, Cairo, Egypt, 2018.

Sayed-Ahmed, E. Y., and Elserwi, A. A., *Limit States Design of Steel Structures,* LAP Lambert Academic Publishing, 2017.

Responsible Design and Delivery of the Constructed Project
Edited by Abdul-Malak, M., Khoury, H., Singh, A., and Yazdani, S.

STRUCTURAL HEALTH MONITORING OF CORROSION POTENTIAL IN CONCRETE BRIDGE DECKS

HANI NASSIF[1], CHAEKUK NA[1], HASAN AL-NAWADI[1], ADI ABU-OBEIDAH[1], and WILLIAM WILSON[2]

[1]*Rutgers Infrastructure Monitoring and Evaluation Group, Rutgers University, Piscataway, USA*
[2] *New Jersey Turnpike Authority, Woodbridge, USA*

Structural Health Monitoring (SHM) of concrete structures during construction, as well as over its service life, has recently become more attractive to owners and consulting engineers. With the introduction of new materials and construction methods, various types of concrete structures are being instrumented with monitoring devices to determine their performance and response to various loading conditions. Among many other objectives, this includes monitoring concrete performance at the serviceability and durability limit states. This paper is an overview of an on-going program for the SHM of concrete bridge decks in the State of New Jersey focusing on field performance. Three types of corrosion sensors are instrumented to monitor the corrosion activities in concrete decks; one is the silver-silver chloride electrode and the other two are multi element probe (MEP) corrosion sensors. Other types of MEPs were also instrumented on bridge decks during reconstruction in late 1990s to monitor the corrosion potential of the bridge decks. Various types of sensors are installed in precast panels during fabrication as well as in-situ cast concrete decks during and after construction. Moreover, a laboratory-based accelerated corrosion testing program is also performed on concrete specimens using various types of rebars. This ongoing study is aimed at correlating laboratory-accelerated corrosion results with long-term performance of the steel in concrete bridge decks under field conditions.

Keywords: Accelerated corrosion testing, Silver-silver chloride electrode, Multi element probe.

1 INTRODUCTION

SHM is mainly applied to evaluate the structural long-term behavior of bridges using various types of sensors (Nassif *et al.* 2003). It can determine the extent of structural deterioration and provide its estimated remaining service life under current environmental conditions. Various factors are involved in the deterioration process of bridges such as material-related factors (i.e., permeability, chloride ingress, and corrosive reinforcement) and load-related factors (i.e., overweight and load-induced fatigue cracks). Among these factors, load induced deterioration has been studied and the phenomenon is well established (Nassif *et al.* 2016). However, corrosion of steel is not generally monitored by SHM protocol because it takes longer time than other monitoring scheme. However, it is of prime importance to monitor the corrosion potential in order to understand the long-term structural behavior as well as to estimate its service life.

Corrosion is a complex phenomenon related to material, structure type, environment, maintenance, etc., but it generally occurs when the reinforcement is exposed to chlorides. Numerous studies have been made to understand the corrosion rate and the on-set time (Arya and Ofori-Darko 2013, Gowripalan *et al.* 2000). Many studies were focusing on the electrochemical process of corrosion and its parameters to predict the corrosion rate and initiation based on results from laboratory testing. When the chloride ions start to ingress within the concrete cover, the reinforcing steel forms a thin layer of its oxide (Bazant 1979). While the thin layer resists the ingress of chloride ions to the steel reinforcement, it is passive and no corrosion will occur. As the concentration of the chloride ions increases, the layer is compromised, and the corrosion is initiated. There are additional factors that also control the corrosion process such as PH, mass transfer and diffusivity of ions and molecules such as Oxygen and ferric hydroxide, and pore water viscosity (ACI 222R-01, 2001).

The ultimate goal of this ongoing study is to find the best model to correlate laboratory-accelerated corrosion results with long-term performance of the steel reinforcement in concrete bridge decks under field conditions. The authors in this paper are presenting and discussing the corrosion monitoring results of four (4) concrete bridges in New Jersey.

2 FIELD STRUCTURAL HEALTH MONITORING (SHM) PLAN

2.1 Corrosion Monitoring Sensors

Two types of corrosion sensors were used; (1) silver-silver chloride electrode (herein Ag electrode) and (2) multi element probe (MEP).

2.1.1 *Silver-silver chloride electrode*

The Ag electrode is a permanent passive electrode used to measure the potential difference between electrode and reinforcement. Figure 1(a) shows the Ag electrode after installation adjacent to a steel rebar. When the chloride ion ingresses in the concrete, the Ag electrode, counter lead, and the concrete works as a battery. It generates a small voltage reading, and the corrosion risk can be determined. Higher chloride ion results in higher voltage readings and therefore higher corrosion potential. Low voltage (< 300 mV) indicates passive corrosion (or not corrosion), high voltage (> 600 mV) indicates active corrosion, and 300~600 mV indicates 50% possibility of corrosion.

2.1.2 *Multi element probe (MEP)*

The MEP measures the corrosion rate as well as on-set of corrosion using different probes. Two types of MEP were used for this study; one is ECI-2 sensor by Virginia Tech., Inc. and the other is CPMP-2 produced by Corrosion Service. Figure 1(b) and (c) shows both MEPs; on the left (b) is ECI-2 and on the right (c) is CPMP-2. ECI-2 MEP integrates of black steel electrode, manganese dioxide (MnO_2) reference electrode, stainless steel base counter electrode, 4 strips of stainless steel (SS) conductive sensors, and Ag electrode. ECI-2 MEP is capable of monitoring the linear polarization resistance (LPR), open circuit potential (OCR), resistivity (R), and chloride ion concentration. The corrosion potential is determined using Table 1 depending on the readings from the MEP. For example, for an LPR over 10 kOhm-cm^2, which is considered high, while the OCP and Chloride content is low (e.g., less than 300 mV) this indicates that there is no corrosion activities. On the other hand, when LPR is less than 10 kOhm-cm^2 and the OCP and Chloride content is high (e.g., more than 600 mV) this indicates high probability of corrosion activity.

Figure 1. Installation example of Ag electrode and two MEPs (ECI-2 and CPMP-2).

Table 1. Application table to judge the corrosion rate.

LPR	OCP	Resistivity	Chloride	Corrosion possibility
Low	Low	Low	Low	High likely
Medium	Low	High	High	Likely
High	Low	Low	Low	Unlikely
High	High	High	High	Unlikely

2.2 Corrosion Monitoring Bridges

2.2.1 *Newark Bay Bridge*

The Newark Bay Bridge (NBB) is a major corridor that connects Newark and Bayonne, NJ. It was originally constructed in 1956 as a part of the New Jersey Turnpike, and the bridge was reconstructed in early 2010 using precast panels to minimize any traffic-induced cracking. This bridge has two types of spans (see Figure 2) – main truss span and floorbeam spans. Two types of rebars were used for the precast panels; the majority of the precast panels were made with epoxy-coated rebars and limited number of panels were made with duplex stainless steel (SS). Ag electrodes and one ECI-2 MEP were instrumented in selected precast panels during casting the concrete in 2012. High-performance concrete (HPC) was used for the entire bridge.

Figure 2. Newark Bay Bridge and typical sensor layout in the precast deck panels.

2.2.2 *Patcong Creek Bridge*

The Patcong Creek Bridge is located on the Garden State Parkway which underwent major widening and rehabilitation that was completed in 2012. The HPC and duplex SS were used for the entire bridge as it is located near the eastern coast ,very close to sea shore. A number of Ag electrodes were installed in multiple spans and one ECI-2 MEP was installed as shown in Figure 3 (a). The corrosion monitoring started in 2013.

2.2.3 *Bridge No. 84.1*

This bridge was reconstructed in 1998 using HPC and galvanized steel. The CPMP-2 MEPS were installed on entire span as shown in Figure 3 (b). The corrosion potential was monitored for the first 2 years by the NJTA engineers, and then by the authors followed up in 2016.

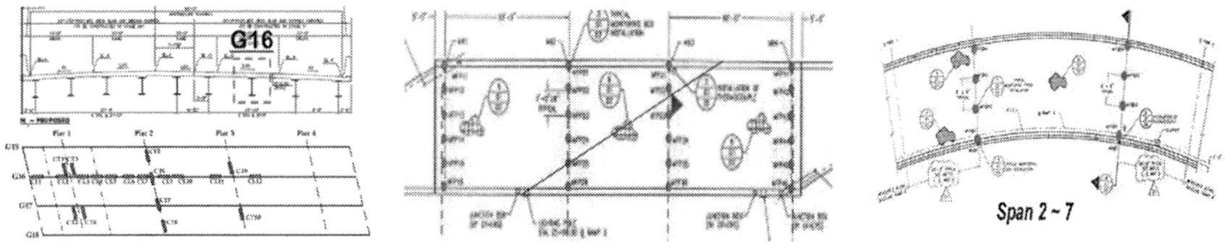

(a) (b) (c)

Figure 3. Sensor layout of Patcong Creek Bridge, Bridge No. 84.1 and Bridge No. 159.7A.

2.2.4 *Bridge No. 159.7A*

This bridge was reconstructed in 1999 using HPC and duplex SS for the new bridge deck. The same engineering consultants instrumented a number of MEPs (CPMP-2) on all spans of the bridge as shown in Figure 3 (c). The corrosion potential was monitored for the first 2 years by the NJTA engineers, and then by the authors since 2016.

3 FIELD MONITORING RESULTS

3.1.1 *Newark Bay Bridge (HPC with stainless steel and epoxy rebar)*

The monitoring has been performed over 5 years since 2013. Figure 4 presents the corrosion potential monitoring results of Ag electrode for two precast panels. Panel 1 has the SS, and Panel 2 has the epoxy rebar. Figure 4 shows that there is no corrosion activity in SS Panel over 5 years, however, there is some corrosion activity in epoxy rebar Panel from year 3 and the corroding area is getting large. Figure 5 depicts the monitoring results of ECI-2 MEP in Precast Panel 1. Based on Table 1, the results confirm that it is unlikely to corrode the stainless steel in Precast Panel 1.

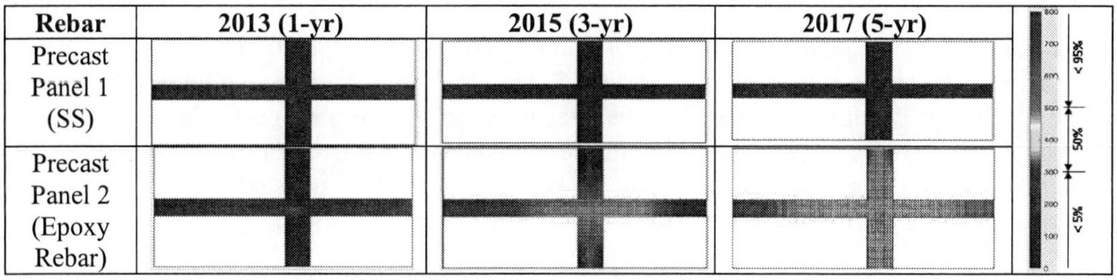

Figure 4. Newark Bay Bridge monitoring results 1 – Ag electrode.

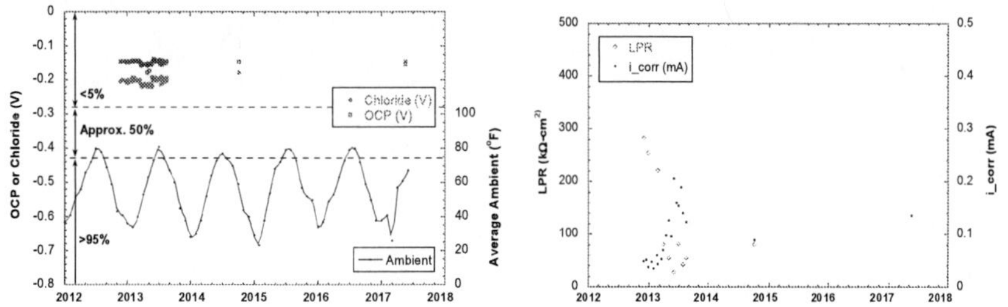

Figure 5. Newark Bay Bridge monitoring results 2 – ECI-2 MEP.

3.1.2 *Patcong Creek Bridge (HPC with stainless steel)*

The corrosion potential has been monitored over 5 years since 2013. Figure 6 and Figure 7 represent the corrosion monitoring results for Ag electrodes and ECI-2 MEP. Similar results as the Precast Panel 1 with SS were observed; no corrosion activity is monitored over 5 years.

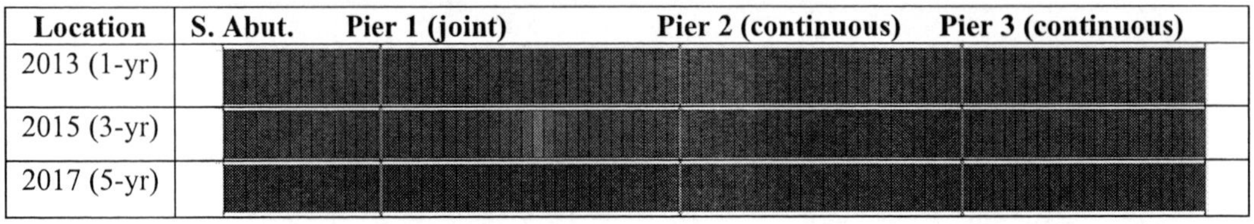

Figure 6. Patcong Creek Bridge monitoring results 1 – Ag electrode.

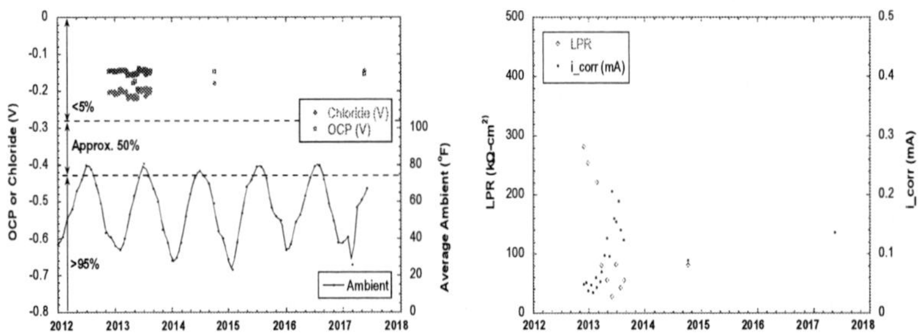

Figure 7. Patcong Creek Bridge monitoring results 2 – ECI-2 MEP.

3.1.3 *Bridge No. 84.1 (HPC with galvanized rebar)*

The bridge was initially monitored in 1999 and more frequently since 2016. Figure 8 shows the data for the corrosion potential and corrosion rate as measured by CPMP-2 MEP at Bridge No. 84.1. After 18 years, some areas (see Figure 8 near locations ① and ②) show some active corrosion in both corrosion potential and rate measurements; Location ① is in the left lane where the major traffic flows, and location ② is in the negative moment region of the bridge. These results indicate that the galvanized steel is in the phase of corrosion initiated within 18 years, and therefore a high probability that the outside Zinc coating layer is undergoing corrosion activity.

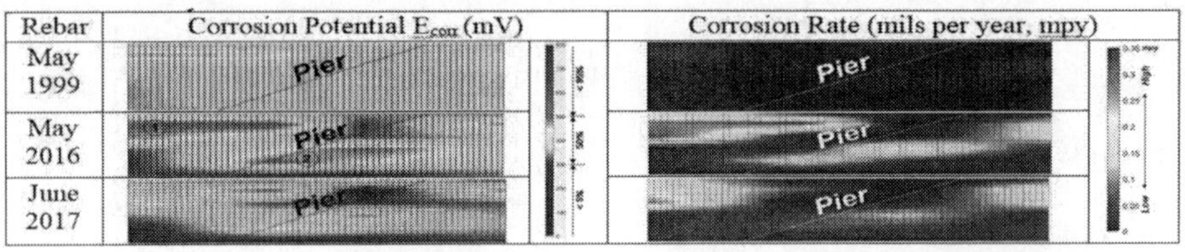

Figure 8. Bridge No. 84.1 monitoring results – CPMP-2 MEP.

3.1.4 *Bridge No. 159.7A (HPC with stainless steel)*

The ramp bridge No. 159.7A was initially monitored in November 1999, and more frequently since 2016. Figure 9 shows the corrosion monitoring results measured by the MEP. Corrosion potential results show that some locations had a 50% chance of corrosion since 1999. However, the data on corrosion rates show no corrosion or a very low corrosion rate. Therefore, it can be deduced that the SS withstands the corrosion activity after 18 years of service life.

Rebar	Corrosion Potential E_{corr} (mV)	Corrosion Rate (mils per year, mpy)
Nov 1999	Span 8 | Span 7 | Span 6 | Span 5 | Span 4 | Span 3 | Span 2 | Span 1 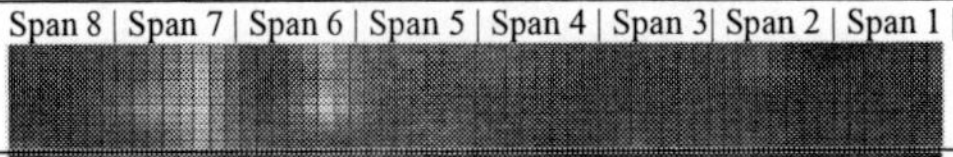	Span 8 | Span 7 | Span 6 | Span 5 | Span 4 | Span 3 | Span 2 | Span 1 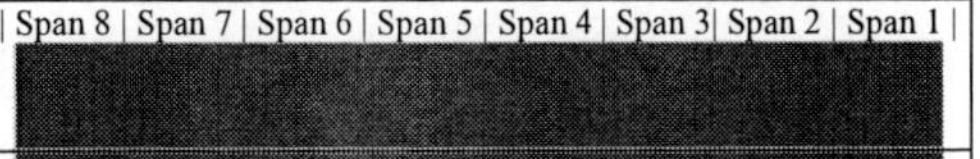
May 2016		
June 2017		

Figure 9. Bridge No. 159.7 monitoring results – CPMP-2 MEP.

4 SUMMARY AND CONCLUSIONS

Based on the field monitoring results, the following conclusions can be made:

1. The Ag electrode can provide the corrosion potential near the reinforcement, while the MEP can provide the corrosion onset as well as the corrosion rate.
2. The field monitoring of the Newark Bay Bridge indicates that the epoxy-coated rebar is not effective in mitigating the corrosion while the SS rebar can withstand the corroding activity. The precast panel with epoxy-coated rebar shows 50% of corrosion probability in 3 years, while the precast panel with SS shows no corrosion in 5 years. The monitoring results of the Patcong Creek Bridge also confirms the effectiveness of SS in resisting corrosion.
3. The monitoring results of Bridge No. 84.1 show that the corrosion onset and high corrosion rate were observed after 18 years for galvanized steel. For Bridge No. 159.7A, no active corrosion was observed after 18 years of service life for SS.
4. There is a need to correlate Lab-based accelerated corrosion results with field monitoring data to help establish these protocols.

Acknowledgments

The authors would like to acknowledge the New Jersey Turnpike Authority for their financial and logistical support for the RIME team during the instrumentation of the sensors and monitoring the bridges.

References

ACI 222R-01, *Protection of Metals in Concrete Against Corrosion*, ACI Committee 222, 2001.

Arya, C., and Ofori-Darko, F. K., Influence of Crack Frequency on Reinforcement Corrosion in Concrete, *Cement and Concrete Research,* Elsevier, 26(3), 345-353, March 1996.

Bazant, Z. P., Physical Model for Steel Corrosion in Concrete Sea Structures, *Journal of the Structural Division,* ASCE, 105(6), 1155-1166, June 1979.

Gowripalan, N., Sirivivatnanon, V., and Lim, C. C., Chloride Diffusivity of Concrete Cracked in Flexure, *Cement and Concrete Research,* Elsevier, 30(5), 725-730, May 2000.

Nassif, H., Suksawang, N., Gindy, M., Abu-Amra, T., and Davis, J., *Instrumentation and Field Testing of the Doremus Avenue Bridge*, Transportation Research Board, Proceedings of the 82nd Annual Meetings, Washington, D.C., 12-17 January 2003.

Nassif, H., Ozbay, K., Wang, H., Noland, R., Lou, P., Demiroluk, S., Su, D., Na, C.K., Zhao, J., and Beltran, M., *Impact of Freight on Highway Infrastructure in New Jersey*, FHWA-NJ-2016-004, NJDOT, 2016.

Responsible Design and Delivery of the Constructed Project
Edited by Abdul-Malak, M., Khoury, H., Singh, A., and Yazdani, S.

AN EFFECTIVE TECHNIQUE UTILIZING FRCM SYSTEM IN STRENGTHENING RC BEAMS IN SHEAR

TADESSE WAKJIRA and USAMA EBEAD

Dept of Civil and Architectural Engineering, College of Engineering, Qatar University, Doha, Qatar

This paper presents the efficacy of a new technique of fabric reinforced cementitious matrix (FRCM) for the strengthening of shear deficient reinforced concrete (RC) beams. This technique involves embedding the FRCM composites in the concrete cover and is referred to as "near surface embedded" FRCM (NSE-FRCM) technique. Five (5) medium scale rectangular RC beams of width × depth × length of 150 × 330 × 2100 mm were constructed and tested under displacement controlled with three-point loading. One beam was kept unstrengthen as a control specimen whereas the other four beams were strengthened with different types and configurations of the NSE-FRCM system. The test parameters were: (a) geometric configuration (intermittent strips of NSE-FRCM versus full NSE-FRCM plate), and (b) fabric types (carbon versus glass fabrics). The test results indicated that the NSE-FRCM technique can successfully be used to significantly enhance the shear capacity of the strengthened beams. The strengthened specimens exhibited an average enhancement in the shear capacity of 62% over the unstrengthen beam. The full NSE-FRCM plate showed higher enhancement in the shear strength compared to that for the intermittent NSE-FRCM configuration. Moreover, the specimens strengthened with carbon FRCM showed higher shear capacity compared to that with glass FRCM counterparts.

Keywords: Rehabilitation, Near surface embedded, Reinforced concrete beams, Fabric reinforced cementitious matrix, NSE-FRCM.

1 INTRODUCTION

In the construction industry, structural rehabilitation is used to restore the capacity of the deteriorated structures and increase their life span. Several structural strengthening methods and materials have been reported in the literature including ferrocement (Ebead 2015), steel plates (Ebead and Marzouk 2002), and fiber reinforced polymer (FRP) (Aly *et al.* 2006, Ebead and Saeed 2014). The latter has widely been used because of its wide range of advantages over the traditional strengthening methods that utilize steel and concrete (Kotynia *et al.* 2008). FRP composites can be applied to existing structural members in different ways; e.g. FRP wraps, externally bonded FRP (Panigrahi *et al.* 2014), mechanically fastened (Elsayed *et al.* 2009a, 2009b), hybrid mechanically fastened and externally bonded FRP (Ebead and Saeed 2014), and near surface mounted FRP (Lorenzis and Nanni 2002). However, the incompatibility of the FRP composites with the concrete substrate becomes a concern. Moreover, FRP composites have poor performance at high temperatures (Raoof and Bournas 2017). Lately, fabric reinforced cementitious matrix (FRCM) composite was introduced as a viable strengthening material that overcomes some of the problems associated with FRP composites (Shrestha *et al.* 2017, Younis *et al.* 2017b). Existing literature showed the successful application of externally bonded FRCM system for flexural and shear strengthening of RC beams (Ebead *et al.* 2017, Elghazy *et al.* 2017, Younis *et al.* 2017a, 2017c), flexural strengthening of RC slabs (Koutas and Bournas 2017), and strengthening of RC columns (Ludovico *et al.* 2010). However, the externally bonded FRCM system is characterized by debonding of FRCM composites from the concrete surface that limits the utilization of the FRCM composite to its full strength. As a remedy to this problem, the authors have recently introduced the hybrid "near surface embedded/ externally bonded" NSE/EB-FRCM technique that has the potential to reduce debonding failure mode (Wakjira and Ebead 2018). With this new technique larger number of FRCM layers can be used with less chance of debonding failure.

Thus, the paper aimed at introducing the NSE-FRCM technique for strengthening of RC beams deficient in shear. The experimental program involved construction and testing of five (5) medium scale RC beams. The test results have been discussed in terms of the shear capacity, deformational characteristics and failure modes. It has been observed that the use of the near surface embedded FRCM system (NSE-FRCM) as a stand-alone technique has the potential to mitigate FRCM/ concrete debonding.

2 EXPERIMENTAL PROGRAM

2.1 Material Properties

Ready-mixed concrete of the same batch, with an average compressive strength of 30 MPa, was used to cast the test beams. Internal reinforcement involved 16 mm diameter bars used as the main tensile reinforcement and 8 mm diameter bars used as the compressive reinforcement and ties. The longitudinal tensile reinforcement bars had an average yield strength of 595 MPa and yield strain of 2660 $\mu\varepsilon$ while the compressive and transverse reinforcement bars had an average yield strength of 535 MPa and yield strain of 2580 $\mu\varepsilon$.

Two different types of fabrics (carbon and glass) were used along with their respective mortar types to form the two types of FRCM systems used in the NSE-FRCM strengthening system. In both fabric types, the textile roving was aligned in two orthogonal directions. For carbon fabrics, both the weft and warp roving are spaced at 10 mm and each roving has a cross-sectional area of 0.047 mm^2 in both directions. For glass fabrics, the warp and weft roving are spaced at 14 mm and 18 mm spacing, respectively. Each roving has a cross-sectional area of 0.066 mm^2 and 0.047 mm^2 in the longitudinal and transverse directions, respectively. Carbon and glass fabrics had respective tensile strength of 4.8 GPa and 2.6 GPa and elastic modulus of 240 GPa and 80 GPa. The carbon and glass fabrics had ultimate elongations of 1.8% and 3.25%, respectively.

2.2 Test Specimens and Test Matrix

The study was carried out on five (5) rectangular RC beams with 150 mm in width and 330 mm in depth. Figures 1a and 1b show the beam dimensions and reinforcement details. One beam was kept unstrengthen to act as a reference specimen whereas the other four beams were strengthened with NSE-FRCM of different geometric configuration.

Table 1 summarizes the test matrix. Two different test parameters have been investigated:

- FRCM type: two different types of FRCM composites were utilized; viz., glass (G) FRCM and carbon (C) FRCM.
- Strengthening configuration: full versus intermittent NSE-FRCM strips. The latter's configuration is 120 mm wide strips spaced at 95 mm within the critical shear span.

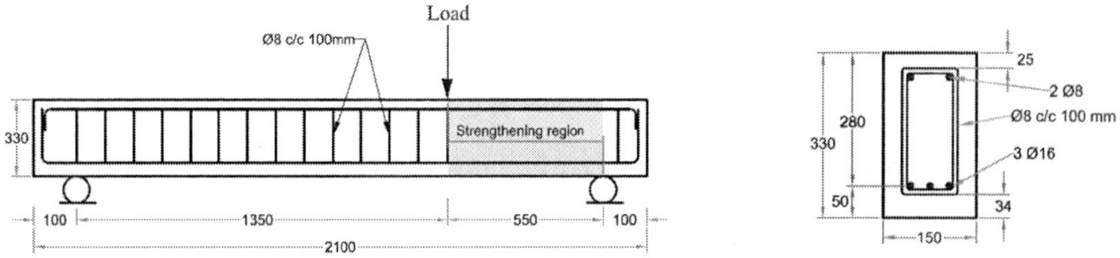

(a) Longitudinal detail (b) Mid-span cross-setion

Figure 1. Typical beam detail.

The NSE-FRCM technique utilized two layers of fabrics in the FRCM composite. The strengthening system involved: (a) preparation of the grooves, (b) application of first layer of the mortar followed by installation of first layer of fabrics, (c) application of second layer of mortar followed by installation of second layer of fabrics and a finishing mortar layer.

Table 1. Test matrix.

Specimen ID	Type of fabric	NSE-FRCM configuration
Control	-	-
C-Full	Carbon	Full
C-Intermittent	Carbon	Intermittent
G-Full	Glass	Full
G-Intermittent	Glass	Intermittent

2.3 Test Setup

The test was executed using Instron static universal testing machine under three-point loading. Linear variable displacement transducers were used to monitor the vertical displacement of the beam under the loading point. Moreover, concrete and steel strains were monitored using strain gauges. The steel strain gauges had 5% strain limit and 5 mm gauge length whereas the concrete strain gauge had a maximum strain limit of 0.2% and 60 mm gauge length.

3 TEST RESULTS

The test results are summarized in Table 2 in terms of the maximum load (P_u), deflection at the maximum load (Δ_u), flexural tensile strain at the maximum load (ε_u), and failure modes.

Table 2. Summary of test results.

Specimen ID	P_u (kN)	Gain in P_u (%)	Δ_u (mm)	Gain in Δ_u (%)	ε_u (%)	Failure mode
Control	104	-	3.25	-	1425	Shear
C-Full	184	77	6.48	99	2711	Shear
C-Intermittent	176	69	5.16	59	1787	Shear
G-Full	174	67	5.98	84	2426	Shear
G-Intermittent	139	34	4.53	39	1762	Shear

3.1. Shear Capacity and Failure Modes

The second and third columns of Table 2 list the load capacity and its percentage increase for the strengthened specimens relative to that of the control specimen, respectively. The NSE-FRCM strengthening system significantly increased the shear capacity of the strengthened beams with a value that ranged from 34% to 77% over the control beam. This result indicates a successful use of the NSE-FRCM system for the strengthening of shear deficient RC beams. The effectiveness of the NSE-FRCM strengthening system varied based on the test parameters.

The C-FRCM system performed better than G-FRCM counterpart in enhancing the strength of the specimens. The highest gain in the shear capacity was recorded for full configuration of carbon NSE-FRCM strengthened beam. This specimen failed at an ultimate load of 184 kN that corresponds to 77% gain in the shear capacity relative to the control specimen as listed in Table 2. Glass NSE-FRCM counterpart of the same specimen failed at an ultimate load of 174 kN that corresponds to 67% gain in shear capacity as listed in Table 2.

With regard to the NSE-FRCM configuration, full configuration of NSE-FRCM showed, as intuitively expected, higher gain in the shear capacity than intermittent NSE-FRCM counterparts of Table 2. For instance, in glass FRCM specimens, the gain in the shear capacity for full NSE-FRCM plate was almost double than the shear capacity for intermittent NSE-FRCM counterpart.

All specimens exhibited a shear failure mode caused by propagation of shear cracks within the critical shear span as listed in Column 8 of Table 2. Generally, no signs of debonding of FRCM system from the substrate concrete were observed demonstrating an effective implementation of the NSE-FRCM technique for mitigating the deboning failure.

3.2. Deformational Characteristics

Figures 2a and 2b show the load versus deflection plots for carbon and glass FRCM, respectively. Moreover, the fourth and fifth columns of Table 2 list the deflection at the maximum load, Δ_u, and the percentage increase in the deflection of the strengthened specimens relative to the control beam, respectively. As can be seen in Figures 2a and 2b, the NSE-FRCM specimens showed higher deflection at the peak load compared to that for the control specimen (Δ_u= 3.25 mm). The highest gain in Δ_u was recorded for carbon FRCM specimen strengthened with full NSE-FRCM plate. Intermittent configuration of the same FRCM type exhibited 59% gain in Δ_u that was 40% lower than the value for the full configuration counterpart. Similarly, full configuration of glass NSE-FRCM plate exhibited 45% higher gain in Δ_u compared to the intermittent configuration counterpart.

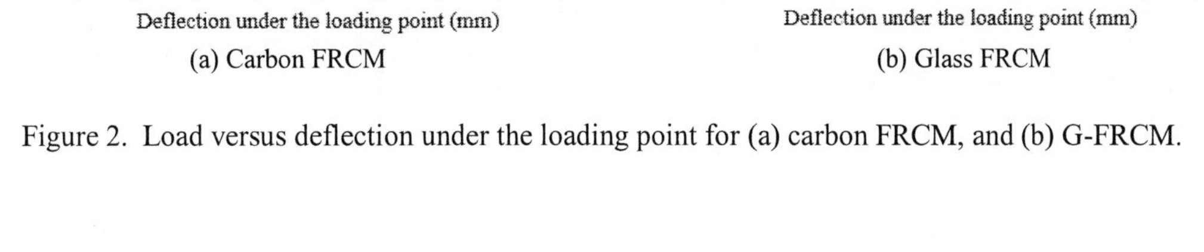

(a) Carbon FRCM

(b) Glass FRCM

Figure 2. Load versus deflection under the loading point for (a) carbon FRCM, and (b) G-FRCM.

As of the FRCM type, carbon FRCM specimens showed higher deflection at the peak load compared to that for glass FRCM counterparts.

With regard to the strain values, all specimens failed in shear before yielding of the longitudinal reinforcement ($\varepsilon_{t,u} < 2660\ \mu\varepsilon$) with an exception of Specimen C-Full as listed in the sixth column of Table 2. Furthermore, the NSE-FRCM technique has increased the strain in flexural reinforcement indicating that brittle shear failure has been delayed for more tensile strains in the tensile bars.

4 CONCLUSION

This paper presented a new form of FRCM strengthening system referred to as NSE-FRCM technique that has the potential to mitigate the debonding failure of FRCM off the substrate concrete. The efficacy of the NSE-FRCM system has been discussed in terms of the increase in the shear strength, deformational characteristics and failure mechanism. Generally, no debonding of FRCM system has been observed attributed to the enhanced FRCM/concrete bond. This result indicated that the NSE-FRCM technique can successfully be used to mitigate the debonding failure mode, which is the governing mode of failure in externally bonded FRCM technique. Prevention of this failure mode leads to better utilization of the FRCM composite. An overall average gain in the shear capacity of 62% has been observed. Carbon FRCM showed higher gains in the shear capacity (73% on average) compared to those of glass FRCM counterpart (51% on average). As expected, specimens strengthened with full configuration of NSE-FRCM showed higher shear capacity compared to that for the intermittent configuration counterparts indicating significance of FRCM continuity.

Acknowledgments

This paper was made possible by Qatar University grant # QUST-CENG-SPR-14/15-15 from Qatar University. The findings achieved herein are solely the responsibility of the authors.

References

Aly, R., Benmokrane, B., and Ebead, U., Tensile Lap Splicing of Bundled CFRP Reinforcing Bars in Concrete, *Journal of Composites for Construction*, 10(4), 287–294, 2006.

Ebead, U., Inexpensive Strengthening Technique for Partially Loaded Reinforced Concrete Beams: Experimental Study, *Journal of Materials in Civil Engineering*, (10), 2015.

Ebead, U., and Marzouk, H., Strengthening of Two-Way Slabs Using Steel Plates, *ACI Structural Journal*, 99(6), 2002.

Ebead, U., and Saeed, H., Flexural and Interfacial Behavior of Externally Bonded/ Mechanically Fastened Fiber-Reinforced Polymer-Strengthened Reinforced Concrete Beams, *ACI Structural Journal*, 111(4), 741–751, 2014.

Ebead, U., Shrestha, K., Afzal, M., Refai, A., and Nanni, A., Effectiveness of Fabric-Reinforced Cementitious Matrix in Strengthening Reinforced Concrete Beams, *Journal of Composites for Construction*, 21(2), 4016084, 2017.

Elghazy, M., Refai, A., Ebead, U., and Nanni, A., Effect of Corrosion Damage on the Flexural Performance of RC Beams Strengthened with FRCM Composites, *Composite Structures*, Elsevier Ltd, 180, 994–1006, 2017.

Elsayed, W., Ebead, U., and Neale, K., Studies on Mechanically Fastened Fiber-Reinforced Polymer Strengthening Systems, *ACI Structural Journal*, 106(1), 49–59, 2009a.

Kotynia, R., Abdel Baky, H., Neale, K., and Ebead, U., Flexural Strengthening of RC Beams with Externally Bonded CFRP Systems: Test Results and 3D Nonlinear FE Analysis, *Journal of Composites for Construction*, 12(2), 190–201, 2008.

Koutas, L., and Bournas, D., Flexural Strengthening of Two-Way RC Slabs with Textile-Reinforced Mortar : Experimental Investigation and Design Equations, *Journal of Composites for Construction*, 21(1), 1–11, 2017.

Lorenzis, L., and Nanni, A., Bond Between Near Surface Mounted FRP Rods and Concrete in Structural Strengthening, *ACI Structures Journal*, 99(2), 123–133, 2002.

Ludovico, M., Prota, A., and Manfredi, G., Structural Upgrade Using Basalt Fibers for Concrete Confinement, *Journal of Composites for ConstructionConstruction*, 14(5), 541–552, 2010.

Panigrahi, A. K., Biswal, K. C., and Barik, M. R., Strengthening of Shear Deficient RC T-beams with Externally Bonded GFRP Sheets, *Construction and Building Materials*, Elsevier Ltd, 57, 81–91, 2014.

Raoof, S., and Bournas, D., TRM versus FRP in Flexural Strengthening of RC Beams : Behaviour at High Temperatures, *Construction and Building Materials*, The Authors, 154, 424–437, 2017.

Shrestha, K. C., Ebead, U., and Younis, A., Effect of Surface Roughening on Concrete/TRM Bond, *Proceedings of the Ninth International Structural Engineering and Construction Conference, Resilient Structures and Sustainable Construction*, ISEC Press, Valencia, Spain, St--43, 2017.

Wakjira, T., and Ebead, U., Hybrid NSE/EB Technique for Shear Strengthening of Reinforced Concrete Beams Using FRCM: Experimental study, *Construction and Building Materials*, 164, 164–177, 2018.

Younis, A., Ebead, U., and Shrestha, K., Different FRCM Systems for Shear-Strengthening of Reinforced Concrete Beams, *Construction and Building Materials*, 153, 514–526, 2017a.

Younis, A., Ebead, U., and Shrestha, K. C., Tensile Characterization of Textile Reinforced Mortar, *Proceedings of the Ninth International Structural Engineering and Construction Conference, Resilient Structures and Sustainable Construction*, ISEC Press, Valencia, Spain, St--28, 2017b.

Younis, A., Ebead, U., and Shrestha, K. C., FRCM Shear Strengthening for Concrete Beams, *Proceedings of the Ninth International Structural Engineering and Construction Conference, Resilient Structures and Sustainable Construction*, ISEC Press, Valencia, Spain, St--27, 2017c.

Responsible Design and Delivery of the Constructed Project
Edited by Abdul-Malak, M., Khoury, H., Singh, A., and Yazdani, S.

EFFECTIVENESS OF FRCM SYSTEM IN STRENGTHENING RC BEAMS IN SHEAR USING HYBRID NSE/EB TECHNIQUE

USAMA EBEAD and TADESSE WAKJIRA

Dept of Civil and Architectural Engineering, College of Engineering, Qatar University, Doha, Qatar

This paper presents an experimental study on the efficacy of a recently introduced hybrid "near surface embedded/ externally bonded" NSE/EB fabric reinforced cementitious matrix (FRCM). Seven medium scale RC beams, deficient in shear, were constructed and tested under three-point loading. Six beams were strengthened using hybrid NSE/EB-FRCM technique and one beam was left un-strengthened as a reference. Two main test variables were considered; viz., strengthening configuration: full versus intermittent strengthening configuration and FRCM type: glass FRCM versus polyparaphenylene benzobisoxazole (PBO). The test results revealed that the hybrid NSE/EB-FRCM strengthening technique is effective in enhancing the shear capacity of the strengthened beams. The strengthened specimens showed an average of 67% higher shear capacity relative to the reference specimen. With regard to the FRCM type, G-FRCM strengthened specimens showed an average of 73% enhancement in the shear capacity, which was 12% higher than the value for PBO-FRCM counterpart. Moreover, the strengthened beams showed higher deflection at the ultimate load than that of the reference specimen. Furthermore, the hybrid NSE/EB-FRCM technique has shown to be a promising alternative to the conventionally used externally bonded FRCM system with the potential to reduce the debonding of FRCM from the concrete substrate.

Keywords: Rehabilitation, Near surface embedded/externally bonded, Fabric reinforced cementitious matrix (FRCM), Reinforce concrete (RC) beam.

1 INTRODUCTION

As a result of structural deterioration caused by different factors (e.g. corrosion of steel bars, end of life, change in service load, and severe environmental conditions), there is a need for developing strengthening methods that can increase their capacity hence extend their life time (Younis *et al.* 2017a). Different methods have been used for the strengthening of the deteriorated structural members. Some of these methods include steel plates (Ebead and Marzouk 2002a, 2002b), ferrocement (Ebead 2015), fiber reinforced polymer (FRP) composites (Ebead and Marzouk 2004, Ebead and Saeed 2014, 2017, Elsayed *et al.* 2009, Kotynia *et al.* 2008), and fabric reinforced cementitious matrix (Elghazy *et al.* 2017, Pino *et al.* 2017, Younis *et al.* 2017b). FRP composites are widely used as a strengthening material due to their favorable behavior over traditional strengthening system. Some of the advantages of FRP composites include the ease and speed of application, resistance to corrosion, high strength to weight ratio, and light weight (Aly *et al.* 2006).

Despite their relative efficiency to the traditional strengthening method FRP composites possess some drawbacks; e.g., incompatibility with the concrete substrate, susceptibility to failure at high temperature and difficulty to apply on wet surface (Raoof and Bournas 2017). To overcome some of these problems, cement-based FRCM composites have recently been developed by replacing the epoxy adhesive with cementitious matrix and fibres with textile fabrics (Trapko *et al.* 2015). FRCM has better adherence with the substrate concrete. In addition to this, FRCM composites possess good resistance to elevated temperature and fire (Raoof and Bournas 2017, Tetta and Bournas 2016). FRCM composites have been used for the strengthening of beams both in flexure and shear (Ombres 2015, Tetta *et al.* 2015, Younis *et al.* 2017c) by externally bonding the material on the surface of the concrete. However, externally bonded FRCM (EB-FRCM) technique is associated with debonding of FRCM off the substrate concrete limiting the utilization of the material (Panigrahi *et al.* 2014, Wakjira and Ebead 2018, Younis *et al.* 2017a). As a remedy to this problem, the authors introduced hybrid "near surface embedded/externally bonded FRCM" technique as a viable alternative to EB-FRCM technique.

Thus, the paper aimed at studying the efficacy of NSE/EB-FRCM technique for the shear strengthening of RC beams. The experimental program involved the construction and testing of seven medium scale RC beams, deficient in shear. The test results are discussed in terms of the shear capacity, deformational characteristics, and failure modes.

2 EXPERIMENTAL PROGRAM

2.1 Material Properties

The beam specimens were cast using ready-mixed concrete with an average compressive strength of 30 MPa. The internal flexural tensile reinforcement used 16 mm diameter steel bars while 8 mm diameter steel bars were used as longitudinal compressive reinforcement and stirrups. Table 1 provides the average mechanical properties of the steel reinforcement.

Table 1. Mechanical properties of the reinforcement steel bars.

Bar diameter (mm)	Yield strength (MPa)	Yield strain (με)	Modulus of elasticity (GPa)
8	535	2580	207
16	595	2660	224

The strengthening technique used two different fabric types; namely, glass and PBO fabric with their associated mortar type as recommended by the manufacturer. The geometric and mechanical properties of the fabrics are summarized in Table 2.

Table 2. Geometric and mechanical properties of the fabrics.

Fabric type	Area per width in the warp direction (mm^2/mm)	Area per width weft direction (mm^2/mm)	Modulus of elasticity (MPa)	Tensile stress (MPa)	Ultimate strain (με)
Glass	0.047	0.066	80000	2600	32500
PBO	0.045	0.0155	270000	5800	21500

2.2 Test Beam

Six specimens were strengthened with different configuration of NSE/EB-FRCM technique, while one beam without strengthening was used as a reference. The dimension and reinforcement details of the reference specimen is shown in Figures 1a and 1b that show the longitudinal detail and cross-sectional detail of the beam, respectively.

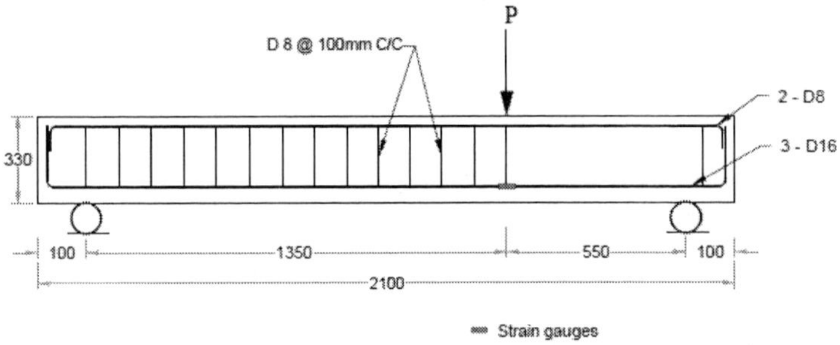

(a)

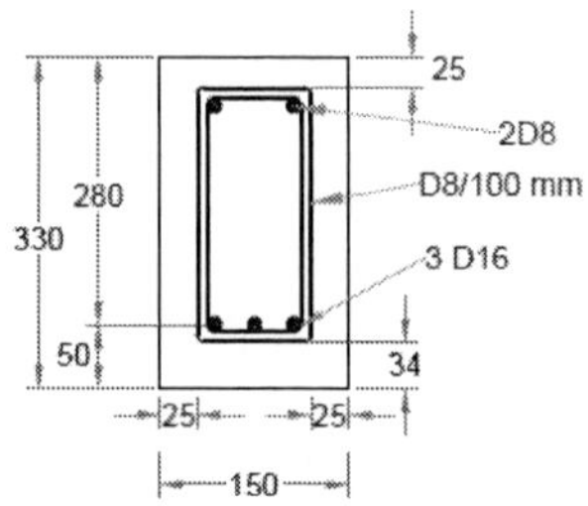

(b)

Figure 1. Specimen detail: (a) longitudinal detail and (b) cross-sectional detail outside the shear span (all dimensions are in mm).

The NSE/EB-FRCM technique utilized two layers of near surface embedded (NSE) FRCM and two layers of externally bonded FRCM system. The strengthening system involved: (a) preparation of the grooves (for the NSE part), and (b) application of first layer of the mortar followed by installation of first, second, third and fourth layer of fabrics by keeping an average of 3 mm thickness of mortar between each layers of fabrics. Finally, surface finishing was done using a trowel and the specimens were cured for a minimum of 28 days prior to testing.

Table 3 lists the test matrix with two main test variables being studied; viz., (a) fabric type: glass versus PBO fabric, and (b) strengthening configuration: both full and intermittent configuration was used for both NSE part and EB part. Intermittent FRCM configuration involved strips of 120 mm wide at 95 mm spacing. The specimen nomenclature for the strengthened beams includes three letters in which the first letter represents the fabric type, the second and last letter represent the strengthening scheme for NSE and EB FRCM part, respectively as listed in Table 3.

Table 3: Test matrix.

Specimen Identifier	Type of fabric	Strengthening configuration	
		NSE part	**EB part**
*R	-	-	-
PFF	PBO	Full	Full
PIF	PBO	Intermittent	Full
PII	PBO	Intermittent	Intermittent
GFF	Glass	Full	Full
GIF	Glass	Intermittent	Full
GII	Glass	Intermittent	Intermittent

*R-reference beam

2.3 Test Setup and Instrumentation

The experimental work was carried out under three-point loading as a simply supported system. The displacement under the loading point was measured using linear variable displacement transducers (LVDT). The strain in the concrete was measured using concrete strain gauge with maximum strain limit of 0.2% and gauge length of 60 mm. Moreover, a strain gauge with 5% strain limit and 5 mm length was used to monitor the strain in the tensile reinforcement bars.

3 RESULTS AND DISCUSSION

Table 4 summarized the test results in terms of the maximum load capacity (P_u), deflection at the peak load (Δ_u), and strain values.

Table 4. Experimental test result summary.

Specimen Identifier	P_u (kN)	Gain in P_u (%)	Δ_u (mm)	Increase in Δ_u (%)	$\varepsilon_{s,u}$ (%)	$\varepsilon_{c,u}$ (%)
R	104	-	3.25	-	0.14	-
PFF	183	76	6.41	97	0.25	0.18
PIF	170	63	6.3	94	0.19	0.13
PII	150	45	5.38	66	0.22	0.13
GFF	187	80	7.44	129	0.26	0.16
GIF	185	78	6.24	92	0.21	0.12
GII	168	61	5.71	76	0.22	0.2

3.1. Shear Capacity

The hybrid NSE/EB-FRCM system has significantly enhanced the shear capacity of the strengthened beams relative to that of the reference specimen (R). The gain in the shear capacity was provided in the third column of Table 4. The mean gain in the shear capacity was 67%, which shows a successful implementation of the hybrid NSE/EB-FRCM technique for strengthening of shear deficient RC beams. The maximum enhancement in the shear capacity was observed in Specimen GFF strengthened with glass NSE/EB-FRCM involving full configuration for both NSE and EB part.

Generally, glass FRCM showed higher enhancement in the shear capacity than that for the PBO FRCM counterparts as listed in Table 4. For instance, the gain in the shear capacity for Specimen GIF was 78%, which was 15% higher than its PBO-FRCM counterpart, PIF (63%) as listed Table 4. Moreover, the full NSE/EB-FRCM configuration showed a higher shear capacity than that for the intermittent FRCM configuration. For instance, the gain in the shear capacity for Specimen GFF, which was strengthened with full NSE/EB-FRCM for both NSE and EB parts, was 76% as listed in Table 4. Altering the NSE configuration to intermittent strips lowered the gain in the shear capacity to 63%. This gain was further reduced to 45% for Specimen PII, strengthened with intermittent configuration for both NSE and EB parts. These results confirmed the significance of the amount of FRCM system in the shear span.

3.2. Deformational Characteristics

The deflection under the loading point at the maximum load (Δ_u) was given in the fourth column of Table 4 while the fifth column of the same table provides the increase in Δ_u relative to the reference beam. The average increase in deflection at the peak load was 99% for the G-FRCM specimen while it was 85% for the PBO-FRCM specimens relative to the reference beam. With regard to the strengthening scheme, the full NSE/EB-FRCM configuration showed higher deflection at the maximum load than that for the intermittent configuration. For instance, the Δ_u value for specimen GFF (7.44 mm) was higher than for Specimen GIF (6.24 mm) which was in turn higher than the value for Specimen GII (5.71 mm).

With regard to the strain values, all beams failed before flexural steel reaches its yielding point as listed in the sixth column of Table 4 and concrete did not reach its crushing point as listed in the last column of Table 4. Thus, the beams failed in shear as expected. Moreover, the hybrid NSE/EB-FRCM system has increased the strain values in flexural steel bars as listed in Column 6 of Table 4, which indicates that the strengthening system has delayed the brittle type of shear failure so that the flexural bars showed more tensile strains.

4 CONCLUSION

In this paper, the efficacy the NSE/EB-FRCM system for strengthening of RC beams in shear has been presented. The experiment work was carried out on seven (7) medium scale RC beams deficient in shear tested under three-point loading. Two different FRCM types were utilized; namely, glass and PBO FRCM. The test parameters were: (a) type of FRCM composite and (b) strengthening configuration. The NSE/EB-FRCM system has successfully increased the shear capacity of the beams. The percentage gain in the shear capacity ranged between 45% and 76% for PBO-FRCM system while it ranged between 61% and 80% for glass FRCM system. Specimens strengthened with G-FRCM system showed higher enhancement in the shear capacity compared to that for the PBO FRCM counterparts. The average increase in the deflection at the peak load were 85% and 99% for G- and PBO-FRCM relative to the reference specimen, respectively. With regard to the strengthening scheme, full FRCM configuration showed better performance in terms of the increase in the shear capacity and deflection at the maximum load compared to that for the intermittent counterparts.

Acknowledgments

This paper was made possible by NPRP grant # NPRP 7-1720-2-641 from the Qatar National Research Fund (a member of Qatar Foundation). The findings achieved herein are solely the responsibility of the authors.

References

Aly, R., Benmokrane, B., and Ebead, U., Tensile Lap Splicing of Bundled CFRP Reinforcing Bars in Concrete, *Journal of Composites for Construction*, 10(4), 287–294, 2006.

Ebead, U., Inexpensive Strengthening Technique for Partially Loaded Reinforced Concrete Beams : Experimental Study, *Journal of Materials in Civil Engineering*, 27(10), 2015.

Ebead, U. and Marzouk, H., Strengthening of Two-Way Slabs Using Steel Plates, *ACI Structural Journal*, 99(6), 23-31, 2002a.

Ebead, U. and Marzouk, H., Strengthening of Two-Way Slabs Subjected to Moment and Cyclic Loading, *ACI Structural Journal*, 99(4), 435–444, 2002b.

Ebead, U. and Marzouk, H., Fiber-Reinforced Polymer Strengthening of Two-Way Slabs, *ACI Structural Journal*, 101(5), 650–659, 2004.

Ebead, U. and Saeed, H., Flexural and Interfacial Behavior of Externally Bonded/Mechanically Fastened Fiber-Reinforced Polymer-Strengthened Reinforced Concrete Beams, *ACI Structural Journal*, 111(4), 741–751, 2014.

Ebead, U. and Saeed, H., FRP/Stirrups Interaction of Shear-Strengthened Beams, *Materials and Structures*, Springer Netherlands, 50(2), 103, 2017.

Elghazy, M., Refai, A., Ebead, U., and Nanni, A., Effect of corrosion Damage on the Flexural Performance of RC Beams Strengthened with FRCM Composites, *Composite Structures*, Elsevier Ltd, 180, 994–1006, 2017.

Elsayed, W., Ebead, U., and Neale, K., Mechanically Fastened FRP-Strengthened Two-Way Concrete Slabs with and without Cutouts, *Journal of Composites for Construction*, 13(3), 198–207, 2009.

Kotynia, R., Abdel Baky, H., Neale, K., and Ebead, U., Flexural Strengthening of RC Beams with Externally Bonded CFRP Systems: Test Results and 3D Nonlinear FE Analysis, *Journal of Composites for Construction*, 12(2), 190–201, 2008.

Ombres, L., Structural Performances of Reinforced Concrete Beams Strengthened in Shear with a Cement Based Fiber Composite Material, *Composite Structures*, Elsevier Ltd, 122, 316–329, 2015.

Panigrahi, A. K., Biswal, K. C., and Barik, M. R., Strengthening of Shear Deficient RC T-beams with Externally Bonded GFRP Sheets, *Construction and Building Materials*, Elsevier Ltd, 57, 81–91, 2014.

Pino, V., Hadad, H., Basalo, F., Nanni, A., Ebead, U., and Refai, A., Performance of FRCM-Strengthened RC Beams Subject to Fatigue, *Journal of Bridge Engineering*, 22(10), 4017079, 2017.

Raoof, S. and Bournas, D., TRM versus FRP in Flexural Strengthening of RC Beams : Behaviour at High Temperatures, *Construction and Building Materials*, 154, 424–437, 2017.

Tetta, Z. and Bournas, D., TRM vs FRP Jacketing in Shear Strengthening of Concrete Members Subjected to High Temperatures, *Composites Part B*, Elsevier Ltd, 106, 190–205, 2016.

Tetta, Z., Koutas, L., and Bournas, D., Textile-Reinforced Mortar (TRM) versus Fiber-Reinforced Polymers (FRP) in Shear Strengthening of Concrete Beams, *Composites Part B: Engineering*, Elsevier Ltd, 77, 338–348, 2015.

Trapko, T., Urbańska, D., and Kamiński, M., Shear Strengthening of Reinforced Concrete Beams with PBO-FRCM Composites, *Composites Part B*, 80, 63–72, 2015.

Wakjira, T. and Ebead, U., Hybrid NSE/EB Technique for Shear Strengthening of Reinforced Concrete Beams Using FRCM: Experimental study, *Construction and Building Materials*, 164, 164–177, 2018.

Younis, A., Ebead, U., and Shrestha, K., Different FRCM Systems for Shear-Strengthening of Reinforced Concrete Beams, *Construction and Building Materials*, 153, 514–526, 2017a.

Younis, A., Ebead, U., and Shrestha, K. C., Tensile Characterization of Textile Reinforced Mortar, *Proceedings of the Ninth International Structural Engineering and Construction Conference, Resilient Structures and Sustainable Construction*, ISEC Press, Valencia, Spain, St-28, 2017b.

Younis, A., Ebead, U., and Shrestha, K. C., FRCM Shear Strengthening for Concrete Beams, *Proceedings of the Ninth International Structural Engineering and Construction Conference, Resilient Structures and Sustainable Construction*, ISEC Press, Valencia, Spain, St-27, 2017c.

Responsible Design and Delivery of the Constructed Project
Edited by Abdul-Malak, M., Khoury, H., Singh, A., and Yazdani, S.

EFFECT OF GRADATION AND REPLACEMENT LEVEL ON ALKALI-SILICA REACTIVITY OF MINERAL FILLERS

NOURA SINNO and MEDHAT SHEHATA

Dept of Civil Engineering, Ryerson University, Toronto, Canada

Mineral fillers are used in concrete for a number of reasons, which include achieving self-consolidation properties. Different fillers are available depending on the type of the original rock. Alkali-silica reaction might occur if the filler is produced from a reactive rock. The present research focuses on evaluating the applicability of the current accelerated mortar bar test, described in ASTM C1260 and CSA A23.2-25A, in evaluating their reactivity. Three different fillers are used: a calcium-carbonate filler, a carbonate silica filler with 27% SiO_2, and a siliceous filler produced from reactive aggregate. Two different gradations of the filler obtained from reactive aggregate are tested: 1) Gradation finer than 75 μm; 2) Gradation with 30% passing 150 μm sieve but is retained on 75 μm sieve and remaining 70% passing 75 μm. Results showed that the calcium-carbonate filler and the carbonate silica filler did not show any expansion higher than the non-reactive sand, while the filler obtained from reactive aggregate did not show expansion higher than the control at 14 days for both gradations, with the expansion showing at later age. This might be due to the high surface area of fillers and the reduced permeability affecting the rate of expansion. Hence, the need to extend the testing period beyond the 14-day specified in the standard to be able to see the expansion.

Keywords: Accelerated mortar bar test, Concrete prism test, ASR, Expansion, Permeability.

1 INTRODUCTION

Alkali-silica reaction (ASR) is a chemical reaction between the silica present in the aggregate and the alkaline pore solution of the concrete. This reaction leads to the formation of a hygroscopic gel that tends to swell in the presence of moisture. The swelling-gel exerts pressure on the concrete leading to cracks and deterioration (Lindgård *et al.* 2012). Many preventive measures are being implemented in order to avoid the damage associated with this reaction, such as the use of supplementary cementing materials and lithium-based compounds (Thomas *et al.* 2006). In addition to their fine size, supplementary cementing materials such as fly ash, slag, and silica fume act as an alkali diluent by substituting a portion of the cement and have a high alkali binding capacity due to their pozzolanic nature (Kandasamy and Shehata 2014).

Several test methods are available to evaluate the reactivity of aggregates. First, the concrete prism test (CPT), described in the American and Canadian standards (ASTM C1293/CSA A23.2-14A), specifies an expansion limit of 0.040% at 1 year above which the aggregate is considered reactive. The accelerated mortar bar test (AMBT), described in ASTM C1260 and CSA A23.2-25A, is another faster test method consisting of soaking mortar bar samples in an alkaline solution and measuring expansion for 14 days with an expansion limit for this test equal to 0.10%. While these two tests are commonly used for testing the reactivity of aggregates, there is no standard test available to evaluate the reactivity of fillers.

Mineral fillers are one of the additives used in self-consolidating concrete (SCC) (Sinno and Shehata 2017) to maintain cohesiveness and stability of the mix (Uysal 2012). SCC is characterized as having high flowability properties and ability to be poured without vibration, while avoiding bleeding and segregation (Khayat 1999). Applications of SCC include its use in areas with congested reinforcement or in the repair of structural elements such as beams and slabs when reaching the repair area is inaccessible (Khayat 1999, Murthy *et al.* 2012). Some types of mineral fillers might have a potential to cause expansion due to alkali-silica reaction (Pedersen 2004). Hence the need of modified test methods to evaluate fillers' reactivity before using them.

2 EXPERIMENTAL PROGRAM

2.1 Materials

Fine aggregates used in this study are of a non-reactive sand type having a fineness modulus equal to 2.64. The cement used is a Type GU cement of 0.99% Na_2O_e having a chemical composition as shown in Table 1.

Table 1. Chemical composition of Type GU cement.

Oxide	GU Cement (Mass %)
SiO_2	19.54
Al_2O_3	5.21
Fe_2O_3	2.16
CaO	62.39
MgO	2.39
SO_3	4.03
Na_2O_e	0.99
Loss on Ignition	2.36

Three mineral fillers are tested in this paper covering different ranges of silica content. The first is a calcium-carbonate filler composed of 94% calcium carbonate and 2.5% magnesium carbonate. The median diameter of this filler is 21 μm having 10% passing 150 μm sieve and retained on 75 μm sieve and the rest is finer than 75 μm. The second is a carbonate silica filler containing 27% SiO_2. This filler has a median diameter of 40 μm with 30% passing 150 μm sieve and retained on 75 μm sieve and the rest finer than 75 μm. These two fillers are commercially available and their particle size distributions are presented in Figure 1.

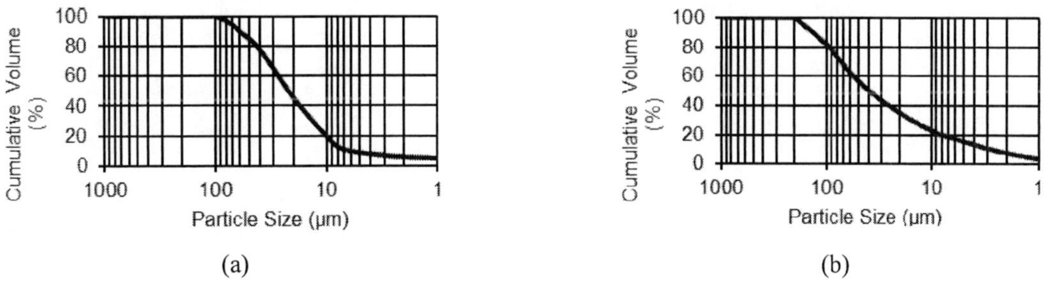

Figure 1. Particle size distribution of (a) carbonate filler and (b) carbonate silica filler.

The third filler is produced from reactive aggregate and is obtained by crushing a highly reactive coarse aggregate with an expansion of 0.22% at 1 year and 0.46% at 14 days using the CPT and AMBT respectively. This filler was crushed and used in two different gradations: the fine filler with all particles passing 75 μm sieve and the coarse filler having 70% passing 75 μm sieve and 30% passing 150 μm sieve and retained on 75 μm sieve. The crushing procedure is explained in next section.

2.2 Experimental Procedure

2.2.1 Crushing of filler obtained from reactive aggregate

Coarse aggregate was crushed using the jaw crusher until all passed 4.75 mm sieve. Then, the crushed aggregate was sieved on the different fine sieves, and each portion, starting with the coarser size, was crushed using the pulverizer and sieved again until all what is on the sieve passed. This is done for all the different sieve sizes until enough filler passing 150 μm sieve and retained on 75 μm sieve and passing 75 μm sieve is obtained.

2.2.2 Sample preparation

The gradation of the mortar bars is as specified in the ASTM C1260. Each set of mortar bars required 990 g of fine aggregate and 440 g of cement. The gradation of the fine aggregate is shown in Table 2.

Table 2. Grading requirements as per ASTM C1260.

Passing	Retained on	Mass (%)
4.75 mm	2.36 mm	10
2.36 mm	1.18 mm	25
1.18 mm	600 μm	25
600 μm	300 μm	25
300 μm	150 μm	15

The fillers were tested at 10% and 20% replacement levels taken from the finest portion of sand. As an example, for 10% replacement level (RL), 10% from the fine aggregate passing 300 μm sieve and retained on 150 μm sieve was replaced by filler leaving 5% sand of this size. The required masses of the sand and fillers are shown in Table 3.

Table 3. Sample preparation for different replacement levels.

Passing	Retained on	Masses required (g)		
		Control	10% RL	20% RL
4.75 mm	2.36 mm	99	99	99
2.36 mm	1.18 mm	247.5	247.5	247.5
1.18 mm	600 μm	247.5	247.5	247.5
600 μm	300 μm	247.5	247.5	198
300 μm	150 μm	148.5	49.5	0
Filler		0	99	198

For the filler produced from reactive aggregate, the amount of filler obtained from the table is divided into 30% passing 150 µm sieve and retained on 75 µm sieve and the rest 70% is passing 75 µm sieve to reach the coarse gradation.

2.2.3 Mixing and placing of mortar bars

A water to cement ratio of 0.47 was used for all the samples which were mixed using the procedure described in ASTM C305. Upon finishing the mixing, the mortar bars were cast in molds of 25 x 25 x 285 mm dimension and cured for 24 hours. The following day, they were put in water preheated at 80°C as required by the standard. Twenty-four hours later, the samples were soaked in 1M NaOH solution and measurements were taken for the next 56 days.

3 RESULTS AND DISCUSSIONS

The expansion of mortar bars was measured and the average of three mortar bars for each set was used in this study. Figure 2 presents the expansion data of the carbonate filler, carbonate filler with silica content and filler obtained from reactive aggregate with the coarse gradation at 10% and 20% RLs. The horizontal line shows the 0.10% limit specified in ASTM C1260 and CSA A23.2-25A.

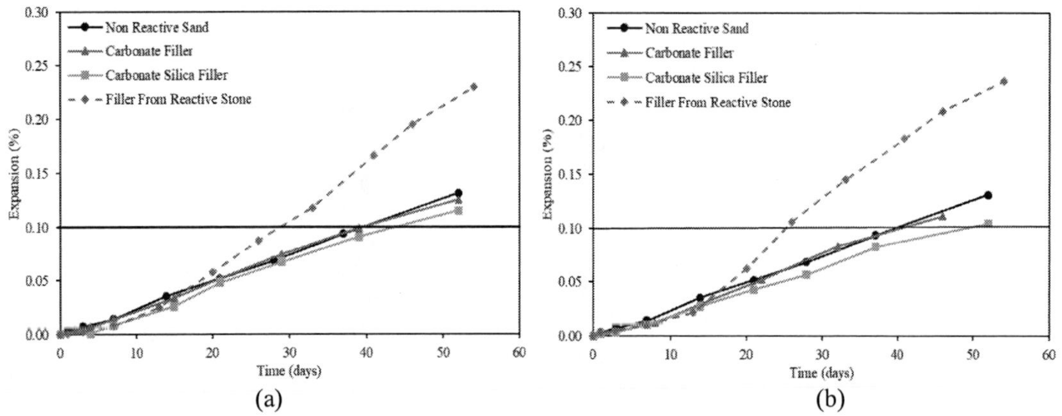

Figure 2. Expansion data of the fillers at (a) 10% RL and (b) 20% RL.

As shown in Figure 2, the carbonate filler and the carbonate silica filler did not show any expansion higher than the non-reactive sand. At 14 days, the expansion was 0.035% for the non-reactive sand compared to 0.034% and 0.026% for the carbonate and carbonate silica fillers. Both fillers had an expansion below the limit of 0.10% specified in the standard. It can be concluded that the silica present in the carbonate silica filler might not be enough to cause expansion or it might not be reactive. When the expansion was measured for 56 days, no expansion higher than the non-reactive sand was obtained. A similar behavior was observed at 20% RL as shown in Figure 2-b). In addition, the filler obtained from reactive aggregate showed expansion higher than the non-reactive sand for 10% and 20% RLs at 56 days. The expansions at 10% and 20% RLs were almost the same with 0.23% and 0.24% respectively. This might be due to the reduced permeability of the increased amount of fine materials at 20% RL hindering the alkalis in solution to reach inside the core which will block further expansion.

Expansion data for the filler obtained from reactive aggregate at 14 days was presented in Figure 3 for the two different gradations.

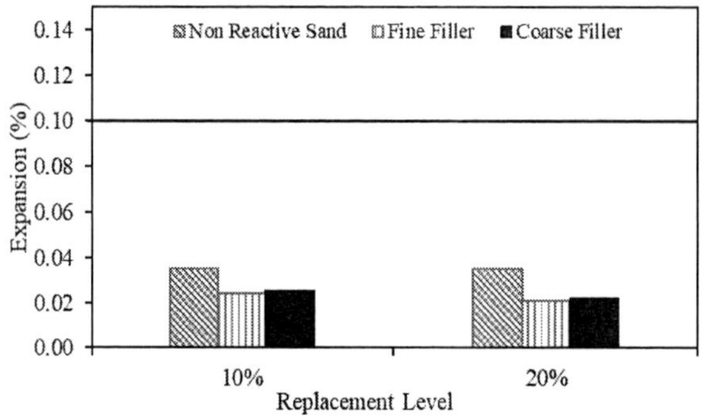

Figure 3. Expansion data of the filler obtained from reactive aggregate at 14 days.

The filler obtained from reactive aggregate did not show any expansion higher than the non-reactive sand at 14 days. The filler with both gradations had an expansion below 0.10% limit specified at 14 days. Measurements were taken for 56 days as shown in Figure 4.

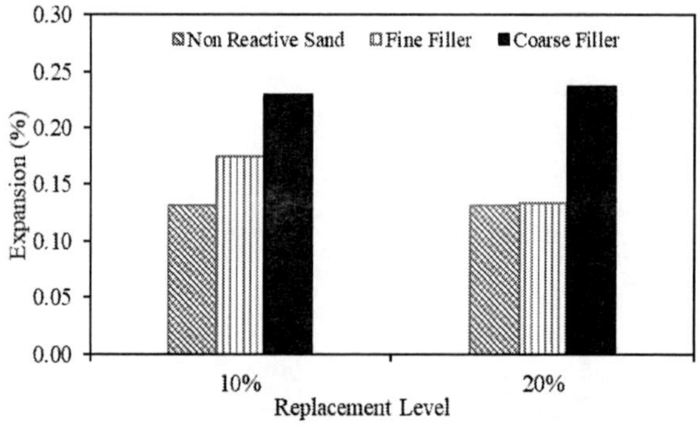

Figure 4. Expansion data of the filler obtained from reactive aggregate at 56 days.

For the filler finer than 75 μm at 20% RL, no expansion higher than the non-reactive sand was observed. However, with the same filler but coarser gradation, an expansion of 0.24% compared to 0.13% for the non-reactive sand was obtained after 56 days. The difference in expansion between the same filler with different gradations might be due to the fact that the larger surface area might result in very low disruptive expansion or it might slow the expansion due to the required large amount of alkali needed to trigger the expansion. The latter reason could be the explanation of why the expansion took place at later age. Moreover, the reduced permeability due to the filler's fine size might hinder the alkalis in solution to reach inside the core.

At 10% RL, the expansion of the coarse filler was 0.23% at 56 days as compared to 0.13% for the non-reactive sand. In addition, the expansion of the finer filler at 10% RL, was lower than the coarse filler at the same RL. This might also be due to the reduced permeability.

In order to better understand whether the filler either lost reactivity when grinded to small size, or its permeability was reduced due to the addition of the filler, concrete prisms were tested using the concrete prism test. At the end of this test, any modifications needed in order to use the accelerated mortar bar test to evaluate fillers, such as extending the testing period to beyond, 14 days will be assessed. At the time of writing this paper, concrete prism tests were still ongoing.

4 CONCLUSION

Potentially reactive fillers need to be tested for expansion due to ASR before implementing them in concrete. With the carbonate silica filler studied in this research, no expansion was observed higher than the control. This might be due to the fact that the size of the filler might have affected its reactivity or the silica present in the filler is not reactive. In addition, the filler obtained from a reactive aggregate did not show any expansion above 0.10% limit at 14 days. At 56 days, the fine filler did not show any expansion higher than the control at 20% RL. However, the coarser filler showed expansion higher than the non-reactive sand at both replacement levels. This suggests that the larger surface area might affect the fillers reactivity and reduce the permeability of the samples. The expansion was not observed at 14 days as specified in the standard, hence the testing period was extended to 56 days. Concrete prims are being monitored to understand whether grinding to very fine size affects reactivity of the filler or the permeability of the mortar bars.

Acknowledgments

This research project is funded by a Discovery Grant from the Natural Sciences and Engineering Research Council of Canada (NSERC). The financial support of this organization is highly appreciated.

References

Kandasamy, S., and Shehata, M., The Capacity of Ternary Blends Containing Slag and High-Calcium Fly Ash to Mitigate Alkali Silica Reaction, *Cement and Concrete Research*, Elsevier, 49, 92-99, May, 2014.

Khayat, K. H., Workability, Testing, and Performance of Self-Consolidating Concrete, *ACI Materials Journal*, American Concrete Institute, 96(3), 346-353, May, 1999.

Lindgård, J., Andiç-Çakir, O., Fernandes, I., Rønning, T., and Thomas, M., Alkali-Silica Reactions (ASR): Literature Review on Parameters Influencing Laboratory Performance Testing, *Cement and Concrete Research,* Elsevier, 42(2), 223-243, February, 2012.

Murthy, K. N., Rao, N. A. V, Reddy, R. I. V., and Reddy, M. V. S, Mix Design Procedure for Self Compacting Concrete. *IOSR Journal of Engineering*, 2(9), 33-41, September, 2012.

Pedersen, B. M., *Alkali-Reactive and Inert Fillers in Concrete*, Norwegian University of Science and Technology, Trondheim, Norway, 2004.

Sinno, N., and Shehata, M., Evaluating the Potential Alkali-Silica Reactivity of Mineral Fillers: a Preliminary Study, in *6th International Conference on Engineering Mechanics and Materials,* Vancouver, Canada, 2017.

Thomas, M., Fournier, B., Folliard, K., Ideker, J., and Shehata, M., Test Methods for Evaluating Preventive Measures for Controlling Expansion Due to Alkali-Silica Reaction in Concrete, *Cement and Concrete Research*, Elsevier, 36(10), 1842-1856, October, 2006.

Uysal, M., Self-Compacting Concrete Incorporating Filler Additives: Performance at High Temperatures, *Construction and Building Materials*, Elsevier, 26(1), 701-706, 2012.

Responsible Design and Delivery of the Constructed Project
Edited by Abdul-Malak, M., Khoury, H., Singh, A., and Yazdani, S.

CONSTRUCTABILITY STUDY OF ASPHALT PAVEMENTS INCORPORATING SHALLOW GEOTHERMAL ENERGY

ALI ZALGHOUT, NOORALHUDA SALEH, SAMIR SARI AD DIN, and GHASSAN CHEHAB

Dept of Civil and Environmental Engineering, American University of Beirut, Beirut, Lebanon

The incorporation of geothermal energy in heating and electricity production has rapidly increased during the last three decades. This paper focuses on the use of shallow geothermal energy in asphalt pavement, herein called Ground Coupled Hydronic Asphalt Pavement (GCHAP) system. GCHAP consists of a series of pipes embedded within the asphalt layer coupled with another network of pipes embedded in the soil. A circulating fluid acts to exchange the heat energy between the soil and the asphalt layer. The system can be used for cooling the pavement, which results in decreasing permanent deformation in the summer. This paper presents a study on the constructability and performance of such systems on a large-scale section. Numerical analyses and a pilot study were conducted to select the system design components. A 9.6 x 4 meter GCHAP section was constructed on a municipal road in addition to another control section. The secondary network of pipes was placed three meters below the ground. Sensors were embedded within the GCHAP system to record and compare the temperature of the pavement to that of the control section. The results showed that GCHAP section can decrease the pavement temperature leading to an increase in resistance to permanent deformation. Moreover, the constructability study showed the importance of performing a pilot study before constructing the system to ensure the resilience of the pipes.

Keywords: Hydronic asphalt pavement, Permanent deformation, PPR pipes, Constructability, GCHAP, Heat energy, Large-scale section.

1 INTRODUCTION

Most roadways around the world are surfaced with asphalt concrete. The condition of these roads highly affects the Gross Domestic Product (GDP), as roads provide the ability to transport goods and services. However, with the increase in transportation activities these roads suffer from various distresses. The two main distresses are permanent deformation and fatigue. The former occurs due to binder softening at elevated temperature under loading conditions, while the latter manifests itself in the form of cracking at the bottom of the asphalt layer due to the repetitive loading and propagates towards the surface. These distresses lead to safety issues, lower the serviceability of roads, and increase the rehabilitation costs. Incorporating long-term strategies to improve the pavement performance is necessary to keep roadways safe and functional. A Ground Coupled Hydronic Asphalt Pavement (GCHAP) presents an alternative pavement system that is both more durable and more sustainable. A GCHAP system consists of a series of pipes embedded in the asphalt layer connected to another series of pipes embedded in the soil. A fluid circulates in this system and transfers the heat from the asphalt layer to the soil during summer, leading to a decrease in rutting. The system also creates an eco-friendlier environment by mitigating the Urban Heat Island (UHI) effect, a phenomenon that cause urban areas to be several degrees warmer than the surrounding regions.

The use of geothermal energy for heating/cooling purposes has been attempted in different applications. In the context of HAP, geothermal application has been implemented in Switzerland (Eugster and Shatzmann 2002) and Japan (Morita and Tago 2005) for pavement deicing and snow melting. Studies have been conducted on the effect of various parameters of HAP on thermal performance. Pan *et al.* (2015) concluded that pipes with larger diameters, closer spacing, and lower depth improve the efficiency of the system. Adl-Zarrabi *et al.* (2016) showed that the effect of pipes spacing on the HAP system is larger than that of pipes depth. Another study revealed that the effect of pipe diameter is more significant than the effect of flow rate in such systems (Mallick *et al.* 2012). Despite the numerous studies, there still remains a need to conduct a comprehensive assessment of, and comparison between, all the parameters of HAP; in addition to investing the performance of HAP coupled to the ground – GCHAP. Moreover, details on constructing the system, issues faced during implementation, and the long-term performance of such systems have not been discussed thoroughly in the literature.

2 OBJECTIVE

The main objective of this study is to evaluate the performance of Hydronic Asphalt Pavements coupled with shallow geothermal energy and present details on the design and construction process of such systems. Moreover, lessons learned behind the construction of such systems are presented as a guide for future GCHAP applications.

3 METHODOLOGY

To determine the optimal parameters of the system, and to make sure that the system is feasible and practical to construct, a numerical study and a pilot study were carried out.

3.1 Simulations

A 3D numerical model was developed in MATLAB to study the sensitivity of the six main parameters of a Hydronic Asphalt Pavement – flow rate, pipe spacing, diameter, depth, pipe conductivity, and pavement conductivity, in order to design the components of the system. On the surface, the pavement receives heat from solar irradiation and exchanges heat by long-wave radiation (with the sky) and convection (with the ambient). The heat is then transferred through the pavement medium by diffusion. The fluid in the pipes transports (advection) and diffuses the heat received from the pavement medium. The model assumes zero-flux boundaries everywhere except the top boundary, in addition to neglecting turbulent flow and viscous dissipation. The initial conditions were calibrated by running an analogous 1D model until steady state conditions were met. Lastly, the model was validated with data found in the literature (Dakessian *et al.* 2016). Figure 1 shows the simulation results.

A set of representative parameters for the pavement were chosen and then individually varied to see the effect in terms of the change in surface temperature of the pavement after 12 hours of simulation. The analysis reveals that that pavement conductivity is the most critical since its corresponding curve exhibits the steepest slope. The spacing and depth of the pipes are mildly sensitive, with pipe diameter and flow rate showing the least sensitivity as shown Figure 1a. The effect of pipe conductivity was investigated separately due to its wide range. Figure 2b shows that the effect of pipe conductivity is limited by the pavement conductivity. The surface temperature decreases steeply as a function of pipe conductivity – until reaching a certain value where the curve begins to approach an asymptote. This value resembles the approximate value of the pavement conductivity. This fact implies that there is no need to invest in highly expensive pipes which have high conductivities, since their effect is limited by the conductivity of the pavement. The analysis done on the influence of pipes spacing on the surface temperature indicated that spacing should be between 17.5 and 25 cm. In this study, a 20 cm pipes spacing was used.

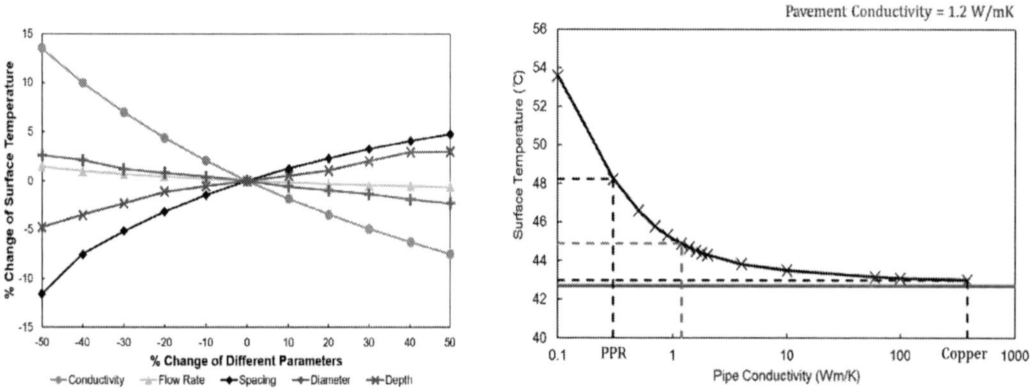

Figure 1. Simulation Results: (a) all design parameters; and (b) pipe conductivity analysis.

3.2 Pipe Material

Copper pipes cost 10 USD/m, while polypropylene (PPR) pipes cost 4 USD/m, according to a local supplier. Class 2 PPR pipes were considered in this study due to financial constraints. The main concern behind using this material is the low conductivity of PPR pipes (0.3 W/mk*) compared to that of copper pipes (385 W/mk); however, pipe materials simulations as discussed earlier showed that the limiting factor to heat transfer is the asphalt – not pipe – conductivity (Watts per meter-Kelvin).

3.3 Pilot Study

During construction, the PPR pipes are subjected to heavy loads under a vibratory steel drum while covered with the hot asphalt mix, which risks their structural integrity. This risk increases when the pipes are closer to the surface. In order to ensure that this problem is avoided during the construction of the large scale GCHAP pavement, a 2mx2m asphalt section with pipes embedded within was constructed (Figure 2). The section involved different types of pipe material (steel, PPR class 2, PPR class 5, PPR with aluminum, and copper), different pipe diameters (20, 25, and 32 mm), different depths from the surface (2 cm and 4 cm), and different paving techniques (laying pipes on un-compacted asphalt layer, laying pipes in subgrade, and using spacers).

Figure 2. Pilot Study Section Construction.

After constructing the 2mx2m section, pressure-testing was done to assess the performance of each pipe. All the pipe types and the laying scenarios presented previously performed well, with no leakage detected, except for the 20-mm diameter plastic pipe that was laid on the subgrade 4 cm below the surface. The reason behind this leakage might have been because the fittings were attached to the pipes by glue that couldn't resist heat.

4 GCHAP LARGE SCALE CONSTRUCTION

The system was constructed on a 9.6mx4m local road section, in addition to a control section. The excavation was carried out to a depth of 3 m, then the polypropylene pipes (PPR) were placed in a serpentine configuration. The spacing between each two consecutive pipes was set to be 20 cm, and steel bars were used to fix the pipes. A fine powder material was placed below and above the pipes to prevent large rocks from breaking the pipes as shown in Figure 3.

Figure 3. Placing fine material below and above the pipes.

After placing the pipes, a pipe connection between the bottom pipes network (3 meters below the surface) and the top pipes network (embedded in the asphalt layer) was established. Two PPR pipes 3 meters in length were welded to the two openings of the pipe network at the bottom of the excavated area. These two PPR pipes were then protected with a 10-cm diameter plastic pipe to prevent coarse material from puncturing the pipes, as shown in Figure 4. Following plastic pipes placement, sensors at three different depths in the soil were set along the pipe. These sensors are located at depths of 1, 2 and 3 meters below the surface. The excavated area was then filled with a good quality base course, and the compactor was used to compact the base course material every 1 meter. The upper piping layer was then placed in a reverse-return configuration, and sensors were placed at different location on the pipes, between the pipes, and at different depth in the asphalt layer. Asphalt was laid down using the bobcat, and mechanical combs. A small compactor was used to compact the sensors located at the surface, to prevent any sensor damage.

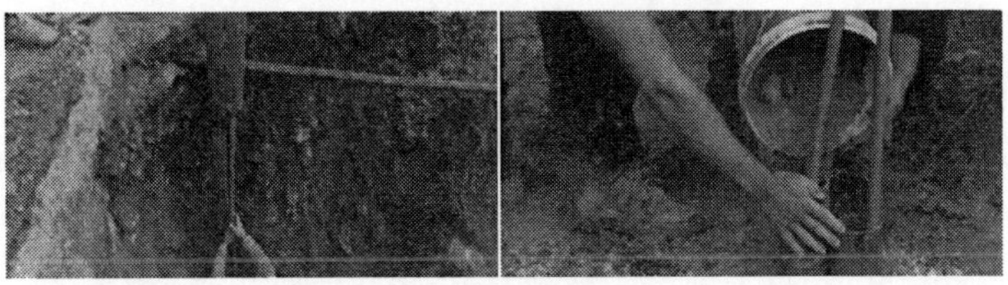

Figure 4. Connecting the pipe network in the asphalt layer with that in the soil.

5 RESULTS

The readings collected for the GCHAP system were affected by the shade of nearby objects at critical times of the day when the temperature reached a peak. The exact performance of the system was thus not well captured. However, inspecting the slopes of the evolution of the pavement temperatures at different depths can still give an idea of how the GCHAP system reduces the pavement temperature as depth, away from the surface, increases as shown in Figure 5. Although not as significant as the change in temperature between the surface and 2.5 cm depth, there exists a decrease in temperature as depth changes from 2.5 cm to 4.5 cm.

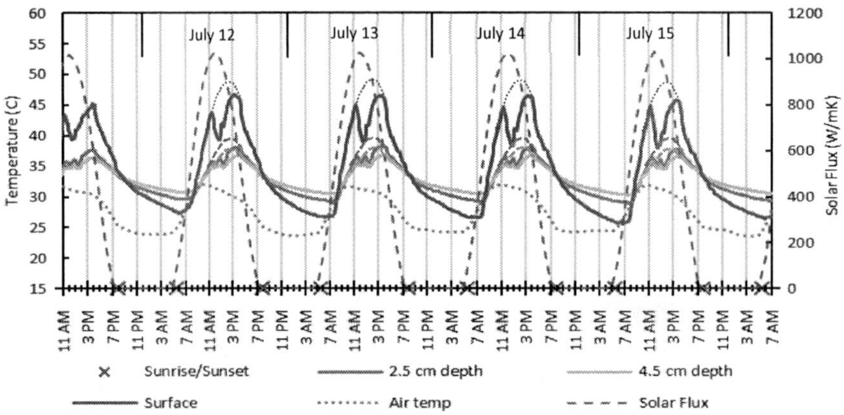

Figure 5. Temperature distribution within pavement incorporating GCHAP system.

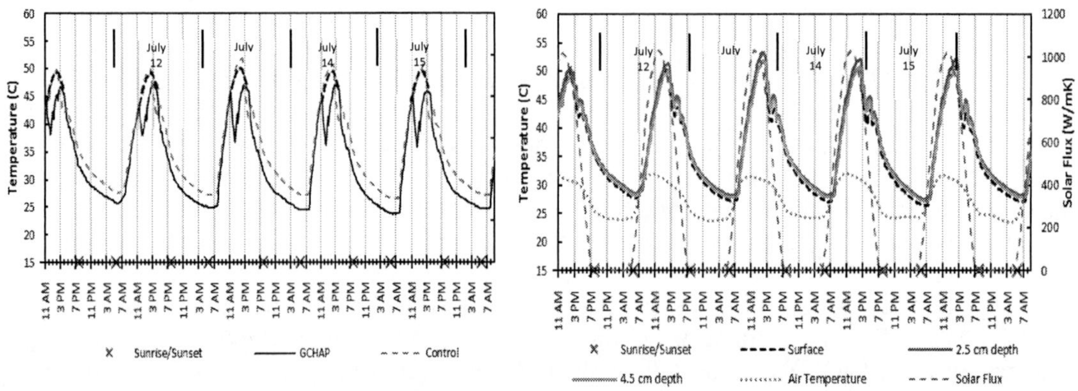

Figure 6. (a) Effect of GCHAP system on surface temperature; and (b) temperature distribution within control section.

Figure 6a shows that the GCHAP system had little impact on decreasing the peak surface temperature (according to an educated extrapolation of the curves) compared to the control section. However, at 2.5 cm and 4.5 cm depth, the GCHAP was able to decrease the temperature by at least 10 °C (Figure 5 and Figure 6b). The ability of the system to decrease the temperature is a function of the location of interest relative to the location of the pipes. Since the surface is the farthest from the pipe, it is the least affected. The direct solar radiation arriving at the surface also plays a major role in counteracting the effects of the system. At 2.5 cm and 4.5 cm depth, the decrease in temperature is more pronounced since they are closer to the pipe but approximately at the same radial distance which explain the smaller difference in temperature between the two. In terms of the cooling curve, the lower temperature of the Soil in the GCHAP system might have played a role in increasing the rate at which the pavement cools and the extent to which it cools as shown in Figure 6(a).

6 LESSONS LEARNED AND CONCLUSIONS

This section points out lessons learned in constructing this unconventional pavement system:

- If unconventional pipes are to be used within the asphalt layer (e.g. use of PPR in this study), it is essential to study the heat and chemical resistance of these pipes against asphalt.
- Covering and protecting both ends of the pipe to prevent obstruction due to contaminants is essential for a more effective operation, in addition to protecting the pipe network embedded in the soil with a layer of fine material before filling and compaction.
- Cool water should be circulated through the pipe network while spreading the asphalt. The pipe network should be pressurized during compaction to better protect the pipes.
- For future research purposes, care should be taken in accounting for the effect of shade, as this would affect the sensors that measure the surface temperature, yielding to temperature values less than the real surface temperature.
- If the test section is conducted on a small section of a minor road, the contractor might not be able to control the asphalt layer thickness. If the thickness of the asphalt layer within which the pipes are embedded is altered, the pipes would be influencing a greater or smaller area and the sensors would be reading temperatures at levels different than what they were meant to.
- In this study, three months after construction and under traffic loading, the pavement started showing a slight deformation, specifically in areas between the pipes. It is possible that the initial deformation the pavement undergoes in its early life was mitigated by the presence of pipes as a support. However, this differential settlement could introduce undesirable shear stresses, thus for future research on GCHAP, more attention should be given to this issue.

As a conclusion, the findings of this research are as follows: (1) The use of PPR in GCHAP is a feasible and practical option, (2) the limiting factor to heat transfer is the asphalt conductivity and not the pipe conductivity, thus it is more economical to invest in increasing the asphalt conductivity rather than in highly conductive pipes such as copper, and (3) GCHAP system decreases the pavement temperature which could lead to a decrease in rutting.

References

Adl-Zarrabi, B., Mirzanamadi, R., and Johnsson, J., Hydronic Pavement Heating for Sustainable Ice-Free Roads. *Transportation Research Procedia*, 14, 704-13, January 2016.

Dakessian, L., Harfoushian, H., Habib, D., Chehab, G. R., Saad, G., and Srour, I., Finite Element Approach to Assess the Benefits of Asphalt Solar Collectors, *Transportation Research Record: Journal of the Transportation Research Board,* 2575, 79-91, 2016.

Eugster, W., and Schatzmann, J., Harnessing Solar Energy for Winter Road Clearing on Heavily Loaded Expressways, *Proc. XIth PIARC International Winter Road Congress*, Sapporo, Japan, January 2002.

Mallick, R., Chen, B. L., and Bhowmick, S., Harvesting Heat Energy from Asphalt Pavements: Development of and Comparison Between Numerical Models and Experiment, *International Journal of Sustainable Engineering*. 5(2):159-169, June 2012.

Morita, K., and Tago, M., Snow-Melting on Sidewalks with Ground-Coupled Heat Pumps in a Heavy Snowfall City, *Proc. World Geothermal Congress*, Antalya, 2005.

Pan, P., Wu, S., and Liu, G., A Review on Hydronic Asphalt Pavement for Energy Harvesting and Snow Melting, *Renewable and Sustainable Energy Reviews,* 48, 624-634, 2015.

Responsible Design and Delivery of the Constructed Project
Edited by Abdul-Malak, M., Khoury, H., Singh, A., and Yazdani, S.

EXPERIMENTAL EVALUATION OF THE FLEXURAL BEHAVIOR OF CONCRETE BEAMS REINFORCED WITH STIFF BIAXIAL GEOGRID

ZAHER AL BASIOUNI AL MASRI, RANA HAJ CHHADE, and GHASSAN CHEHAB

Dept of Civil Engineering, American University of Beirut, Beirut, Lebanon

Limited research exists on the benefits of using geogrids to reinforce Portland cement concrete structural and non-structural members. Preliminary research findings have shown that the inclusion of geogrids in Portland cement concrete leads to a definitive improvement in the post-cracking behavior in terms of ductility, load capacity, and crack propagation control. However, the geogrid sheet caused partial discontinuity in the concrete section which in some cases lead to complete separation of the bottom concrete layer. Through a rigorous experimental program, this study aims to verifying and further enhancing the current knowledge regarding the behavior of geogrid-reinforced concrete elements as well as proposing a geogrid reinforcing configuration that is expected to maintain the continuity throughout the concrete section while achieving or improving the aforementioned benefits. A total of nine concrete beam specimens were prepared and tested under four-point bending. Testing results and observations re-confirmed the reinforcing benefits of geogrids and proved that the proposed configuration is very effective as it resulted in better post-cracking behavior while overcoming the concrete separation problem.

Keywords: Post-cracking behavior, Four-point bending test, Crack opening, PCC, Reinforcement.

1 INTRODUCTION

Geogrids are geosynthetic materials made from geotextiles or polymers such as polypropylene, polyethylene or polyester. They are commonly used for soil stability purposes in road embankments or reinforced earth walls due to their tensile reinforcing capability. They can be classified as uniaxial, biaxial, or triaxial based on the number of directions they reinforce. Each of these types can be further classified as stiff or flexible geogrids based on their physical and mechanical properties (Tang *et al.* 2008).

The use of geogrids in concrete is not as common as their use in soils. However, efforts are recently being invested to assess the feasibility of using geogrids to reinforce Portland cement concrete (PCC) in order to benefit from their tensile strength and ductility. Studies conducted so far have resulted in promising findings as concrete gained both post-cracking ductility and load capacity (Tang *et al.* 2008, El Meski and Chebab 2014, Itani *et al.* 2016).

The objective of this study is to verify and enhance the current knowledge regarding the behavior of geogrid-reinforced concrete elements and investigate the effectiveness of circular geogrid reinforcement which is expected to maintain the continuity throughout the concrete section while providing more ductility and post-cracking load capacity. The experimental program encompasses testing nine plain and geogrid-reinforced beam specimens. The structural behavior of each sample type was analyzed and compared to that of the control samples based on load-deflection patterns and the maximum load capacity attained.

2 LITERATURE REVIEW

While geogrids are widely used in geotechnical applications, and significant research exists on the use of geogrids as reinforcements of pavement structures, only few studies investigated the advantages and limitations of using geogrids to reinforce Portland cement concrete elements.

Itani *et al.* (2016) introduced uniaxial geogrids in thin concrete overlays to study their effectiveness in preventing reflective cracking. Two experimental setups were adopted: the direct tension and the flexure tests (both monotonic and cyclic) to simulate thermal and traffic loads respectively. The results of both tests indicated that geogrids provided concrete with post-cracking ductility and extra load capacity. However, it was observed that geogrid-reinforcement weakened the concrete section during the pre-cracking phase and, in some cases, lead to the separation of the upper and lower concrete layers along the geogrid sheets. El Meski and Chehab (2014) used uniaxial, biaxial, and triaxial geogrids with different physical and mechanical properties to reinforce normal and high strength concrete beams. The specimens were subjected to four-point monotonic bending until failure. Comparing the load-deflection patterns of geogrid-reinforced and plain concrete beams, a much larger deflection was observed for all geogrid-reinforced samples indicating a ductile post-cracking behavior. Tang *et al.* (2008) also investigated the behavior of geogrid-reinforced PCC members by comparing the effect of introducing one or two layers of stiff and flexible biaxial geogrids. Similar benefits of using geogrid reinforcement were observed in terms of improved post-cracking ductility and load capacity. Stiff geogrids were found to achieve better overall results compared to flexible geogrids, which implies that the physical and mechanical properties of the geogrids are key factors in the effectiveness of geogrid reinforcement.

Chidambaram and Agarwal (2014) tested the effectiveness of confining concrete specimens with geogrids under compressive, flexural and tensile loading. It was concluded that the use of geogrids as a confinement mechanism for concrete resulted in a significant improvement in the behavior of concrete compared with conventional confinement techniques. Furthermore, it was found that the number of layers of geogrids used for confinement as well as their mechanical properties had a major effect on the load-deformation behavior of concrete. In a separate study, Chidambaram and Agarwal (2015) used geogrids for shear reinforcement in steel-reinforced PCC beams. Specimens were tested under single point monotonic loading. Testing results showed that geogrid shear reinforcement significantly enhanced the post-cracking behavior of the beams.

Al Hedad *et al.* (2017) studied the effect of using geogrids on the drying shrinkage behavior of concrete pavements. To do that, beam and slab samples were prepared and reinforced with a sheet of biaxial geogrid at different locations along the specimen depth. They were cured for 7 days and then placed in a drying chamber until day 56. Results showed that geogrids tended to decrease the drying shrinkage strains by 7-28 % compared to plain concrete specimens.

3 TESTING PROGRAM

A total of nine simply supported concrete beams were prepared and tested under monotonic four-point bending until they failed in flexure. The flexure test was conducted following the ASTM Standard Test Method for Flexural Strength of Concrete (ASTM C78/C78M-16 2016). Three of the specimens were reinforced with one sheet of biaxial geogrid placed at 3 cm from the bottom. Another three samples were reinforced with the proposed circular-shaped geogrid while maintaining a minimum concrete cover of 3 cm. The remaining three plain concrete beams served as control samples for comparison. All the specimens had the same length of 56 cm and a cross section of 15x15cm. Beam dimensions and geogrid reinforcing details are shown in Figure 1. The plain concrete, one-layer geogrid, and circular-shaped geogrid specimens were labeled as P, S, and O respectively. A 100-ton closed-loop, servo-hydraulic universal testing machine was used to apply a monotonic load at a constant crosshead displacement rate of 0.02 mm/sec to each simply supported beam. A vertical spring-loaded linear variable differential transducer (LVDT), with a range of +/- 25mm, was used to measure the vertical displacement at the middle of the bottom surface of each specimen. Data was then collected using a 16-bit data acquisition system.

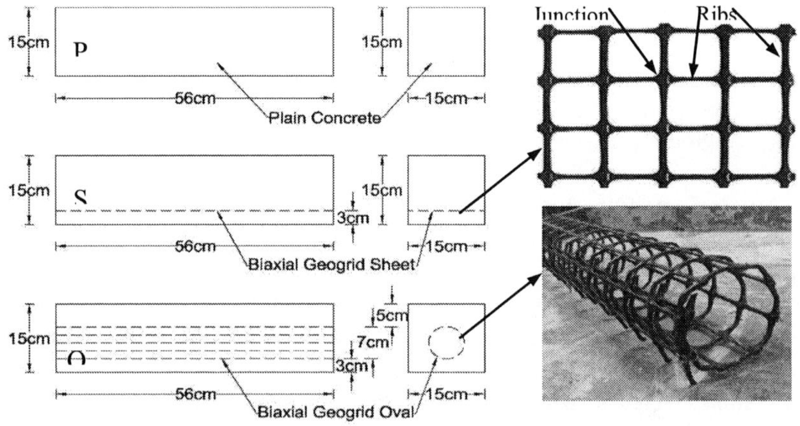

Figure 1. Specimen dimensions and reinforcing details.

4 MATERIAL PROPERTIES

4.1 Geogrids

Biaxial geogrids consist of two-directional thin ribs joined together at thicker junctions; hence, they provide tensile reinforcement in both longitudinal and transverse directions. The geogrids used for this study are non-woven stiff geogrids made up of polypropylene. Table 1 presents the properties of the used geogrid as obtained from the manufacturer.

Table 1. Physical and mechanical characteristics of the biaxial geogrids used.

Property	Unit	Value
Load at 2% strain	KN/m	14
Load at 5% strain	KN/m	28
Ultimate Tensile Strength T_{ult} (L/T)	KN/m	40/40
Strain at T_{ult} (L/T)	KN/m	11/10

Note : L = Longitudinal direction ; T = Transverse direction

4.2 Portland Cement Concrete

In brief, the concrete mix adopted in this study was prepared using medium-size coarse aggregates having a nominal maximum aggregate size of 9.5 mm, fine aggregates consisting of natural sand, Type-I Portland cement and a water to cement ratio of 0.5. For 1 cubic meter of concrete, the weights of cement, sand, aggregates and water are respectively: 395, 815, 1080 and 217 Kg. The given mixture proportion is corrected based on the absorption capacity of the aggregates. After several trial mixes, the aggregate gradation and the water to cement ratio which resulted in the best mix workability in the presence of geogrids were adopted.

Three standard concrete cylinders were prepared and tested for compressive strength measurement. The average 28-day concrete compressive strength was 25 MPa. The relatively low value of the strength is expected due to the adopted water to cement ratio, which is needed for flowable concrete to easily penetrate the apertures of the geogrids. It should be noted that the use of superplasticizers wasn't necessary since the needed workability was achieved by increasing the water to cement ratio while maintaining an acceptable strength. Concrete for all beams and cylinders was cast in three successive layers with proper vibration. The reinforcement was positioned in place after laying the first 3cm concrete cover layer. All specimens were subjected to soaked curing conditions. Some pictures showing the specimen fabrication process as well as the concrete texture and slump obtained are presented in Figure 2.

Figure 2. Concrete slump/texture, and specimen fabrication.

5 RESULTS AND ANALYSIS

The load-displacement curves for the different specimens tested with related images showing the observed modes of failure in the unreinforced and geogrid-reinforced samples are presented in Figure 3. The three curves for each beam type refer to the three replicates. The plot in Figure 3-a reveals the brittle behavior of the plain concrete specimens (P), with a minimal displacement of 0.8 mm obtained at the failure load of 15 KN.

As for the beams with one layer of geogrid "S", initial cracking occurred at a similar value of load as that for the plain concrete specimens. However, the crack growth was limited and was arrested by the geogrid layer. Its further propagation slowed down significantly as the section gained post-cracking ductility. Upon failure of the geogrids, the crack instantly reached the top surface leading to complete beam failure. Since the initial crack occurred at a similar value of load as that for the plain concrete specimens, this suggests that the geogrids were not engaged and thus did not contribute mechanically during the pre-cracking phase; their main contribution was limited to providing post-cracking ductility and extra load capacity. Figure 3-b shows the load-displacement curves for the three replicates of beam specimens with one geogrid layer. Upon careful observation, it appears that geogrids failed in two modes, junction failure and rib failure. Within the post-cracking phase, the curves exhibited multiple steep drops in the load; each drop corresponds to the failure of one or more ribs or junctions. The specimen that experienced rib failure exhibited relatively higher ductility but lower load capacity than the specimens that exhibited junction failures. This was indicated by the higher displacement (9mm) and lower load attained (1.5KN) just before failure. It is believed that when the initial crack hits the geogrid ribs first, they start to fail one after another resulting in the observed shape in Figure 3-b of the load-displacement curve for the rib failure mode. However, for specimens in which the initial crack propagated and reached the geogrid junctions first, the post-cracking load capacity is sustained without any noticeable drops in load until sudden failure at higher load (14KN) and slightly lower displacement values (7mm). It should be noted that the maximum load reached in the post-cracking phase (approximately 14KN) was more than 90% of the load that initiated the first crack in the concrete section.

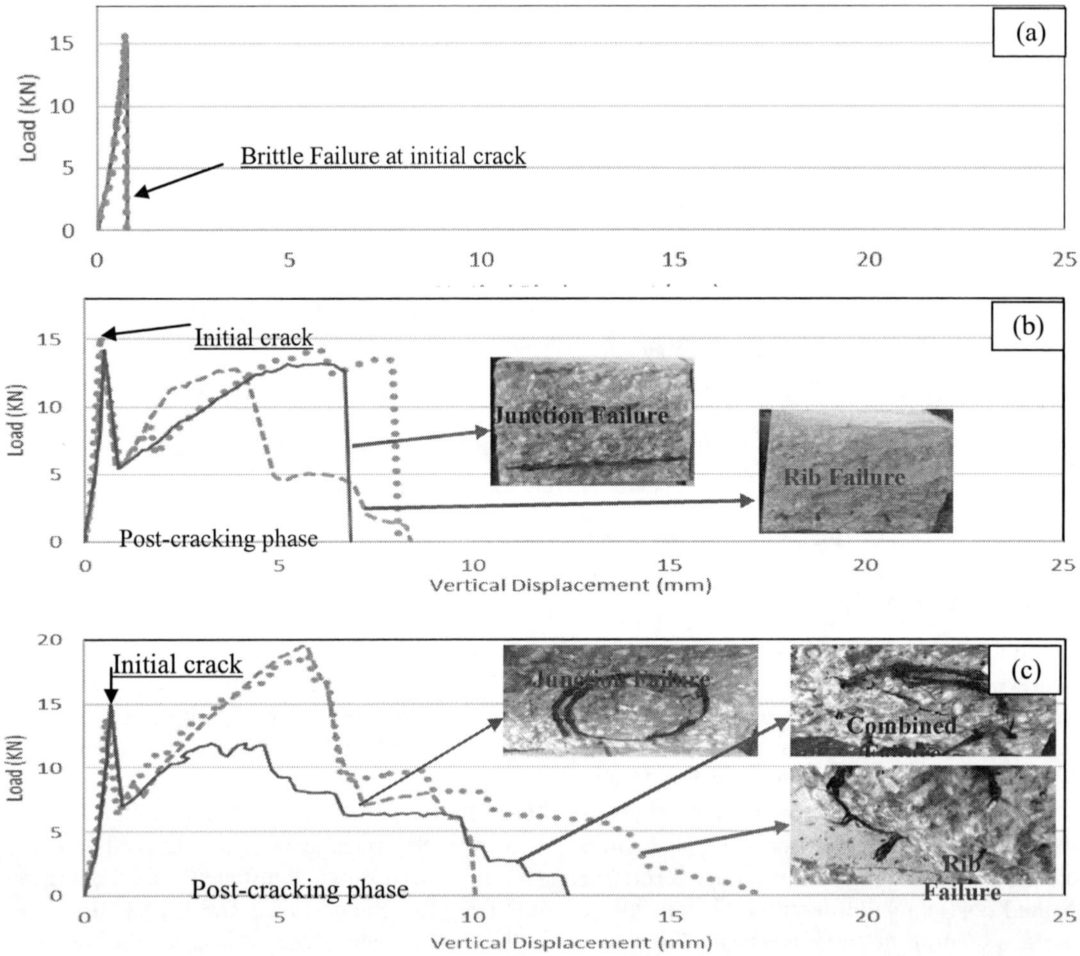

Figure 3. Load vs. vertical displacement patterns and failure modes for beam types: (a) P, (b) S, and (c) O.

As for the "O" category, where the geogrids were rolled and placed to form a circular cross section, the behavior of each replicate varied based on the observed failure mode, rib vs. junction. Similar to the "1-G" samples, geogrids did not contribute much in resisting the load during the pre-cracking phase. After the initial crack, geogrids start to resist the bottom tensile stresses while controlling crack propagation. During the post-cracking phase, the section significantly gained load capacity until it peaked at 20 KN, which is around 130 % of the load at the initial crack, before it started to decrease gradually as some ribs and junctions started to fail progressively. It should be noted that the "O" configuration yielded larger displacements before failure and attained a higher load capacity during the post-cracking phase than the "1-G" category. The load-displacement curves of the "O" specimens as well as the observed failure modes are presented in Figure 3-c. Similar to the "1-G" category, the location and trajectory of the initial crack, and thus the location at which the crack hits the geogrid, is the main factor that controls whether the geogrid will fail along its ribs, junctions, or a combination of both. It can be seen from the load-displacement patterns that specimens failing along ribs exhibited higher ductility with displacement at failure reaching 17mm which can be explained by the relatively stiffer nature of the junctions.

6 CONCLUSIONS AND RECOMMENDATIONS

Reinforcing concrete beams with geogrids, either in the form of a sheet or a circular cross section, provided the samples with post-cracking ductility and load capacity. Furthermore, the circular-shaped geogrid reinforcement resulted in a better performance compared to the geogrid sheet. The circular-shaped geogrid reinforcement was more effective in the control of crack opening and propagation compared to one-layer geogrid reinforcement, which is evident from the lower release in energy (drop in the load) at the initial crack. It also resulted in higher post-cracking load capacity and endured larger displacement before failure. The circular-shaped geogrid addressed the issues of concrete section weakening and concrete separation along the geogrids which were observed in earlier studies.

Furthermore, three different types of failure were observed: Junction failure, rib failure, and a combination of both. It was noticed that ribs exhibited a more ductile behavior where they failed gradually one after another until the section failed at a low applied load. The junctions, on the other hand, provided ductility in terms of the displacement at failure, but they remained intact, maintaining a relatively high load capacity, until they suddenly failed altogether. The results of this study proved the expected benefits of the use of the circular shaped geogrid as a reinforcing tool in concrete beams and validated the behavior of stiff biaxial geogrids in concrete elements.

References

Al Hedad, A. S., Bambridge, E., and Hadi, M. N., Influence of Geogrid on The Drying Shrinkage Performance of Concrete Pavements, *Construction and Building Materials*, 146, 165-174, 2017.

ASTM C78/C78M-16, American Society for Testing and Materials, Committee C-9 on Concrete and Concrete Aggregates, *Standard Test Method for Flexural Strength of Concrete (Using Simple Beam with Third-Point Loading)*. ASTM International, (2016).

Chidambaram, R. S., and Agarwal, P., The Confining Effect of Geo-Grid on The Mechanical Properties of Concrete Specimens with Steel Fiber Under Compression and Flexure, *Construction and Building Materials*, 71, 628-637, 2014.

Chidambaram, R. S., and Agarwal, P., Flexural and Shear Behavior of Geo-Grid Confined RC Beams with Steel Fiber Reinforced Concrete. *Construction and Building Materials*, 78, 271-280, 2015.

El Meski, F. and Chehab, G. R., Flexural Behavior of Concrete Beams Reinforced with Different Types of Geogrids, *Journal of Materials in Civil Engineering*, 26(8), 04014038, 2014.

Itani, H., Saad, G., and Chehab, G., The Use of Geogrid Reinforcement for Enhancing the Performance of Concrete Overlays: An Experimental and Numerical Assessment, *Construction and Building Materials*, 124, 826-837, 2016.

Tang, X., Kim, S., and Chehab, G. R., Laboratory Study of Geogrid Reinforcement in Portland Cement Concrete, *RILEM International Conference on Cracking in Pavements,* Chicago, USA, 2008.

Responsible Design and Delivery of the Constructed Project
Edited by Abdul-Malak, M., Khoury, H., Singh, A., and Yazdani, S.

USE OF RECYCLED GLASS IN CONCRETE PRODUCTION

DANIELA IONESCU, ANDREW KILPATRICK, and JOE PETROLITO

Dept of Engineering, La Trobe University, Australia

Worldwide, rapid population growth causes an increasing production of solid waste, and sustainable waste management solutions are required to deal with this. Waste glass is a significant component of the solid waste stream, and most of the recycled glass cullet is used for new glassware. However, waste containers with mixed types of glass, such as windows, ovenware, and crystal, which have higher melting points and are of different colors, cannot be used for new glassware. Hence, the contaminated load is sent to landfill. As landfill capacity is limited, it is critical to consider using this waste material for other applications. While the effects of the substitution of fine or coarse aggregates have been extensively researched, information on the effects of partial replacement of coarse aggregates and total replacement of fine aggregate with glass cullet is scarce. This paper reports on continuing research at La Trobe University on using recycled aggregates in the production of concrete. The effects of partial or total replacement of coarse crushed rock aggregates with recycled glass aggregates on the strength and durability of concrete were investigated. It was found that concrete mixes produced with recycled glass exhibited lower performance than the mixes produced with crushed rock aggregates. However, the production costs were lower and the appearance of concrete was improved, making it applicable for finishing products, noise barriers and light-trafficked footpaths.

Keywords: Durability, Environmentally friendly, Glass cullet, Glasscrete, Laboratory investigation, Mechanical properties, Strength development, Waste glass.

1 INTRODUCTION

Modern societies require massive quantities of construction materials to build and maintain infrastructure. One of the preferred construction materials is concrete, due to its design and construction adaptability, mechanical properties, durability and economic viability (Kosmatka and Wilson, 2011). However, the extraction and processing of natural resources to produce aggregates and cement production are both energy-intensive and costly processes. Hence, the use of recycled aggregates for concrete production has become common practice to conserve natural resources, minimize environmental impacts and reduce solid waste (Ionescu 2005, Ionescu *et al.* 2015, Ionescu *et al.* 2016).

Glass is an inert material that can be recycled and reprocessed indefinitely without any loss in quality, although the recycled glass must satisfy strict specifications for a high-quality product. Glass recovery from mixed waste loads is a difficult and costly process that includes sorting by color and crushing the separated glass to form a product called cullet, which is then sent to the glass beneficiation plant. Other types of glass (flow glass, window and mirrors glass, ovenware,

Pyrex, crystal, etc.), which are normally present in very small amounts (around 5 g/ton), cause contamination, and an entire load of recovered glass can end up in landfill (Randell *et al.* 2014). Moreover, automated sorting equipment tends to break the glass, and it ends up in smaller and smaller pieces until it is not readily recoverable (Randell *et al.* 2014). Hence, it is imperative to find an alternative use for the considerable quantities of broken, mixed color and contaminated glass to minimize the environmental impact.

Total or partial replacement of natural aggregates with glass cullet, as well as supplementing cementing materials with glass powder during concrete production, results in an environmentally-friendly concrete that is known as glasscrete (CCAA 2008). Successful strategies for using waste glass can be implemented if its engineering properties and its effects on the performance of concrete are established. To date, research on glasscrete performance has investigated the discrete effects of either glass powder supplement for cementitious materials and/or the substitution for fine aggregates (Omran e*t al.* 2017), and the replacement of coarse aggregates (Serniabat *et al.* 2014). Despite the positive outcomes of using recycled glass for concrete production, the processes of separating glass cullet into coarse aggregate or fine aggregate fractions, as well as grinding it down to powder to supplement the cementing materials, adds to the production costs (Serniabat *et al.* 2014). In addition, total replacement of coarse aggregate with glass cullet results in reduced concrete durability, due to the alkali-silica reaction (Omran e*t al.* 2017). Furthermore, there are no accepted specifications for the physical and chemical properties of glass cullet when used as a substitute of aggregates in concrete (CCAA 2008). Hence, this paper is focused on the performance of glasscrete produced with the partial replacement of coarse aggregates and total replacement of fine aggregate with a readily-available glass cullet.

2 MATERIALS REQUIREMENTS

Concrete comprises about 75% by volume of aggregates, which have a crucial effect on concrete properties (Neville and Brooks 2010). Past research into the possible uses of glass cullet (Spitty 2003) suggested that crushed glass has most of the required properties for concrete aggregates, i.e., it is inert, has angular grains and a broad gradation to produce a minimum voids space in the concrete matrix. Hence, it may be used as a partial or total replacement for naturally-occurring aggregates in glasscrete production.

3 EXPERIMENTAL PROGRAM

The supplied materials were subjected to physical characteristic tests in accordance with the relevant Australian Standards (primarily AS 1012, AS 1289, AS 2758) to acquire a reasonable indication of their mechanical properties. Compliance with the current specifications and the economic feasibility of obtaining the optimum material were the prevailing factors during the study.

3.1 Concrete Aggregate

The aggregates were used as supplied (without additional crushing, sieving or washing), to maximize production cost savings using glass cullet. It was reported that finer-sized, green glass cullet minimizes the alkali-silica reaction (Omran *et al.* 2017). Visy supplied a green glass cullet with a nominal size of 6 mm from its glass and cardboard resource recovery plant in Melbourne. Hymix Quarry in Axedale (HQA) supplied the coarse basalt in fractions and the washed river sand in a blended state. Gradation 1 (20 mm) and Gradation 2 (10 mm) for aggregates for concrete production (Standards Australia 2014b) were used to combine the supplied aggregates as follows:

- HQA 20 mm coarse basalt and blended washed river sand (20B-CA + BWRS-FA)
- HQA 20 mm coarse basalt blended with finer than 6 mm Visy glass cullet (20B&GC-CA + GC-FA)
- HQA 10 mm coarse basalt and blended washed river sand (10B-CA + BWRS-FA)
- HQA 10 mm coarse basalt blended with finer than 6 mm Visy glass cullet (10B&GC-CA + GC-FA).

This resulted in a total replacement of natural sand with glass cullet, whereas the coarser aggregates were replaced in proportions of 31.2% to 34.3% by waste glass for the 10 mm and 20 mm maximum grain size gradations, respectively.

Figure 1(a) presents the gradation for the materials supplied for this research, compared with the current specifications for both natural and manufactured sand. The blended washed river sand (BWRS) sand fits well between the boundaries of both natural and manufactured sand, whereas the glass cullet (GC) is parallel to the coarser boundary for sand, and comprises 45% by weight of fractions coarser than 4.75 mm. Figure 1(b) shows the CA + FA combined gradations against the ideal Gradation 1 for a maximum aggregate size 20 mm and the ideal Gradation 2 for a maximum grain size 10 mm, respectively. Both gradations containing glass cullet (20B&GC-CA + GC-FA and 10B&GC-CA +GC-FA) are below the ideal gradations, indicating that they lacked fines, particularly the fractions between 2.36 mm and 0.6 mm. The chemical composition of the glass cullet is presented in Table 1. The BWRS fines (grains < 0.425 mm) were tested for any clay content. A summary of the consistency tests (Standards Australia 2009) and the clay and silt contents (Standards Australia 2015) is presented in Table 2. It was found that the finer fractions classify as non-plastic/low plasticity silts. In addition, the physical properties of both coarse and fine aggregates were determined in accordance with the relevant specifications (AS 1141.6.1 2000), and they are summarized in Table 3. As expected, the glass cullet was inert to water absorption, and it was only 4% to 11% lighter than the basalt and blended washed river sand, resulting in a concrete with comparable properties to concrete produced with natural aggregates.

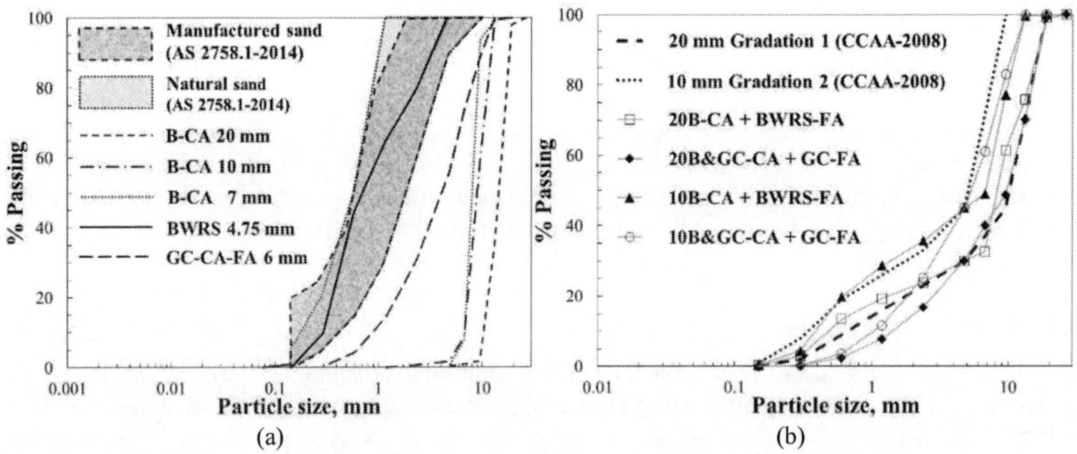

Figure 1. Particle size distribution of aggregates used to prepare the concrete specimens: (a) supplied aggregates; (b) combined gradations (CA + FA) for both nominal sizes 20 mm and 10 mm, respectively.

Table 1. Chemical composition of the glass cullet supplied from Visy.

SiO_2 (wt%)	Al_2O_3 (wt%)	Fe_2O_3 (wt%)	CaO (wt%)	Na_2O (wt%)	MgO (wt%)	K_2O (wt%)	Cr_2O_3 (wt%)	MnO (wt%)	TiO_2 (wt%)	SO_3 (wt%)	P_2O_5 (wt%)
67.8	1.53	0.36	10.93	11.09	1.58	0.55	0.1	0.018	0.062	0.03	0.023

Table 2. Consistency characteristics and silt content in fine aggregates.

Material	Liquid limit (%)	Plastic limit (%)	Plasticity index (%)	Linear shrinkage (%)	Silt content (%)
BWRS-FA	18	14	4	2	6
GC-FA	1	NP	-	0	4

Table 3. Physical properties of aggregates used.

Material	Water absorption (%)	Moisture content (%)	Particle density SSD (t/m³)
B-CA	2.23	3.0	2.72
GC-CA	-	-	2.41
BWRS-FA	0.6	7.5	2.61
GC-FA	-	-	2.51

3.2 Concrete Strength

The quantities for the four concrete mixes were based on a nominal characteristic compressive strength of 32 MPa and are listed in Table 4. Concrete was produced with Portland cement and aggregates in as-supplied conditions. The mix design used a 50-mm slump, and the water quantities for concrete preparation were corrected for the moisture content of the natural aggregates.

Table 4. Proportions used for the mix design.

Mix type	Cement (kg)	Water (kg)	Fine aggregates(kg)	Coarse aggregates (kg)
20B-CA + BWRS-FA	370	158	550	1270
20B&GC-CA + GC-FA	370	196	870	790
10B-CA + BWRS-FA	390	160	780	950
10B&GC-CA + GC-FA	390	212	1240	300

Table 5 summarizes the strength at 7 days and 28 days. Glasscrete produced a lower strength when compared with the control batches, with strength reductions of 29.7% (20B&GC-CA + GC-FA) and 40.8% (10B&GC-CA + GC-FA) at 7 days. At 28 days, the strength of the glasscrete was 35.1% (20B&GC-CA + GC-FA) and 46.6% (10B&GC-CA + GC-FA) lower than that of the control concrete. The trend was similar for the flexural and splitting tensile strengths.

Table 5. Average strength of hardened concrete.

Mix type	7 days			28 days		
	f_{cm} (MPa)	$f_{ct.f}$ (MPa)	$f_{ct.sp}$ (MPa)	f_{cm} (MPa)	$f_{ct.f}$ (MPa)	$f_{ct.sp}$ (MPa)
20B-CA + BWRS-FA	45.5	4.1	2.4	57.0	5.3	4.3
20B&GC-CA + GC-FA	32.0	3.4	2.0	37.0	4.0	2.5
10B-CA + BWRS-FA	51.7	4.3	2.6	65.5	6.2	4.5
10B&GC-CA + GC-FA	30.5	3.3	2.0	35.0	3.7	2.4

Note: measured f_{cm} = mean compressive strength, $f_{ct.f}$ = flexural tensile strength, $f_{ct.sp}$ = splitting tensile strength

3.3 Durability of Concrete

An indirect correlation exists between the durability of concrete and the water permeable voids present in the hardened concrete. The water absorption and the apparent volume of permeable voids were determined from tests performed in accordance with Australian Standards (Standards Australia 2014a), and the results are summarized in Table 6. The 20B&GC-CA + GC-FA mix showed a reduction, whereas the 10B&GC-CA + GC-FA mix showed a slight increase in the apparent volume of permeable voids.

Table 6. Apparent volume of permeable voids for 28 days old concrete.

Mix type	Immersed absorption (%)	Boiled absorption (%)	Apparent volume permeable voids (%)
20B-CA + BWRS-FA	7.0	7.1	16.1
20B&GC-CA + GC-FA	7.3	7.7	15.2
10B-CA + BWRS-FA	6.8	6.9	15.7
10B&GC-CA + GC-FA	6.7	7.2	15.2

4 CONCLUSIONS

The effects of the partial replacement of coarse aggregates and the total replacement of natural sand with glass cullet for concrete production were discussed in this paper, and the following conclusions can be drawn.

The use of waste glass resulted in a significant reduction in the strength of the design batches, although they satisfied the design compressive strength requirements. This trend was also observed for the flexural and splitting tensile strength of glasscrete. The durability of the glasscrete was also affected, with both measurements for water absorption showing increased values. The apparent volume of permeable voids results was inconclusive, as they contradict the trends observed for water absorption.

Partial replacement of coarse natural aggregates and total replacement of natural sand with glass cullet appear to be suitable solutions for concrete production, although, their use may result in a less-durable concrete, due to a slightly higher volume of permeable voids. However, the cost of production of concrete with recycled glass aggregates is lower than that of natural aggregate concrete, while providing environmental benefits. Furthermore, the appearance of glasscrete makes it attractive for architectural applications such as finishing products, noise barriers and low trafficked footpaths.

The strength of glasscrete may be improved if partial replacement of fine aggregates is used instead of total replacement. Additionally, the durability of concrete may possibly be improved if water-repellent admixtures are added to the mix. Future research is required to study these aspects.

Acknowledgements

The authors gratefully acknowledge the help provided by All Stone Quarry, Bendigo and Visy Melbourne in supplying materials free of charge. The assistance with the preparation and testing of specimens provided by Mr. David Osborne is also acknowledged.

References

Cement Concrete and Aggregates Australia (CCAA), *Use of Recycled Aggregates in Construction*, CCAA Publications, Sydney, 2008.

Ionescu, D., Construction Potential of a Mining By-Product, *Achieving Sustainability in Construction*, Dhir, R. K., Dyer, T. D. and Newlands, M. D. (eds), 169-176, Dundee, United Kingdom, 2005.

Ionescu, D., Petrolito, J., Forrest, J., German, S. and Pinnuck, S., Effects of Alternative Aggregates on Concrete Strength, *Interaction between Theory and Practice in Civil Engineering and Construction*, Komurlu, R., Gurgun, A. P., Singh, A., Yazdani, S. (eds), ISEC Press, 2016.

Ionescu, D., Petrolito, J., Mainali, B., Sweet, J. and Robinson, D., Alternate Aggregates for Concrete, *Science, Engineering and Environment*, Vol. 1, Hossain, Z. and Kaneco, S. (eds.), 318-323, Tsu, Mie, Japan, 2015.

Kosmatka, S. H. and Wilson, M. L., *Design and Control of Concrete Mixtures: The Guide to Applications, Methods and Materials*, Portland Cement Association, USA, 2011.

Neville, A. M. and Brooks J.J. *Concrete Technology*, Pearson, United Kingdom, 2010.

Omran A. F., Morin, E. D., Harbec, D., Tagnit-Hamou, A., Long-Term Performance of Glass-Powder Concrete in Large-Scale Field Applications, *Construction and Building Materials*, 135, 43–58, January, 2017.

Randell, P., Pickin, J. and Grant, B., Waste Generation and Resource Recovery in Australia, *Reporting Period 2010/11*, 4 February 2014.

Serniabat T. S., Khan, M. N. N. and Zain, M. F. M., Use of Waste Glass as Coarse Aggregate in Concrete: A Possibility towards Sustainable Building Construction, *International Journal of Civil, Environmental, Structural, Construction and Architectural Engineering,* 8(10), 1071-1078, 2014.

Spitty, A., *The Use of Recycled Glass as Aggregate in New Concrete*, La Trobe University, Bendigo, Australia, 2003.

Standards Australia, *Methods of Testing Soils for Engineering Purposes, AS 1289*, Standards Australia, Sydney, Australia, 2009.

Standards Australia, *Methods of Testing Concrete – Determination of water absorption and apparent volume of permeable voids in hardened concrete, AS 1012.21*, Standards Australia, Sydney, Australia, 2014a.

Standards Australia, *Aggregates and Rock for Engineering Purposes - Concrete Aggregates, AS 2758.1*, Standards Australia, Sydney, Australia, 2014b.

Standards Australia, *Methods of Sampling and Testing Aggregates Methods for sampling and testing aggregates, AS 1141.6.1, AS 1141.33*, Standards Australia, Sydney, Australia, 2015.

Responsible Design and Delivery of the Constructed Project
Edited by Abdul-Malak, M., Khoury, H., Singh, A., and Yazdani, S.

GLASS FIBER POLYMER ON THE FRESH AND HARDENED PROPERTIES OF CONCRETE

MOHAMMED JOBAER UDDIN[1], ARIFUL HASNAT[1], MOHAMMAD S ISLAM[2], and SYED F AHMAD[3]

[1]*Dept of Civil Engineering, University of Asia Pacific, Dhaka, Bangladesh*
[2]*Dept of Civil Engineering, University of Tabuk, Tabuk, Saudi Arabia*
[3]*Lievense, Arabia Consulting Engineers, Riyadh, Saudi Arabia*

The study describes an experimental study into the relationship among incorporation of Glass Fiber on the fresh and hardened properties of concrete. The effect of fiber fractions on the slump, compressive strength, splitting tensile strength and flexural strength of Glass Fiber Reinforced Concrete (GFRC) were investigated for volume fraction (VF) of 0%, 1%, 2% and 4%. In order to serve the purpose, concrete cylindrical and prism specimens were made with various rates of fiber-glass polymer. The cylinder specimens were tested at 7 and 28 days for compressive strength and at 28 days for splitting tensile strength, whereas the prism specimens were tested at 28 days for flexural strength. The experimental test results show that the additions of glass fiber polymer to concrete decrease slump, but increase compressive strength, splitting tensile strength and flexural strength. In addition, an analytical model has been proposed to predict slump of fresh concrete.

Keywords: GFRC, Workability, Compressive strength, Tensile strength, Flexural strength, Slump prediction model.

1 INTRODUCTION

Concrete is the most consumed construction material. From its invention in roman era, its applications in different industries have ever been rising. This phenomenon is mainly attributed by its usefulness over other types of construction materials in respect of strength, durability and economy (Yoo *et al.* 2013). However, susceptibility to weak in tension and low fracture strain capacity has severely censored its application (Chandramouli *et al.* 2010). To compensate this brittleness nature of concrete, fiber reinforced concrete is often considered as an alternative. Fiber has been used to reinforce brittle matrix from as early as Biblical period (Brandt 2008). Fibers classify as steel, glass, synthetic and natural, has been incorporated in concrete to strengthen concrete in tension (Coenen *et al.* 1977, Kosmatka *et al.* 2011). Among all these, steel fiber is the most widely used in concrete (Brandt 2008, Islam and Alam 2013, Islam 2017). However, vulnerability of steel fiber to corrosion limits its applications. To resolve this problem glass fiber is often inferred as an alternative. Earlier researches have revealed that incorporation of glass fiber in concrete reduces workability, i.e. slump of fresh concrete. However, the effect of fiber inclusion on compressive strength of concrete is still an issue of argument (Kizilkanat *et al.* 2015). In addition to this, previous studies reported that addition of glass fiber in concrete improves tensile strength, flexural strength, toughness, ductility of plain concrete.

2 MATERIALS

2.1 Cement

Blended hydraulic Portland composite cement, CEM II/BM (S-V-L) was used as cementing material in the study.

2.2 Aggregates

Bholagonj stone and Sylhet sand, locally available materials were used respectively as coarse aggregate (CA) and fine aggregate (FA) in the study. The specific gravity of the coarse aggregate and fine aggregate are respectively 2.66 and 2.45, and unit weight are 1543.27 kg/m^3 and 1667.92 kg/m^3. ASTM C136 was followed to obtain grain size distribution, and represented in Figure 1.

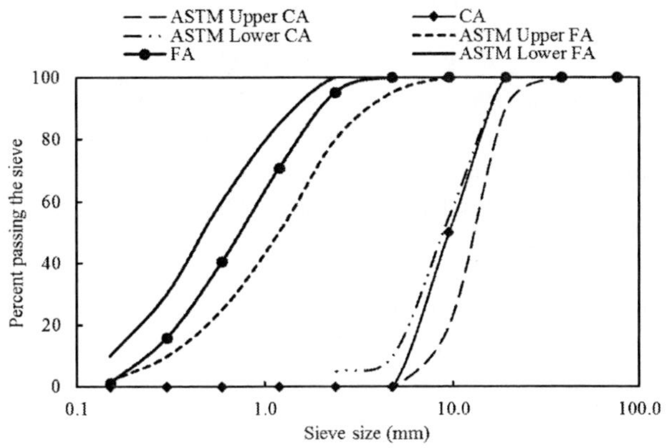

Figure 1. Gradation curves of the aggregates used in this study.

2.3 Glass Fiber Polymer

Glass fiber of density 1700-2250 kg/m^2, compressive and tensile strength of 490-840 kg/cm^2 and 210-320 kg/cm^2 respectively were utilized in this study. The macro fiber possesses young's modulus of 104 MPa, elongation at break was of 0.6% to 1.2%.

2.4 Mix Design and Test Specimens

Four cases of mix proportions were designed by volumetrically varying the fiber content to concrete at different fractions to perform the study. The glass fiber mesh was added at a fraction of 0%, 1%, 2% and 4% respectively. The water to cement ratio (w/c) of 0.45, and the sand to aggregate ratio (s/a) of 0.42 was kept constant for all mixes. The mixture constituents for concrete cylinder and beam specimens are documented in Table 1.

Table 1. Mix proportions for concrete cylinder and beam specimens.

Volume Fraction (VF) (%)	Coarse Aggregate (kg/m^3)	Fine Aggregate (kg/m^3)	Cement (kg/m^3)	Water (kg/m^3)
0	1534.15	1023.23	7.05	3.17
1				
2				
4				

2.5 The Experiment's Methods

For each of the four cases shown in Table 1, a total of 1.07 ft^3 (0.03 m^3) of concrete was prepared to conduct slump test as per ASTM C143, and thirty-two 4 in X 8 in (10 cm X 20 cm) concrete cylinders and eight 4 in X 4 in X 11.5 in (10 cm X 10 cm X 29 cm) beams were also prepared to determine the mechanical properties of hardened concrete. Concrete cylinders were tested as per ASTM C39 at 7 and 28 days to determine the compressive strength of concrete and for splitting tensile cylinders test as per ASTM C496 at 28 days. The stress-strain responses of specimens also recorded for compressive strength at 28 days. The beam specimens were subjected to center point loadings as per ASTM C293 to measure the flexural strength at 28 days.

3 RESULTS & DISCUSSIONS

3.1 Compressive Strength and Splitting Tensile Strength

Compressive strength test result, at 7 days and 28 days has been demonstrated in Figure 2. At 7 days, reduction in average compressive strength was attributed for VF of 1% and 4%, as shown in Figure 2. Nevertheless, VF of 4% showed an increase in compressive strength to control at 28 days alone. For rest of the investigated cases, a reduction in compressive strength with glass fiber incorporation. However, it was noticed that presence of glass fiber prevented development of crack width in concrete.

Influence of fiber addition on tensile strength of hardened concrete has been demonstrated in Figure 3. Overall, an increase in tensile strength is observed at 28 days, 1.67% and 14.45% enhancement in tensile strength respectively for VF 2% and VF 4%. However, VF 1% shows a decline in tensile strength by 10.89%. It was also noticed that presence of glass fiber limits complete split of concrete by abridging the crack and transferring stress. These eventually prevent complete rupture of concrete, and shows numerous micro crack in concrete.

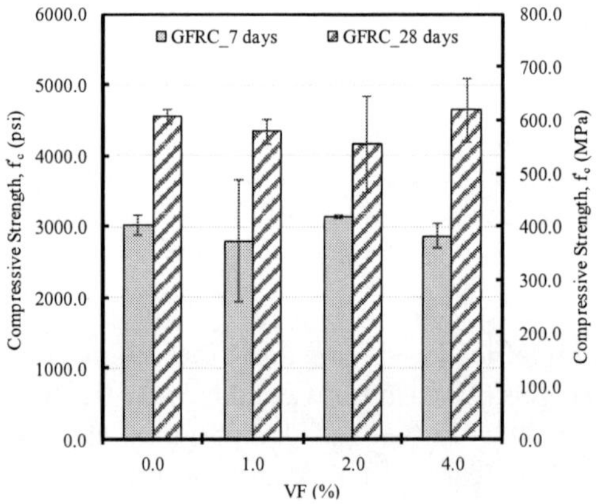

Figure 2. Influence of fiber incorporation on compressive strength of concrete.

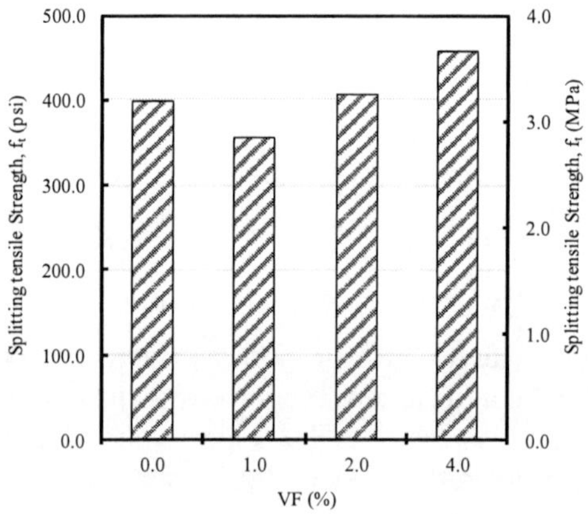

Figure 3. Influence of fiber addition on splitting tensile strength of concrete.

3.2 Stress-Strain Response

Stress-strain responses recorded at 28 days has been illustrated in Figure 4. The slopes of stress-strain responses are quite similar, however an increase in strain capacity with increase in fiber fractions is also depicted from Figure 4. In case of VF 1% and VF 2%, there are not noticed any significant enhancement in terms of both stress and strain capacities. However, for VF of 4% there were significant increase in both compressive strength and strain, as shown in Figure 4.

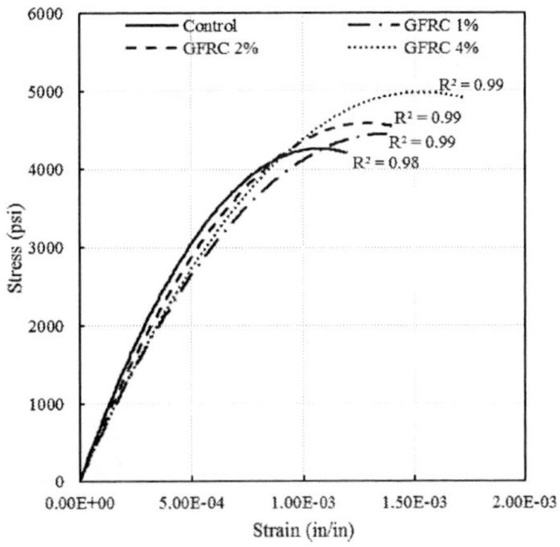

Figure 4. Stress-strain response at 28 days.

3.3 Flexural Strength

Beam test recorded results at 28 has been illustrated in Figure 5. Overall an enhancement in flexural strength is overserved with fiber addition, as shown in Figure 5. Figure 5 shows approximately 33% enhancement in flexural strength capacity for both VF of 1.0% and 2.0%, whereas about 53% gain in flexural strength was achieved for VF 4.0%.

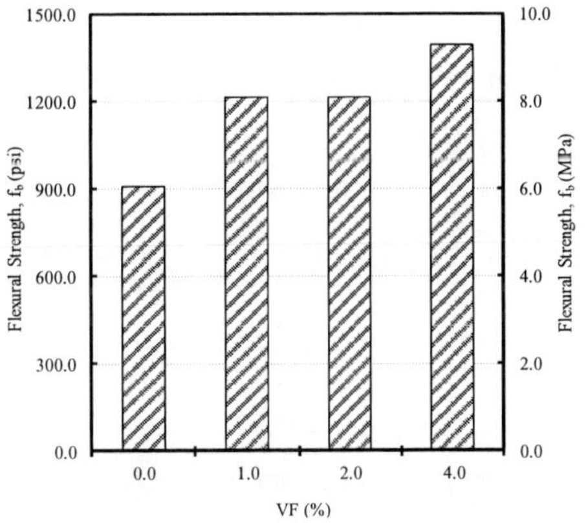

Figure 5. Influence of fiber addition on flexural strength of concrete.

3.4 Proposed Model for the Prediction of Slump of GFRC

Figure 6 shows the effect of the glass fiber content on the slump of GFRC. As can be shown, the slump value decreased linearly with an increase in fiber content. The simple model that described the data set with R^2 value of 85% is shown in Eq. 1. The Prob (F) and F-ratio values of Eq. 1 shown to be 0.078 and 11.23, respectively. The t-ratio for the regression parameters a and b were found to be more than 1.0. These showed a very good evidence that the propose model is more likely to predict the nearly accurate slump value of the GFRC. The proposed model was used to determine the analytical slump value of GFRC and compared with the experimental values obtained from this study. The results are shown in Figure 7. As can be seen, the predicted slump values obtained from the proposed model showed a very good alignment having R^2 values of 84% with the actual slump values generated from the experimental procedures.

$$S_f = a\,V_f + b \tag{1}$$

Where: S_f is the slump value of GFRC in cm; V_f is the fiber content in percentage; a and b are the coefficients and their values are -1.0486 and 5.36, respectively.

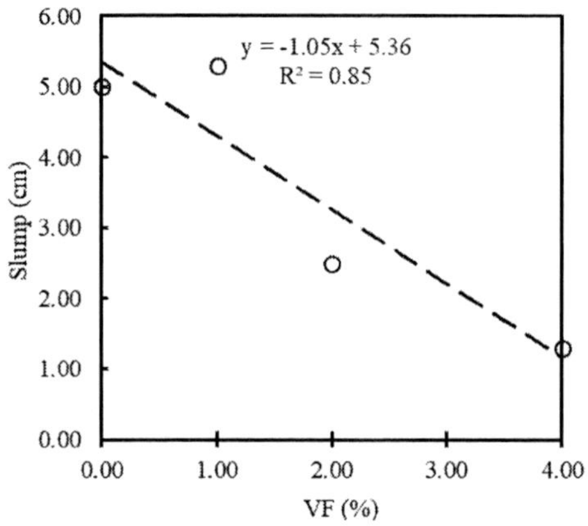

Figure 6. Influence of fiber addition on slump of concrete.

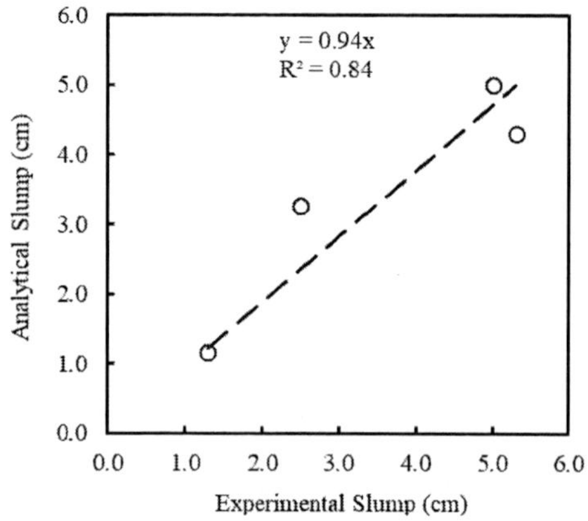

Figure 7. Relationship between analytical slump and experimental slump.

CONCLUSIONS

The findings of this study can be summarized as follows:

(a) Incorporation of glass fiber increases viscosity of concrete and thus reduces flowability.

(b) Effect of addition of fiber is not noteworthy on concrete compressive strength and modulus of elasticity.

(c) Effect of fiber incorporation on tensile strength of concrete is not quite appreciable.

(d) Fiber incorporation increase flexural strength of plain concrete remarkably.

Acknowledgements

This research was supported by University of Asia Pacific, Bangladesh. The authors would like to show their gratefulness to AuspiciousBD Co. Ltd for facilitating this research by supplying glass fibers.

References

Brandt, A. M., Fibre Reinforced Cement-Based (FRC) Composites After Over 40 Years of Development in Building and Civil Engineering, *Composite Structures,* Elsevier, 86(1), 3-9, November, 2008.

Chandramouli, K., Srinivasa, R. P., Pannirselvam, N., Seshadri, S. T., and Sravana, P., Strength Properties of Glass Fiber Concrete, *ARPN Journal of Engineering and Applied Sciences*, 5(4), 1-6, 2010.

Coenen, M. C., *Glass Compositions Suitable for Incorporation into Concrete*, Jenaer Glaswerk Schott & Gen., assignee, United States Patent US 4,002,482, January, 1977.

Islam, M. S., Simplified Shear Strength Prediction Models for SFRC Beams Using Genetic Algorithm, *Proceedings of the Institution of Civil Engineering (ICE)-Construction Materials,* https://doi.org/10.1680/jcoma.16.00073, 2017.

Islam, M. S., and Alam, S., Principal Component and Multiple Regression Analysis for Steel Fiber Reinforced Concrete (SFRC) Beams, in *International Journal of Concrete Structures and Materials,* 7(4), 253-266, 2013.

Kizilkanat, A. B., Kabay, N., Akyüncü, V., Chowdhury, S., and Akça, A. H., Mechanical Properties and Fracture Behavior of Basalt and Glass Fiber Reinforced Concrete: An Experimental Study, *Construction and Building Materials,* Elsevier, 100, 218-224, December, 2015.

Kosmatka, S. H., Panarese, W. C., and Kerkhoff, B., *Design and Control of Concrete Mixtures*, Portland Cement Association, Illinois, 2002.

Yoo, D. Y., Lee, J. H., and Yoon, Y. S., Effect of Fiber Content on Mechanical and Fracture Properties of Ultra High Performance Fiber Reinforced Cementitious Composites, *Composite Structures*, Elsevier,106, 742-753, December, 2013.

Responsible Design and Delivery of the Constructed Project
Edited by Abdul-Malak, M., Khoury, H., Singh, A., and Yazdani, S.

RESPONSIBLE DELIVERY OF THE BUILT PROJECT: WHOSE RESPONSIBILITY IS IT?

MAYSSA KALACH, MOHAMED-ASEM ABDUL-MALAK, and ISSAM SROUR

Civil and Environmental Engineering, American University of Beirut, Beirut, Lebanon

Several procurement approaches have evolved over the years to suit the delivery of the ever increasing complexity and size of construction projects, coupled with the increasing pressure to embrace green methodologies as part of any such approach. Many of these available approaches reflect various levels of integration that may be desired to achieve in respect of critical project aspects, including: design, construction, operation, and finance. Decisions as to the most suitable delivery approach are to be made by project owners/developers, while being often assisted by market analysts, project management professionals, and other specialty consultants. While it is established that such a decision-making process does not follow an exact science, this paper tackles the issue of where responsibility truly rests for ensuring that the preferred or adopted project procurement strategy is consciously argued for with the interests of all involved stakeholders in mind. It touches on (1) the roles of the public and development policies and regulations prevailing in the locality of the intended project as well as (2) the capabilities and readiness of concerned project participants in enforcing, or at least, promoting more environmental, social, and economic consciousness into the procurement approach decision-making process.

Keywords: Integration, Project procurement, Green policies, Sustainability.

1 BACKGROUND

The construction industry makes a vital contribution to the social and economic development of every country and has a major impact on its environment. To set the elements of sustainable development in an engineering context, twelve guiding principles were published by the Royal Academy of Engineering (RAE). Such principles advocate forward looking advices such as: "Innovate and be creative", "Seek engagement from all stakeholders", "Make sure you know the needs and wants", and "Do things right, having decided on the right thing to do" (Dodds and Venables 2005). Aiming for a successful built project, owners/developers are increasingly realizing that focusing on the delivery process of the project may be as important as focusing on its technical aspect. A project delivery system is defined as a system that determines (1) the relationships between the different project stakeholders and (2) their timing of engagement in the process of providing the intended built facility (El Asmar *et al.* 2013). To suit the delivery of the ever-increasing complexity and size of construction projects, several procurement approaches have evolved over the years to replace the conventional Design-Bid-Build (DBB) method. These methods include the fast-track oriented methods (i.e., the Construction Manager at Risk (CMR) and the Design-Build (DB) method) and the Integrated Project Delivery (IPD). As for the DBB and the CMR, in both methods the owner contracts separately with the designer and the contractor for design and construction services. However, the CMR approach is characterized by an early involvement of the general contractor (GC) during the design stage to enhance constructability, in contrast to the DBB where the contractor is involved upon 100% design completion. The DB method, on the other hand, is characterized by a single contractual relationship between the owner and the design builder for design and construction services. In fact, the old concept of a master builder is resurfacing with the partial integration of design and construction under the DB approach and the more contemporary methodology of IPD. The latter approach, which is a relatively new concept is distinguished by a multiparty agreement including the owner and the very early involvement of key participants (El Asmar *et al.* 2013), i.e., at the pre-design stage.

The traditional comparison of cost, time and quality metrics of the different project delivery methods (PDMs) have been the subject of many construction research studies. For instance, El Asmar *et al.* (2013) evaluated the performance of 35 IPD projects in comparison to projects delivered using other PDMs and showed statistically significant performance outcomes in the case of IPD. Sullivan *et al.* (2017) analyzed two decades of literature covering different PDMs comparison (namely DBB, CMR and DB). The results showed that: (a) no single delivery method consistently performs better on unit cost, (b) CMR and DB were the most accurate in controlling the schedule variation of a project, and (c) DB was superior in delivery speed in all explored studies and continues to increase its advantage over time. However, such comparison metrics govern the traditional decision-making process regarding the most suitable delivery method of the built project (Sullivan *et al.* 2017).

Unlike conventional building development where the environmental effects of a built project are often disregarded, green building (GB) strategies focus on the improved environmental performance. A recent study by Ahmad *et al.* (2017) explored the effect of different PDMs on the green performance of GB projects using a systematic research review. To this end, the study defined GB projects as innovative projects and their level of green performance as an indication of their level of innovation. The authors argued that coordination among project team members is strongly influenced by the PDM which by itself impacts project innovation. Depending on the extent of innovative features incorporated, each PDM was found to have the capacity to produce successful results (Ahmad *et al.* 2017).

As for a responsible delivery process, it is believed to address what is beyond embracing green methodologies, to manifest as a holistic decision-making process addressing the constraints of economic, social and environmental factors and embracing the needs of all involved stakeholders, resulting in a sustainable investment. As such, more criteria than what are often adopted in the traditional decision-making process need to be considered and evaluated. Based on how the owner chooses to put into effect integration/concurrency/etc., the choice made can end up better serving the social, economic, and/or environmental dimension(s) of sustainable (responsible) development. On one hand, different project delivery approaches reflect various levels of integration that may be desired to achieve in respect of critical project aspects. On the other hand, different approaches cater for different needs, constraints and priorities, all affecting final project outcomes, particularly sustainability goals. While it is established that a decision-making process regarding the most suitable delivery approach does not follow an exact science, this paper tackles the issue of where responsibility truly rests for ensuring that the preferred or adopted project procurement strategy is consciously argued for with the interests of all involved stakeholders in mind.

2 RESEARCH SCOPE AND METHODOLGY

While it is evident that the project owner (whether public or private) has the upper hand as to selecting the project delivery approach that better suits his interests, this paper sheds light on how such choices could be made responsibly. To this end, considerations for a responsible investment are first identified. Then, an extensive review on the impact of different attributes of PDMs on project performance is carried out and different relationships are identified accordingly. As such, when weighing the options as to the delivery approach coupled with associated strategies (partnering, alliancing, full integration, other relational contracting choices, green building, etc.), this paper explores how the owner should take into account social, economic, and environmental issues/factors/benefits, in order to end up delivering the intended facility with sustainability in mind. Consequently, a conceptual framework is presented illustrating the responsibility share/assignment for the different participants illustrating (1) the roles of the public and development policies and regulations prevailing in the locality of the intended project as well as (2) the capabilities and readiness of concerned project participants in enforcing or, at least, promoting more environmental, social, and economic consciousness into the procurement approach decision-making process.

3 CONSIDERATIONS FOR A RESPONSIBLE INVESTMENT

For a responsible investment, several considerations, as presented in Figure 1, should be evaluated. Starting by the underlying drivers (i.e., owner, project, and market drivers) behind such an investment, these are presented as driving the selection of potential procurement strategies. In fact, the selection process of the most suitable PDM often follows a "process of elimination", i.e., paring away obviously inappropriate methods until reasonable alternatives prevail. To eliminate inappropriate organizations, Gordon (1994) specified three types of drivers that must be assessed: (a) project drivers (e.g., time constraints, flexibility needs, preconstruction service needs, design process interaction and financial constraints), (b) owner drivers (e.g., construction sophistication, current capabilities, risk aversion, restriction on methods and other external factors) and (c) market drivers (e.g., availability of appropriate contractors, current state of the market and package size of the project). The type of investment - be it public, private, or a combination of both (i.e., public private partnership, PPP), and irrespective of the underlying driver - is conceptualized to be serving either the priorities of a national development or that of a private investment portfolio. For instance, in addressing the complexity of sustainable development, Fenner *et al.* (2006) emphasized that sustainability is observed differently when viewed from individual perspectives or from national or global levels, and a holistic approach in respect of the financial consideration requires that the interests of individual parties are at least viewed in the light of the wider project interests and wider community needs. However, a responsible investment, while it is normally expected to be overshadowed by the consideration of meeting the desired financial viability and regardless of the priorities it serves, should be satisfying the three pillars of sustainability, i.e., the environmental, the social and the economic aspects. As for the environmental considerations, Mukherjee and Muga (2009) studied the decision-making driving the individual adoption of sustainable technology and its diffusion in the architecture, engineering and construction (AEC) industry by presenting a combined “top-down” and “bottom-up” approach. The “top-down” perspective, i.e., that of the individual stakeholder, aids in the decision-making process (individual level) by providing comparisons between alternative designs; it develops a definition of value that can aid individual stakeholders to adopt sustainable alternatives. The “bottom-up” perspective investigates how individual adoption of green technology drives diffusion of green thinking (systems level) through different industry sectors and stakeholder groups. On one hand, the rate at which the industry is shifting towards the adoption of green technology and more sustainable practices is shaped by the diversity of priorities and differing perceptions of risk among industry stakeholders (Mukherjee and Muga

2009). On the other hand, risk management in construction projects is influenced by the procurement options (Osipova and Eriksson 2011). To this end, Osipova and Eriksson (2011) clarified how to improve risk management by adopting appropriate procurement options in terms of project delivery method, form of payment, and use of collaboration or partnering arrangements. As for the economic considerations, several studies from around the world demonstrated a growing interest in green building and sustainable design; this can be attributed to the recognition by clients that there are direct economic benefits from green buildings (e.g., higher rents or sale prices) (Ochieng *et al.* 2014). In contrast, the industry's size and fragmentation and the increased first cost associated with sustainable buildings were identified as major barriers for owners to pursue and promote sustainable construction (Ochieng *et al.* 2014). As for the social aspect of a development and its impact, Esteves *et al.* (2012) defined the social impact assessment (SIA) as the process of managing the social issues of a development. A 'good' SIA practice was therefore defined as one that: (a) is participatory, (b) supports affected peoples, proponents and regulatory agencies, (c) increases understanding of change and capacities to respond to change, (d) seeks to avoid and mitigate negative impacts and to enhance positive benefits across the life cycle of developments, and (e) emphasizes enhancing the lives of vulnerable and disadvantaged people (Esteves *et al.* 2012).

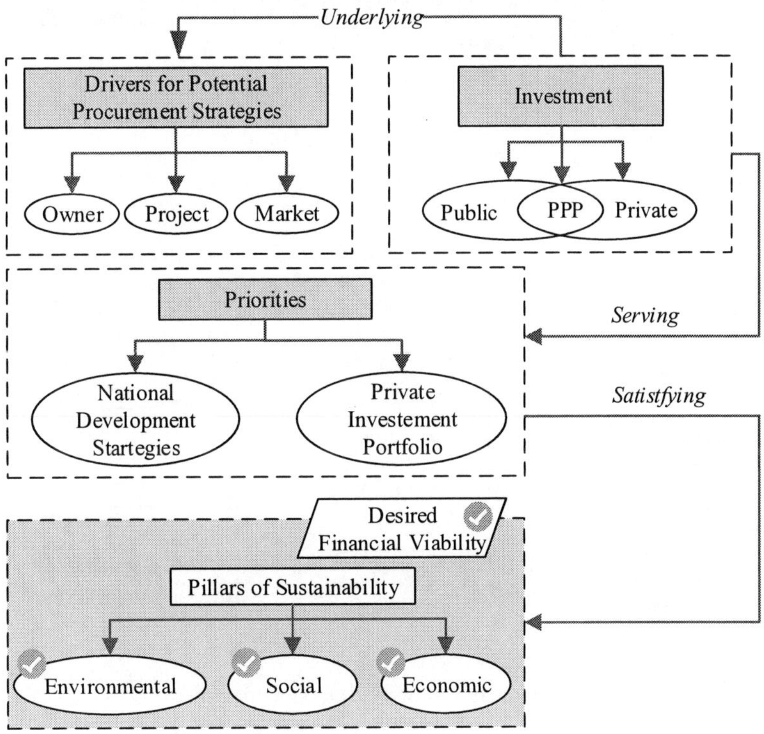

Figure 1. Considerations for a responsible investment.

4 RESPONSIBLE DELIVERY PROCESS

Different pooling of functions can be used to deliver a built facility (Figure 2). This pooling defines the basket of functions that a specific team is held responsible for. For instance, pooling design and construction under one contract refer to the DB method. Pooling construction, operation and finance designates the Build-Operate-Transfer (BOT) approach. Moreover, a team could be assigned early on to be responsible for some extent of residual planning in addition to other functions. As for financing the job, it could be either short-term or long-term financing. As such, it is presented as dashed, as it could encompass the whole process from planning till operation or stop at any level. In defining attributes for a responsible delivery process, decision makers are advised to weigh the options as to the pooling of functions coupled with associated strategies (partnering, alliancing, full integration, other relational contracting choices, green building, etc.) in order to end up delivering the intended facility with sustainability in mind. The strategies, as presented in Figure 2, are defined by: (a) the concurrency of functions, (b) the timing of involvement of the general contractor (GC) or the construction manager (CM) and (c) the level of participants' integration through the type of contract adopted.

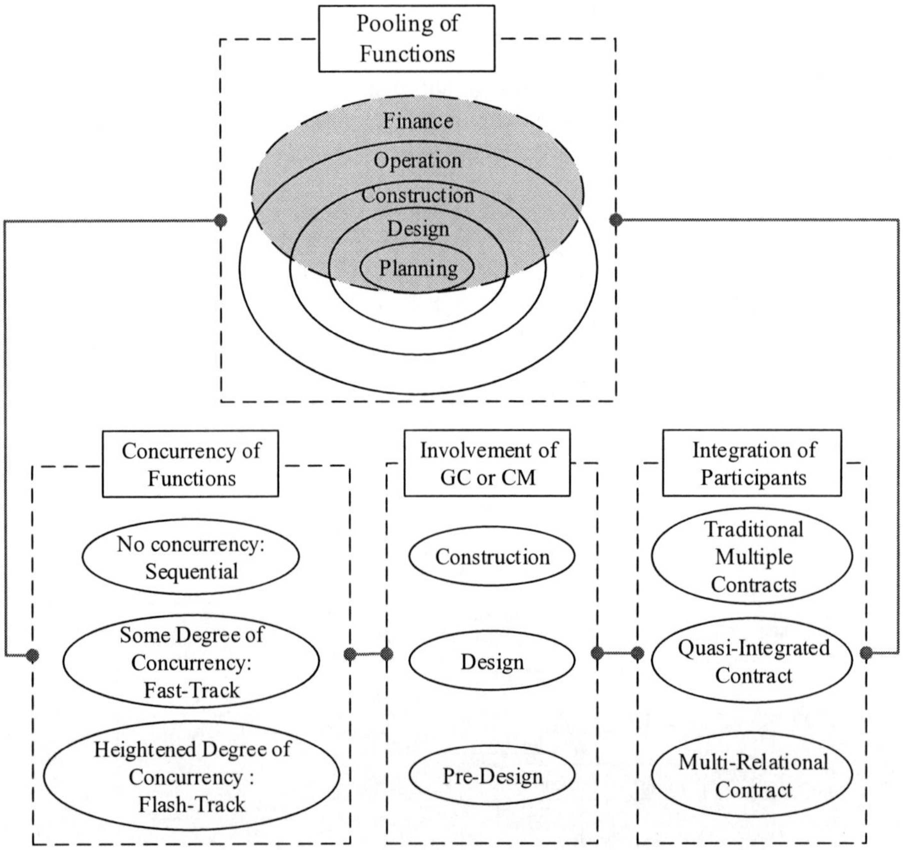

Figure 2. Selecting attributes of a responsible delivery process.

An improved environmental performance could be achieved using any delivery method. For instance, Molenaar *et al.* (2009) found that all PDMs could lead to all levels of LEED certification. However, success differs for each PDM. This is supported by the research findings of Ahmad *et al.* (2017), concluding that early involvement of key project participants, collaboration, and the use of technology are important attributes responsible for enhanced environmental performance. However, although these attributes can be incorporated into any PDM, IPD facilitates the use of technology and incentivizes early participant involvement and collaboration. Thus, using IPD tends to provide higher assurance that project performance targets (e.g., LEED certification) will be met or even exceeded (Ahmad *et al.* 2017).

As for the concurrency of functions, the sequential or "linear approach to procure, design, build, operate, (decommission) can lead to a failure to recognize the wider context in which engineering takes place as part of a series of complex systems with feedback loops involving society and the environment" (Fenner *et al.* 2006). Inherent with the fast-track alternative is the time saving resulting from the compressed schedule due to the degree of overlap between design and construction providing the salient advantage over the sequential option. Moreover, the pursuit of faster fast-track or "flash-track" alternative could be deemed necessary in certain circumstances such as emergency rebuilds, competitive market advantage, and regulatory compliance (Austin *et al.* 2015). Whereas fast tracking can be defined as a time-driven process that by necessity requires some degree of concurrency between engineering, procurement, and construction, flash tracking is defined as a time-driven project, which by necessity requires a heightened degree of concurrency between these functions (Austin *et al.* 2015). To be noted that not all organizations can pursue executing flash-track projects. That is, readiness assessment algorithms were developed by Pishdad-Bozorgi *et al.* (2016) to enable an organization to assess its readiness to execute time-critical or flash-track projects.

On one hand, through focusing on the relationships and interdependencies between 47 essential flash-track practices, it was identified that personnel selection, contractually aligning project participants and establishing fully integrated teams, as being from the most central and core flash-track enablers (Pishdad-Bozorgi *et al.* 2017). On the other hand, integration of participants (a) holds the potential to mend construction industry fragmentation, which often results in poor performance (Mesa *et al.* 2016) and (b) leads to innovative solutions thereby presenting a better project environmental performance (Ahmad *et al.* 2017). In contrast, traditional lump-sum contract is the most detrimental to innovation, involving the highest cost risk, the highest occurrence of adversarial relationships, lowest integration level across the supply chain, and poorest innovation outcomes (Ahmad *et al.* 2017). Moreover, disintegration of the construction process and the resulting adversarial relationship associated with the traditional multiple contracts have caused construction professionals to advocate for more relaxing relational contracts that incorporate higher levels of cooperation and integration. Collaboration under a relational contracting arrangement involves an equitable sharing of risks and rewards in order to reduce adversarial relationships and align the interests of different parties. The establishment of long-term partnership agreements has been an example of developing trust between organizations who work repeatedly together (Fenner *et al.* 2006, Ahmad *et al.* 2017). Lahdenperä (2012) compared the three-different relational contracting (RC) arrangements: 1) Project Partnering (PP), 2) Project Alliancing (PA), and 3) Integrated Project Delivery (IPD). It was concluded that all incorporate common features but to a varying degree, such as the early involvement of the main parties, transparent financial system, shared risks and awards, joint decision-making, and collaborative agreement among multiple parties. Moreover, the core philosophy in these types of relational contracting is to generate a cooperative and trustful climate for the benefit of the project. Besides, early involvement of participants and availability of opportunities for open dialogue and collaboration was proved to enhance the risk management process (Osipova and Eriksson 2011). However, while PDMs define risk allocation formally, the use of incentives and collaboration or partnering arrangements helps in establishing a collaborative approach to risk management (Osipova and Eriksson 2011). To this end, Boukendour and Hughes (2014) suggested a fair risk-sharing formula that incentivizes the partners to truthfully propose their target costs. The importance of such incentive formula is to remove any suspicion as well as increase trust and collaboration among the contracting parties during advanced stages of the project (Boukendour and Hughes 2014).

As such, integration of participants, as shown in Figure 2, covers the full spectrum from the lack of integration using traditional multiple contracts (e.g., DBB) to the quasi-integration when signing relational contracts but excluding the owner, to the full integration using multi-relational contracts, where all the key participants, including the owner, signs one contract and agree to gain/pain sharing formulas. However, project delivery methods influence an owner's ability to achieve its sustainability goals mediated through the level of integration achieved in the delivery process. Research by Mollaoglu-Korkmaz *et al.* (2013) involving 12 in-depth case studies showed that the level of integration in the delivery process affects final project outcomes, particularly sustainability goals. However, although DB and CMR have better chances of facilitating integration, results show that DBB also has the potential to provide higher levels of integration if it informally involves the constructor in the earlier phases of the project.

As such, involvement of the GC or CM at a stage earlier than construction, either at the design stage like in the DB approach or at the pre-design stage like in IPD will lead to an increased potential for achieving integration and leading to improved performance. For instance, Swarup *et al.* (2011) emphasized that early involvement of the constructor and the early inclusion of green strategies are attributes crucial for the delivery process and can have altering effects on project outcomes, especially on sustainability goals. Design charrette, compatibility of team members, and commitment to project sustainability goals were also found to be crucial in achieving team integration and overall project success and innovation (Swarup *et al.* 2011, Mollaoglu-Korkmaz *et al.* 2013, Ahmad *et al.* 2017).

5 RESPONSIBILITY SHARE/ASSIGNMENT FRAMEWORK

Shifting the construction sector away from its current mode toward a more responsible one necessitates a clear identification of the responsibility share/assignment of each party of the process as per the assigned duties/participation. "Clearly, every decision does not have to address the global issues of climate change, global poverty, and so on. However, where the boundaries are drawn must be acknowledged as having a profound effect on the decisions taken" (Fenner *et al.* 2006). Addressing the term sustainability requires new ways of thinking, practices and attitude. Therefore, it requires practical change in our ethics and behavior (Fenner *et al.* 2006, Ochieng *et al.* 2014). To move towards sustainable construction, key drivers such as commitment by leadership (Swarup *et al.* 2011, Ochieng *et al.* 2014), incentives, strategies and stricter environment policies are needed (Ochieng *et al.* 2014). To this end, the framework presented in Figure 3 highlights the responsibility share/assignment of each party including: (a) the responsibility share of the government, (b) the responsibility share of the owner/developer (c) the responsibility shared between the owner and other participants, and (d) the responsibility share of other project participants. It also shows the actions deduced to be made by each party.

Starting by the roles of the public and development policies and regulations prevailing in the locality of the intended project, governments have responsibilities through mandating and/or enforcing policies, legislations, regulations and incentives/penalties, at least in respect of environmental compliance. Fines and penalties for non-compliance with regulations lead to more cautious attitude (Ochieng *et al.* 2014).

Having the ultimate power over the project, owners/developers carry a substantial level of responsibility. Through their involvement in the building delivery process, they can lead innovation practices, incentivize and foster positive interaction among team members (Kilinc *et al.* 2015, Olanipekun *et al.* 2017). They undertake an important role in terms of: (a) creating and promoting the right project conditions for the realization of innovation and (b) understanding and sharing the needs of both end-users and stakeholders (Kilinc *et al.* 2015). Their involvement in the delivery of green building projects is regarded as owner commitment (OC) (Olanipekun *et al.* 2017). Through carrying out a systematic literature in the area of project delivery of green building projects, Olanipekun *et al.* (2017) identified nine indicators of OC. As these indicators were found to influence project performance such as cost, time, quality and sustainability rating metrics- thereby ensuring successful delivery of green building projects - they are added to the presented framework. By virtue of this commitment, project owners can ensure not only successful projects in term of green performance, but a responsible delivery of the built project.

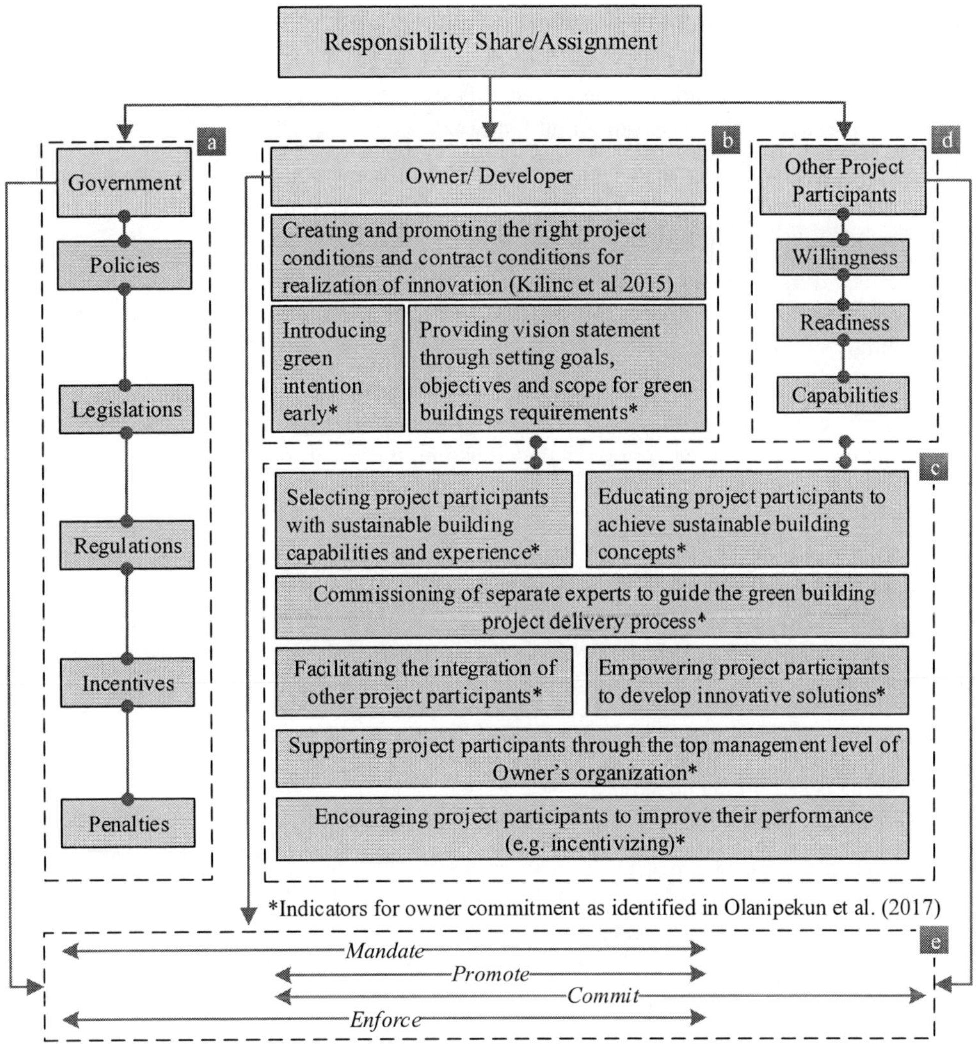

Figure 3. Responsibility share/assignment framework.

As for other project participants, their share of responsibility should be prevalent at least through their commitment. Their willingness, capabilities, and readiness to commit to injecting more environmental, social, and economic consciousness into their share of duties in the construction project, are regarded as forming a responsible attitude. Therefore, the deduced actions for acting responsibly are presented in the last box in Figure 3 (i.e., box "e"). As such, mandating and enforcing are the responsible actions related to the government and the owner, committing is the responsible action related to the owner and other participants, and promoting is the responsible action regarded as solely that of the owner.

6 CONCLUDING REMARKS

The following concluding remarks can be stated:

- Responsible project delivery thinking and strategies, as illustrated throughout this paper, address what is beyond embracing green building methodologies.
- The paper argues for a holistic decision-making process which starts by identifying the underlying drivers behind the concerned investment, all while satisfying the constraints of economic, social and environmental factors.
- While it is evident that project owner (whether public or private) has the upper hand as to selecting the project delivery approach that better suits his interests, this paper identified how such choices could be made responsibly.
- The paper offers a process model encompassing the attributes of PDMs, whose impact of on project performance has been identified from the literature and different relationships accordingly emphasized.
- The paper finally presents a conceptual framework that addresses where responsibility ultimately rests, and it illustrates the responsibility share/assignment for the different participants along with the actions deduced as being critical to be fulfilled by each.

References

Ahmad, T., Aibinu, A. A., and Stephan, A., September. Green Building Projects: Process Innovation Leading to Project Innovation, *Proceeding of the 33rd Annual ARCOM Conference*, 299-308, 2017.

Austin, R. B., Pishdad-Bozorgi, P., and de la Garza, J.M., Identifying and Prioritizing Best Practices to Achieve Flash Track Projects, *Journal of Construction Engineering and Management*, 142(2), 04015077, 2015.

Boukendour, S., and Hughes, W., Collaborative Incentive Contracts: Stimulating Competitive Behavior without Competition, *Construction Management and Economics*, 32(3), 279-289, 2014.

Dodds R., and Venables R., *Engineering for Sustainable Development: Guiding Principles*, Royal Academy of Engineering, London, 2005.

El Asmar, M., Hanna, A. S., and Loh, W. Y., Quantifying Performance for the Integrated Project Delivery System as Compared to Established Delivery Systems, *Journal of Construction Engineering and Management*, 139(11), 04013012, 2013.

Esteves, A. M., Franks, D., and Vanclay, F., Social Impact Assessment: The State of the Art, *Impact Assessment and Project Appraisal*, 30(1), 34-42, 2012.

Fenner, R. A., Ainger, C. M., Cruickshank, H. J., and Guthrie, P. M., Widening Engineering Horizons: Addressing The Complexity of Sustainable Development, *Engineering Sustainability*, 159(1), 145-151, 2006.

Gordon, C. M., Choosing Appropriate Construction Contracting Method, *Journal of Construction Engineering and Management*, 120(1), 196-210, 1994.

Kilinc, N., Ozturk, G. B., and Yitmen, I., The Changing Role of The Client in Driving Innovation for Design-Build Projects: Stakeholders' Perspective, *8th Nordic Conference on Construction Economics and Organization*, 279-287, 2015.

Lahdenperä, P., Making Sense of the Multi-Party Contractual Arrangements of Project Partnering, Project Alliancing and Integrated Project Delivery, *Construction Management and Economics*, 30(1), 57-79, 2012.

Mesa, H. A., Molenaar, K. R., and Alarcón, L. F., Exploring Performance of the Integrated Project Delivery Process on Complex Building Projects, *International Journal of Project Management*, 34(7), 1089-1101, 2016.

Mollaoglu-Korkmaz, S., Swarup, L., and Riley, D., Delivering Sustainable, High-Performance Buildings: Influence of Project Delivery Methods on Integration and Project Outcomes, Journal of Management in Engineering, 29(1), 71-78, 2011.

Mukherjee, A., and Muga, H., A Decision-Making Framework to Assess Stakeholder Value in Adoption of Sustainable Practices in Construction, *Construction Research Congress 2009: Building a Sustainable Future*, 548-557, 2009.

Ochieng, E. G., Wynn, T. S., Zuofa, T., Ruan, X., Price, A. D., and Okafor, C., Integration of Sustainability Principles into Construction Project Delivery, *Journal of Architectural Engineering Technology*, 3(1), 2014.

Olanipekun, A. O., Chan, A. P., Xia, B. P., and Ameyaw, E. E., Indicators of Owner Commitment for Successful Delivery of Green Building Projects, *Ecological Indicators*, 72, 268-277, 2017.

Osipova, E., and Eriksson, P. E., How Procurement Options Influence Risk Management in Construction Projects, *Construction Management and Economics*, 29(11), 1149-1158, 2011.

Pishdad-Bozorgi, P., de la Garza, J. M., and Austin, R. B., Readiness Assessment for Flash Tracking, *Journal of Construction Engineering and Management*, 142(12), 06016005, 2017.

Pishdad-Bozorgi, P., Austin, R. B., and de la Garza, J. M., Network Analysis of Flash-Track Practices, *Journal of Management in Engineering*, 33(1), 04016024, 2016.

Sullivan, J., Asmar, M. E., Chalhoub, J., and Obeid, H., Two Decades of Performance Comparisons for Design-Build, Construction Manager at Risk, and Design-Bid-Build: Quantitative Analysis of the State of Knowledge on Project Cost, Schedule, and Quality, *Journal of Construction Engineering and Management*, 143(6), 04017009, 2017.

Swarup, L., Korkmaz, S., and Riley, D., Project Delivery Metrics for Sustainable, High-Performance Buildings, *Journal of Construction Engineering and Management*, 137(12), 1043-1051, 2011.

Responsible Design and Delivery of the Constructed Project
Edited by Abdul-Malak, M., Khoury, H., Singh, A., and Yazdani, S.

A SYSTEM-OF-SYSTEMS FRAMEWORK FOR PUBLIC-PRIVATE PARTNERSHIP PROJECTS

RANA KHALLAF[1] and MAKARAND HASTAK[2]

[1]*Dept of Structural Engineering and Construction Management, Future University in Egypt, Cairo, Egypt.*
[2]*Division of Construction Engineering and Management, Lyles School of Civil Engineering, Purdue University, W. Lafayette, USA*

Public-private partnership projects are a complex form of megaprojects since they include a close-knit long-term contractual relationship between a public and private entity. Studying a PPP project as a monolithic system often leads to disregarding the emergent properties that occur from the interdependency between the subsystems. Hence there is a need for providing a framework for analyzing PPPs using the systems engineering approach. This paper proposes the adoption of a System-of-Systems (SoS) approach to analyzing PPP projects and proposes a framework for it. A top-down approach is used to identify the key components and their interrelationships. The analysis presented in this paper reveals the SoS taxonomy and lexicon for a PPP project and the hierarchical levels under it. The findings provide a step towards effectively analyzing the relationship within and across the various systems. They are expected to assist researchers in understanding and simulating the dynamics between the different subsystems.

Keywords: Systems engineering, Megaprojects, Taxonomy, Methodology.

1 INTRODUCTION

Public-private partnership projects have been in use for many types of projects such as roadways, water projects, and universities. They have many forms with combinations of build/operate/maintain/lease/transfer, but common among these diverse projects is their complexity and the multitude of parties involved (Khallaf *et al.* 2016). From this stems a need to understand and analyze PPPs using a systems engineering perspective. Systems engineering allows for integration of the multiple diverse systems and their interactions to better model them and ensure that their separate properties as well as combined effects are captured. Previous literature studied PPPs at one level, for example, project risk (Chung *et al.* 2010, Attarzadeh *et al.* 2011, Li *et al.* 2017), project bidding (Jang 2010, Ho 2009) or success/failure factors (Abdul-aziz and Kassem 2011, Liu *et al.* 2015) but did not consider the different systems and their interdependencies thus failing to realize the non-monolithic nature of PPPs. Hence this paper uses the single level studied as a point of departure to propose a framework for PPPs by studying them as a complex SoS. This will guide in the PPP design and help achieve synergy among its subsystems.

This systems engineering approach has been applied mainly to the aerospace and defense industries but has also spread to healthcare, transportation, and other industries. In the context of SoS applications in civil engineering, several studies applied the SoS technique to model and simulate topics under infrastructure projects. Some of the previous applications of SoS in civil engineering have been to model project performance (Zhu and Mostafavi 2014), plan national infrastructure (Otto *et al.* 2016), and propose megaproject systems integration (Zhu *et al.* 2017). However, to the authors' knowledge, no previous studies have looked into how PPPs can be studied and modeled as a SoS.

2 PUBLIC-PRIVATE PARTNERSHIP AS A SYSTEM-OF-SYSTEMS

2.1 System-of-Systems (SoS)

A system-of-systems can be defined as a complex large-scale system consisting of autonomous subsystems with diverse frameworks under it. These subsystems work independently but also interact in order to deliver the needed service (DeLaurentis and Crossley 2005). SoSs have common characteristics such as operational and managerial independence and the manifestation of emergent properties from the internal complexity emanating from their interactions. A SoS's components must also have a common objective or purpose, which in the case of PPPs is the delivery of a service. PPPs can be considered a collaborative SoS because the components work together for the common goal of service delivery. Similar to national transportation systems (DeLaurentis and Callaway 2004), PPPs are considered complex because of the heterogeneity of their subsystems, the physically-dispersed systems, and uncertainty of the conditions. One of the important challenges facing megaprojects and SoSs is systems integration and studying interactions among systems (DeLaurentis and Crossley 2005 and Zhu *et al.* 2017).

2.2 SoS Traits

Public-private partnerships are commonly used for the delivery of megaprojects. These projects are composed of multiple systems working together for a common goal, the delivery of a service. The growing application of PPPs along with the failures that have occurred in some of them enforces that there is a current issue facing PPPs that needs to be studied. Studying a PPP as a monolithic system leads to the neglecting of a PPP's emergent properties. According to DeLaurentis and Crossley (2005), there is a need to identify SoS examples and create a taxonomy of SoS types to solve these problems.

According to Maier (1998), there are five traits of a SoS: operational independence (each entity in a PPP such as the public/private entity is autonomous in its operations), managerial independence (each entity has its own hierarchy of management), evolutionary development (transition toward wireless technology and cloud storage of data), emergent behavior (interaction between the entities), and geographical distribution (the entities/services are not located in the same area). A SoS generally has three characteristic dimensions: type, control, and connectivity of systems (DeLaurentis *et al.* 2011). In a PPP, the system type is neither wholly technological nor wholly human; instead it is a hybrid of both types, since it depends on people and technology to deliver the service. In general, a PPP has a high level of control and is not controlled by any authority; on the other hand, it is governed by the contract between the public and private entities. So the concessionaire would have power in this case that emerges from its contractual relationship with the public entity. Finally, the connectivity of systems under a PPP leads to the presence of emergent behavior. This emergent behavior can arise from the dynamics between the public and private parties as well as from external non-contractual entities. It can also arise due to the interdependency between a PPP's subsystems.

2.3 Top Down Approach

A top-down approach is applied in this paper to view the PPP holistically and identify its components and how they interact. Systems engineering introduced this concept to optimize an entire system rather than optimize its components separately (Ender 2006). This top-down approach helps transform high-level needs and requirements to lower-level processes that are planned, managed, and tracked more easily and efficiently. Functional decomposition is performed on the SoS to reach the lower level components (Ender 2006). Similar to megaprojects, a PPP project can be decomposed into its socio-technical constituents (Bygstad 2010). The social component includes the interaction between the parties involved whether contractual or not, and the technical component is the service to be provided or project to be built.

2.4 Abstraction and Taxonomy

Abstraction of the PPP is needed to explore the systems and emergent properties. SoS is used to analyze the relationship between systems and to suggest problem-solving techniques (Jackson 1990). The lexicon categories in a SoS are: resources, operations, policy, and economics. Resources are the physical entities that manifest in the SoS; operations are the processes that guide the movement of the resources; economics are the non-physical financial-related aspects that guide the resources of the SoS; and finally policies are the guidelines/procedures that guide the activity of the entities (DeLaurentis and Crossley 2005). According to Mostafavi *et al.* (2014), effective policies are needed for the expansion of infrastructure projects. In the United States, some states have developed policies for PPP projects such as Virginia and California, and thus have a higher number of PPP projects than other states like Montana and Wyoming that do not have PPP-enabling legislation. Table 1 shows the hierarchical classification of the lexicon under three distinct hierarchical levels. These three levels (project, market, and macro) are adapted from Khallaf (2016) based on the proposed risk classification. Within each of these categories is a hierarchy of subsystems. The lowest level under a resource, α, would be an individual entity, which then aggregates to the next level, β, which finally aggregates to the highest level γ (an aggregation of β). The level of aggregation chosen for an analysis would depend on what is to be studied. For example, if the level of focus were on a specific PPP project to study the effect of the policies and economics on it, the project (α) level would be chosen and the lexicon identified for that level only. Computational models can be used to simulate these interactions and show their effects.

Table 1. PPP project ROPE.

Level	Resources	Operations	Policy	Economics
α Project	Project resources	Activities for operating a project	Policies relating to a single PPP project (by the public or private entity)	Economics of constructing/operating a single resource for a PPP project
β Market	Resources in a market	Activities for operating a project at a market level	Policies relating to multiple resource use for PPP projects (by the state/country)	Economics of constructing/operating resource networks for PPP projects
γ Macro	Global resources	Activities for operating a project at a global level	Policies relating to the global PPP projects (by global entities)	Economics at a global level

Under the abstraction phase the entities and their roles are identified within a system. Figure 1 shows the entities involved in a PPP project at the base level (α). They are classified into two categories based on their power and influence on a project: endogenous vs. exogenous and explicit vs. implicit. Explicit entities are those that manifest physically while implicit entities are usually implied. Exogenous entities are found outside the system while endogenous entities are internal within the system. The stakeholder group consists of entities that can affect or can be affected by the project including the users, concessionaire, owner, politicians, and media. Disruptors have a negative impact on the project such as financial issues or resource delay or unavailability. Drivers include advancements or policies made within a project to enable its success. The entities can also be identified and mapped out for the market level (e.g. stable market conditions (drivers), competitors (disruptors)) as well as for the global level (e.g. policies enabling PPP projects (drivers).

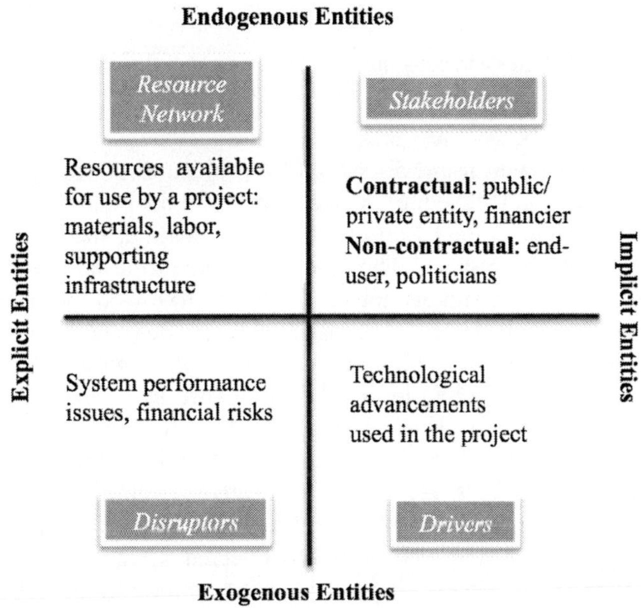

Figure 1. System-of-systems entities in a PPP project at the base level.

Figure 2 shows the policies and economics in a PPP project for the three hierarchical levels (macro, market, and project) based on the classification by Khallaf (2016). Although each level has its own policies, the lower level policies can be affected by the higher-lever policies. Examples of policies in a country are conditions for which to bail out a concession. The concession toll road program that started in the beginning of the 1990s in Mexico was bailed out by the government in 1997 with an approximate cost of 1 to 1.7 per cent of GDP (Guasch *et al.* 2007). Three out of four projects awarded in France in the 1970s went bankrupt and were rescued by the government (Guasch 2003). A mixed-use building at the University of Quebec at Montreal was $200 million over budget, which doubled its cost to a total of $400 million (Sanger and Crawley 2009). The high cost and dire consequences from PPP failure have driven the need for studying PPPs.

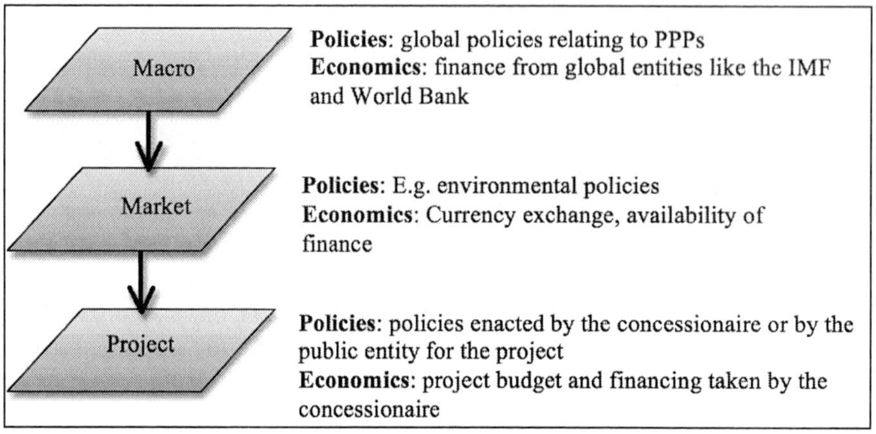

Figure 2. Policies and economics under the three hierarchical levels

3 CONCLUSION

This paper presented an integrated framework for PPP projects as a system-of-systems. A top-down approach was used to identify the hierarchical levels and the resources, operations, policies, and economics under each level. Based on the SoS classification, the three hierarchical levels identified are macro, market, and project. The first step in the proposed framework is identifying the resources, operations, policies, and economics (Table 1) followed by identifying the level of analysis (α, β, γ). Based on the level chosen, the entities (endogenous/exogenous and explicit/implicit) can then be identified and mapped as shown in Figure 1. The interrelationships between the hierarchical levels can also be identified (Figure 2) to show the cascading effects between them. This framework can be used to model the intra-level relationships such as the dynamics between the hierarchical levels or the inter-level relationships such as those between the different subsystems (under one level). Using the systems engineering approach also enables the design of the subsystems themselves. PPPs need to be designed effectively and operated both as separate autonomous systems and also as interconnected systems to capture the emergent dynamics from their interactions. Connecting the distinct and dissimilar systems together through SoS enables the modeling and discovery of vulnerabilities in a system. This will help in the conceptual phase of system design by minimizing vulnerabilities and maximizing systems connections. DeLaurentis and Crossley (2005) suggest that underperformances in complex SoS are the result of inadequate attention to the interaction among the subsystems. This can be solved by following the framework for PPP SoS and taking into consideration the emergent dynamics that occur from the systems.

References

Abdul-aziz, A. R. and Kassem, P. S. J., Objectives, Success and Failure Factors of Housing Public-Private Partnerships in Malaysia, *Habitat International*, 35(1), 150-157, 2011.

Attarzadeh, M., Risk Management of Long Term Infrastructure Projects "PPP-BOT Projects" by Using Uncertainty, Probabilistic and Stochastic Methods and Models, *First International Symposium on Uncertainty Modeling and Analysis and Management (ICVRAM),* Maryland, *USA,* 2011.

Bygstad, B., Nielsen, P. A., and Munkvold, B. E., Four Integration Patterns: A Socio-Technical Approach to Integration in IS Development Projects, *Journal of Information Systems,* 20, 53-80, 2010.

Chung, D., Hensher, D. A., and Rose, J. M., Toward the Betterment of Risk Allocation: Investigating Risk Perceptions of Australian Stakeholder Groups to Public-Private Partnership Tollroad Projects, *Research in Transportation Economics*, Vol.30, 43-58, 2010.

DeLaurentis, D., Crossley, W. A., and Mane, M., Taxonomy to Guide Systems-of-Systems Decision-Making in Air Transportation Problems, *AIAA Journal of Aircr*aft, 48(3), 760–770, 2011.

DeLaurentis, D. and Callaway, R. K., A System-of-Systems Perspective for Future Public Policy Decisions, *Review of Policy Research*, 21(6), 829–837, 2004.

DeLaurentis, D. and Crossley, A Taxonomy-based Perspective for Systems of Systems Design Methods, *IEEE System, Management and Cybernetics Conference*, Hawaii, USA, 2005.

Ender, T. R., A Top-Down, Hierarchical, System-of-Systems Approach to the Design of an Air Defense Weapon, *PhD Thesis*, School of Aerospace Engineering, Georgia Institute of Technology, Atlanta, USA, 2006.

Guasch, J. L., Granting and Renegotiating Infrastructure Concessions: Avoiding the Pitfalls, *The World Bank,* Washington DC, USA, 2003.

Guasch, J. L., Laffont, J. J., and Straub, S., Concessions of Infrastructure in Latin America: Government-led Renegotiations, *Journal of Applied Econometrics*, 22, 1267-1294, 2007.

Ho, S. P., Government Policy on PPP Financial Issues: Bid compensation and financial renegotiation. In: Akintoye, A., Beck, M. (Eds.), *Policy, Finance and Management for Public–Private Partnerships*. Blackwell, West Sussex, UK, 267–300 (Chapter15), 2009.

Jang, G., The Bids-Evaluation Decision Model Development and Application for PPP Transport Projects: A Project Risk Modeling Framework, *PhD Thesis*, Dept of Mechanical Engineering, Colorado State University, Fort Collins, Colorado, USA, 2010.

Khallaf, R., Naderpajouh, N., and Hastak, M., Interactional Analysis for Two-Party Risk Assessment in Public Private Partnerships (PPP), *The 5th World Construction Symposium 2016: Greening Environment, Eco Innovations & Entrepreneurship, Colombo, Sri Lanka,* 2016.

Khallaf, R., *Interactional Risks Associated with Parties to PPP Project,* PhD Thesis, Purdue University, Indiana, USA, 2016.

Li, Y., Wang, X., and Wang, Y., Using Bargaining Game Theory for Risk Allocation of Public-Private Partnership Projects: Insights from Different Alternating Offer Sequences of Participants, *Journal of Construction Engineering and Management,* 143(3), 04016102, March, 2017.

Liu, J., Love, P., Smith, J., and Regan, M., Life Cycle Critical Success Factors for Public-Private Partnership Infrastructure Projects, 31(5), 04014073, 2015.

Maier, M. W., Architecting Principles for System-of-Systems, *Systems Engineering*, 1(4), 267–284, 1998.

Mostafavi, A., Abraham, D., and DeLaurentis, D., Ex ante Policy Analysis in Civil Infrastructure Systems. *Journal of Computing in Civil Engineering*, 28(5), A4014006, 2014.

Otto, A., Hall, J. W., Hickford, A. J., Nicholls, R. J., Alderson, D., and Barr, S., A Quantified System-of-Systems Modeling Framework for Robust National Infrastructure Planning, *IEEE Systems Journal*, 99, 1-2, 2014.

Sanger, T. and Crawley C., The Problem with Public-Private Partnerships. Canadian Centre for Policy Alternatives Publications, 2009. Retrieved from https://www.policyalternatives.ca/publications/monitor/problem-public-private-partnerships on January 30, 2018.

Zhu, J. and Mostafavi, A., Integrated Simulation Approach for Assessment of Performance in Construction Projects: A System-of-Systems Framework. *Proceedings of the 2014 Winter Simulation Conference, Savannah, USA,* 2014.

Zhu, J., Mostafavi, A., and Whyte, J., Towards Systems Integration Theory in Megaprojects: A System-of-Systems Framework. *LC3 2017*: *Proceedings of the Joint Conference on Computing in Construction (JC3)*, Greece, 1, 313-320, 2017.

Responsible Design and Delivery of the Constructed Project
Edited by Abdul-Malak, M., Khoury, H., Singh, A., and Yazdani, S.

EVALUATING PPP PROJECTS USING A LEAN FRAMEWORK

ZEINA MALAEB and FAROOK HAMZEH

Dept of Civil and Environmental Eng., American University of Beirut, Beirut, Lebanon

A Public Private Partnership (PPP) describes an arrangement between the public and the private sectors for providing a public asset or service. PPPs are characterized by their long-term nature, complex contractual agreements, and distinct risk allocation formulas, all which distinguish them from traditional procurement routes. Performance evaluation is a prerequisite of performance improvement and is vital to the realization of project success, particularly in PPP projects. In fact, the absence of an effective performance evaluation framework in PPPs serves as a trigger for generating below optimum service quality. Although several performance evaluation frameworks exist, they prove more suitable for traditional projects, which highlights the need to account for the unique features of PPPs. This paper first introduces the various perspectives, from the literature, used to classify project success and evaluate performance. The paper then addresses the aforementioned need by introducing a new framework for evaluating project success, using the lean concepts of transformation, flow, and value. Key Performance Indicators (KPIs) are suggested as part of the framework to serve as performance measures on PPP projects. The introduced lean framework is the first of its kind and fills the gap in PPP research by being one of the few performance evaluation tools customized for PPPs.

Keywords: Key performance indicators (KPIs), Lean construction, Performance evaluation, Public private partnerships (PPPs), Special purpose vehicle (SPV), TFV.

1 INTRODUCTION

A Public Private Partnerships (PPP) describes an arrangement between the public and the private sectors for providing a public service. PPPs are characterized by their long-term nature, bundling of project functions, complex contractual agreements, and distinct risk allocation formulas (Grimsey and Lewis 2004), all of which set them apart from traditional procurement routes. The public sector's reasons for adopting PPPs are twofold: (1) they allow the public sector to benefit from private financing which relieves the stress on limited public budgets, and (2) they allow the public sector to utilize the private sector's skill and management expertise in project delivery (Osei-Kyei and Chan 2015). In fact, PPPs are claimed to offer a range of benefits such as accelerated infrastructure provision, timely project implementation, reduced whole life cycle costs, reduced pubic risk, and improved service quality and innovation (Leiringer 2006).

Performance evaluation is the prerequisite of performance improvement and is vital to the realization of project success, particularly in PPP projects. In fact, the absence of an effective performance evaluation framework in PPPs serves as a trigger for generating below optimum service quality (Liu *et al.* 2014a). Several frameworks and perspectives for measuring project success have been developed over the years. However, the unique characteristics of PPP projects, in addition to the complex nature of their stakeholder interfaces, separate them from traditionally procured projects and prove the existing frameworks that serve the latter insufficient (Liu *et al.* 2014b). Additionally, changes in project structures obligate respective changes in performance metrics, driven by the altered characteristics (Tranfield *et al.* 2005).

Comprehensive evaluation of project success necessitates investigating several fronts: the transformation of inputs to outputs in a process, the flow within the process, and the value generation to the customer at the end of the process. This transformation-flow-value (TFV) theory is sourced from the lean philosophy, which attempts to generate a rounded outlook of a project and its processes. Of the few frameworks that attempt to evaluate PPP project success in specific, none focus on measuring success from these three aforementioned fronts. The above highlights the need for a holistic performance evaluation framework, specifically designed for PPP projects, which takes into account the transformation, flow, and value perspectives of project processes. This research attempts to address this need by introducing such a framework and suggesting a number of Key Performance Indicators (KPIs) that can be used to measure PPP project success. The introduced lean framework is the first of its kind and fills the gap in PPP research by being one of the few performance evaluation tools customized for PPPs.

This paper first describes the main characteristics of PPP projects. Then, it introduces different ways of classifying project success, as found in the literature. The shortcoming of these perspectives is highlighted next, which signifies the need for a new holistic framework for measuring PPP success. Finally, the TFV framework developed for this research is introduced, and a set of KPIs generated for evaluating the success of PPP projects.

2 CHARACTERISTICS OF PPP PROJECTS

The PPP project structure involves two main project stakeholders: the public authority and the private entity. The private sector is represented by a project company called the Special Purpose Vehicle (SPV). This SPV enters into a contractual agreement with the public authority for the financing, designing, building, management and operating of a public facility. This contract is of a long term nature, usually for 30 or more years (Sarmento and Renneboog 2016). Being a distinct procurement route, PPPs are characterized by several features that differentiate them from traditionally procured projects. Some of these are: uncertainties and risks stemming from the long-term agreements, major rearrangements in the roles of the multiple project stakeholders, increased responsibilities and risks for the private sector, and complex contractual arrangements between different stakeholders (Zhang 2005). Another major characteristic is the bundling of project functions, from design to construction to operation, in one and handing them over to one private entity (the SPV), which assumes the responsibility of the whole bundle. The structure of the SPV is also distinct, as this project company combines the designer, contractor, and service provider under one umbrella. This aligns stakeholder goals and interests and generates a whole-life cycle approach for the project. PPPs involve stakeholders early on in the process, which leads to the formation of collaborative environments. Furthermore, their use of output specifications in the design is claimed to drive the potential for innovation on the project (Grimsey and Lewis 2004, Leiringer 2006, Tawiah and Russell 2008, World Bank 2009).

3 CLASSIFICATIONS OF PROJECT SUCCESS

Performance evaluation of projects encompasses two parts: defining success and measuring success. As perceived in Table 1 below, there exists numerous ways and perspectives of describing success on projects. There might not be one correct way to define success and delineate its measures; however, it is certain that this definition should be in consistency with the subject project and its distinct characteristics. To elaborate, evaluating the success of a project should take into consideration the different features stemming from its procurement route, stakeholder structure, commercial and contractual frameworks, and operating systems and processes (Thomsen et al. 2009). None of the various perspectives defined in Table 1 below are customized to account for the unique features of PPP projects, which reflects the need for a new holistic framework for evaluating such projects.

Table 1. Different classifications of project success.

No	Classification of Success Dimensions	Source
1	Project characteristics; contractual arrangements; project participants; interactive processes	Elattar (2009); Kwofie *et al.* (2016)
2	Project management perspective; stakeholder perspective; contract management perspective	Liyanage and Villalba-Romero (2015)
3	Technical KPIs; operational and functional KPIs; financial KPIs	Mladenovic *et al.* (2013)
4	Project management success; project success; consistent project success	Cooke-Davies (2002)
5	Macro viewpoint vs. micro viewpoint	Lim and Mohamed (1999)
6	Process domain and performance domain	Toor and Ogunlana (2010)
7	Quantitative KPIs vs. qualitative KPIs	Cox *et al.* (2003)
8	Project efficiency; impact on the customer; business and direct success; preparing for the future	Shenhar *et al.* (1997)
9	Process: doing it right; system: getting it right; benefits: getting them right	Atkinson (1999)
10	Meeting design goals; benefits to the end user; benefits to the developing organization; benefits to the economy	Sadeh *et al.* (2000)
11	Success efficiency and success effectiveness	Pinto and Slevin (1988)
12	Stakeholder satisfaction; strategies; processes; capabilities; stakeholder contribution	Liu *et al.* (2014a)
13	Physical characteristics of projects; requirements of stakeholders including financing and marketing, innovation and learning, and stakeholders; project process	Yuan *et al.* (2009)

4 DEVELOPED LEAN PERFORMANCE EVALUATION FRAMEWORK

4.1 TFV Theory

The optimal perspective for evaluating PPP project success would be one that takes into account a holistic description of the different elements of a project and its processes. To elaborate, it is not sufficient to merely evaluate the external outputs of a process, while ignoring the occurrences within the processes, and the creation of value at different steps. This line of thought is compatible with the Transformation-Flow-Value (TFV) theory of the lean philosophy. In short, the TFV theory introduces the concept of studying the transformation, flow, and value views of a process. It considers that traditional theories, which encompass only the "transformation" view, are not reflective of the overall project fundamentals. Figure 1 describes these three views. The transformation view merely measures the inputs and outputs of each process individually, and disregards the internal interactions between processes, which reduces the project to a black box. It is important to study the "flow" between processes, in order to minimize the waste occurring within. "Flow" pertains to the flow of material, labor, information, and space in the project while "waste" refers to wasted time, cost, and, material. Finally, the overall objective of a process is to generate value to the customer, through meeting his requirements. Consequently, this generation of value should also be evaluated when observing a project to ensure that the customers' objectives are met (Koskela *et al.* 2002). Table 2 clarifies these concepts.

The distinct characteristics of PPP projects, identified in Section 2 above, separate these projects from traditionally procured projects, and add an inherent level of complication to the former. Consequently, evaluating PPP project success would require a much more comprehensive and holistic study than evaluating the success of traditional projects. Therefore, when considering PPP projects, success should be appraised from a well-rounded perspective to encompass the unique character of these arrangements. The TFV viewpoint seems to satisfy these requirements, and thus, this research projects concepts from the TFV theory into the developed evaluation framework to measure PPP project success.

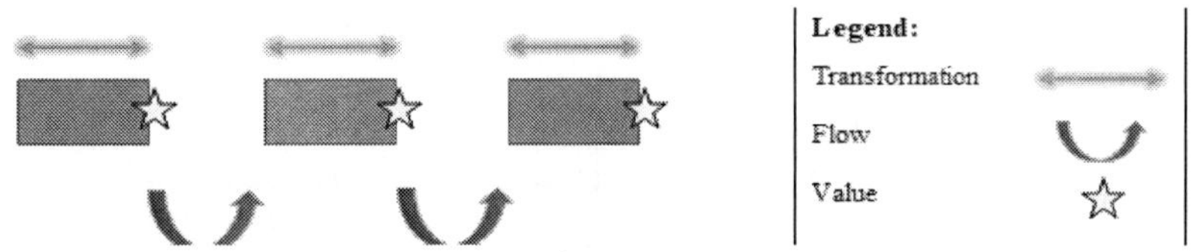

Figure 1. Depiction of the transformation, flow, and value views.

Table 2. Concepts of the transformation, flow, and value views (Amended from Koskela *et al.* 2002).

Viewpoint	Transformation	Flow	Value
Conceptualization of a Process	Transformation of inputs to outputs	Flow of material and information from its initial state to the final product	Generation of value to the customer
Function	Making individual tasks efficient	Minimizing waste (non-value adding activities)	Minimizing value loss (Ensuring customer requirements are met)
Application	Task Management	Flow Management	Value Management
Evaluation	The "input" and "output" of each internal process on its own	The flow between the different processes	The product (intermediate and final) vs. customer

4.2 TFV Framework

As discussed, the first step of performance evaluation is deciding on the perspective used to measure success that forms the basic skeleton of the framework. The second step is developing performance measures, or KPIs, to evaluate success. The KPIs are first identified and filtered from a thorough literature review of studies on the performance evaluation of construction projects, shown in Table 1, specifically detailing project success indicators and factors. The KPIs are then segmented into the three perspectives: transformation (T), flow (F), and value (V). "T" KPIs measure only the process inputs and outputs, "F" KPIs measure the efficiency within the processes, and "V" KPIs measure the generation or loss of value at the product or bi-product level. Although some KPIs may fall under more than one perspective, the most fitting one is chosen. Table 3 presents the identified KPIs, 18 in total, along with their descriptions.

5 CONCLUSIONS AND FUTURE RESEARCH

An efficient performance evaluation framework that accounts for the specific characteristics of PPP projects is critical to their success (Liu *et al.* 2014a). A performance evaluation framework based on a lean background, along with KPIs, adopting a rounded transformation, flow, and value perspective, ensures a holistic representation of PPP project success. Future research can develop further performance measures and apply the framework to a case study for validation of results.

Table 3. TFV framework and key performance indicators (KPIs).

KPI Title	Description	T	F	V
1. Cost / Financial	Cost Growth = (actual incurred cost – initial planned cost) /initial planned cost	x		
2. Project Duration	Time Growth = (final completion time – planned completion time) / planned completion time	x		
3. Level of Dispute	Number of claims and disputes on the project; Size of disputes on the project; Incidents of imposing penalties or damages	x		
4.Environmental Impact	Utilization of non-renewable resources; Greenhouse gas emissions; Carbon and embodied energy; Disturbance to flora and fauna; Consideration for endangered and protected species; Water use (quantity and quality); Promotion of alternative greener or renewable energy; Types of material used			x
5.Health and Safety	Accident rate; Number of injuries or deaths			x
6.Construction Efficiency	Efficiency = actual productivity rate / planned productivity rate	x		
7.Construction Delays	Delays to activity start or completion ; Delay in supplying materials, tools, and/or equipment		x	
8.Operation quality	Dependability of operation service including availability and reliability; Management of health and safety in operation stage			x
9.Maintenance quality	Speed and frequency of maintenance processes; Management of disruption of services			x
10. Information and Communication Systems	Type and quality of intra and inter organizational information systems		x	
11. Monitoring and Management Mechanisms	Quality and efficiency of the monitoring PM systems used		x	
12. Life-Cycle efficiency	Cost of operating and maintaining the facility	x		
13. SPV Stakeholder Involvement	Input of contractor during the design; Input of operator during the design; Number and size of value engineering efforts		x	
14. SPV Collaboration Efficiency	Number of VOs in the construction phase; Incidents of design rework; Incidents of construction rework		x	
15. SPV Communication Efficiency	Means of communication.; Frequency and density of communication; Parties involved in the communications		x	
16. Problem Solving Efficiency	Average time taken to solve problems; Parties involved in solving problems		x	
17. Level of Innovation	New technologies and processes used; Value engineering proposals; Input of parties to the design process			x
18. Innovation Efficiency	Cost and time savings on the project	x		

References

Atkinson, R., Project Management: Cost, Time and Quality, Two Best Guesses and a Phenomenon, It's Time to Accept Other Success Criteria, *Int. Journal of Project Management,* 17(6), 337-342, 1999.

Cooke-Davies, T., The "Real" Success Factors on Projects, *International Journal of Project Management,* 20(3), 185-190, 2002.

Cox, R. F., Issa, R. R., and Ahrens, D., Management's Perception of Key Performance Indicators for Construction, *Journal of Construction Engineering and Management,* 129(2), 142-151, 2003.

Elattar, S. M. S., Towards Developing an Improved Methodology for Evaluating Performance and Achieving Success in Construction Projects, *Scientific Research and Essays,* 4(6), 549-554, 2009.

Grimsey, D. and Lewis, M., *Public Private Partnerships: The Worldwide Revolution in Infrastructure Provision and Project Finance*, Edward Elgar Publishing, 2004.

Koskela, L., Howell, G., Ballard, G., and Tommelein, I., The Foundations of Lean Construction, *Design and Construction: Building in Value,* 211-226, 2002.

Kwofie, T. E., Afram, S., and Botchway, E., A Critical Success Model for PPP Public Housing Delivery in Ghana, *Built Environment Project and Asset Management,* 6(1), 58-73, 2016.

Leiringer, R., Technological Innovation in PPPs: Incentives, Opportunities and Actions, *Construction Management and Economics*, 24(3), 301-308, 2006.

Lim, C. S., and Mohamed, M. Z., Criteria of Project Success: an Exploratory Re-Examination, *International Journal of Project Management,* 17(4), 243-248, 1999.

Liu, J., Love, P. E., Davis, P. R., Smith, J., and Regan, M., Conceptual Framework for The Performance Measurement of Public-Private Partnerships, *Journal of Infrastructure Systems,* 21(1), 04014023, 2014a.

Liu, J., ED Love, P., Smith, J., Regan, M., and Sutrisna, M., Public-Private Partnerships: A Review of Theory and Practice of Performance Measurement, *International Journal of Productivity and Performance Management,* 63(4), 499-512, 2014b.

Liyanage, C., and Villalba-Romero, F., Measuring Success of PPP Transport Projects: A Cross-Case Analysis of Toll Roads, *Transport Reviews,* 35(2), 140-161, 2015.

Mladenovic, G., Vajdic, N., Wündsch, B., and Temeljotov-Salaj, A., Use of Key Performance Indicators for PPP Transport Projects to Meet Stakeholders' Performance Objectives, *Built Environment Project and Asset Management,* 3(2), 228-249, 2013.

Osei-Kyei, R., and Chan, A. P., Review of Studies on the Critical Success Factors for Public–Private Partnership (PPP) Projects from 1990 to 2013, *International Journal of Project Management,* 33(6), 1335-1346, 2015.

Pinto, J. K., and Slevin, D. P., Project Success: Definitions and Measurement Techniques, *Project Management Institute*, 1988.

Sadeh, A., Dvir, D., and Shenhar, A., The Role of Contract Type in the Success of R&D Defense Projects under Increasing Uncertainty, *Project Management Journal,* 31(3), 14-14, 2000.

Sarmento, J. M., and Renneboog, L., Anatomy of Public-Private Partnerships: Their Creation, Financing and Renegotiations, *International Journal of Managing Projects in Business,* 9(1), 94-122, 2016.

Shenhar, A. J., Levy, O., and Dvir, D., Mapping The Dimensions of Project Success, *Project Management Journal,* 28(2), 5-13, 1997.

Tawiah, P. A., and Russell, A. D., Assessing Infrastructure Project Innovation Potential as a Function of Procurement Mode, *Journal of Management in Engineering,* 24(3), 173-186, 2008.

Thomsen, C., Darrington, J., Dunne, D., and Lichtig, W., *Managing Integrated Project Delivery,* Construction Management Association of America (CMAA), McLean, VA, 105, 2009.

Toor, S. and Ogunlana, S. O., Beyond the 'Iron Triangle': Stakeholder Perception of Key Performance Indicators (Kpis) for Large-Scale Public Sector Development Projects, *International Journal of Project Management,* 28(3), 228-236, 2010.

Tranfield, D., Rowe, A., Smart, P. K., Levene, R., Deasley, P., and Corley, J., Coordinating for Service Delivery in Public-Private Partnership and Private Finance Initiative Construction Projects: Early Findings from an Exploratory Study, *Proceedings of the Institution of Mechanical Engineers, Part B: Journal of Engineering Manufacture,* 219(1), 165-175, 2005.

World Bank, *Good Governance in Public Private Partnerships: A Resource Guide for Practitioners.* Washington DC: World Bank, 2009.

Yuan, J., Zeng, A. Y., Skibniewski, M. J., and Li, Q., Selection of Performance Objectives and Key Performance Indicators in Public–Private Partnership Projects to Achieve Value for Money, *Construction Management and Economics,* 27(3), 253-270, 2009.

Zhang, X., Criteria for Selecting the Private-Sector Partner in Public–Private Partnerships, *Journal of Construction Engineering and Management,* 131(6), 631-644, 2005.

Responsible Design and Delivery of the Constructed Project
Edited by Abdul-Malak, M., Khoury, H., Singh, A., and Yazdani, S.

A DECISION SUPPORT SYSTEM FOR THE RENEGOTIATION OF PPP CONTRACTS

AMIRA SHALABY and AMR HASSANEIN

Construction Engineering Dept, The American University in Cairo, Cairo, Egypt

Public Private Partnership (PPP) contracts tend to have longer contract durations compared to other conventional procurement methods. A contract renegotiation becomes inevitable in most of the cases. The renegotiation process usually develops a number of scenarios in order to regain the contract equilibrium. The common renegotiation scenarios used are: increasing the service charges, increasing the concession period, or paying a lump sum amount to the party of concern in order to maintain a fixed rate of return and keep the return on equity constant. In this paper, a method of selecting the optimum scenario among the different scenarios is proposed. The aim of this research is to facilitate the renegotiation process by offering an automated system to select the optimum renegotiation scenario that preserves the rights and the interests of the project stakeholders. This is done using a weighted sum model to calculate the weights and ranks of a number of factors influencing the stakeholders' decisions. A Decision Support System (DSS) is developed with the aid of Microsoft Excel 2013, Visual Basic for Applications (VBA) programming language, and the Precision Tree 5.5 for Excel add-in.

Keywords: Dispute, Negotiation, Financial model, Contract re-equilibrium, Spreadsheet.

1 RESEARCH MOTIVATIONS

Public Private Partnership is a delivery method which depends on co-operation between the public sector and the private sector in order to be able to finance mega infrastructure projects. The choice between the Public Private Partnership delivery method and other conventional delivery methods depends on many factors that should be looked at thoroughly. PPP projects are very challenging in terms of attracting investors to enter the bidding process. This is due to several factors; one of which is the private sector's fear of the long-term nature of PPP projects, which makes it extremely difficult to anticipate contingencies along the projects' lifecycles. Due to the lengthy contract durations, the principles and bases upon which the original PPP contract was made may no longer apply; they may be simply altered or totally changed as the project evolves. A different set of conditions and situations may appear later in the project, making the need for a contract re-equilibrium inevitable, and with re-equilibrium comes renegotiation. During the renegotiation stage, conflicts may arise between the different stakeholders of the project. These lengthy renegotiations become full of conflicts that, in some cases, may lead to contract terminations and major losses for several parties. However, those conflicts can be avoided through a pre-agreed renegotiation mechanism. Tools are needed to ease the lengthy renegotiation process. This research provides a tool that facilitates a renegotiation process in which the interests of all parties are considered in the final decision. This tool will help in attracting the private sector to enter PPP contracts and ensure project continuity and stability of transactions.

2 LITERATURE REVIEW

The time value for money is the difference between the project value that is incurred by the private sector in case of choosing the PPP option and the project value if executed by the public sector. Although the cost of borrowing for the private sector is usually much higher than the cost of borrowing for the public sector, the PPP option, in many cases, may have a higher value for money for several reasons, such as the ability of the private sector to provide better and more efficient services than the public sector (The Construction Management Association of America, 2012). Private sector know-how is another reason for choosing a PPP approach. A value for money study should be conducted to determine whether the PPP option is more efficient in the long run than other procurement methods. In other words, the public sector should only choose the PPP option over conventional procurement methods when the private sector is providing a better service with more efficiency, or when the private sector is providing a higher value for money (VFM). Almost all PPP advantages lead to the same conclusion: a better VFM (Skanska 2004).

Archer and Cameron (2003) published results of a survey conducted by the National Audit Office (NAO) of the United Kingdom. The survey included 37 PPP projects. Cost overruns and time delays were estimated. The results indicated that almost 75 percent of conventional procured projects suffer from cost overruns while less than 25 percent of PPP projects suffer from cost overruns. The study also found that almost 24 percent of PPP projects suffer from time delays. The conclusion showed that PPPs are more efficient in terms of time and cost than conventional procurement methods.

Sarmento and Renneboog (2016) have explained another aspect which is called "off-balance sheet" approach. In their paper, the concept was explained as an advantage to the governments using PPP delivery method. PPP value is not part of the government expenditures listings; in other words, it is off their balance sheet. Hence, theoretically, it will not affect the public debt.

Katz (2006) states that due to the length of PPP contracts, such contracts are usually prone to renegotiation as a result of various unexpected contingencies that might occur along the project lifecycle. In addition, it is very difficult to control the performance of the private sector and its quick response to growing demand, especially when the private sector is paid by the government rather than the user fees method. This may lead to political issues due to the dissatisfaction of the service users. A PPP contract may lose efficiency with time due to the fact that the project cannot be re-tendered, which diminishes competition and the incentive to provide a better service.

Cruz and Marques (2012) has investigated the reasons of contract renegotiation in infrastructure projects. They claim that contract renegotiation is usually a result of contract incompleteness. Cruz and Marques states that in order to have a complete contract, this will result in a very high transaction costs. Hence, instead of working on a complete contract, some rules and conditions are to be agreed in case of renegotiation.

Cruz *et al.* (2014) has stated that the main reasons of renegotiation vary between changes initiated by the government and changes in the expected return of the project. The event which occurs after the award of the project leads to a completely new risk matrix with different allocation of the risks and risk allocation.

In summary, the idea of a PPP is to combine the expertise and resources of the public and the private sectors in order to reach efficiency and VFM, yet PPP if not carefully managed, it can increase the cost of procurement services more than any other traditional procurement method,

(Carbonara *et al.* 2016). The above disadvantages along with insufficient feasibility studies and poor decisions can turn a PPP from an advantageous method of delivering services to the public sector into leading to undesired results.

3 THE METHODOLOGY

This framework is a fragment of a comprehensive research which aimed to developing the re-equilibrium scenarios of PPP, and selecting the optimum scenario which fits both the public and the private sector. However, for the purpose of this paper, the research shall focus on the scenario selection process.

The purpose of this research is to facilitate the PPP renegotiation process by choosing the optimum scenario to satisfy all parties. Stakeholders have different interests and concerns in the renegotiation process; hence, the proposed model will work on finding common ground between all parties in order to reach an optimum renegotiation outcome. This will ensure a fair approach and enhance mutual trust between the stakeholders, which will benefit the project's overall progress. The following framework is to be of great benefit to decision makers in public and private sectors, yet the objective of the proposed framework is not to be a substitute for decision makers.

In PPP projects, events occur that disturb the re-equilibrium of the financial model; in other words, it affects the cash flow of the project which may lead to changing the agreed rate of return of the private sector. In order to fulfill the purpose of PPP projects and return the equilibrium to the PPP financial model, there are commonly used four re-equilibrium scenarios which are: paying a lump sum amount to the private sector, increasing the service charges, increasing the concession period, or a combination of one or more of the above. The scenarios selection process aims to choice between the four re-equilibrium scenarios is made. The goal of the scenarios selection process is to account for the interests of the stakeholders of the PPP project, mainly the public sector and the private sector, when selecting the renegotiation outcome scenario.

The criteria and their corresponding sub-criteria are defined, based on which the selection is to be made. In order to evaluate each scenario, a score shall be given to each scenario. The scenario score is calculated using the weighted sum model. This weighted sum model works by allocating different criteria to each scenario. The criteria will have relative importance to each other. The scenario score shall equal the summation of the weight of importance for each criterion multiplied by its value as shown in Equation 1. (x) is the number of alternatives of scenarios, (i) is the number of main criteria, (j) is the number of sub-criteria of a certain main criterion, (W) is the weights and (R) is the rankings. The following equation should be within where $x=1$ to 4, $i=1$ to n, $j=1$ to m.

$$ScenarioScore(SS_x) = \sum_{(j=1)}^{m} W_{ci} W_{sj} R_{sj} \qquad (1)$$

After inputting the criteria, the model provides the scenarios selection process with the importance of each criteria with respect to the rest of the criteria for each of the four scenarios. The model then determines the weights of the criteria and sub-criteria with respect to the different re-equilibrium scenarios. The approach is to draw matrices for the main criteria and each of the sub-criteria categories in which the top row is the criteria or the sub-criteria and the rest of the matrix is a mirror image. Decision matrices were invented by Stuart Pugh (Pugh 1993). An example of the decision matrix is shown in Figure 1.

The above exercise is done for the criteria and the sub-criteria for each one of the four scenarios, as the weights may differ when dealing with making a lump sum payment or just extending the concession period. Separate rankings are required from both the public and the private sector in order to reflect their interests and preferences. Using the weighted sum model, eight scenario scores are calculated: four for the public sector, and the other four from the perspective of the private sector. The outputs of this step are the best scenarios or the scenarios with the highest scores from the private sector perspective and the public sector perspective. Then, averages of the private sector rankings and the public sector rankings are taken to calculate the four re-equilibrium scenarios scores from the perspective of both parties. Figure 2 shows the integration between the inputs and outputs of the model. An example of which shall be illustrated in the following section.

	Criteria 1	Criteria 2	Criteria 3	Criteria 4
Criteria 1		Criteria 2	Criteria 1	Criteria 4
Criteria 2			Criteria 3	Criteria 2
Criteria 3				Criteria 4
Criteria 4				

Figure 1. Example of a decision matrix.

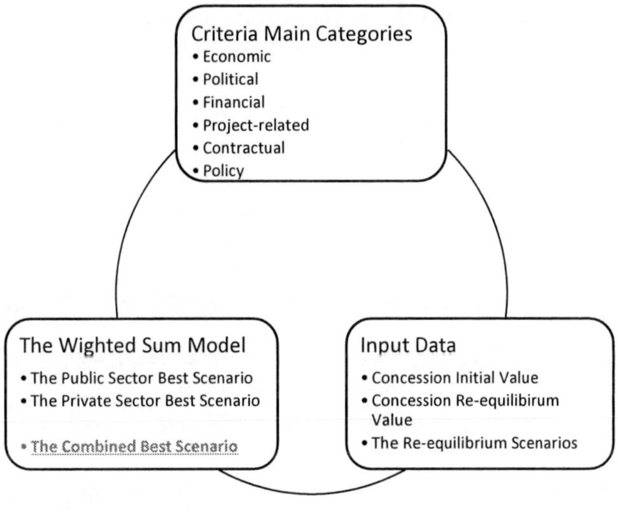

Figure 2. The Integration between the model inputs and outputs.

4 DISCUSSIONS AND ANALYSIS

The previous framework can be applied using computer software. The model is developed using Microsoft Excel 2013, Visual Basic for Applications (VBA) programming language, and the Precision Tree 5.5 for Excel add-in. In order to demonstrate the application of the framework developed, dummy data were used.

The aim of the model is to maximize the satisfaction of both the private and the public sectors. In order to achieve this goal, a set of criteria should be defined in order to evaluate this degree of satisfaction. The criteria main categories are economic, political, financial, project related, contractual, and policy. The economic criteria have to do with the economic situation of the country where the project is located. The sub-criteria for this category include inflation, general conditions of the country, stability of the exchange rates, and how the IRR is affected by all of the above. This is followed by the political criteria. The sub-criteria include long-term stability of the country, the current political situation, and the justice system in general. In addition, the financial criteria have to do with the party's financial standing. It includes any current financial obligations required to be paid by the party, additional financial obligations, liquidity, the ratio between debt and equity, and the transparency of financial data. On the other hand, the project-related criteria are specific to the project nature, which includes original concession period, preferable concession period, and the level of complexity of the project. Moreover, the contractual criteria include existence of regulator, risk sharing agreement, and how clear the termination clauses and re-equilibrium clauses are. Finally, the policy criteria are the general manner in which a certain party usually reacts to a certain situation. It includes the long-term business strategy, the likelihood of repeated business with a certain partner who is able to recognize the claim, and the experience of other partners.

The weights of each criteria and sub-criteria change depending on its relation to a certain scenario, as explained in the methodology; hence, the model user shall select which is more important among combinations of two criteria with respect to the selected scenario. The model user has the right to select both criteria if he believes that they are equally important. The weights of the criteria or the sub-criteria are calculated by counting the number of occurrences of the criteria or the sub-criteria in the decision matrix and dividing it by the total number of occurrences of all the criteria or the sub-criteria to get a percentage.

There existed separate weights for the criteria and the sub-criteria for each one of the four scenarios. However, the model user is to enter only one rank for all the four. There exist two separate sheets for the private sector and the public sector in order to decide whether these sub-criteria is in favor of the party or not. The following column is related risk, where the model user chooses the risk that affects his decision from a dropdown menu. The ranks are very suitable, suitable, neutral, unsuitable, and very unsuitable. Finally, the scenarios with the highest scores are identified both separately and combined for the public and the private sectors, Figure 3.

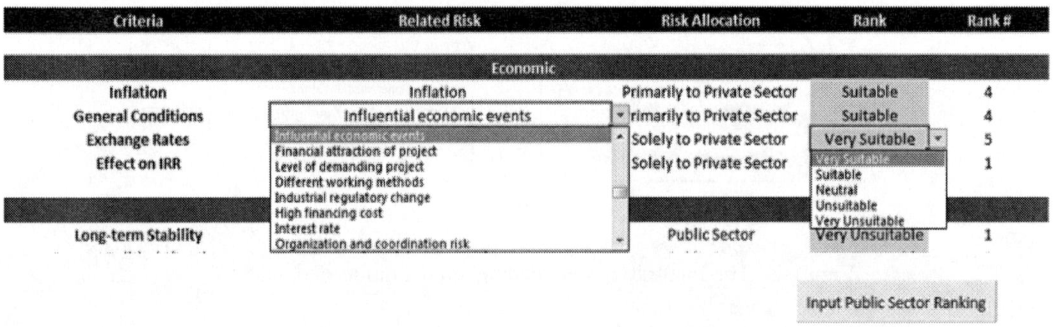

Figure 3. Example of sub-criteria ranking.

The model presents three decision tree reports, each presenting the expected monetary value (EMV), which is another way of calculating the scenario score. The trees start with a decision node, with the four alternatives of scenarios branching from it. Each scenario has branches presenting the criteria, and each criteria branches into the sub-criteria. The branches have the weights, and the rankings are assigned at the end of the tree branches.

The model then develops the risk profile graphs as it shows the risks and opportunities of choosing one scenario over the other. The model produces a probability chart, which shows the effect of changing the weights of the criteria and sub-criteria on the scenario scores, reflecting the probability of choosing a certain scenario over the other. A cumulative graph is also developed, which presents the ranges of values that the scenario can yield versus the corresponding probabilities or weights. A statistical summary report is then formulated to provide some statistics about all the ranges of possible outcomes of the decision tree.

The model also develops policy suggestion reports, which help the decision maker look at the broader picture when making his decision. The decision table report provides the "benefit of correct choice," which is basically the difference between the value of the highest scenario score and the lowest one. This helps in showing whether the scenario scores are close or not. The policy suggestion reports also include an optimal tree report that shows the path of the optimal decision only.

5 CONCLUSIONS

The purpose of this research is to develop a tool which will assist the re-equilibrium process of the financial model of PPP contract. Re-equilibrium is the process of adjusting the cash flow of the PPP project to return the rate of return of the private sector to the original rate of return at the contract commencement. Contract renegotiation at this stage is necessary to protect the rights of different stakeholders of the project. The model is developed using Microsoft Excel 2013, Visual Basic for Applications (VBA) programming language, and Precision Tree 5.5 for Excel add-in.

The main contribution of this research is that it develops a framework that facilitates the PPP contract renegotiation process. The methodological framework is applied as a Decision Support System prototype model. The purpose of the suggested framework is to enable all contract stakeholders to agree on a unified method of developing the different re-equilibrium scenarios and choosing the optimal scenario that suits all parties. This will facilitate the PPP renegotiation process, which will, in turn, encourage investors to enter PPP projects. The developed framework is of great benefit to project stakeholders, including the private sector, the public sector, and the users of the service. It saves time and money invested in lengthy negotiations, and it enforces transparency and mutual trust between the different parties by providing a tool that significantly minimizes conflicts during the renegotiation process and defines clear steps to be followed in order to reach an agreement that will maximize the benefits for both private and public sectors.

References

Archer, D., and Cameron, A., Making Public Private Partnerships Work, *Socia Partnerships Applied*, 2003.

Carbonara, N., Costantino, N., and Pellegrino, R., A Transaction Costs-Based Model to Choose PPP Procurement Procedures, *Engineering, Construction and Architectural Management*, 23(4), 491-510, 2016.

Cruz, C. O., Marques, R. C., and Cardoso, P., Empirical Evidence for Renegotiation f PPP Contracts in the Road Sector, *Journal of Legal Affairs and Dispute Resolution in Engineering and Construction*, 7(2), 05014003, 2014.

Cruz, C. O., and Marques, R. C., Using The Economic and Financial Reequilibrium Model to Decrease Infrastructure Contract Incompleteness, *Journal of Infrastructure Systems*, 19(1), 58-66, 2012.

Katz, D., Financing Infrastructure Projects: Public Private Partnerships (PPPs), *New Zealand Treasury*, 2006.

Pugh, S., *Total Design: Integrated Methods for Successful Product Engineering*, Wokingham, England, 1993.

Sarmento, J. M., and Renneboog, L., Anatomy of Public-Private Partnerships: Their Creation, Financing and Renegotiations, *International Journal of Managing Projects in Business*, 9(1), 94-122, 2016.

Skanska, *European Commission Second International Workshop on PPPs*, European Commission, 2004.

The Construction Management Association of America, *An Owner's Guide to Project Delivery Methods*, CMAA, 2012.

Responsible Design and Delivery of the Constructed Project
Edited by Abdul-Malak, M., Khoury, H., Singh, A., and Yazdani, S.

MULTI-METHOD MODEL FOR INFRASTRUCTURE PORTFOLIO MANAGEMENT

GREGORY J. FITCH[1], IBRAHIM ODEH[1], and FREDERICK KAUTZ[2]

[1]*Dept of Civil Engineering & Engineering Mechanics, Columbia University, New York City, USA*

[2]*Dept of Civil & Environmental Engineering, Worcester Polytechnic Institute, Worcester, USA*

Management practices have a direct impact on the cost of capital used for financing the operations and maintenance of existing infrastructure on a concurrent basis with planning and constructing new projects. This paper identifies best management practices that can improve credit rating, which in-turn theoretically lowers the interest paid on debt. As a result, these best management practices lower the weighted average cost of capital used to maintain existing assets on a concurrent basis with the construction of new infrastructure. As a practical example, this paper identifies the best management practices from the perspective of municipal credit rating agencies in the United States. The current research then constructs a deterministic integer programming model based on these best management practices. The deterministic model is part of multi-method model that will be mathematically embedded into a system dynamics model that can then be used as a facilities management plan. The facilities management plan aims at lowering the weighted average cost of capital for maintaining existing assets while planning and constructing new infrastructure. While the multi-method model is based on municipal credit rating in the U.S., the research goal is to nevertheless show how a system dynamics model is used to implement best management practices that are generally accepted as economically sustainable worldwide.

Keywords: Sustainability, System dynamics, Finance, Economics, Best management practices, Strategic and operational planning.

1 INTRODUCTION

The purpose of this research is to advance the economic sustainability of the infrastructure procurement process by hedging the risk of financial loss to private equity sponsors and commercial lenders. To achieve this primary goal, the objectives of the current research are threefold. The first objective is the construction of an integer programming (IP) model to optimize the management of a portfolio of infrastructure assets by meeting a set of specific economic and financial constraints while ensuring all user demands are met. The second objective is the construction of a system dynamics (SD) model to analyze and simulate the causal effects related to asset deterioration, rehabilitation processes, cost accumulation and inflexibility of resources across a portfolio of infrastructure assets. By jointly considering the first two objectives, a third and primary objective will be to present a holistic method for supporting a process to optimize decision-making during the procurement of new assets within a portfolio of existing infrastructure. The decision-making process is part of an overarching management plan that incorporates economic and financial targets while ensuring all user demands are met.

2 RESEARCH METHODOLOGY

The investigation presented in this first part of a series of research papers is intended to elucidate a methodology where SD models can be used to manage a portfolio of infrastructure assets from multiple perspectives. The methodology outlined in this paper is intended to be adaptable and used by any one of several stakeholders in the infrastructure procurement process. For practicing civil engineers and concerned governments, the SD model will be used essentially as a facility management plan. Conversely, this same methodology with a modified SD model is equally intended to be used by equity sponsors and commercial lenders for optimizing the economic return and mitigating financial risk arising from a portfolio of infrastructure investments. The methodology begins with the current research that identifies the best management practices (BMP) from the perspective of credit rating agencies in the U.S. Several of these BMPs can be mathematically modeled and are presented in an IP model in this paper. The IP model includes constraints that capture the BMPs from the perspective of credit rating agencies in the U.S. While the model is based on municipal credit rating in the U.S., the research goal is to investigate and demonstrate how IP and SD models can be constructed and used in tandem to manage an investment portfolio subject to a variety of optimizing constraints.

The IP model is presented in detail in this current research paper. The IP model includes an objective function to minimize all costs associated with managing a portfolio of infrastructure assets while ensuring all user demands are met, subject to a set of financial and economic practices that are considered highly favorable from the perspective of the credit rating agencies. This current research will be followed by an investigation that presents an illustrative case study that demonstrates how the objective function can be optimized. By doing so, the maintenance of existing assets and the procurement of new assets can be prioritized as a function of user demands, costs and the optimization constraints included in the IP model. The third and final stage of the overarching research will integrate the optimization results into a system dynamics framework similar to the "aging chain" SD model that is presented in cursory form in this current paper. Simulation results from the SD model will be used to analyze how the portfolio of infrastructure performs in meeting all user demands while satisfying the optimization constraints.

3 MANAGEMENT IMPACT ON CREDIT

For purposes of brevity, a complete discussion of management's impact on credit is precluded from this research paper. The integer programming model presented, however, is largely based on Larkin (2001) in which the BMPs have been identified as highly favorable to concerned governments and other issuers of debt. Table 1 includes those BMPs and also indicates the respective constraint(s) within the IP model that captures the mathematical representation of each practice.

4 INTEGER PROGRAMMING MODEL

The IP modeling method is concerned with optimization problems in which some of the variables are required to take on discrete values. Rather than allow a variable to assume all real values in a given range, only predetermined discrete values within a prescribed range are permitted. In most cases, these values are integers, giving rise to the name of this class of models (Jensen and Bard 2003). While a full discussion explaining the methodology of IP modeling is beyond the scope of this paper, the current research presents an IP model that mathematically represents the BMPs from the perspective of municipal credit rating agencies in the U.S.

Table 1. Management practices and integer programming model constraints.

Management Practice	Constraint No.
Fund balance reserve policy & working capital reserves	No. 4
Multiyear financial forecasting	No.'s 4, 5, 6, & 7
Contingency planning policies	No.'s 4, 5, 6, & 7
Policies regarding nonrecurring revenue	No. 5
Depreciation of general fixed assets	No. 6
Debt affordability reviews and policies	No.'s 4, 5, 6, & 7
Pay-as-you-go capital funding policies	No. 4
Five-year capital improvement plan integrating operating costs of new facilities	No. 6
Rapid debt retirement policies (greater than 65% of debt retired in 10 years)	No. 7

It is noteworthy that for academic purposes, IP models are often presented as a method for optimizing a warehouse or other facility location-type problem. There are several notable similarities and distinctions between the facility location-type problem and the application of IP modeling method in the current research. To begin, in the facility location-type problem the IP model includes constraints that are a function of the costs for building a facility at a specific location in order to meet all customer demands. Likewise, the IP model presented in this research aims to optimize the expenditure of funds while meeting all demands required of a portfolio of infrastructure assets. Furthermore, in the facility location problem constraints are related to shipping within geographical boundaries. Conversely, the IP model in the current research is bound within a range of economic and financial constraints established by credit rating agencies. In addition and as the name suggests usage of the IP model in the facility location problem is used to select a location for a facility prior to construction. In the current research the IP model is used to structure a management plan for a portfolio that consists of already existing infrastructure and other assets that have yet to be designed and/or constructed.

4.1 IP Model for BMPs

The following IP model includes constraints 1, 2 & 3 which are required to establish that all user demands are met and to account for annual operational costs for a portfolio of existing facilities on a concurrent basis with significant upgrades and the construction of new projects.

The integer programming model follows:

$$X_{ijk} = \begin{cases} 1 & \text{If calendar year } j \text{ is the } k^{th} \text{ year of upgrading or constructing an asset } i \\ 0 & \text{Otherwise} \end{cases} \quad (1)$$

where c_{ijk} equals the cost in year k for upgrading an existing infrastructure asset or a procuring new facility i in calendar year j.

Plant i take on values of A, B, C, D etc.

For a facility management plan of N years starting in year 1: $1 \leq j \leq N$.

Let S be the set of variables defined as the following:

- $X_{Aj,1}$..........................$X_{Aj,k}$ for all j
- $X_{Bj,1}$..........................$X_{Bj,k}$ for all j

For A, B, C, D etc.

The objective is to minimize the total cost subject to the requirements that all demands be met subject to the capacity of current and future assets contained within the infrastructure portfolio.

$$\text{Minimize} \sum_i \sum_j \sum_{k=1}^{k(i)} c_{ijk}\, X_{ijk(i)} \tag{2}$$

where *K(i)* equals the number of years to operate and maintain or upgrade existing facilities *or* construct and add new assets into the infrastructure portfolio; subject to the following constraints:

Constraint No. 1 – All demands must be met over *N* years.

$$\sum_j^N X_{ijk} = d_i \tag{3}$$

Constraint No. 2 – For upgrading and/or constructing infrastructure asset *i* in *n* number of years, the nth year follows $n^{th} - 1$ year etc.

$$X_{i,j+1,2} \geq X_{i,j,+1};\ X_{i,j+2,3} \geq X_{i,j,+2};\ X_{i,j+3,4} \geq X_{i,j,+3}\ etc. \tag{4}$$

Constraint No. 3 – For each infrastructure asset *i* and each calendar year *j*, *j* is at most 1 year of operations and maintenance and / or new construction.

$$X_{i,j,1} + X_{i,j,2} + X_{i,j,3} + X_{i,j,k(i)} \leq 1 \tag{5}$$

Constraint No. 4 – Maintaining a fund balance reserve policy and working capital reserves.

$$\sum_i \sum_j \sum_{k=1}^{k(i)} c_{ijk}\, X_{ijk(i)} \leq U_j + \sum y_j \tag{6}$$

where U_j is defined as the upper bound defined by conservatively forecasted revenue less the amount placed into working capital reserves and pay-as-you-go capital funding policies and y_j is defined as the sum of allowable debt for year *j*.

Constraint No. 5 – A policy in place regarding nonrecurring revenue.

$$\sum_j (C_{A,j,1}X_{A,j,1} + C_{A,j,2}X_{A,j,2} + C_{A,j,3}X_{A,j,3}) \leq non - recurring\ revenue + \%\, U_j + \% \sum y_j \tag{7}$$

Constraint No. 6 – Implementation of a five (5) year capital improvement plan integrating operating costs.

$$\text{For each year j the} \sum c_{i,,j,k} X_{i,j,k} \leq U_j + \sum y_j \tag{8}$$

Constraint No. 7 – A policy for the rapid retirement of debt with at least 65% of debt retired in the next 10 years.

$$.65 \sum y_j \geq \sum y_j\ in\ year\ j + 10 \tag{9}$$

5 SYSTEM DYNAMICS MODELS

Of particular relevance to the current research is the work of Rashedi and Hegazy (2015), in which a SD model simulates the "aging chain" of the deterioration and rehabilitation processes as part of an infrastructure management plan. A Markovian process along with a transition probability matrix (TPM) is used to capture the nonlinear progression of infrastructure assets evolving toward worsening states of condition before receiving major rehabilitation or being completely replaced. The SD model is used to preclude budget shortfalls on a concurrent basis with the preservation of existing infrastructure. The TPM is formulated as follows:

$$\text{TPM} = \begin{bmatrix} P_{11} & P_{12} & 0 & 0 & 0 \\ 0 & P_{22} & P_{23} & 0 & 0 \\ 0 & 0 & P_{33} & P_{34} & 0 \\ 0 & 0 & 0 & P_{44} & P_{45} \\ 0 & 0 & 0 & 0 & 1 \end{bmatrix} \qquad \text{Eq. (10)}$$

where P_{ii} is the probability of an asset in state i remaining in state i while $\boldsymbol{P}_{ij}$ is the probability of the same asset deteriorating to state j ($P_{ii} + P_{ij} = 1$).

5.1 Developing the SD Model

The current research begins to construct the SD model (Figure 1) consistent with Rashedi and Hegazy (2015) but for purposes of brevity only a simple stock-and-flow diagram is presented. The stock-and-flow diagram presented within this research was developed with the computer software iThink 10.1.2® in order to demonstrate how a facility management plan can be developed within a SD model platform. The SD model employs the industry-accepted general structure for stock-and-flow diagrams including stocks, flows, converters, and connectors. Rectangles represent stocks and signify both physical and non-physical accumulations resulting from an activity. Flows represent activities or actions that transport quantities into or out of a stock over time. Inflows and outflows are represented by pipes or arrows pointing into (adding to) and out of (subtracting from) stocks, respectively. Unless there is a net value of zero, material stock exists at a given point in time and will remain even when the processes of inflows and outflows are complete.

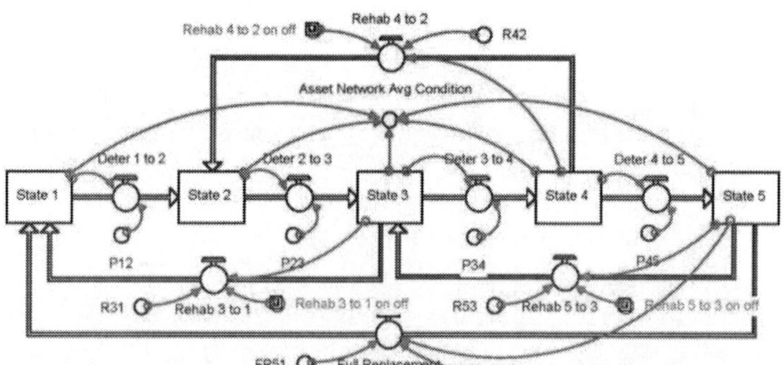

Figure 1. System dynamics model for an "Aging Chain" of infrastructure assets.

In the current research, the SD model includes the rate of deterioration based on a time constant between states that captures the nonlinear progression of infrastructure assets evolving toward worsening states of condition before receiving major rehabilitation or being completely replaced. Figure 2a shows an example for a case of 100 assets in State 1 (which could be interpreted as the "new", as built state), wherein the rate of deterioration is exponential for transitions between each successor state given the characteristic time constants for transition between these states. Figure 2b shows a reduced rate of deterioration after a rehabilitation process is introduced, which transitions assets from state #3 to the original state #1. In the follow-on research the costs for rehabilitating assets back to earlier states of condition will be constrained by the optimization results from the IP model's objective function.

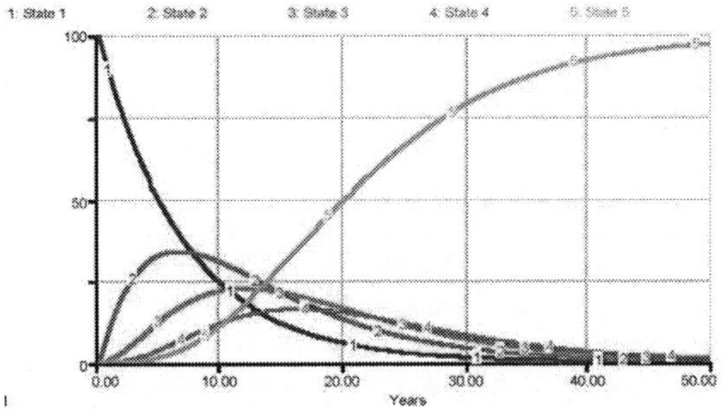

Figure 2a. Assets with no rehabilitation.

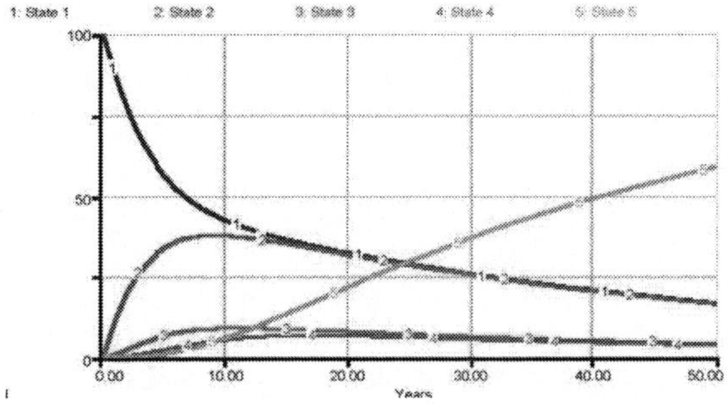

Figure 2b. Assets with rehabilitation.

6 CONCLUSION AND FURTHER RESEARCH

The research presented in this paper identifies BMPs that have been determined to improve a government credit rating. An improved credit rating translates into a lowered cost of capital, which in-turn makes future projects more feasible. The primary objective of the current research is to construct a multi-method model that can be used to manage a portfolio of infrastructure assets while meeting these BMPs. This investigation investigates a first step to facilitate a more manageable and streamline SD model. It includes the construction of an IP model in which the objective function is to minimize all costs for maintaining a portfolio of assets while meeting all user demands subject to the aforementioned BMPs. Follow-on research will include solving the IP model and optimizing the objective function at the appropriate level of abstraction to classify infrastructure assets by type while ensuring all user demands are met within the framework of the SD model. The objective of this evolving research is to develop a multi-method model to manage a portfolio of infrastructure assets.

References

ISEE Systems iThink version 10.1.2 [Computer software]. Lebanon, NH

Jensen, P. A. and Bard, J. F., *Operations Research Models and Methods*, John Wiley & Sons Inc., Danvers, Massachusetts, USA, 2003.

Larkin, R. P., Impact of Management Practices on Municipal Credit, *Government Finance Review,* 23-26, February, 2001.

Rashedi, R. and Hegazy, T., Holistic Analysis of Infrastructure Deterioration and Rehabilitation Using System Dynamics, *ASCE Journal of Infrastructure Systems,* 10.1061/(ASCE)IS.1943-555X.0000273, September, 2015.

Responsible Design and Delivery of the Constructed Project
Edited by Abdul-Malak, M., Khoury, H., Singh, A., and Yazdani, S.

PRIORITIZING SIDEWALK UPGRADE PROJECTS TO MAXIMIZE COMPLIANCE WITH ACCESSIBILITY REQUIREMENTS

AYMAN HALABYA and KHALED EL-RAYES

Dept of Civil and Environmental Engineering, University of Illinois at Urbana-Champaign, Urbana, United States of America

People with disabilities form 18.7% of the United States population, as reported by the U.S. Census Bureau in 2012. To avoid discrimination against this significant portion of the population, state and local governments are required by federal and state laws to provide and maintain accessibility for people with disabilities on their sidewalks and pedestrian facilities. To achieve compliance with these laws, state and local governments need to conduct self-evaluations to identify inaccessible pedestrian facilities and develop transition plans to schedule upgrade projects for these inaccessible pedestrian facilities. The federally-mandated transition plan requirements include the development of a schedule that displays, in detail, deadlines for all upgrade projects needed to achieve full compliance with accessibility requirements. To prepare this schedule, public entities are required to rank and prioritize pedestrian facilities upgrade projects. This paper presents the development of a novel methodology to quantify the impact of upgrading inaccessible pedestrian facilities on people with disabilities. The developed methodology considers several factors related to pedestrian facilities' conditions and location to estimate the number of expected pedestrians with disabilities impacted by upgrading each inaccessible pedestrian facility. This methodology is designed to assist decision makers in state and local governments in the process of ranking and prioritizing inaccessible pedestrian facilities upgrade projects.

Keywords: ADA, Self-evaluation, Transition plan, Pedestrian facilities, Ranking.

1 INTRODUCTION

People with disabilities form 18.7% of the United States population (U.S. Census Bureau 2012). This significant portion of the population requires special accommodations in order to be able to fully participate in society. These special accommodations help create accessible environments that enable people with disabilities to carry out their day-to-day activities independently. Several federal and state accessibility laws and regulations have been enacted to prevent discrimination against people with disabilities, including (1) the Architectural Barriers Act (ABA) (U.S. Congress 1968), (2) the Rehabilitation Act (U.S. Congress 1973), and (3) the Americans with Disabilities Act (ADA) (U.S. Congress 1990). Federal agencies such as the U.S. Department of Justice and the U.S. Department of Transportation are appointed by the aforementioned accessibility laws to develop and enforce federal regulations such as the 2010 ADA Title II regulations (U.S. Department of Justice 2010). Failure to comply with these accessibility laws and regulations often forces people with disabilities to travel in dangerous conditions, navigate difficult environments, or suffer painful injuries. In addition, state and local governments have incurred costly settlements due to non-compliance with accessibility laws and regulations. For example, the City of Los Angeles agreed in 2016 to pay $1.4 billion to upgrade its sidewalks and pedestrian facilities and achieve full compliance with accessibility requirements in order to avoid litigation (United States District Court 2016).

To avoid these costly settlements and guarantee a higher level of service for their citizens, state and local governments are required to conduct self-evaluations to identify non-compliant pedestrian facilities and develop transition plans to explain in detail the actions and steps required to achieve full compliance with accessibility laws and regulations (U.S. Department of Justice 2010). Public entities are also required to make these self-evaluations and transition plans available to the public and keep them in record for at least three years. This federally mandated transition plan requirement proved to be one of the most difficult tasks for municipalities and local governments because federal and state laws, regulations, standards, and guidelines did not offer any guidance on how to rank and prioritize pedestrian facilities upgrade projects, which leaves state and local governments with the challenging task of developing methodologies to perform this ranking process (CCRPC 2016).

Several studies have been conducted to assist state and local governments in ranking and prioritizing the upgrade projects of non-compliant pedestrian facilities. Most of these studies adopted a conceptual framework that utilizes two main factors: (1) *Pedestrian Potential Index* (PPI), that represents travel demand; and (2) *Pedestrian Deficiency Index* (PDI) that incorporates infrastructure and safety factors. Each of these two indices can include multiple indicators that are utilized to calculate a weighted index that can be used to rank pedestrian facilities (CCRPC 2016, City of Bellevue 2008, City of Clayton 2014, City of Fishers 2015, Loewenherz 2010). Higher PPI and PDI values means higher upgrade priority.

PPI considers factors that affect the degree of use of pedestrian facilities and pedestrian traffic volume at each of those facilities such as (1) proximity to public facilities, (2) population density, (3) percentage of people with disabilities, and (4) land use near pedestrian facilities. Due to the lack of pedestrian traffic data, all the aforementioned studies utilized heuristics to represent the need to upgrade non-compliant pedestrian facilities (Guensler *et al.* 2015). Other studies focused on measuring and estimating the number of pedestrians using sidewalks and pedestrian facilities during a specific time period (Davis *et al.* 1988, Ryus *et al.* 2017, Sanders *et al.* 2017, Seyfried *et al.* 2006). These studies however focused on the analysis of a small area for the purpose of extracting design guidelines and best practices rather than accurately estimating the expected traffic at each pedestrian facility in a large urban area which is essential to provide a reliable ranking of upgrade projects for non-compliant pedestrian facilities. To overcome this limitation, this paper presents a novel methodology for ranking pedestrian facilities based on their level of use.

2 PEDESTRIAN LEVEL OF USE

To rank the upgrade projects of pedestrian facilities, the present model utilizes a novel metric that is named "Pedestrian Level of Use" (PLU) to represent the total number of pedestrian trips that utilize each pedestrian facility in a specific period of time. This metric reflects the varying levels of use for pedestrian facilities and can be used to rank these facilities based on how often they are used by pedestrians. The proposed methodology calculates the PLU for each pedestrian facility in four major steps that calculate (1) the average number of pedestrian trips for each pedestrian facility based on the available federal, state, and local statistics and surveys; (2) the preliminary Expected Pedestrian Trips (EPT) for each pedestrian facility that considers the level of service index LOSI that this facility can accommodate; (3) an adjusted EPT for each pedestrian facility that reflect the connectivity of sidewalk network; and (4) the final PLU that considers proximity to Pedestrian Trip Generators (PTG) in addition to all previous factors (see Figure 1). The following four sections provide concise description of each of these steps.

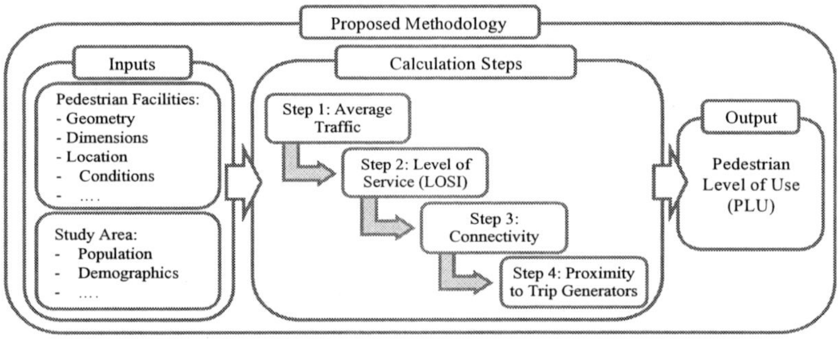

Figure 1. Proposed methodology.

2.1 Step 1: Average Pedestrian Traffic

The National Household Travel Survey (NHTS) reported in 2009 that the average number of walking trips per person in the United States is 362 (Santos *et al.* 2011). It also reported that the average walking trip length is 0.98 miles and the average trip time is 16.15 minutes in 2009 (NHTS 2009). In this step, the model utilizes these statistics to calculate the total number of pedestrian trips, miles travelled, and time spent walking for the entire study area by multiplying these national averages by the population of the study area.

2.2 Step 2: Level of Service Index

Level of Service Index (LOSI) for sidewalks quantifies the amount of pedestrian traffic that can be accommodated by that sidewalk, and is calculated in the present model based on the measurements and geometry of that sidewalk. For example, sidewalk width is one of the main factors that determines the level of service for sidewalks. Wider sidewalks can accommodate higher pedestrian traffic with acceptable LOSI, while narrower sidewalks provide much lower LOSI for the same number of pedestrians (Landis *et al.* 2001). In this step, the model utilizes LOSI calculations to find the maximum number of pedestrians that can be accommodated by each sidewalk to provide the highest possible levels of service, and then uses this value as relative importance weight to find the weighted average EPT for each pedestrian facility. For example, Sidewalk A in Figure 2 can accommodate twice the number of pedestrians that sidewalk B can accommodate at the same level of service because Sidewalk A is twice as wide as Sidewalk B (see Figure 2). Accordingly, Sidewalk A will have an EPT that is twice the value of Sidewalk B.

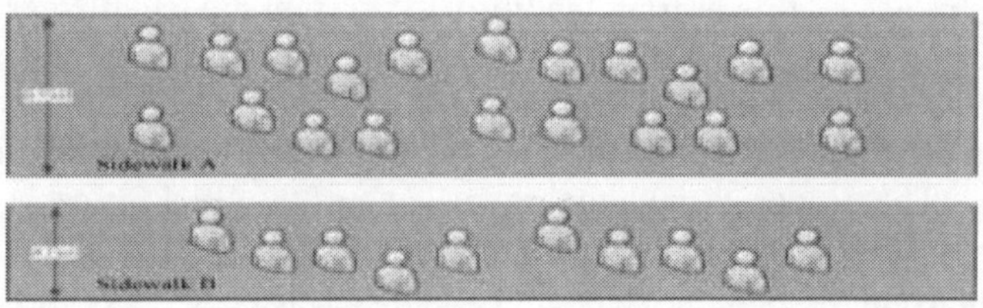

Figure 2. Level of service index.

2.3 Step 3: Connectivity

The main purpose of building sidewalks is to connect different points in the city in a way that allows pedestrians to use this interconnected network of sidewalks to travel from one point to the other with ease. In this step, the model utilizes a novel metric that is named *Connectivity Index (CI)* to consider the continuity of LOSI across the sidewalk network. This index identifies pedestrian facilities that disrupt the flow of pedestrians and provide a way to avoid that disruption in the calculations of PLU. For example, if we have a sidewalk that includes three connected segments A, B, and C with the widths of 6 feet, 4 feet, and 6 feet respectively (see Figure 3). In this example, the calculated LOSI for sidewalk segments A and C in the previous step is higher than that of sidewalk B due to their larger width. In reality, the LOSI in all three segments should be equal since they are all connected. To achieve this, the present model adjusts the LOSI of sidewalk B to match those of sidewalks A and C to maintain the same level of service throughout the entire sidewalk network.

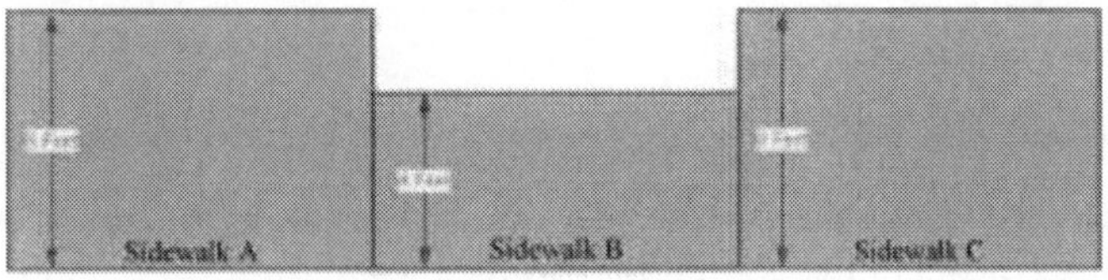

Figure 3. Connectivity.

2.4 Step 4: Proximity to Pedestrian Trip Generators (PTGs)

Pedestrian facilities near Pedestrian Trip Generators (PTGs) such as public facilities, transit stations, hospitals, schools, and shopping malls are exposed to higher volumes of pedestrian traffic due to the additional pedestrian trips to visit these PTGs (CCRPC 2016, City of Bellevue 2008, City of Clayton 2014, Frackelton and Guensler 2014). In this step, the model utilizes a novel index named *Block Index (BI)* to represent the proximity to PTGs. BI reflects the distance in urban street blocks between pedestrian facilities and trip generators. For example, sidewalks on the same block as the trip generator will have the highest BI values, while sidewalks in the adjacent blocks will have lower BI values, and the further a sidewalk is from the trip generator, the lower their BI value will be. This step calculates BI values for each block, and then uses them as relative importance weights to re-adjust the values of the previously calculated EPT and generate the final PLU values for each pedestrian facility in the case study. It should be mentioned that this step takes into account the potential latent travel demand (traffic that is currently latent due to inaccessible sidewalks) by adjusting the value of all sidewalks based on their proximity to PTGs regardless of their current accessibility conditions.

3 CASE STUDY

A case study of a small town with the population of 3830 was used to test the developed methodology. The case study includes 864 sidewalks that have the collective length of 22.52 miles, as shown in Figure 4. The case study requires calculating PLU for all sidewalks to rank and prioritize pedestrian facilities' upgrade projects. The required input data includes (1) existing dimensions, geometry, and conditions of all sidewalks in the case study, (2) locations of all sidewalks in the case study, (3) locations and types of all pedestrian trip generators in the case study area, and (4) demographic and statistical information about the case study area. The methodology utilizes this input data to calculate PLU for each pedestrian facility. An example sidewalk from the case study has received EPT of 1044 as an initial value in Step#1, 1153 due to its high capacity for pedestrian traffic in Step#2, 1215 to ensure the continuity of traffic with the two adjacent sidewalks, and 1258 due to its proximity to a hospital.

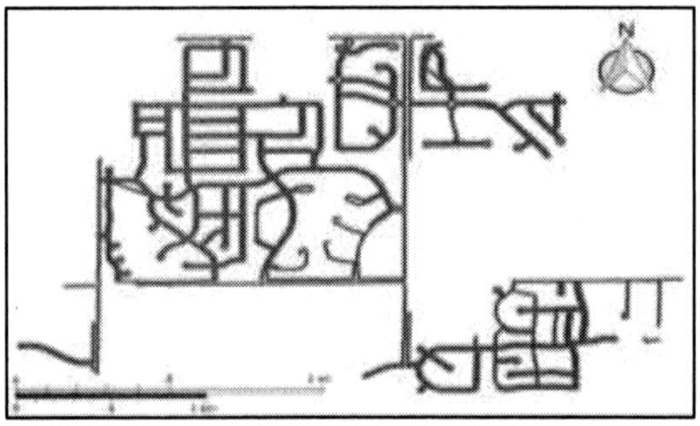

Figure 4. Map of the study area.

The capability of the developed methodology to efficiently calculate PLU for each pedestrian facility in the case study can be illustrated by its ability to calculate (1) the average pedestrian trips per pedestrian facility per year, (2) LOSI for each pedestrian facility, (3) EPT for each pedestrian facility that considers the continuity of the sidewalk network, and (4) final PLU value for each pedestrian facility that considers proximity to PTGs. In the case study the average number of pedestrian trips for each pedestrian facility calculated in Step 1 was 1044. After Step 2, the minimum number of pedestrian trips was 2 while the maximum was 2188. After Step 3, the maximum and minimum numbers have remained the same, while the median increased from 1044 to 1072. Finally, after Step 4 the total number of expected pedestrian trips was distributed over all pedestrian facilities considering proximity to trip generators, as shown in Figure 5.

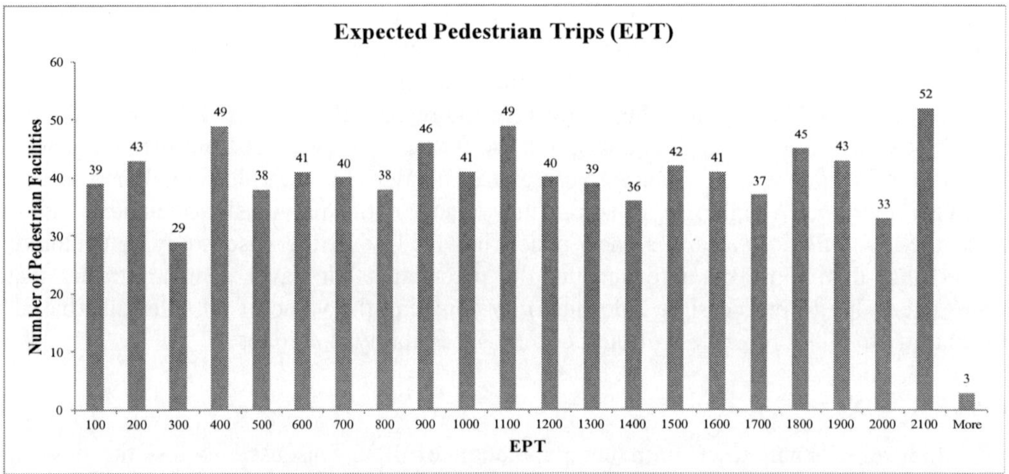

Figure 5. EPT distribution for all sidewalks in the case study.

4 CONCLUSION

A methodology was developed to calculate Pedestrian Level of Use (PLU). The PLU values were calculated based on available dimensional, geometric, demographic, and geospatial data.

The resulting PLU considered (1) total number of pedestrian trips in the study area, (2) level of service of pedestrian facilities, (3) connectivity and continuity of the sidewalk network, and (4) proximity to pedestrian trip generators. This methodology enables decision makers to rank and prioritize pedestrian facilities upgrade projects to maximize the impact of these upgrades and achieve full compliance with accessibility requirements.

References

CCRPC, *Sidewalk Network Inventory and Assessment*, Champaign, IL, 2016.

City of Bellevue, *City of Bellevue, Washington Americans with Disabilities Act (ADA) Sidewalk & Curb Ramp Inventory*, City of Bellevue, Washington, 2008.

City of Clayton, *City of Clayton ADA Transition Plan*, Clayton, MO, 2014.

City of Fishers, *Americans with Disabilities Act Transition Plan for Public Rights-of-Way*. City of Fishers, Indiana, 2015.

Davis, S. E., King, L. E., and Robertson, H. D., Predicting Pedestrian Crosswalk Volumes, *Transportation Research Record: Journal of the Transportation Research Board*, 1168, 25–30, 1988.

Frackelton, A. and Guensler, R., Pedestrian Project Prioritization: Incorportating App-Collected Sidewalk, *Transportation Research Board 93rd Annual Conference*, The National Academies of Sciences, Engineering, and Medicine, Washington, D.C, 2014.

Guensler, R., Frackelton, A., Grossman, A., Elango, V., Xu, Y., Toth, C., Akanser, A., Castrillon, F., Khoeini, S., Palinginis, E., and Sadana, R., *Automated Sidewalk Quality and Safety Assessment System (Regional UTC)*. Atlanta, GA, 2015.

Landis, B., Vattikuti, V., Ottenberg, R., McLeod, D., and Guttenplan, M., Modeling the Roadside Walking Environment: Pedestrian Level of Service, *Transportation Research Record: Journal of the Transportation Research Board*, 1773(1), 82–88, 2001.

Loewenherz, F., Asset Management for ADA Compliance Using Advanced Technologies, *Institute of Transportation Engineers Annual Meeting and Exhibit 2010: Vancouver, BC, Canada, 8-11*, Vancouver, BC, Canada, 654-667, 2010.

NHTS, *National Household Travel Survey*, 2009.

Ryus, P., Butsick, A., Proulx, F. R., Schneider, R. J., and Hull, T., *Methods and Technologies for Pedestrian and Bicycle Volume Data Collection: Phase 2*, The National Academies Press, 2017.

Sanders, R. L., Frackelton, A., Gardner, S., Schneider, R., and Hintze, M., Ballpark Method for Estimating Pedestrian & Bicyclist Exposure in Seattle : a Potential Option for Resource-Constrained Cities in an Age of Big Data, *Transportation Research Board 96th Annual Meeting, The National Academies of Sciences, Engineering, and Medicine,* Washington, D.C., 1–25, 2017.

Santos, A., McGuckin, N., Nakamoto, H. Y., Gray, D., and Liss, S., *Summary of Travel Trends: 2009 National Household Travel Survey*, 2011.

Seyfried, A., Steffen, B., and Lippert, T., Basics of Modelling the Pedestrian Flow, *Physica A: Statistical Mechanics and Its Applications*, 368(1), 232–238, 2006.

U.S. Census Bureau, *Americans with Disabilities: 2010 Household Economic Studies*, *US Census Bureau*, 2012.

U.S. Congress. *Architectural Barriers Act*. 42 U.S.C. § 4151 et seq., U.S.A., 42 U.S.C. §§4151 et seq. (1968).

U.S. Congress. *Rehabilitation Act*. 29 U.S.C. § 701 et seq., U.S.A., 29 U.S.C. § 701 et seq. (1973).

U.S. Congress. *Americans With Disabilities Act*. 42 U.S.C. § 12101 et seq., U.S.A., 42 U.S.C. § 12101 et seq. (1990).

U.S. Department of Justice. *Americans with Disabilities Act Title II Regulations. 28 CFR PART 35*, Washington, DC, USA (2010).

United States District Court, *Willits v. City of L.A.*, C.D. California, Los Angeles, 2016.

Responsible Design and Delivery of the Constructed Project
Edited by Abdul-Malak, M., Khoury, H., Singh, A., and Yazdani, S.

FAST-PACED CONSTRUCTION OF THE NEW KHALIFA PORT BRIDGE PROJECT

ANTOINE GERGESS

Civil and Environmental Engineering, University of Balamand, El-Koura, Lebanon

The significant growth and development in the United Arab Emirates in recent years necessitated considering advanced construction techniques for bridge structures. This paper presents features of the New Khalifa Port Project which consisted of bridge structures with a combined length of 3.64 km. The strict project design criteria, schedule and budget represented significant challenges which required considering specific designs using efficient precast elements and construction methods. The bridge structural system consisted of 39.8 m long precast, pretensioned modified AASHTO (American Association of State Highway and Transportation Officials) girders with cast-in-place concrete deck slabs. The main challenge was to optimize the size of the precast girder as the self-weight of the initially proposed elements exceeded the lifting capacity of the available equipment. The other challenge was to correlate between the British Standards and AASHTO LRFD (load and resistance factor design) code to design the girders for static and dynamic loads. The casting yard was set at 70 km from the project site and delivery roads were assigned to accommodate the special tandems carrying girders weighing 80 tons each. Lessons learned during design, fabrication, transportation and construction of the precast girders add on to the efficiency of the Accelerated Bridge Construction (ABC) processes.

Keywords: Girder, Precast, Prestressed, AASHTO, LRFD, Challenges.

1 INTRODUCTION

Precast elements are often used for the construction of bridge structures. Standard AASHTO girders can accommodate span lengths from 12 m up to 70 m. The girder size depends on the span length, load design criteria, service load stress criteria and most importantly the capacity of available lifting equipment. Design is usually performed using an internationally recognized design code such as the British Standards (1990, 2001, 2006) or AASHTO LRFD (AASHTO 2017). Nevertheless, constructability of precast structures can present major challenges that should be addressed by relying on the ingenuity of the project team.

This paper provides insight into the design and construction of the Khalifa Port bridge project in the United Arab Emirates (UAE) to satisfy stringent structural design and durability requirements. The Khalifa Port project comprises three structures with a total length of 3.64 km. The bridges connect islands made from reclaimed seabed material to the mainland. The main Port of Khalifa Island comprises two bridge structures, 1 km long each, labelled as the Highway Bridge and the Utility Bridge. The other island which serves as a marine terminal consists of a 1.64 km bridge structure, labelled as the Trestle Bridge. (Karapiperis *et al.* 2010).

The casting yard was set at 70 km from the project site and delivery roads were assigned to accommodate the special tandems carrying girders weighing 80 tons each. Lessons learned during design, fabrication, transportation and construction of the precast girders that add on to the efficiency of the Accelerated Bridge Construction (ABC) process are presented in this paper.

2 OBJECTIVE AND OUTLINE

The main objective of the paper is to present the accelerated bridge design and construction process for the 3.64 km long bridge structures. Description of the bridge structures is first provided. Gaps in the originally submitted construction scheme that led to presenting alternative design and construction processes are then discussed. These range from using precast elements for substructure pier columns and pier caps to reducing the weight of the superstructure precast girders while maintaining its flexural rigidity. Emphasis is placed on proper coordination that led to the fast-track construction of the 3.64 km of concrete bridges within approximately 12 months.

3 DESCRIPTION OF THE BRIDGE STRUCTURES

The two-bound Highway Bridge is 28.9 m wide with three - 3.65 m traffic lanes and a 2 m wide walkway in each direction. It connects the Khalifa Port Industrial Zone to the mainland loading and unloading terminals (Figure 1a). The parallel 27.5 m wide Utility Bridge accommodates pipelines and conveyors and one lane for maintenance vehicles (Figure 1a). The 19.5 m wide trestle bridge accommodates a 10 m wide vehicle carriageway and a heavy-duty dry bulk conveyor gallery supported on protruding cantilevers from the pier caps (Figure 1b). The Highway and Utility bridges are supported by 14 AASHTO girders each while the Trestle Bridge is supported by 5 girders only. The concrete deck slab thickness is 225 mm for the Highway and Utility bridges and 250 mm for the Trestle Bridge. Continuity for the three bridges was achieved in the deck slabs along four consecutive spans with expansion joints provided every 160 m.

(a) Utility (left) and Highway (middle & right). (b) Trestle with protruding pier caps.

Figure 1. Khalifa Port bridges.

4 DESIGN

The design of the superstructure girders and slabs was performed using British Standards for static loads (BS 5400-2 2006, BS 5400-4 1990) and the American Association of State Highway and Transportation Officials load and resistance factor design (LRFD) method for the seismic loads (AASHTO 2009). For the Highway and Utility bridges, the AASHTO precast, prestressed girders were designed for HA and HB45 traffic loadings (BD *37/01*) with HB45 being the heaviest design truck and HA representing lane loading. The Trestle Bridge girders were designed for HA traffic loading only (Gergess and Tepavcevic 2011).

The original layout for precast girders consisted of 2,300 mm deep girders that were spaced at 2.05 m for the main highway bridge and at 1.98 m for the utility bridge (Figure 1a). The 2,300 mm deep girders spaced at 2.56 m were also used for the trestle bridge (Figure 1b).

Based on value engineering, a 2,200 mm deep girder with a 1,070 mm wide top flange and a 200 mm web was used. The 200 mm web thickness reduced the self-weight by 20% without affecting its stiffness (no prestressing strands were provided in the web). The top flange width of 1,070 mm (compared to 800 mm) enhanced the stability of the girder as the weak axis moment of inertia for the 2,200 mm deep girder was larger compared to the original girder.

The final design resulted in using 15.2 mm diameter prestressing strands with a tensile strength of 1,860 MPa. Ordinary reinforcement in the girders was grade 460B, high yield deformed steel (type 2) with a minimum yield strength of 460 MPa (Gerges and Gergess 2012).

4.1 Durability

The design criteria emphasized on durability, mainly the need to achieve a 100-year service life under the extremely high temperatures of the Persian Gulf region (temperatures up to 55°C) over harsh marine environment (salt water and humidity up to 100 percent).

Durability measures involved increasing the concrete cover for the steel rebar, imposing strict limitations on crack width and using corrosion inhibitors within the concrete mix. For the substructure elements, a 100 mm cover was used and a maximum crack width of 0.125 mm was allowed. For the pretensioned AASHTO girders, a cover of 50 mm was used and no cracking was allowed as per the design criteria for limiting tensile stresses in prestressed Class 2 elements (BS 5400-4 1990). For slabs and diaphragm, a cover of 50 mm was used as well and limits for the crack width were set at 0.150 mm for exposed surfaces and 0.175 mm for non-exposed surfaces.

4.2 Concrete Mix

The minimum 28-day compressive cube strength for the precast, prestressed girders was set at 50 MPa. The aggregates size did not exceed 20 mm and for AASHTO girders it was supplemented with an increased dosage of corrosion inhibitor. The concrete mix was subjected to tests (NT Build 492) to determine the chloride penetration in the concrete.

5 CONSTRUCTABILITY

The pier columns and caps were precast onshore, transported on floating crane barges, and lowered into their final positions (Figure 2a). Once in position, they formed the working surface on which the precast girders were laid and from which the cast-in-place slab was constructed. Stay-in-place corrugated galvanized iron sheets were used for casting the slab (Figure 2b). Precast- construction was highly valued due to the repetitive construction cycles anticipated in the project.

5.1 Bridge Bearings

Laminated elastomeric bearings placed on the reinforced concrete pier caps were used to support the girders (Figure 3). Bearings were sized to accommodate expansion/contraction and at the same time minimize the thermal forces applied to the substructure based on its low horizontal stiffness. They were also designed for seismic loads (AASHTO 2000). Fixity to the adjacent concrete surfaces was ensured using top and bottom checkered steel plates that were glued using epoxy mortar to the concrete.

(a) Precast pier columns. (b) Corrugated metal sheets.

Figure 2. Construction process for precast piers and girders.

Experimental testing was performed to confirm the ability of the bearings to deform and transfer the design loads. The maximum shear deformation was measured at 155 mm due to seismic, thermal and time-dependent creep and shrinkage deformations.

Lateral stability between the precast girders was ensured by monolithically casting in-situ 1,950 mm deep by 500 mm wide diaphragms centered with the transverse axis of the bearings.

Figure 3. 2.2 m deep, 39.6 m long AASHTO girder.

6 CHALLENGES

The tight construction schedule and the need to keep costs within budget represented significant challenges during design, fabrication, transportation and erection of the precast elements.

6.1 Selection of Precast Girders

Selection of the girder type was governed by cost and constructability. The use of longer spans can reduce the number of intermediate piers, but the size and weight of the girders was limited by the availability of suitable lifting equipment. To reduce weight, girders were first designed as continuous over the supports using in-situ concrete diaphragms between the girder ends allowing negative moments to develop at the supports and limiting positive moments at the centers of the spans. This scheme was not adopted due to potential cracking of the solid end diaphragms which affects durability and service life of the bridge structures. Consequently, the girder size was based on the conventional single span design method where the precast girders are laid out as simply supported and continuity is achieved only in the deck slabs over the supports.

To reduce weight, an optimized 2,200 mm deep girder with 800 mm bottom flange, 1070 mm top flange and 200 mm web was used without changing the total number of girders (Figure 3). This also increased the compression axial stress due to prestress which allowed reducing the number of prestressing strands to 48 compared to 52 in the original design.

6.2 Fabrication

Fabrication of the girders was done at a casting yard distant 70 km from the project site (Figure 4a). 906 precast girders were produced during a period of 9 months at a peak rate of 36 girders per week (6 girders per day). 10 – 140 m long prestressing lines were set for production with a 1,200 tons bulkhead capacity (for the jacking prestress force). Three sets of steel and GRP molds were used for casting. 16 tons of 15.24 mm high strength prestressing strands were consumed daily.

(a) Casting yard. (b) Construction site.

Figure 4. Fabrication and erection of precast girders.

6.3 Transportation

Transporting the 80 tons girders to site was carefully planned to ensure that the associated roads and junctions can accommodate the special tandems. 200 tons multi-axial hydraulic trailers were used to deliver up to 13 precast girders per day. Two gantry cranes with a capacity of 70 tons each were used for handling the precast girders at the casting yard (Figure 4a).

6.4 Erection

Precast girders were loaded on barges and lifted using heavy-duty floating cranes for offshore erection (Figure 4b). Erection of the girders required high precision in operation of cranes during lifting. They had to be accurately placed on the supporting bearings within the setting time of the epoxy mortar applied to the checkered plates of the bearing pads (Figure 3). For this purpose, temporary steel guide frames were fixed to the pier cap to guide the lifted girders into place (Figure 3). These frames could also resist possible impact when placing the precast girder in place.

7 CONCLUSIONS

This paper described the process of accelerating the construction of 3.64 km of bridges. Value engineering led to changing the construction scheme from cast-in-situ to precast for substructure piers and adopting the conventional simple span methodology with continuity in the topping slab for superstructure precast girders. Proper coordination between the contractor's construction engineering consultants, precast fabricators and the client facilitated implementation of the modified design and construction schemes which led to the fast-track construction of the 3.64 km of concrete bridges within approximately 12 months. This was in line with the construction schedule and within the project budget.

Based on findings from the design and construction processes, the following conclusions were made:

- Value engineering resulted in reducing the weight of the precast girders by 20% while maintaining its flexural rigidity and stability.
- Precast construction for substructure pier caps and columns and superstructure girders was highly valued due to the repetitive construction cycles.
- The concept of using simple spans instead of making the girders continuous for composite loads and live loads was preferred for durability and maintainability.
- The use of elastomeric bearing pads contributed to reducing the magnitude of the seismic forces transmitted to the foundations.

Acknowledgments

The author acknowledges the contribution of Archirodon Construction, Abu Dhabi Ports Company (ADPC) (client for Khalifa Port Bridges project) and Al-Meraikhi Industrial Complex in the United Arab Emirates.

References

AASHTO, *Guide Specifications for Seismic Isolation Design,* American Association of State Highway & Transportation Officials, 2nd edition, Washington, DC., 2000.

AASHTO, *Guide Specifications for LRFD Seismic Bridge Design,* American Association of State Highway & Transportation Officials, 1st edition, Washington, DC., 2009.

AASHTO, *LRFD Bridge Design Specifications,* American Association of State Highway & Transportation Officials, 8th edition, Washington, DC., 2017.

BS 5400-2, *Steel, Concrete and Composite Bridges—Part 2: Specification for Loads*, United Kingdom: British Standards Institute, 2006.

BS 5400-4 *Steel, Concrete and Composite Bridges – Part4: Code of Practice for Design of Concrete Bridges*, United Kingdom, British Standards Institute, 1990.

BS BD37/01, *Loads for Highway Bridges. Design Manual for Roads and Bridges*, v1, section 3, part 14. United Kingdom: British Standards Institute, 2001.

Gergess, A. N., and Tepavcevic, M., *Implication of Using the British Standards on the Design of Precast, Prestressed Concrete AASHTO Girders,* PCI 57th Annual Convention and National Bridge Conference, Salt Lake City, Utah, 2011.

Gerges, N., and Gergess, A. N., Implication of Increased Live Loads on the Design of Precast Concrete Bridge Girders. *PCI journal,* 57(4), 78-95, 2012.

Karapiperis, D., Lykidis, G., Savvopoulos, K., El-Sayed, K., and Loukakis, K., Combined Effort. *ASCE Civil Engineering Magazine*, 80(9), September, 2010.

Responsible Design and Delivery of the Constructed Project
Edited by Abdul-Malak, M., Khoury, H., Singh, A., and Yazdani, S.

THE COST IMPACT OF FRONT END ENGINEERING DESIGN (FEED) ACCURACY FOR LARGE INDUSTRIAL PROJECTS

MOUNIR EL ASMAR and G. EDWARD GIBSON Jr.

School of Sustainable Engineering and the Built Environment, Arizona State University, Tempe, AZ, USA

Assessing the accuracy of front end engineering design (FEED) for large industrial projects is a critical task with the potential to considerably impact overall project success. At this early phase of the project, the owner's expectation is to be able to make informed decisions including cost predictions to determine whether the project should proceed to the next phase. The primary objective of this paper is to evaluate FEED accuracy and measure its impact on project cost performance. The authors collected and analyzed data from 33 completed large industrial projects representing over $8.83 billions of total installed cost. A primary finding is that projects with high FEED accuracy significantly outperformed projects with low FEED accuracy in terms of cost growth in relation to the approved budget at Phase Gate 3. This finding assists owners and contractors focusing on the front end planning phases of their large industrial projects to considerably improve cost performance.

Keywords: Front end planning, Construction industry institute, Project definition, Rating index, Cost growth, Performance.

1 BACKGROUND AND OBJECTIVE

For many projects, front end planning (FEP) is considered to be the most important process within the project lifecycle (Construction Industry Institute (CII) 2012). Planning efforts conducted during FEP can have significantly more influence on project success than efforts undertaken after detailed design and construction have begun (Gibson *et al.* 1993). While addressing FEP of projects in general, past research efforts have not focused specifically on assessing the accuracy of the engineering component of front end engineering design (FEED) for large industrial projects. A consistent understanding of FEED will allow the owner, engineer, and contractor to be better aligned as the project design process moves forward.

This paper quantifies the accuracy of FEED deliverables at the end of the detailed scope phase. It provides details about the research background and methodology, data collection efforts, project data analysis, and conclusions of the work performed by the research team. The research team set forth the following objectives: (1) identifying the enabling factors that drive effective engineering design during FEED (e.g. capability of the engineers, turnover of key design team members, time allowed for FEED, etc.); and (2) testing the impact on project performance by comparing the level of accuracy definition at the end of FEED versus corresponding cost performance factors for a sample of completed large industrial projects.

1.1 FEP Definition

FEP is defined as the process of developing sufficient strategic information with which owners can address risk and make decisions to commit resources in order to maximize the potential for a successful project. FEP is also known as front end loading, pre-project planning, feasibility analysis, conceptual planning, programming/schematic design, and early project planning (Construction Industry Institute (CII) 2012). FEP Phase 1 is known as "feasibility," Phase 2 is known as "concept," and Phase 3 is known as "detailed scope" as seen in Figure 1. Each of these phases is followed by a Phase-Gate (diamonds in the figure) that marks a decision to move forward to the next stage of the project. The focus of this research effort is the detailed scope phase or Phase 3, the gateway to detailed design and construction.

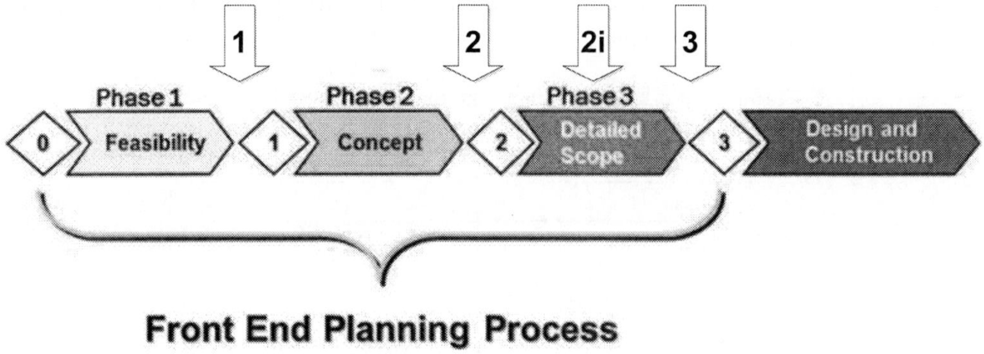

Figure 1. Typical front end planning process.

1.2 Research Scope

Within the FEP body of knowledge, the scope of this research specifically deals with the accuracy of FEED engineering deliverables. For maximum impact, the study also concentrates on large industrial projects, namely ones with the following characteristics:

- Projects completed within industrial facilities such as (or similar to): Oil/gas production facilities; refineries; chemical plants; pharmaceutical plants; paper mills; steel/aluminum mills; power plants; manufacturing facilities; food-processing plants; textiles mills.
- Total installed cost greater than USD 10 million (in some cases, respondents felt that their projects were complex and met the remaining criteria, despite being slightly below $10 million) and construction duration greater than nine months.
- More than ten core team members (e.g., project managers, project engineers, owner representatives) and a combination of part-time and full-time to complete full-time availability of core team members.

2 RESEARCH METHOD

The specific research hypothesis is as follows: FEED accuracy impacts project cost growth. To test this hypothesis, a comprehensive literature review was completed to identify all the relevant accuracy factors. These factors were provided and prioritized through a series of four industry-sponsored workshops with engineering and construction professionals experienced in the front end planning of large industrial projects. Specific project data regarding (1) the FEED development effort, (2) measures for accuracy factors, (3) cost budgets at the beginning of detailed design, and (4) project cost at the completion of the projects, were collected and analyzed. FEED accuracy scores were calculated for each project and compared to project performance data through statistical analysis.

The authors identified a total of 37 accuracy factors from the literature review and the survey and completed focus group exercises to identify any missing factors that needed to be added or any similar factors that needed to be combined. The authors and their industry research team of 20 FEED experts also developed detailed definitions for each of the accuracy factors based on the literature, and verified them based on industry experience. Through the workshops with 48 industry experts, the number of factors was narrowed down to a final list of 27, and the research team developed weights for each of these factors through a modified Delphi ranking approach. Four geographically dispersed workshops were hosted at various locations across the United States and Canada, as shown in Table 3-2. Overall, 48 industry professionals representing 31 organizations (14 owners and 17 contractors) attended the workshops as shown in Table 1.

Table 1. Workshop locations and number of participants.

Location	No. of Participants
Houston, Texas	14
Seal Beach, California	6
Cherry Hill, New Jersey	9
Calgary, Alberta, Canada	19
Total: 4 Workshops	**Total: 48 Participants**

The authors used several statistical methods to analyze the data collected at the workshops. Statistical analysis allowed the authors to interpret the data and provided a basis to offer recommendations regarding the efficacy of measuring FEED accuracy and predicting project outcomes. The methods employed by the authors include boxplots, histograms, normality tests, variance tests, regression analyses, t-tests, Mann-Whitney-Wilcoxon ranked sum tests and step-wise sensitivity analyses. Microsoft Excel™, Minitab™, and the statistical R package were the primary software platforms used to analyze data.

3 INITIAL FINDINGS: FEED ACCURACY FACTORS

The accuracy of FEED is not studied in literature. Therefore, the authors started by studying the accuracy of other project requirements, such as cost and schedule estimates, as there are established criteria for evaluating accuracy for these types of estimates in construction projects (e.g.; Skitmore *et al.* 1990, Construction Industry Institute (CII) 1998, Oberlender and Trost 2001, Heinemann and Zeiss 2002, Chen *et al.* 2005, Construction Industry Institute (CII) 2006, Rigby and Bilodeau 2015, Lim *et al.* 2016). Next, literature and past CII research efforts regarding factors that impact accuracy are evaluated, including those related to the project team (leadership and execution team) and project resources. The studies reviewed were conducted over four decades, span various industries, and were mined by the authors for accuracy factors that may apply to FEED. The discovered FEED accuracy factors are organized in four distinct types as shown in Figure 2.

In summary, no research was found in the literature that focused on FEED accuracy for large industrial projects. Thus, accuracy factors from previous research were studied and adapted in developing the FEED accuracy factors. The final list of 27 accuracy factors was developed with input from 23 literature references, the survey, 20 research team members, and 48 industry experts that participated in the workshops.

	1. PROJECT LEADERSHIP TEAM
1.a	Leadership team's previous **experience executing** a project of similar size, scope, and/or location, including FEED
1.b	**Stakeholders** are appropriately represented on the project leadership team
1.c	Project **leadership** is defined effective and accountable
1.d	**Leadership team and organizational culture** fosters trust, honesty, and shared values
1.e	Project leadership team's attitude toward **change**
1.f	Key personnel **turnover** (e.g., how long key personnel stay with the leadership team)
	2. PROJECT EXECUTION TEAM
2.a	**Technical capability and relevant training/certification** of the execution team
2.b	**Contractor/Engineer's team experience** with the location, with similar projects, and with the FEED process
2.c	**Stakeholders** are appropriately represented on the project execution team
2.d	Level of **involvement** of design leads or managers in the engineering process
2.e	Key personnel **turnover** including the **stability/commitment** of key personnel on the owner side through the FEED process
2.f	**Co-location** of execution team members to one another
2.g	Team culture or history of the execution team **working together**
	3. PROJECT MANAGEMENT PROCESS
3.a	**Communication** within the team is open and effective; a communication plan with stakeholders is identified
3.b	**Priority** between cost, schedule, and required project features is clear
3.c	Organization implements and follows a **front end planning** process (e.g., phase gates, clear requirements) and a **formal structure** or process to prepare FEED
3.d	Significant **input of construction knowledge**
3.e	Adequate process for **coordination** between key disciplines
3.f	**Alignment** of FEED process with **available project information,** including the existence of peer reviews and a standard procedure for updating FEED
3.g	**Documentation** of information used in preparing FEED
3.h	**Review and acceptance** of FEED by appropriate parties
	4. PROJECT RESOURCES
4.a	**Commitment** of key personnel on the project execution team
4.b	**Calendar time** allowed for preparing FEED
4.c	**Quality** and level of detailed of engineering data available
4.d	**Amount of funding** allocated to perform FEED
4.e	**Local knowledge** (e.g., institutional memory, understanding of laws and regulations, understanding of site history)
4.f	**Availability of standards and procedures** (e.g., design standards, standard operating procedures, and guidelines)

Figure 2. Accuracy TYPES and factors.

Workshop participants were also asked to rank order the top five accuracy factors in each accuracy type and to prioritize each accuracy type by allocating percentage values that represented the relative importance of each accuracy type to the accuracy of FEED. The final factor numbers and weights are based on the rank ordering of factors per accuracy type, and distributed among the following five possible ratings for each factor:

- High Performing = 100%
- Meets Most = 75%
- Meets Some = 50%
- Needs Improvement = 25%
- Not Acceptable = 0%

4 THE IMPACT OF FEED ACCURACY ON PROJECT COST PERFORMANCE

A total of 48 industry professionals from 31 different organizations provided project data, professional comments, and suggestions, during the 4 workshops. These participants have a combined engineering/project management experience of 962 years with an average of 23 years of experience per participant. Complete project data were collected for 33 projects with sizes ranging from US $8 million to US $14 billion, and an average project cost of approximately US $276 million. These 33 completed projects represent over US $8.83 billion in total installed cost and are geographically dispersed across six countries and nine states of the US. Table 2 provides some descriptive statistics for the completed project sample.

Table 2. Descriptive statistics (N=33).

Metrics	Avg.	Median	Std. Dev.	Min	Max
Total Installed Cost ($M)	276.01	114.20	455.80	7.78	1,939.00
Total Project Duration (Days)	924.39	750.00	472.02	240.00	2,340.00
Cost Change (%)	9.47	7.16	18.67	-27.27	53.44

The authors calculated accuracy scores for each completed project. Scores ranged from 24 to 97. A regression analysis was conducted to investigate any correlations between accuracy and cost growth. For the linear regression of cost change vs. accuracy, the resulting p-value was 0.015 indicates that there is a statistically significant relationship, e.g.; the slope of the regression equation is non-zero. This suggests that changes in the predictor variable (accuracy score) are significantly correlated with changes in the response variable (cost change). However, the R-squared value of 0.181 indicates that only 18.1% of the variance is explained by the predictor (accuracy score), so more variables are also impacting cost growth as expected. These other variables can be related to design, construction, scope, and other factors throughout the project. The data points and regression line are shown in the fitted line plot in Figure 3. Looking at the trend, one can see how accuracy scores starting around 75 and higher are associated with smaller cost change compared to projects with low FEED accuracy scores. In fact, several of these high accuracy projects have negative cost growth which implies savings in cost.

On projects that average millions of dollars, even small percentage savings are considerable. These quantitative results were also tested qualitatively on a dozen in-progress large industrial projects that used this method to gauge their FEED accuracy and confirmed the significant value added to their projects and FEP process.

5 CONCLUSION

This paper evaluated FEED accuracy and measured its impact on project cost performance. A primary finding is that projects with high FEED accuracy significantly outperformed projects with low FEED accuracy in terms of cost growth. Ongoing work is also measuring the effect of

FEED maturity on project performance, in an effort to understand the combined impacts of FEED accuracy and maturity on a number of key performance metrics.

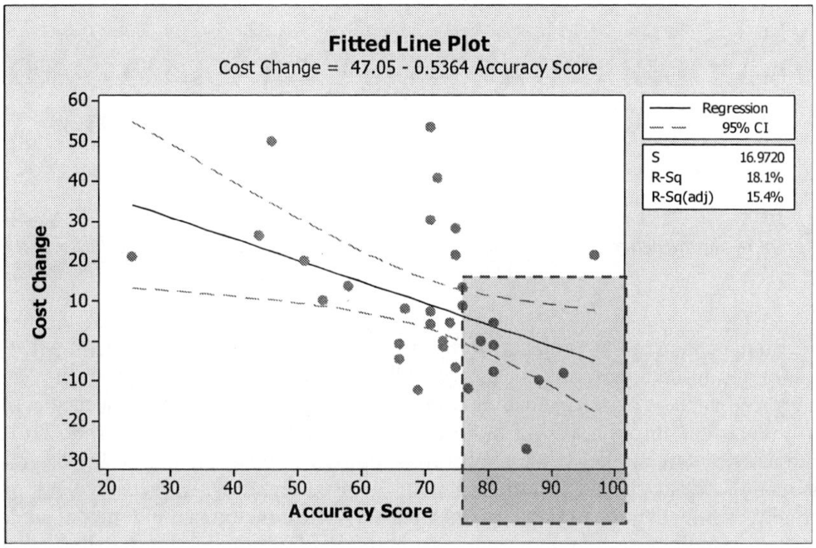

Figure 3. Regression results for cost change vs. FEED accuracy score.

Acknowledgments

This work was funded by the Construction Industry Institute (CII) under Research Team 331 "Assessing the Maturity and Accuracy of Front End Engineering Design (FEED) to Support Phase-Gate Approvals."

References

Chen, H. L., O'Brien, W. J., and Herbsman, Z. J., Assessing the Accuracy of Cash Flow Models: The Significance of Payment Conditions, *Journal of Construction Engineering and Management*, 131(6), 669–676, 2005.

Construction Industry Institute (CII), *Improve Early Estimates*, Research Summary 131-1, University of Texas at Austin, Austin, TX, 1998.

Construction Industry Institute (CII), *Front End Planning: Break the Rules. Pay the Price*, Research Summary 213-1, Austin, TX, 2006.

Construction Industry Institute (CII), RR256-11 - *Project Site Leadership Role in Improving Construction Safety*, Construction Industry Institute (CII), Austin, Texas, 2012.

Gibson, G. E., Kaczmarowski, J. H., and Lore, H. E., *Modeling Pre-Project Planning for the Construction of Capital Facilities*, Source Document 94 Prepared for Construction Industry Institute, University of Texas at Austin, Austin, TX, 1993.

Heinemann, G. D., and Zeiss, A. M., A Model of Team Performance, *Team Performance in Health Care*, Springer, 28–42, 2002.

Lim, B., Nepal, M. P., Skitmore, M., and Xiong, B, Drivers of the Accuracy of Developers' Early Stage Cost Estimates in Residential Construction, J*ournal of Financial Management of Property and Construction*, 21(1), 4–20, 2016.

Oberlender, G. D., and Trost, S. M., Predicting Accuracy of Early Cost Estimates Based on Estimate Quality, *Journal of Construction Engineering and Management*, 127(3), 173-182, 2001.

Rigby, D., and Bilodeau, B., *Management Tools & Trends 2015*, Bain & Company; 2015.

Skitmore, M., Stradling, S., Tuohy, A., and Mkwezalamba, H., *The Accuracy of Construction Price Forecasts*, University of Salford, 1990.

Responsible Design and Delivery of the Constructed Project
Edited by Abdul-Malak, M., Khoury, H., Singh, A., and Yazdani, S.

ON MODELING ACTIVITY CRASHING AND OVERLAPPING: A FIRST ALGORITHM

KAMEL MOHAMED ELAMROUSY[1] and PABLO BALLESTEROS-PÉREZ[2]

[1]*Dept of the Built Environment, University of Reading, Reading, United Kingdom*
[2]*Dept of Architecture, Building and Civil Engineering, Loughborough University, Leicestershire, United Kingdom*

Accelerating construction projects is a commonly used method for meeting the project deadlines and/or compensating for current delays. There are several approaches to speed-up a project, such as activity crashing and overlapping. Activity crashing means reducing activity durations through adding more resources. Activity overlapping means executing certain activities in parallel when they were supposed to be sequential in the original plan. Construction management literature is mostly focused on studying each acceleration mode separately, while the focal point of this paper is to develop a joint model and a first algorithmic implementation that involves both acceleration methods. Particularly, on reviewing the project management and scheduling literature, a mathematical model combining activity crashing and overlapping is reformulated. Also, a Genetic Algorithm is implemented on a fictitious case study. Preliminary findings of the model and algorithmic implementations identify that activity crashing alone and a mixed approach are preferred when significant compression is required, whereas activity overlapping is recommended for minimizing compression costs when only a small compression is required.

Keywords: Compression, Acceleration, Delay, Schedule, Stochastic, Productivity, Risk.

1 RESEARCH METHODOLOGY

Many construction projects suffer delays (Ballesteros-Pérez *et al.* 2015, 2016). One of the most common approaches used by project managers to deal with those delays and meet the deadlines is project acceleration (expedition) (Baker 1991). Haga (1998) defined project acceleration (schedule compression) by "shortening the normal duration of the project schedule without reducing the original scope of work."

There are three methods for project acceleration: activity crashing, activity overlapping (fast-tracking) and activity substitution (Eduardo *et al.* 2008). Crashing is a method for reducing the project duration for the least possible cost through increasing the number of resources. The most common approaches to crash an activity are to add more laborers, use multiple shifts, offer overtime and add more resources; which will incur more cost (Baker 1991, Eduardo *et al.* 2008, Sahu and Sahu 2014). Activity overlapping involves starting the activity earlier than its planned start date so that it would commence before its predecessor(s) has finished; that is, converting some sequential activities into parallel ones (Meier *et al.* 2015). Similarly, this would require additional costs due to inevitable rework. This paper will deal simultaneously with both methods.

Based on the previous models found in the literature, a new merged model will be created and a Genetic Algorithm (GA) adopted due to the ability of this evolutionary technique to provide near-optimal and quick solutions. Furthermore, a case study of fictitious construction project will also be examined.

Crashing was the first approach to be studied and discussed in the context of project management (Eduardo *et al.* 2008). The most popular method for crashing PERT/CPM networks is based on the average duration of the activities overlooking any uncertainties that may lead to higher durations (Haga 1998, Sahu and Sahu 2014). One common approach to solve crashing problems is to use linear programming which was introduced by Fulkerson (1961). Also, goal programming was found to be quite efficient too when dealing with secondary objectives (Moore *et al.* 1978). The crashing part of the model proposed here will follow a deterministic approach with a linear relationship between duration decrement and cost increment similar to the overwhelming majority of models found in the literature.

Activity overlapping has been a subject for many researches since the early 1990s. Krishnan *et al.* (1997) were the first to introduce a model that explains how activity overlapping works. Terwiesch and Loch (1999) studied the effectiveness of activity overlapping on product development. Xiao and Si (2003) introduced an approach to calculate the total execution times of the upstream and downstream activities. Wang and Lin (2009) pointed out the importance of the process structure for more efficient overlapping using simulation algorithm similar to the approach followed in this paper. Following the research done by Bogus *et al.* (2011) and Ballesteros-Pérez (2017), this paper employed Monto Carlo simulations to calculate the average compression in the overlapping operations.

Although there is a need to study the two acceleration modes simultaneously, there are very few papers that have already handled this topic. Roemer and Ahmadi (2004) studied both crashing and overlapping concurrently in the modeling framework for the first time in the literature. Their model was based on an upstream and downstream approach that was suggested by Krishnan *et al.* (1997). Moreover, they studied the impact of crashing and overlapping solely and jointly on a multi-stage example. Eduardo *et al.* (2008) conducted a significant set of research that is relevant to this paper, yet their scope is much wider; they focused on combining the three modes of project acceleration. However, the two significant models neglected the stochastic nature of the activity duration in their study which is taken care of in this paper. In other words, the overlapping part was stochastically calculated while the deterministic approach was used to calculate the crashing part of the model.

2 MODEL FORMULATION

2.1 Overlapping Model

Assume two activities: a predecessor activity (g) whose duration is d_g and its cost is C_g; and a successor activity (a) whose duration is d_a and its cost is C_a. To initiate the successor activity, two types of costs may be incurred: the upfront cost C_{ai} and the operational cost C_{aj}. The overlap that would occur between the two activities can be denoted by o. The delay can be represented by k and the float is denoted by f. By its very definition, fast-tracking requires the successor activity to be brought forward by a certain number of time units u so that it would start earlier than originally planned. The actual overlap would start when the activity's float vanishes, such that ($u > f$) and the difference between them is the overlap, such that ($o = u - f$). The occurrence of fast-tracking may incur some costs. Eq. (1) represents the average incremental cost Δc_a.

$$\Delta c_a = Risk(R) \times IMPACT(I) = \left(\frac{u-f}{d_g+k}\right)^{\alpha} \cdot \left(C_{ai} + \left(\frac{u-f}{d_a}\right)^{\beta} C_a\right) = \left(\frac{o}{d_g+k}\right)^{\alpha} \cdot \left(C_{ai} + \left(\frac{o}{d_a}\right)^{\beta} C_a\right) \quad (1)$$

when ($\Delta c_a \leq C_a$), ($0 \leq \alpha$), ($\beta \leq +\infty$) and $0 \leq o \leq MIN\,(d_a, d_g + k)$.

The predecessor's sensitivity is represented by the variable α, while variable β represents how quickly C_a is being spent while the successor is being executed. The model developed here assumes a case of probability which can be measured by multiplying the risk by the impact, as per Eq. (1). The proportional relationship between the risk (R), and overlap o suggests that the risk would be 0% when the overlap is zero, and the risk would be 100% when the overlap is maximal. The impact (I) is proportional to that amount of money (cost) allocated to the portion of the activity that would be overlapped. Also, it can be assumed for many construction projects on average conditions that there would be no upfront costs and both variables α and β are linear (i.e., $\alpha = \beta = 1$). This scenario is described by Eq. (2).

$$\Delta c_a = (R) \times (I) = \left(\frac{u-f}{d_g+k}\right) \cdot \left(\frac{u-f}{d_a} \cdot C_a\right) = \frac{C_a \cdot (u-f)^2}{d_a \cdot (d_g+k)} = \frac{C_a \cdot o^2}{d_a \cdot (d_g+k)} \quad (2)$$

when there is more than one predecessor for a certain successor (a), this scenario can be calculated using Eq. (3).

$$\Delta c_a = (R) \times (I) = \left(1 - \prod_{i=1}^{n}\left(1 - \left(\frac{o_i}{d_{g_i}+k_i}\right)^{\alpha}\right)\right) \cdot \left(C_{ai} + \left(\frac{o_{\max}}{d_a}\right)^{\beta} C_a\right) \quad (3)$$

with ($\Delta c_s \leq C_a$), ($0 \leq \alpha$), ($\beta \leq +\infty$) and $0 \leq o_{max} = \text{MAX}\ o_i \leq MIN\,(d_s, d_g + k)$

The number of days that a successor can be brought forward u can be calculated such that ($u = o + f$). The main concern now is to calculate the average effective amount of time u_{av} by which a successor activity can be brought forward using Eq. (4).

$$u_{av} = f \cdot Risk + (o+f) \cdot (1 - Risk) = f + o \cdot (1 - Risk) = f + o \cdot \left(1 - \left(\frac{o}{d_g+k}\right)^{\alpha}\right); u \geq f \quad (4)$$

The overall effective (expected) overlapping can be calculated for the project through the summation of the effective overlaps u_e of critical activities which have zero float (lies on the critical path CP). This can be summarized by equation (5).

$$u_e = \sum_{i \in CP} o_i \times (1 - Risk_i) \quad (5)$$

2.2 Crashing Model

As noted earlier, when it comes to crashing, the relationship between time and cost is deemed to be linear following previous models like Moore *et al.* (1978). Let us assume there is an activity (a); its normal duration, which is a subjective input by the user, is d_a^N, crashed duration is d_a^C, normal cost is C_a^N and the crashed cost is C_a^C. Having calculated the number of crashing days Cr, the crashed duration can be calculated using Eq. (6).

$$d_a^c = d_a^N - Cr \tag{6}$$

when $Cr_{min} \leq Cr \leq Cr_{max}$

Having calculated the crashed duration of activity (a), the actual crashed cost C_a^C can be calculated using Eq. (7). Similar to the normal duration, the normal cost of each activity as well as the maximum crashed cost are subjective inputs by the user.

$$C_a^C = v_a \cdot Cr = v_a \cdot (d_a^N - d_a^C) \rightarrow \; v_a = \frac{C_a^C - C_a^N}{d_a^N - d_a^C} \tag{7}$$

2.3 Concurrent Crashing and Overlapping

Having calculated the number of crashed days Cr, and the average effective amount of time u_{av}, the total project acceleration Ac can be calculated using Eq. (8).

$$Ac = u_{av} + Cr = f \cdot Risk + (o + f) \cdot (1 - Risk) + (d_a^N - d_a^C) \tag{8}$$

The additional total cost C_T incurred by an activity due to both crashing and fast-tracking can be calculated using equation (9).

$$C_T = \left(1 - \prod_{i=1}^{n}\left(1 - \left(\frac{o_i}{d_g + k_i}\right)^{\alpha}\right)\right) \cdot \left(C_{ai} + \left(\frac{o_{\max}}{d_a}\right)^{\beta} C_a\right) + \left(v_a \cdot (d_a^N - d_a^C)\right) \tag{9}$$

2.4 Case Study

The two most important parameters that are being measured by the proposed model are the maximum acceleration that can be achieved and the additional cost associated with this compression. Thus, the mathematical model was applied to a case study consisting of 12 activities, a total duration of 100 days, and a total cost of $1000 as shown in Figure 1.

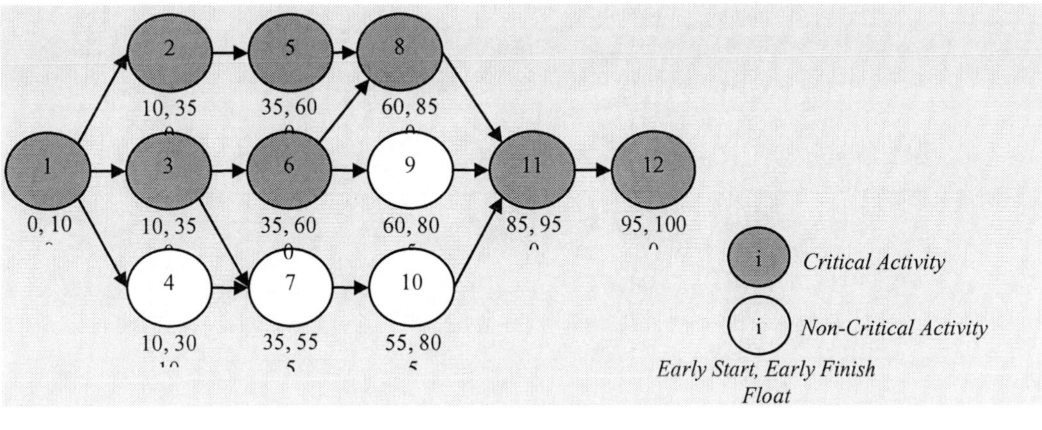

Figure 1. Network diagram for the case study.

A Genetic Algorithm (GA) model was developed using Excel Solver, as based on the mathematical models proposed earlier in the methodology section. The inputs for this model is shown in Table 1. The user shall decide on the crashed cost per day and the crashing limits which are assumed here to be 25% of the normal duration. Also, the amount of overlap allowed would be limited to 25%, which corresponds to a maximum risk of 25%.

Table 1. Input parameters for the Genetic Algorithm model.

ID	Inputs											
	Duration			***Cost (c)***			***Relationships***		***Model Limits***			
	ND	CD	Slack	Cost (c)	CC	CC/day	IPs	IPs durations	Crashing		Overlapping	
	d_a^N	d_a^C	f	C_a^N	C_a^C	v_a	g_i	d_g	Cr_{min}	Cr_{max}	o_{min}	o_{max}
1	10	8.20	CA	50	63	5.00	-	-	0	2.5	-	-
2	25	18.76	CA	50	63	2.00	1	8.2	0	6.3	0	4.1
3	25	18.81	CA	100	125	4.00	1	8.2	0	6.3	0	4.1
4	20	16.51	5	50	63	2.50	1	8.2	0	5.0	0	4.1
5	25	18.84	CA	100	125	4.00	2	18.8	0	6.3	0	9.4
6	25	18.84	CA	100	125	4.00	3	18.8	0	6.3	0	9.4
7	20	15.11	5	50	63	2.50	3,4	18.8 16.5	0	5.0	0	5.1
8	25	18.80	CA	100	125	4.00	5,6	18.8 18.8	0	6.3	0	5.5
9	20	15.59	5	50	63	2.50	6	18.8	0	5.0	0	9.4
10	25	18.80	5	100	125	4.00	7	15.1	0	6.3	0	7.6
11	10	7.51	CA	100	125	10.00	8,9,10	18.8 15.6 18.8	0	2.5	0	3.6
12	5	3.76	CA	150	188	30.00	11	7.5	0	1.3	0	3.8

ND: Normal Duration CD: Crashed Duration CC: Crashed Cost IP: Intermediate Predecessor

Maximum crashing and maximum overlapping were then calculated as shown in Table 2. Also, Monte Carlo simulation was utilized; the values were simulated 10,000 times. The costs were calculated using the mathematical model explained earlier.

Table 2. Output parameters for the Genetic Algorithm model.

Activity ID	Outputs					
	Optimum Crashing	***Active overlaps***	***Active overlaps cost***	***Actual crashing cost***	***Overlap Risk***	***Effective overlap***
	(Solution) Cr	***(Solution)*** o	Eq. (3) (ΔCa)	Eq. (7) C_a^C	Eq. (3) R	Eq. (5) u_e
1	**1.80**	-	-		-	-
2	***6.24***	***3.76***	4.58	12.47	0.46	2.04
3	***6.19***	***3.87***	9.69	24.78	0.47	2.04
4	***3.49***	***2.46***	2.23	8.74	0.30	1.72
5	***6.16***	***5.89***	9.82	24.64	0.31	4.04
6	***6.16***	***5.48***	8.47	24.65	0.29	3.88
7	***4.89***	***3.41***	3.95	12.24	0.35	2.21
8	***6.20***	***4.07***	8.34	24.80	0.39	2.50
9	***4.41***	***5.32***	4.81	11.04	0.28	3.82
10	***6.20***	***5.13***	9.26	24.78	0.34	3.39
11	***2.49***	***3.42***	21.71	24.86	0.48	1.79
12	***1.24***	***3.76***	74.94	37.19	0.50	1.88

The results summarized in Table 3 show that at the same limits, activity fast-tracking led to lower compression than activity crashing yet cheaper. It can be said that fast-tracking is preferred for low-to-moderate delays while activity crashing and the concurrent approach is preferred for extreme delays. Also, a combination of the two approaches shall be considered.

Table 3. Outcome summary Table.

Initial project cost	$1,000.0	Effective crashing (***Cr***); Eq. (6)	24.1
Total overlap cost	$157.8	Effective overlapping u_e;(Deterministic); Eq. (5)	12.1
Total crashing cost	$230.2	Effective acceleration Ac_e (Deterministic)	**36.2%**
Compression cost (***Ct***); Eq. (9)	$388	Average acceleration actually achieved ***Ac***; Eq. (8)	32.1
Percentage of cost increase	**39%**	**(Average from project duration simulations)**	**32.1%**

3 CONCLUSIONS

The model proposed in this paper followed previous crashing and fast-tracking models, and it combined them in one model that is based on both deterministic and stochastic approaches. One advantage of this model is its ability to calculate the total compression achieved, as well as extra costs that result from using the two scheduling acceleration modes either separately or jointly. Also, the model not only considered the case of the single predecessor, but also the multi-predecessor scenarios.

References

Baker, R. D., *Time-Cost Relationships in Construction*. Springfield, Virginia, 1991.

Ballesteros-Pérez, P., del Campo-Hitschfeld, M. L., Mora-Melià, D., and Domínguez, D., Modeling Bidding Competitiveness and Position Performance in Multi-Attribute Construction Auctions, *Operations Research Perspectives*, 2, 24-35. doi: 10.1016/j.orp.2015.02.001, December, 2015.

Ballesteros-Pérez, P., Skitmore, M., Pellicer, E., Zhang, X., Scoring Rules and Competitive Behavior in Best-Value Construction Auctions, *Journal of Construction Engineering and Management*, 142(9), 04016035. doi: 10.1061/(ASCE)CO.1943-7862.0001144, 2016.

Ballesteros-Pérez, P., Modelling the Boundaries of Project Fast-Tracking, *Automation in Construction*, 84, 231-241, doi: 10.1016/j.autcon.2017.09.006, 2017.

Eduardo, J., Gerk, J. E. V., and Qassim, R. Y., Project Acceleration via Activity Crashing, Overlapping, and Substitution. *IEEE Transactions on Engineering Management*, 55(4), 590-601, 2008.

Fulkerson, D. R., A Network Flow Computation for Project Cost Curves, *Management Science*, 7(2), 167-178, 1961.

Haga, W. A., *Crashing PERT networks*, PhD Thesis, University of Northern Colorado, ProQuest Dissertations Publishing, Greeley, Colorado, USA, 1998.

Krishnan, V., Eppinger, S. D., and Whitney, D. E., A Model-Based Framework to Overlap Product Development Activities. *Management Science*, 43(4), 437-451, 1997.

Meier, C., Browning T. R., Yassine A. A., and Walter, U., The Cost of Speed: Work Policies for Crashing and Overlapping in Product Development Projects, *IEEE Transactions on Engineering Management*, 62(2), 237–255, 2015.

Moore, L. J. et al., Analysis of a Multi-Criteria Project Crashing Model, *AIIE Transactions*, 10(2), 163-169, 1978.

Roemer, T. A., and Ahmadi, R., Concurrent Crashing and Overlapping in Product Development, *Operations Research*, 52(4), 606–622, 2004.

Sahu, K., and Sahu, M., Cost & Time and also Minimum Project Duration Using Alternative Method, Presented at *International Review of Applied Engineering Research*, 4(5), 403-412, 2014.

Terwiesch, C., and Loch, C. H., Measuring the Effectiveness of Overlapping Development Activities, *Management Science*, 45(4), 455–465, 1999.

Wang, J., and Lin, Y., An Overlapping Process Model to Assess Schedule Risk for New Product Development, *Computers & Industrial Engineering*, 57(2), 460-474, 2009.

Xiao, R., and Si, S., Research on the Process Model of Product Development with Uncertainty Based on Activity Overlapping, *Integrated Manufacturing Systems*, 14(7), 567-574, 2003.

Responsible Design and Delivery of the Constructed Project
Edited by Abdul-Malak, M., Khoury, H., Singh, A., and Yazdani, S.

SCHEDULE RISK ASSESSMENT IN GREEN BUILDING PROJECTS

ASLI PELIN GURGUN[1], HASAN GOKBERK BAYHAN[1], GUL POLAT[2], and HARUN TURKOGLU[2]

[1]*Dept of Civil Engineering, Yildiz Technical University, Istanbul, Turkey*
[2]*Dept of Civil Engineering, Istanbul Technical University, Istanbul, Turkey*

Construction projects are subjected to a variety of risks eventually affecting the overall duration and green building projects are no exception. Use of risk assessment means in such projects is also critical to complete them within aimed schedule. In this study, risks that have potential schedule impacts in certified green building projects are identified and their impact and likelihood of occurrence values are researched. For this purpose, a survey is administered to construction professionals with green building experience. Based on the responses, risks are plotted to draw their potential impact on schedule/likelihood of occurrence in a graph to identify the most significant risk factors. This tool may help the practitioners to identify and prioritize potential risks and allocate resources in projects accordingly to prevent or minimize their possible impacts.

Keywords: Sustainable construction, Certified buildings, Duration, Project planning, LEED.

1 INTRODUCTION

As the interest in sustainable construction expands, the number of certified green building projects also increases. Green building means and methods are becoming contract requirements and the parties involved in green projects are subjected to risks that they did not encounter before (Coglianese 2009). Failures in achieving schedule objectives can be encountered frequently as the results of undesired risks inherent in such projects. Construction projects are normally subjected to several risks due to a variety of sources. There are additional risk factors in green building projects causing challenges for timely completion of them. For this reason, it is essential to understand these risks and factor them in project management processes to maintain schedule boundaries.

This study presents the findings of a survey study carried out in the U.S. that seek risks specific to LEED certified green buildings and their potential impacts on project schedule. Likelihood of occurrence and impact values are plotted to present analysis results. It is aimed to facilitate multiple insights to practitioners about schedule risk management in green building projects by providing a tool for:

- Prioritizing risks and allocate project resources accordingly during planning phase
- Bringing project stakeholders to a common understanding about project details including specific risks

- Organizing project control mechanisms.

2 RISKS IN GREEN BUILDING PROJECTS

Green buildings distinguish from conventional buildings in terms of risks. Green building construction processes may accommodate different set of risks tending to result with bottlenecks, cost and time overruns related to their meticulous selection of the construction processes, participants, systems and materials (Iqbal *et al.* 2015, Kibert 2013, Tserng *et al.* 2009). The common risks identified after a diligent literature survey and included in this study can be seen in Table 1. Gurgun *et al.* 2016 studied impacts of risks on projects costs provided in this table. On the other hand, this study focuses on analyzing same risks but for schedule impacts.

Table 1. Risk factors (Gurgun *et al.* 2016).

Risk Category	Description
Consultant, contractor and subcontractor related issues	CCSI1 - Lack of green construction experience and qualification CCSI2 - Contractors and subcontractors agreeing to standards that are not within their expertise and competence
Material related issues	MPPI1 - Doubts about long-term viability and performance of new and untested products, materials and technologies MPPI2 - Faulty performance of HVAC/electrical/plumbing systems and alternative water systems/alternative power generating equipment MPPI3 - Failure to receive materials/products in a timely fashion causing delays MPPI4 - Lack of expertise in new products/technologies
Legal and contractual issues	LCI1 - Inadequate definition of project parties' contractual roles and responsibilities LCI2 - Inconsistencies between formal regulations (e.g., existing federal, state and local legislation) and LEED LCI3 - Concern that project owners and participants lose potential benefits because of the stringent standards of LEED
Financial and economical issues	FEI1 - High cost of certification process FEI2 - Scarcity of insurance solutions FEI3 - Rental or resale value loss due to delay related to green construction procedures and conditions FEI4 - Failure to use of financial incentives (tax/loan discounts, low financing rates) because of delays or lower certification levels than expected

2.1 Consultant, Contractor and Subcontractor Related Issues

Consultants are most of the time the key members of the green construction organization because of their important missions of offering appropriate suppliers, fulfilling the knowledge gap between parties, helping to create an effective communication web and directing the project to desired certification level. After deciding on desired LEED certification level according to the specified resource limitations, first thing is to develop a team and create an effective communication web.

Experience and knowledge of green building systems are important components for green certification (Gurgun *et al.* 2016). The possibility of delays in projects decreases, if project participants such as subcontractors and consultants are familiar with activities and requirements of green building certification processes.

As the demand in green buildings and sustainable construction grows, contractors' interest also increase whether or not they are experienced. Contractors and subcontractors agreeing to standards that are not within their expertise and competence may cause delays and as well as cost overruns in projects.

2.2 Material Related Issues

Availability and selection of the appropriate environmental-friendly materials is important in green building projects. Use of new and recently developed materials and products is common and some of these materials and products are not analyzed over long-term assessments and tests (Gurgun *et al.* 2016). As buildings are designed for certain design periods, project participants should consider the uncertainties related to the performance and long term durability of new and untested products, materials and technologies (Hwang *et al.* 2017).

One of the major objectives of sustainable buildings is to improve energy and water efficiency in the structures. Green building certification systems encourage use of advanced systems for energy and water efficiency, which contributes to the minimization of adverse environmental impacts. Credit points regarding energy efficiency particularly share the largest portion of total collectable points. There are many alternative implementations such as occupant monitoring systems, renewable energy technologies, replacement of plumbing and power generating systems, waterless urinals, and sensor-activated faucets and electrical systems improve energy and water efficiency in operating buildings. Faulty performance of such systems can cause delays in project schedule, as certification systems need additional check-ups before their usage.

On the other hand, meeting schedule objectives requires careful planning of sustainable materials that are to be used for sustainability goals. Utilization of environmental-friendly materials, products and technologies is encouraged and rated in certification systems. Projects can qualify for substantial amount of credit points by including such materials and products. The uncertainties related to availability and on-time delivery of materials can be much more difficult depending on the local conditions.

Lack of expertise has been reported as another source of liabilities in green building projects (BCCA 2011, Ofori-Boadu 2012). Therefore, insufficient expertise and being unfamiliar in new products, materials and technologies is another factor that should be considered as a risk in green building projects.

2.3 Legal, Regulatory and Contractual Issues

As in conventional projects, defining project party roles and responsibilities in contract prevents task conflicts in green building projects. Today, it is easier than ever to decide on these definitions with the help of contract standards. For example, American Institute of Architects (AIA) publishes numerous standards. AIA B214-2007 includes preparation of a LEED certification plan, conducting a pre-design workshop where the LEED rating system to be reviewed and LEED points to be targeted. Preparing and monitoring a LEED certification plan, providing LEED specifications for inclusion in the contract documents, preparing a LEED certification report, detailing the LEED rating the project achieved is also available in this document (Dingwell 2010). B214-2012 is the newer version of "Standard Form of Architect' Services: LEED Certification". In this contract LEED certification services and agreements, drawings and specifications, LEED workshop, certification plan, project registration and submission of LEED documentation to GBCI, certification services during bidding or negotiation, construction for prime and consultant architect, additional services, owner's responsibilities, compensation and miscellaneous provisions are included. A detailed contract is the best option to avoid delays due to some complex processes of green building construction and certification that can be time challenging for parties.

2.4 Financial and Economical Issues

Financial and economical concerns are inherent in building projects including green ones. These concerns may depend on a variety of reasons. Decreasing bureaucratic procedures (license, approval, permit) priorities/fees, adjustment of tax rates based on achieved certificate level, grants and low interest loans some of the financial are factors encouraging projects owners investing in sustainable construction projects.

Schedule planning of projects is strongly related with cost estimations. Decision of building green construction might impose specific set of risks affecting both schedule and cost calculations. Insurance solutions that would help mitigation or minimization of results of undesired events are important as a financial instrument. Yet, policies developed for green building projects can still be scarce and the market is still improving.

The concept of being green is commonly used as marketing factor that might positively increase the perception of the buildings as in higher quality of living (Zhang *et al.* 2018, Kok *et al.* 2012). It is not unlikely to rent or sell sustainable buildings at rates higher than usual. Appraisal Institute reported that rental premiums of green commercial buildings raised ranging from 2% to 27% and sale premiums raised ranging from 9% to 25% (IMTAI 2013). When these numbers combined with lower operating expenses, investors are more likely to build green. But the unavailable materials, systems, paperwork and processes mixed with inexperience resulting delays can be encountered.

In the US, there are many financial incentives ranging from green density bonuses to priority to green buildings in bureaucratic procedures. To get use of these incentives, certain level of certification systems should be achieved. The construction design must qualify for a certain certification value and receive the certification level expected in order to be able to get the desired incentive. For example, in Cincinnati, 100% tax abatement is applicable for 15 years for new residential construction, and 10 years for renovation. Another incentive program was initiated in Washington D.C. that requires achievement of certification through LEED NC and LEED CS in new construction and major renovation of privately-owned non-residential buildings over 50,000 square feet starting in 2012 (Department of Energy 2012). Also, North Carolina provides a corporate or personal income tax credit in the amount of 35% of the cost of renewable energy systems, Indiana provides a property tax assessment exemption in the amount of 100% of the cost of systems that generate energy using solar, wind, hydropower or geothermal resources in New Jersey, the tax exemption amount is %100 of the property tax due and in Chicago, it is possible to take building license as short as 15 days (Garciano 2014). It is clear that project delays could cause loss of incentives resulting with extra time overruns on planned schedules.

3 ANALYSIS AND FINDINGS

According to answers collected from 402 different respondents, the results of likelihood of occurrence, impact on schedule and order of importance values are given in Table 2. Analysis results indicate that top five risks that have the highest effect on project schedule are:

i) contractors and subcontractors agreeing to standards that are not within their expertise and competence
ii) lack of expertise in new products/technologies

iii) lack of green construction experience and qualification
iv) high cost of certification process, and
v) doubts about long-term viability and performance of new and untested products, materials and technologies

Table 2. Results of the study.

Risk ID	Likelihood of occurrence	Rank	Impact on Schedule	Rank	Likelihood*Impact Rank
CCSI2	3.32	2	3.11	1	1
MPPI4	3.23	3	2.89	2	2
CCSI1	3.07	5	2.83	4	3
FEI1	3.28	1	2.65	8	4
MPPI1	3.14	4	2.66	7	5
LCI1	2.94	6	2.71	6	6
LCI2	2.93	7	2.57	9	7
MPPI3	2.66	8	2.88	3	8
MPPI2	2.61	9	2.75	5	9
FEI2	2.59	10	2.23	11	10
FEI4	2.50	11	2.27	10	11
LCI3	2.41	12	2.17	12	12
FEI3	1.68	13	1.67	13	13

In Figure 1, impact and likelihood of occurrence values of risks are shown using a graph.

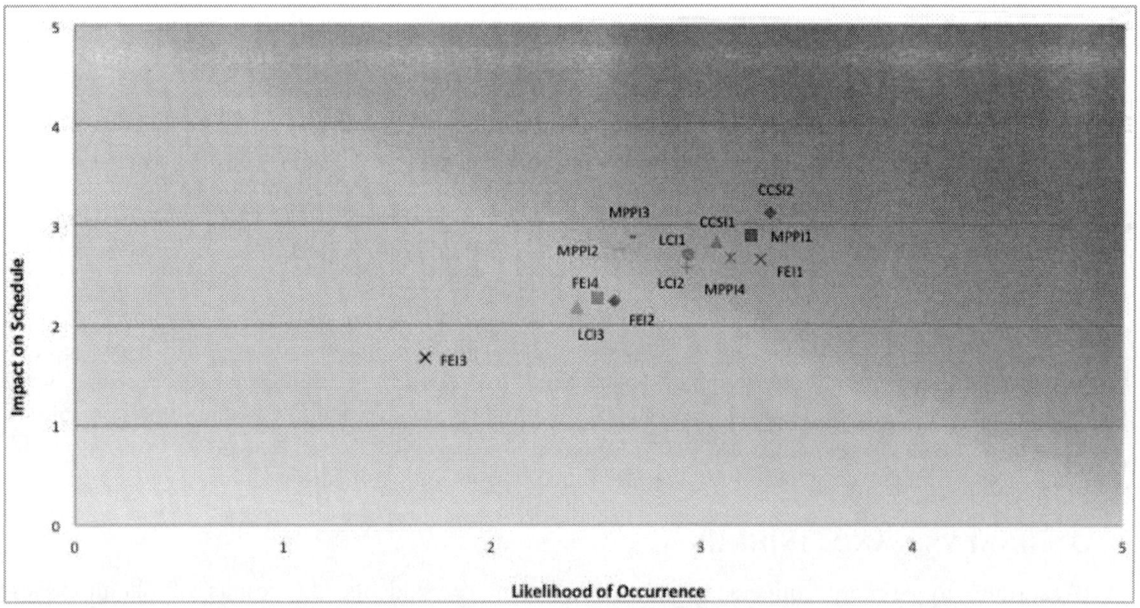

Figure 1. Impact of risks in green building projects on schedule.

This study aimed to reveal risks in green building projects for potential schedule impacts and facilitate insights to practitioners about schedule risk management in green building projects by providing a tool for prioritizing risks and allocate project resources accordingly during planning phase and organizing project control mechanisms.

References

BCCA, A Study on the Risks and Liabilities of Green Building, *British Columbia Construction Association*, I. Edition, Victoria, 7-8, 2011.

Coglianese, M. P., Construction's Legal Risks in the New Green Paradigm, *Construction Law*, 9(12), 43, 2009.

Department of Energy, *Incentives*, 2012. Retrieved from http://www.dsireusa.org/incentives/incentive.cfm?Incentive_Code=OH22F on 11 November 2016.

Dingwell, J., Green Economy Post, *An Introduction to Green Building Construction Contracts*, 2010. Retrieved from http://greeneconomypost.com/green-buildingconstruction-contracts-8059.htm, 31 Oct 2017.

Garciano, J. L., *Energy Efficiency and Renewable Energy Tax Incentives Federal and State Energy Tax Programs*, KLEIN HORNIG LLP, I. Edition, Washington, 2014.

Gürgün, A. P., Arditi, D., and Casals Vilar, P., Impacts of Construction Risks on Costs in Leed-Certified Projects, *Journal of Green Building*, 11, 163-181, 2016.

Hwang, B. G., Shan, M., Phua, H., and Chi, S., An Exploratory Analysis of Risks in Green Residential Building Construction Projects: The Case of Singapore, *Sustainability*, 9, 1116-1137, 2017.

Institute for Market Transformation and Appraisal Institute, *Green Building and Property Value: A Primer for Building Owners and Developers*, 2013. Retrieved from https://www.appraisalinstitute.org/assets/1/7/Green-Building-and-Property-Value.pdf, 03 Nov 2017.

Iqbal, S., Choudhry, R. M., Holschemacher, K., Ali, A., and Tamošaitienė, J., Risk Management in Construction Projects, *Technological and Economic Development of Economy*, 21(1), 65-78, 2015.

Kibert, C. J., *Sustainable Construction: Green Building Design and Delivery*, Hoboken, N.J.: Wiley, 2013.

Kok, N., and Jennen, M., T*he Impact of Energy Labels and Accessibility on Office Rents,* Energy Policy, 46, 489-497, 2012.

Ofori-Boadu, A., Owusu-Manu, D. G., Edwards, D., and Holt, G., Exploration of Management Practices for LEED Projects: Lessons from Successful Green Building Contractors, *Structural Survey,* 30(2), 145-162, 2012.

Tserng, H. P., Yin, S. Y. L., Dzeng, R. J., Wou, B., Tsai, M. D., and Chen, W. Y., *A Study of Ontology-Based Risk Management Framework of Construction Projects Through Project Life Cycle*, *Automation in Construction*, 18(7), 994–1008, 2009.

Zhang, L., Wu, J., and Liu, H., Turning Green into Gold: A Review on the Economics of Green Buildings, *Journal of Cleaner Production*, 172, 2234-2245, 2018.

Responsible Design and Delivery of the Constructed Project
Edited by Abdul-Malak, M., Khoury, H., Singh, A., and Yazdani, S.

DELAY IN CONSTRUCTION INDUSTRY: CAUSES, EFFECTS AND MITIGATION MEASURES

YIANNIS VACANAS and CHRIS DANEZIS

Dept of Civil Engineering and Geomatics, Cyprus University of Technology, Limassol, Cyprus

The main objectives of this paper are to determine the main causes of delays that occur in construction projects in Cyprus and their effects, and suggest measures that can assist in their effective control and reduction. Seeing that no relevant work has been undertaken for the case of Cyprus, a detailed literature review was carried out to identify the most significant causes of delay at the international level. Based on the review, a survey questionnaire was distributed to members of the construction industry of Cyprus. The analysis of the results indicated variations requested by the client, mistakes and missing information from consultants' drawings, and low productivity and inadequate programming of works from the contractor's side to be the main causes of delay. By considering the delay causes analysis specific mitigation measures are proposed that are mainly related to the programming of the works, work progress and productivity monitoring, and the appointment process of the parties involved in a construction project.

Keywords: Disputes, Time, Completion, Disruption, Projects, Arbitration.

1 INTRODUCTION

In the construction industry, the completion of projects within the agreed time constraints is considered an important contractual issue. In the case of delayed completion, all the main parties involved will potentially attempt to recover the financial losses they suffered because of the occurred delay by claiming the payment of financial losses by other parties of the project.

Depending on the size of the project and the agreement amount, the reimbursement sums, which are claimed for alleged delay or disruption, may be extremely high. Consequently, parties aim to avoid liability, which in turn result in frequent disputes (Vacanas *et al.* 2016a). Disputes, despite the various alternative dispute resolution methods available nowadays, will always cause further costs and losses to all parties. Additionally, delayed infrastructure projects (roads, hospitals, airports etc.) affect public's quality of life. For these reasons, researchers and institutions attempted to examine the causes of delay in construction projects.

Considering the above, the objectives of this paper are set to: (a) present an overview of the main causes delay in construction projects at the international level, (b) investigate the causes of delay in the Cypriot construction industry, and (c) examine measures that can aid the delay, and subsequently dispute, avoidance in the Cypriot construction industry.

2 CAUSES OF DISRUPTION AND DELAY IN CONSTRUCTION PROJECTS

2.1 Studies at the International Level

Studies regarding the causes of delay have been carried out by a number of researchers for various for various countries and geographical areas (Abd El-Razek *et al.* 2008, Ahmed *et al.* 2002, Chan and Kumaraswamy 1996, Chua *et al.* 1999, CIOB 2009, El-Sayegh 2008, Frimpong *et al.* 2003, Memon *et al.* 2010, Muhwezi *et al.* 2014, Odeh and Battaineh 2002, Orangi *et al.* 2011, Polat *et al.* 2014, Xiao and Proverbs 2003, Zou *et al.* 2007). The main causes of delay that were identified are summarized ranked from most frequently identified to the least frequently identifies in Table 1.

Table 1. Causes of delay in ranked from most to the least frequently mentioned in literature.

Cause of Delay
Changes and Variations by Consultants
Delay in payments to Contractor
Contractor's poor management ability
Intervention and changes to design by Client/ Owner
Inadequate contractor experience / poor technical performances
The financing by contractor during construction
Delay in issuing permits and approvals by government
Incomplete Documents/ design errors
Inadequate programme scheduling/ improper planning
Price inflation of construction materials
Suppliers' incompetency to deliver materials on time
Subcontractors' poor performance and management
Shop Drawings Approval/ delays in approvals
Design Development
Tight programme schedule
Shortage in manpower supply and availability/ shortage of site workers
Inadequate site investigation
Poor communication
Inspections
Changes in Laws and Regulations
Lack or departure of qualified staff
Weather conditions
Rework
Low speed of decision making involving all project teams
Labour productivity
Decision During Development stage/ slow decision making
Procurement delays

As it can be seed in Table 1, the 'Changes and Variations by Consultants', 'Delay in payments to Contractor', 'Contractor's poor management ability', 'Intervention and changes to design by Owner', 'Inadequate contractor experience / poor technical performances' and 'The financing by contractor during construction' are the most important causes of delay in construction projects identified at the international level.

2.2 Causes of Delay in Cyprus

In order to identify the most important causes of delay in construction projects in Cyprus we designed a questionnaire that was asking the participants to value the impact of twenty identified causes on a scale of 1 to 5; (1) corresponds to a very high level of impact, (2) to a high level, (3) indicates an average level, (4) a low level and (5) shows a very low level of impact, respectively. The twenty causes of delay proposed by the experts, and their respective significance as experienced by the participants are illustrated in Table 2.

Table 2. Causes of delay in Cyprus construction industry in ranked order.

Cause of Delay
Changes by the Client
Mistakes and Missing Information form Consultants drawings
Low productivity by Contractor
Inadequate experience by Consultants on specialised projects
Delayed Instructions by Consultants
Difficulties in Financing of the works by the Contractor
Inadequate programming of works
The involvement of a large number of parties
Payments Delay
Problems between Contractor and his Subcontractors
Delay to responses by Consultants to requests of information
Delay to responses by Client to requests of information
Bad communication
Delay to material approval
Suspension of Works by Contractor
Problems between Contractor and his Suppliers
Works required by public authorities (Electricity, Water, etc.)
Delay to approval of final drawings
Bad weather
Accidents

Based on the survey results, it was concluded that delay causes that are identified as the ones with the highest impact on projects' completion are changes by client, mistakes and missing information from consultants drawings', low productivity by contractor, inadequate experience by Consultants on specialized projects', 'late instructions by consultants, difficulties in financing of the works by the contractor, and inadequate programming of works.

2.3 Effects of Delay in Cyprus

To identify the effects of delay in projects the participants were asked to identify the percentage of projects where delay occurred. The participants' replies are summarized in Table 3. Only 17% of the participants state that delay occurred up to 20% of the projects they worked on, where 30% of the participants experienced delay on more than 60% of the projects they worked on.

Table 3. Delay occurrence in projects.

Delay Occurrence in Projects	%
Delay occurred in 0-20% of projects	17%
Delay occurred in 21-40% of projects	28%
Delay occurred in 41-60% of projects	26%
Delay occurred in 61-80% of projects	17%
Delay occurred in 81-100% of projects	13%

With regards to the duration of the delay, 28% of the participants experienced delay of duration up to 10% of the original contract duration, and 80% of the participants experienced delay with duration up to 30% of the original contract duration (Table 4).

Table 4. Delay duration in projects.

Delay Duration in projects	%
0-10 % of original duration	28%
11-20 % of original duration	26%
21-30 % of original duration	26%
31-40 % of original duration	9%
41-50% of original duration	7%
> 50% of original duration	4%

Regarding the agreement on EOT award duration, 53% of the participants stated that always or usually there is an agreement until the delivery of the project; and 47% that rarely or never is there an agreement on EOT duration until the delivery of the project (Table 5).

Table 5. Agreement of EOT award duration until the delivery of a project.

Answer	Percentage
Always	7%
Usually	46%
Rarely	39%
Never	7%

Depending on the contract, disputes in construction contracts are resolved via arbitration or court proceedings. The standard form contracts for construction projects in Cyprus are based on JCT 63 (private building projects) or FIDIC 4 (public works). More than 60% of the participants replied the arbitration proceeding has duration up to 2 years, whereas most of the court proceedings last more than 2 years without taking into account time spent on appeals (Table 6).

Table 6. Duration of arbitration proceedings.

Answer	Arbitration Proceedings	Court Proceedings
1 - 12 months	28%	2%
13 - 24 months	33%	2%
25 - 36 months	15%	26%
more than 36 months	9%	41%
I do not know	15%	29%

The enforcement of an Arbitral Award is an issue that has not been investigated thorough so far. According to the participants' experience usually there is requirement for the Court's intervention for the enforcement of the Arbitrator's award (Table 7). This is possibly because the parties that have the obligation to pay the amounts of the award know that they can delay their payment without serious consequences. Therefore, it is preferable to retain the amounts in their accounts for as long as possible, taking into advantage the slow judiciary system of Cyprus.

Table 7. The Arbitrator's award requires the court's aid to be enforced.

Answer	Percentage
Always	9%
Usually	54%
Rarely	11%
Never	4%
I do not know	22%

3 DELAY AND DISPUTE MITIGATION MEASURES

By considering the causes of delay and the effects of delay the following measures could be applied for mitigating delay and its effects including the generation of disputes:

3.1 Programming of the Works Related Measures

Programming is important in construction projects for organizing and scheduling the construction activities as well as controlling the various time constraints agreed. For this reason, there are various measures and practices related to programming that if applied could aid delay mitigation and avoidance:

- Preparation of realistic programme of works with critical path method
- Periodical (weekly, monthly) update of the works programme according to progress of works, with all changes highlighted to all parties
- When a disruption event occurs the programme to be updated and distributed

3.2 Work Progress and Productivity Monitoring Measures

Early warning for delays can make the parties act in a responsible way towards delay control and avoidance. Concordantly, there has to be frequent, clear and transparent updating of the parties on the works progress. The combined use of traditional methods (meetings, site reporting, frequent schedule updating etc.) and new technologies (e.g. use of BIM technology with UAV and Laser scanning for works progress recording and updating via 3D images) (Vacanas *et al.* 2016b) can provide clear updating of all the interested parties regarding the progress of works.

3.3 Assessment of Consultants' Experience During Appointment Process

One of the most important causes of delay is the changes and variations by Consultants and Clients. The reasons for these variations and changes are mainly the mistakes and missing information on consultants' drawings and the incomplete design because of restrictive timeframes and budgets. The appointment process of Consultants must include assessment of their ability to fulfill any special project requirements. Furthermore, there must be an audit methodology to ensure complete design preparation by the Consultants prior the tender period. Apparently, the Clients do not fully understand the end result of the design until they see it being constructed. BIM technology may be used to provide to Clients 3D visualization of the end project so that changes by the Clients during construction are avoided.

3.4 Assessment of Contractor's Experience and Financial State During Tendering Process

Every construction project is different but there are similarities between projects, which some contractors have experienced more than others. For this reason, the experience and ability of Contractors, especially for projects with certain particularities, have to be taken into consideration during tendering process. Difficulties in works financing by the Contractor and Late Payments are two major causes of delay and for this reason the financial state of the Contractors has to be investigated to avoid awarding the contract to a Contractor that will not be able to finance the works in the case there are late payments or any other payment issues.

3.5 Arbitration

Arbitration is clearly the dispute resolution method that offers fast resolution of disputes. However, this particular method for dispute resolution is not provided by all construction contracts. This is an issue that needs to be looked at. Nevertheless, the Arbitration Act of Cyprus was enacted around the middle of the previous century. The Cyprus Arbitration Act needs to be updated and certain disadvantages have to be improved.

4 CONCLUSIONS

Timely completion is one of the basic components that can set a construction project as successful. The main causes of delay have been analyzed at international and local level. Certain measures for delay mitigation have been suggested and discussed. The data acquired from the questionnaire research will be analyzed and the results presented in the future via journal papers.

References

Abd El-Razek, M. E., Bassioni, H. A., and Mobarak, A., Causes of Delay in Building Construction Projects in Egypt, *J. Constr. Eng. Mgmt.*, 134, 831–841. doi:10.1061/(ASCE)0733-9364, 134:11(831), 2008.

Ahmed, S. M., Azhar, S., Castillo, M., and Kappagantula, P., Construction Delays in Florida: An Empirical Study, Final Report, Florida International University, 2002.

Chan, D. W. M. and Kumaraswamy, M .M., An Evaluation of Construction Time Performance in The Building Industry, *Build. Environ.,* 31, 569–578. doi:10.1016/0360-1323(96)00031-5, 1996.

Chua, D. K. H., Kog, Y. C., and Loh, P. K., Critical Success Factors for Different Project Objectives, *J. Constr. Eng. Manag.,* 125, 142–150, doi:10.1061/(ASCE)0733-9364(1999)125: 3(142), 1999.

CIOB, Managing The Risk of Delayed Completion in the 21st Century, 2009.

El-Sayegh, S. M., Risk Assessment and Allocation in the UAE Construction Industry, *Int. J. Proj. Manag.,* 26, 431–438, doi:10.1016/j.ijproman.2007.07.004, 2008.

Frimpong, Y., Oluwoye, J., and Crawford, L., Causes of Delay and Cost Overruns in Construction of Groundwater Projects in A Developing Countries; Ghana as a Case Study, *Int. J. Proj. Manag.,* 21, 321–326, doi:10.1016/S0263-7863(02)00055-8, 2003.

Memon, A. H., Abdul Rahman, I., Abdullah, M. R., and Abdu Azia, Ade, A., Factors Affecting Construction Cost in Mara Large Construction Project: Perspective of Project Management Consultant, *Int. J. Sustain. Constr. Eng. Technol.,* 1, 41–54, 2010.

Muhwezi, L., Acai, J., and Otim, G., An Assessment of the Factors Causing Delays on Building Construction Projects in Uganda, *Intnl. J. Constr. Eng. Mgmt.,* 3, 13–23, doi:10.5923/j.ijcem.20140301.02, 2014.

Odeh, A. M. and Battaineh, H. T., Causes of Construction Delay: Traditional Contracts, *Int. J. Proj. Manag.,* 20, 67–73, doi:10.1016/S0263-7863(00)00037-5, 2002.

Orangi, A., Palaneeswaran, E., and Wilson, J., Exploring Delays in Victoria-Based Australian Pipeline Projects, *Procedia Eng.,* 14, 874–881, doi:10.1016/j.proeng.2011.07.111, 2011.

Polat, G., Okay, F., and Eray, E., Factors Affecting Cost Overruns in Micro-Scaled Construction Companies, *Procedia Eng.,* 85, 428–435, doi:10.1016/j.proeng.2014.10.569, 2014.

Vacanas, Y., Themistocleous, K., Agapiou, A., and Danezis, C., Methodology for Infrastructure Project Management, Dispute Avoidance and Delay Analysis, *An. Edif.,* 2, 12–19, doi:10.20868/ade.2016.3193, 2016a.

Vacanas, Y., Themistocleous, K., Agapiou, A., and Hadjimitsis, D., The Combined Use of Building Information Modelling (BIM) and Unmanned Aerial Vehicle (UAV) Technologies for the 3D Illustration of the Progress of Works in Infrastructure Construction Projects, *Proceedings of SPIE - The International Society for Optical Engineering,* doi:10.1117/12.2252605, 2016b.

Xiao, H. and Proverbs, D., Factors Influencing Contractor Performance: An International Investigation, *Eng. Constr. Archit. Manag.,* 10, 322–332, doi:10.1108/09699980310502937, 2003.

Zou, P. X. W., Zhang, G., and Wang, J., Understanding the Key Risks in Construction Projects in China, *Int. J. Proj. Manag.,* 25, 601–614, doi:10.1016/j.ijproman.2007.03.001, 2007.

Responsible Design and Delivery of the Constructed Project
Edited by Abdul-Malak, M., Khoury, H., Singh, A., and Yazdani, S.

A SIMPLIFIED METHOD FOR PHYSICAL PROGRESS MEASUREMENT

RUVEYDA KOMURLU[1] and AKIN ER[2]

[1]*Dept of Architecture, Kocaeli University, Kocaeli, Turkey*
[2]*TUPRAS General Management, Kocaeli, Turkey*

Project stakeholders frequently have problems while monitoring the progress of the projects during execution. Since an unbiased measurement system could not be established, progress figures are mainly based on project stakeholders' own experiences or subjective judgements. Considering the size of the projects, small to medium scale projects suffer the most from this problem. Usually, planning and site organizations of such scale projects are not set up adequately due to tight budgets or limited number of qualified personnel. Therefore, progress calculation method could not be defined efficiently or the data required for calculating project progress could not be gathered accurately. In order to overcome aforementioned difficulties and measure the progress correctly, simplified progress measurement techniques can be a solution. This article intends to describe how a simplified and convenient physical progress measurement method can be established. By separating the project into main elements and defining the most indicative commodities of them, number of parameters, which should be followed in the project can be decreased to a manageable level. Project weights of the elements can be calculated by taking into account standard man-hours required to complete each of them. Finally, construction milestones of the elements can be scheduled in a table, which helps to track and calculate the progress. A footbridge example is simply illustrated in the article to describe the method to readers explicitly.

Keywords: Activity weighting, Incremental milestones, Equivalent unit, Standard unit man-hour, Indicative commodity, Unbiased measurement system.

1 INTRODUCTION

Project stakeholders are mostly paying attention to three subjects during the execution of projects. They wonder to know;

(i) whether the project activities are progressing in line with the target work schedule or not,

(ii) whether the project cost is within the estimated project budget or not, and

(iii) how much is the physical progress of the project?

Usually, the answer to the last question is the most inconsistent one. Since an unbiased measurement system could not be established at the beginning, progress figures are mainly based on project stakeholders' own experiences or subjective judgements (Lee 1997). Thus, it will be better to concentrate on the meaning of physical progress in a project in advance.

Physical progress is "the status of a task, activity or discipline, based on pre-established guidelines related to the amount or extent of work completed" (AACE International 2017). It is well understood from the definition that measurement method which is going to be applied during the project, should be defined at the beginning. Considering the size of the projects, small to medium scale projects suffer from this problem the most since planning organizations and procedures in such projects are not set up properly. On the contrary, in large scale projects, the strategy and the procedures for progress measurement are in place at the beginning of the project, because owners and contractors are well aware about the importance of planning.

Besides, a measurement method should also be designed to follow the amount or extent of work completed against the total amount of work. This may be another disadvantage for small to medium scale projects since the data required for calculating project progress could not be gathered accurately due to insufficient site organizations. In order to overcome aforementioned difficulties, simplified progress measurement techniques can be preferred.

2 COMMON METHODS FOR MEASURING PROJECT PROGRESS

Various progress measurement methods have previously been discussed in several studies (Construction Industry Institute (CII) 1987, Thomas and Mathews 1996, Fleming and Koppleman 1996). Most commonly accepted methods (Del Pico 2013) are listed below:

(i) Units Completed: This method could be used for repeated works where each of the unit work requires almost the same resources and duration. For example, if it is planned to plant 1.000 trees to a park and 250 of them were already planted, the completion percentage of the work equals to 25%.

(ii) Incremental Milestones: If a work includes several steps that should be followed in an order, incremental milestones method could be useful for measuring the project progress. For example, construction of a foundation basically includes steps such as lean concrete, formwork, reinforcement and concrete pouring. Taking into account the total effort required for the whole foundation, the ratio of total effort required for each step can be calculated as a weighting in the total scope. Scalar product of progress ratio of each step and its weighting will give the contribution of that step to the project progress.

(iii) Start / Finish: Some activities have only start and finish points and no other activity has been defined in between. Leakage test of a pipeline can be taken as an example. Progress ratios of each point achievement can be defined as 30% - 70%, 50% - 50%, 0% - 100% and progress of the work can be calculated accordingly.

(iv) Cost Ratio: This method can be used when the project scope is limited with supply of labor only. Cost will be a linear function of man-hours spent. Thus, cost ratio will be almost same as spent labor ratio of the project.

(v) Experience / Opinion: This subjective method should be preferred as the last option since progress figures are mainly based on personal experiences of those who makes the judgement.

(vi) Weighted or Equivalent Units: This method is accepted as the most reliable and consistent technique. Considering the scope of the project, one of the production units is selected as the standard unit, and all others production units are converted to the selected one. For example, welding of a 6" standard sketch carbon steel pipe joint in a pipeline project could be accepted as the standard welding unit. Then, all other joints on the pipeline should be converted to the predefined standard welding unit by using coefficients previously determined for standardization of welding parameters such as diameter, sketch thickness and material type. Progress ratio of the welding activity is the ratio of total welded 6" equivalent joints to sum of all 6" equivalent joints in the subject pipeline project.

3 A SIMPLIFIED METHOD SUGGESTION

Combination of two methods, namely incremental milestones and weighted/equivalent units, can be used as a simplified method for progress measurement. Weighted/equivalent units method will maintain the accuracy of the results. Incremental milestones method will ensure simplicity while monitoring and calculating the project progress. Work flow sequence of the method is as follows:

(i) Separate the project into main elements,

(ii) Define the most indicative commodity of each element,

(iii) Calculate the project weight of each element by converting indicative commodity quantities to the equivalent unit,

(iv) Tabulate all elements in the rows of a chart versus their incremental milestones and their weightings at the columns,

(v) Calculate the progress of the project considering the site progress of each element.

3.1 Case Study

A footbridge is selected as a case study to describe the approach explicitly.

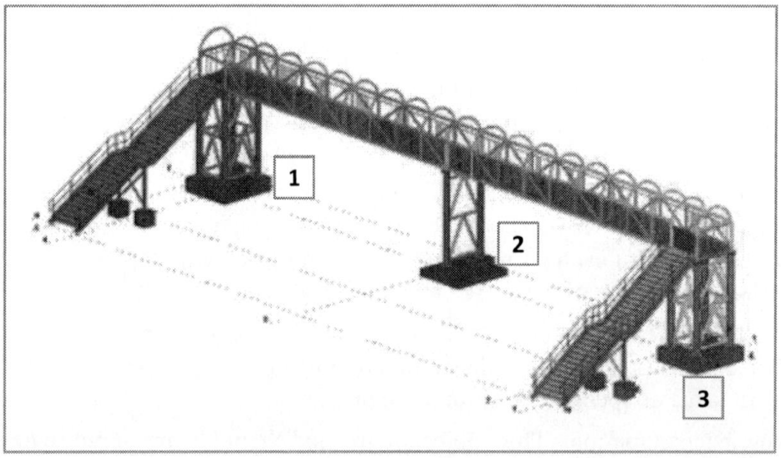

Figure 1. Footbridge.

Step 1: Structure is divided into nine elements as shown in Table 1. The number of elements depends on the monitoring capacity of planning and site organizations of the project. In a detailed study, each activity of level-three schedule can be considered as an element.

Step 2: Most indicative commodity of each element is determined and presented in Table 1. Quantities of the commodities are estimated considering the sketch given in Figure 1.

Step 3: Converting quantities to equivalent standard man-hours is the most convenient way to calculate equivalent unit of the elements. Scalar product of commodity quantity and standard unit man-hour gives the equivalent total man-hours of each element. By this way, weighting of each element could be easily reckoned as shown in Table 1. Standard unit man-hours should include all construction steps of the elements. Common industrial figures could be utilized while defining the standard unit man-hours. However, it is recommended to calculate those figures according to previous construction experiences like the authors did in Table 1.

Table 1. Summary of Step 1 to Step 3.

Structural Elements	Most Indicative Commodity	Quantity	Standard Unit Man-hours (incl. all steps)	Total Man-hours	Weight in the Project
Foundation-1	m3 of concrete	20	15	300	15%
Foundation-2	m3 of concrete	12	15	180	9%
Foundation-3	m3 of concrete	20	15	300	15%
Column-1	tons of steel	3	70	210	11%
Column-2	tons of steel	1	70	70	4%
Column-3	tons of steel	3	70	210	11%
Ladder-1	tons of steel	0,5	120	60	3%
Ladder-3	tons of steel	0,5	120	60	3%
Deck	m2 of deck	50	12,2	610	31%
			TOTAL	2.000	100%

Step 4: The findings are gathered in Table 3 in which elements are placed in the rows versus their incremental milestones, and their weightings at the columns.

Step 5: The project progress is calculated according to the site progress report given in Table 2. Sum-product of milestone weightings and site progress ratios of the milestones give the progress of each element. By multiplying that figure with the project weight of the element, contribution of each element to the project progress is calculated. The sum of all element contributions equals to the total progress of the project.

Table 2. The summary of site progress.

Structural Elements	Site Progress
Foundation-1	Completed
Foundation-2	Completed
Foundation-3	Completed
Column-1	Completed
Column-2	Completed
Column-3	Completed
Ladder-1	Supporting structure completed. Stairs and handrails partially completed. (80% - 50%)
Ladder-3	Supporting structure completed. Stairs and handrails partially completed. (60% - 30%)
Deck	Supporting structure completed. Stairs and auxiliary partially completed. (60% - 10%)

Table 3. Calculation of project progress (Step 4 and Step 5).

		MILESTONES							
Structural Elements	**Weight in the Project**	**1**	**2**	**3**	**4**	**5**	**6**	**Progress of the Element**	**Progress of the Project**
		Excavation	*Lean Concrete*	*Formwork*	*Reinforce-ment*	*Anchor Bolts*	*Concrete Casing*		
		10%	5%	25%	40%	10%	10%		
Found.-1	15%	100%	100%	100%	100%	100%	100%	100%	15%
Found.-2	9%	100%	100%	100%	100%	100%	100%	100%	9%
Found.-3	15%	100%	100%	100%	100%	100%	100%	100%	15%
		Base Plate	*Structure Preassembly*	*Structure Erection*	*Alignment Tightening*				
		10%	20%	40%	30%	0%	0%		
Column-1	11%	100%	100%	100%	100%			100%	11%
Column-2	4%	100%	100%	100%	100%			100%	4%
Column-3	11%	100%	100%	100%	100%			100%	11%
		Base Plate	*Structure Erection*	*Stairs*	*Hand Rails*				
		10%	50%	20%	20%	0%	0%		
Ladder-1	3%	100%	100%	80%	50%			86%	3%
Ladder-3	3%	100%	100%	60%	30%			78%	2%
		Platform Preassembly	Platform Erection	Hand Rails	Auxiliary				
		20%	40%	20%	20%	0%	0%		
Deck	31%	100%	100%	60%	10%			74%	23%
								TOTAL	**91%**

3.2 Achievements of the Method

The following points are achieved through this method:

(i) The procedure which is defined in the work flow is not complex. It is easy to follow.

(ii) Calculation of the quantities for all project activities is not required. Quantities of the most indicative commodities should be enough. It is less time consuming.

(iii) Standard man-hours, experienced by the contractors in previous projects, can be used while converting quantities to the equivalent unit. It is easy to adopt.

(iv) Following the progress by quantities is not required. Following milestones with completion percentages ensures simplicity. It is easy to gather progress information.

(v) This method ensures a consistent and reliable progress calculation technique.

(vi) It is applicable with a reasonable number of personnel.

4 CONCLUSION

The project progress rate is one of the indicative parameters that enable the project management to follow the project performance concurrently. However, it is usually difficult to set up an appropriate method to measure the project progress, especially for small scale projects. Due to this concern, similar studies to this article have been conducted by many researchers up to now (Chin *et al.* 2004, De Marco *et al.* 2009, Daneshyar and Walker 2014). The aim of this study is to provide a guideline for individuals who would like to follow project progress in a simple and an accurate way. Two common progress measurement methods are combined, namely incremental milestones and weighted/equivalent units. Weighted/equivalent units method maintains the accuracy of the results. Incremental milestones method ensures the simplicity while reporting and calculating the project progress. Thus, the calculation method is uncomplicated, reliable and easy to follow by less number of personnel. Therefore, the method can be utilized by the contractors which have difficulties to establish sufficient planning and site organizations for small to medium scale projects due to tight budgets or limited number of qualified personnel.

References

AACE International, *Cost Engineering Terminology*, Recommended Practice 10S-90, Morgantown, WV, 2017.

Chin, S., Yoon, S., Kim, Y., Jung, Y., Park, S. and Chung, M., A Project Progress Measurement and Management System, *Proceedings of the 21st ISARC*, Jeju, South Korea, 2004.

Construction Industry Institute (CII), *Project Control for Construction*, Publication No. 6-5, 1987.

Daneshyar S., and Walker P., Construction Progress Measurement Tool for Small Projects, *2014 AACE Technical Paper*, PS.1597, 1-18, 2014.

De Marco, A., Briccarello, D., and Rafele, C., *Cost and Schedule Monitoring of Industrial Building Projects: A Case Study*, Journal of Construction Engineering and Management, Vol. 135 No. 9, 853-862, 2009.

Del Pico, W. J., *Project Control: Integrating Cost and Schedule in Construction*, John Wiley & Sons, Inc., Hoboken, New Jersey, 2013.

Fleming, Q. W., and Koppleman, J. M., *Earned Value Project Management*, Project Management Institute (PMI), Upper Darby, Pa., USA, 1996.

Lee, B., *Improvements of Determination Methods for Work Progress and Progress Payment in Construction Projects*, Construction & Economy Research Institute of Korea, Korea, 1997.

Thomas, H. R., and Mathews, C. T., *An Analysis of the Methods for Measuring Construction Productivity*, Construction Industry Institute (CII) Source Document 13, Austin, TX, 1996.

Responsible Design and Delivery of the Constructed Project
Edited by Abdul-Malak, M., Khoury, H., Singh, A., and Yazdani, S.

ON THE LIMITATIONS OF THE EARNED VALUE MANAGEMENT TECHNIQUE TO ANTICIPATE PROJECT DELAYS

PABLO BALLESTEROS-PÉREZ[1] and KAMEL MOHAMED ELAMROUSY[2]

[1]*School of Architecture, Building and Civil Engineering, Loughborough University, Leicestershire, United Kingdom*

[2]*School of the Built Environment, University of Reading, Reading, United Kingdom*

A significant proportion of projects across the construction industry fail to meet their planned completion dates, being this a recurrent topic in the project management literature. Multiples causes of project delays have been proposed, however, hardly any attention has been paid to the fact that the most celebrated project monitoring and control technique – the Earned Value Management (EVM) – may not be as fit for purpose as it seems. It is proposed that because EVM ignores activity duration variability it always results in optimistic completion dates which may be very difficult to meet in the real projects. This research offers a fresh and long overdue critique of EVM in its most common implementation (assuming deterministic activity durations and costs), while highlighting its shortcomings. Particularly, Monte Carlo simulations are implemented to exemplify how the merge event bias phenomenon is inadvertently impacting the schedule in both time and cost dimensions. A fictitious case study is used to demonstrate the connection between these shortcomings and what is then conceived as a delay in project completion.

Keywords: Project management, Monitoring, Scheduling, Merge event bias, Stochastic network analysis.

1 INTRODUCTION

Earned Value Management (EVM) is, probably, one of the most prominent monitoring techniques for tracking project progress and also one that has received extensive research attention (Vanhoucke 2009, Vanhoucke 2011, Vanhoucke 2012).

EVM, in its classical form and using the most recent metrics denominations consists of the Planned Value (PV), the Actual Cost (AC) and the Earned Value (EV). There is a fourth metric named 'Earned schedule (ES)' (Lipke 2003) which measures in time units what EV measures in money units. These four metrics allow the calculation of a series of time and cost variances, performance indicators and forecasting indicators which (theoretically) represent how the project is doing in either or both time and cost dimensions.

The main advantage of this technique is its relative simplicity and that it only requires the kind of information (activity percentages of completion and actual costs mostly) that is actually gathered for other purposes during the project execution. The biggest disadvantage, which has apparently remained out-of-sight for most practitioners and researchers alike is that EVM is a technique that measures 'amount of work performed', NOT time deviations.

Put it in simple words, EVM measures the differences between EV and PV (for time deviations) or between EV and AC (for cost deviations). If EV is greater that either PV or AC, the project is doing well in time or cost, respectively. But EV increases as more activities are finished, irrespective of whether the order of execution was the most appropriate.

The reasonable question then is: are the activities being finished so far the ones that will lead to achieving the project duration we planned originally? EVM cannot tell. Indeed, if we deviate significantly from the project schedule, complementary indices like the p-factor (Lipke 2004), which measure schedule adherence, inform us about the risk of rework but they still fail to anticipate what (negative) time deviations can be expected.

EVM is therefore, a technique that works perfectly in the cost dimension (Ballesteros-Pérez et al. 2015, Ballesteros-Pérez et al. 2016), as all activities contribute to the total project cost according to their respective budgets. But that is not the case for time deviations. Activities need to happen in specific order to achieve a particular duration.

In this paper, a case study showing the execution of a project with paths in parallel will be exemplified. Detailed calculations of all (even the latest) EVM metrics will evidence how, despite this project is experiencing delays, the metrics cannot inform the project manager until it is too late (the situation cannot be remediated). Discussions will show how further research in EVM is still required to come up with alternative metrics that take care of situations where an imbalanced project progress among paths is occurring and at least one path is falling behind.

2 CASE STUDY

Let us present an example highlighting relevant shortcomings of the EVM technique. Figure 1 represents a fictitious project (schedule) consisting of three paths with four activities each. For the sake of simplicity it will be assumed that each activity has the same cost (100 monetary units) and the same duration (modeled by the cast of a six-sided die). This example could have also been illustrated with just one activity in each path, but for the sake of not using decimal places when casting dice, four activities have been considered per path.

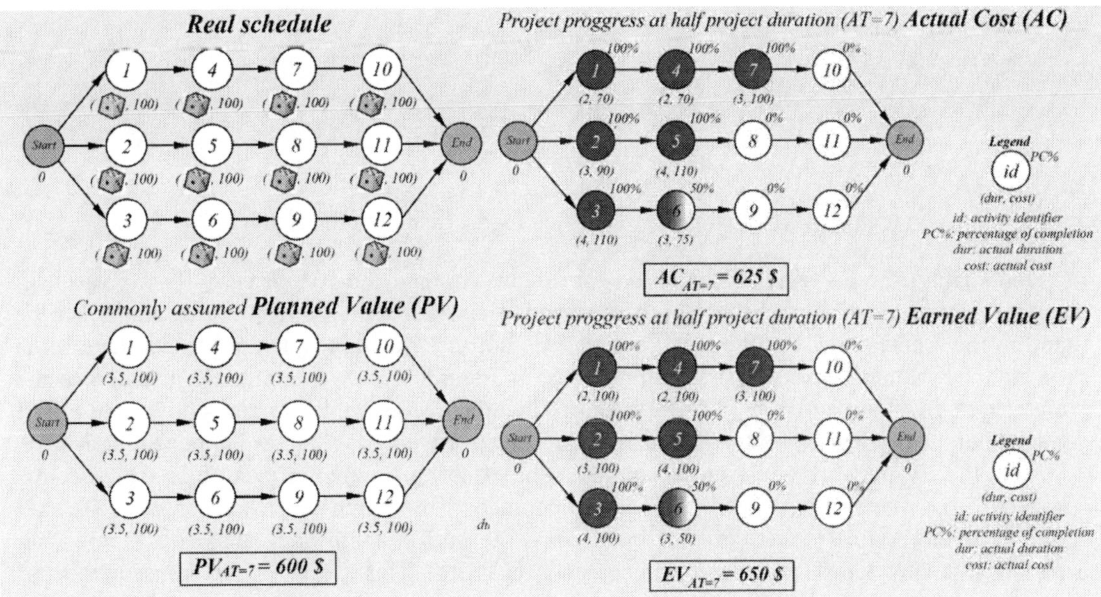

Figure 1. Activity on Node (AoN) networks and EVM calculations for the fictitious project.

Figure 1 represents at the top left the 'real model' of this schedule: each activity is variable concerning its duration, but not in the cost dimension. The bottom left network represents the abstraction (simplification) normally assumed by most Project Managers (PMs) when deciding which is the most representative (mistakenly believed to be the average) scenario. This is the Planned Value (PV) network, which could also have been represented, and indeed it usually is in real settings, by means of a Gantt chart. According to all specified data, this PV diagram tells that the project should last 14 days and cost 1,200 monetary units. The top right and bottom right networks represent the Actual Cost (AC) and the Earned Value (EV) with the progress at half of (what is believed to be) the project duration (Actual Time, AT, equal 7 days or other time unit).

By having a look at the percentages of completion (top right from each activity node) the path located in the center has advanced exactly according to the plan: two activities out of four (activities 2 and 5) have been completed at AT=7. The path at the top has progressed beyond what was planned: three activities out of four (activities 1, 4 and 7) have finished, only activity ten is left. Conversely, the path at the bottom has underperformed and only one and a half activities (100% of activity 3 and 50% of activity 6) are finished as of AT=7.

At this moment it was expected (planned) that the project would have had an EV of 600 monetary units, as the PV anticipated that 6 activities should have been finished at AT=7. However, whichever six activities is not something that the EV metric differentiates. Let us observe in Figure 2 the evolution of the PV, AC and EV metrics since this project started.

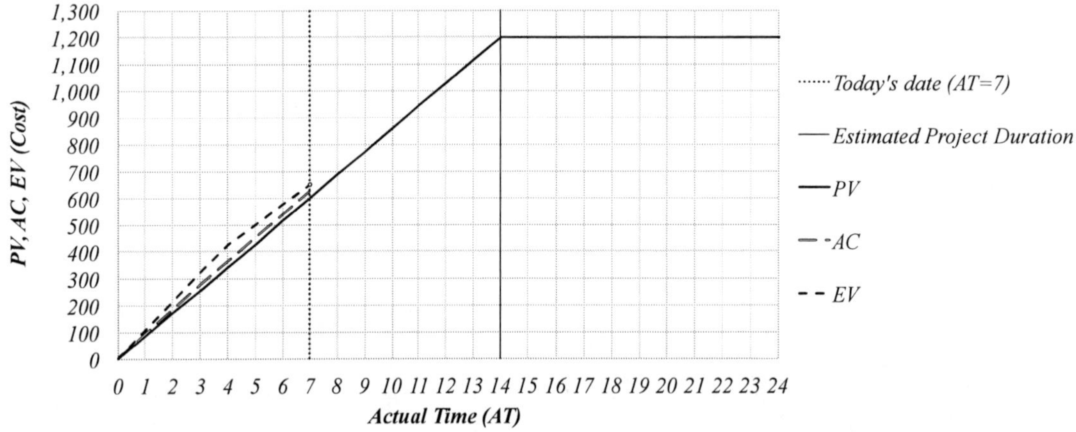

Figure 2. EVM metrics representation for the fictitious project at AT=7.

Undeniably, the EV curve remains on top of the AC and the PV curves. This project, in theory, is doing great: it is ahead of schedule and it cheaper (details of all calculations can be found in the Appendix). But the 'real' situation is not that favorable. If the project duration is estimated by Monte Carlo (MC) simulations as in Figure 3 at two moments: just before the project started (AT=0) and now (AT=7), both say that, on average, this project is not going to last 14 days, but much longer (around 17 and 16.5 days, respectively). What has happened then?

First, the PV network could not account for the effect of the merge event bias (Ballesteros-Pérez 2017a). The merge event bias is the phenomenon by which when two or more parallel activities with variable (uncertain) durations converge into a single one cause (on average) an extra local delay before the single successor can start. This delay is the consequence of computing the maxima of multiple duration probability distributions, unlike the average, which is the common operation when activities are in series (Ballesteros-Pérez 2017b). Therefore, from the very beginning this project was not going to last 14 days on average, but 17. At AT=7, the project progress has actually outperformed the expectations (the probabilistic duration curve is actually slightly located to the left compared to the curve at AT=0), but, when there are several paths, we just need one to underperform (fall behind) in order to cause a project delay (not meet the initial completion date). From the statistical point of view, irrespective of how many activities each parallel path has, it is likely that, as the number of parallel paths increases, there will be at least one which will not progress as planned.

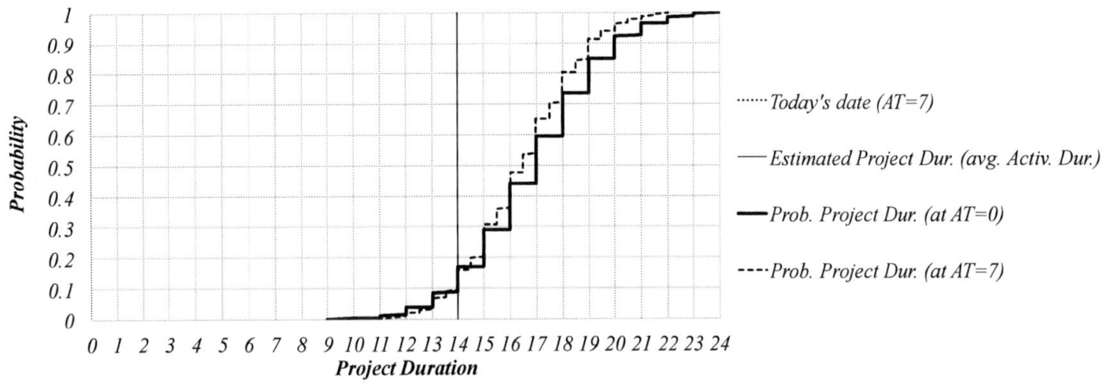

Figure 3. Project duration estimation for the fictitious project at AT=0 and AT=7.

Initially, this does not seem to be a problem: during the first stages of any project execution, there are plenty of activities to carry out. Actually, during the initial stages of a project, most paths are opening. This is mathematically obvious: for any series of paths to converge, first, those paths should have been opened (branched out from other activities). But later, the project progresses and the paths start to converge and, as the merge event bias effect prescribes, delays keep cumulating at each converging node, and finally translated to the overall project duration. That is why many projects do not seem to experience a delay initially (during the stages in which paths opening are the majority against paths converging), and then, all of a sudden the PM starts losing track of them gradually and they end irremediably late. At that advanced stage, hardly anything can be done by the PM when the project is already late as, normally, resources are not that transferable and/or activities cannot be crashed beyond certain limits (Ballesteros-Pérez 2017b). Therefore, no signs of alert came from the EVM technique on time.

3 DISCUSSION

A smart way of addressing this situation might have been then, to simulate multiple times the possible planned expenditure by MC simulations for example, that is, the potential PV or EV curve, as they represent the same here. Then take the average of all those curves so that we can have a representative 'average' PV curve against which the future EV curve can be fairly compared (the AC curve does not have any bias, therefore ignored from now on). These curves have been named 'Average EV simulations' and 'Median EV simulations' in Figure 4, but, they could have also been named 'Average PV simulations' and 'Median PV simulations'.

The results are quite surprising in two ways. First, near the end, both the average and the median EV curves remain below the commonly assumed (average, and straight in this case) PV line because of the merge event bias. The most representative curve is actually the median, as that is the curve that reaches the PV of 1,200 monetary units at day 17 (when half of the project simulations have finished the project). The average EV curve will not reach the cost of 1,200 until all the simulations have finished the project, that is, with just one project simulation unfinished, the average will not reach 1,200. This means that the average EV curve is not a good alternative for the PV as theoretically the average will always be an asymptote of the BAC.

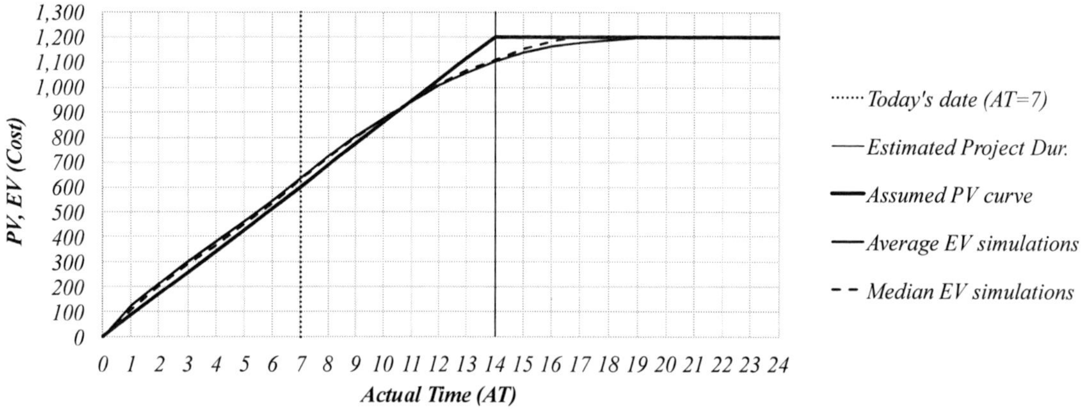

Figure 4. 'Commonly assumed' PV versus the 'actual' average and median EV.

Second, both the average and the median EV curves, are initially 'above' the 'assumed PV curve'. This means that, at initial stages, the EVM technique is going to make the scheduler think that the, despite the progress might be just 'average', it is doing better than expected (planned) by comparing to the wrong 'commonly assumed' PV line. Only later (at AT=11 in this project) both curves will move 'below' the 'commonly assumed' PV curve and the project will be late. This happens whenever there are more than two parallel paths (virtually in any real project).

Therefore, the EVM technique is, at least as it is conceived nowadays, dangerously misleading concerning its time dimension: it will frequently make the project manager believe that everything might be going well initially, but later, probabilistically speaking, the project will still have high chances experiencing delays when it is too late to correct. What is worse, this flaw cannot be corrected by just simply adopting a 'median EV simulation'-like curve instead of the straight PV curve as in Figure 4. This is because, as it was stated initially, the EVM only measures 'amount of work' without differentiating whether the activities performed so far have been the ones that will keep the optimum (shorter) sequence of project execution (duration).

4 CONCLUSIONS

The major limitations of the EVM technique have been highlighted. This technique is used by many Project Managers around the world and it is one of the most referenced project monitoring and control tool by core Project Management guides (e.g. PMBoK, IPMA competence Baseline, APMBoK). However, without exception, by neglecting activity duration variability and the merge event bias, project duration monitoring tasks with this technique are significantly optimistic in presence of multiple parallel paths. Furthermore, even in cases with no activity duration variability, EVM could still provide misleading interpretations about the project progress. This as EVM measures 'amount of work performed' without considering if that work has been performed in the right order of execution.

The influence of this unaccounted duration variability is different for each industry, though. The manufacturing industry, in general, tend to work indoors (lower climate influence), with semi-repetitive products (higher opportunity of learning from the past), with relatively stable project teams and clients. This might limit the activity duration variability and, hence, cause a lower impact of the optimistic biases described here. On the other side of the spectrum, the construction industry needs to deliver projects with more variable conditions (works are carried out mostly outdoors, projects are generally unique, constant changes in project team is the norm, different project locations, etc). It may not be that surprising then, that construction projects have traditionally been one of the sectors suffering more from missing the original completion dates.

Acknowledgments

This research was supported by the CIOB Bowen Jenkins Legacy Research Fund at the University of Reading (H5405400).

Appendix. EVM calculations

Summary of EVM metrics calculations at AT=7 with PV=600, AC=625 and EV=650:
SV = EV – PV = 50 > 0   SPI = EV / PV = 1.08 > 1 → Project ahead of schedule (Wrong!)
CV = EV – AC = 25 > 0   CPI = EV / AC = 1.04 > 1 → Project with cost underrun (Correct)
Metric calculations with Earned Schedule at AT=7 with ES=7.58:
SV(t) = ES – AT = 0.58 > 0   SPI(t) = ES / AT = 1.08 > 1 →Project ahead of schedule (Wrong!)
Calculations with p-factor at AT=7 with p=0.93, $EV(e)_{min}$=EV·p=604.5 and $ES_{EV(e)}$=7.05:
SV = $EV(e)_{min}$ – PV = 4.5 > 0   SPI = $EV(e)_{min}$ / PV = 1.01 > 1 → Ahead of schedule (Wrong!)
SV(t) = $ES_{EV(e)}$ – AT = 0.05 > 0   SPI(t) = $ES_{EV(e)}$ / AT = 1.01 > 1→ Ahead of schedule (Wrong!)

Calculations of all the Time Forecasting indicators (EAC(t) either with PV, ED or ES would also give optimistic Project Durations (<14) since all the variances and performance indicators are above 0 and 1, respectively.

Calculations of all the Cost Forecasting indicators (EAC) will be accurate only if the Performance Factor (PF) used is the CPI, otherwise (PF=SPI, SPI(t), SCI or SCI(t)) it will be not.

References

Ballesteros-Pérez, P., del Campo-Hitschfeld, M.L., Mora-Melià, D., and Domínguez, D., Modeling Bidding Competitiveness and Position Performance in Multi-Attribute Construction Auctions, *Operations Research Perspectives*, 2, 24-35, December, 2015. doi: 10.1016/j.orp.2015.02.001

Ballesteros-Pérez, P., Skitmore, M., Pellicer, E., and Zhang, X., Scoring Rules and Competitive Behavior in Best-Value Construction Auctions, *Journal of Construction Engineering and Management*, 142(9), 2016. 04016035. doi: 10.1061/(ASCE)CO.1943-7862.0001144

Ballesteros-Pérez, P., M-PERT: A Manual Project Duration Estimation Technique for Teaching Scheduling Basics, *Journal of Construction Eng. and Management*, ASCE, 143(9), 4017063, 2017a. doi: 10.1061/(ASCE)CO.1943-7862.0001358

Ballesteros-Pérez, P., Modelling The Boundaries of Project Fast-Tracking, *Automation in Construction*, Elsevier, 84, 231–241, September, 2017b. doi: 10.1016/j.autcon.2017.09.006

Lipke, W., Schedule is Different, *The Measurable News*, 31–34, Summer, 2003.

Lipke, W., Connecting Earned Value to the Schedule, *The Measurable News*, 1, 6–16, Winter, 2004.

Vanhoucke, M., Measuring Time - Improving Project Performance Using Earned Value Management, *Springer Berlin Heidelberg*, Berlin, Heidelberg, 2009.

Vanhoucke, M., On the Dynamic Use of Project Performance and Schedule Risk Information During Project Tracking, *Omega*, Elsevier, 39(4), 416–426, 2011.

Vanhoucke, M., Project Management with Dynamic Scheduling. *Springer Berlin Heidelberg*, Berlin, Heidelberg, 2012.

Responsible Design and Delivery of the Constructed Project
Edited by Abdul-Malak, M., Khoury, H., Singh, A., and Yazdani, S.

RESEARCH TRENDS IN EVM LIMITATIONS AND PROPOSED IMPROVEMENTS

FARAH DEMACHKIEH and MOHAMED-ASEM ABDUL-MALAK

Dept of Civil and Environmental Engineering, American University of Beirut, Beirut, Lebanon

The earned value method (EVM) is a recognized management technique that is used for monitoring, evaluating, and controlling the progress of projects. The method's employability challenges have consistently been addressed in the literature along with proposed enhancements that aim at improving its application. The overall objective of this paper is to synthesize those limitations reported to be encumbering EVM employability and to examine the main enhancements proposed for the EVM traditional set of metrics. To this effect, this paper initially validates the importance of project controls, as the top factor cited as being critical to the success of projects. Secondly, it presents a classification of the shortcomings reported to be impeding EVM employability, along with their corresponding improvement studies. As such, the paper looks at the temporal distribution of these limitation and improvement research studies, while shedding light on the trend of such interconnected research efforts over the last three decades. As such, the paper presents a synthesis of the improvements found to be affecting the representation, usage and measurements of the EVM set of metrics, with the aim of gaining perceptiveness on how the reported EVM shortcomings may probably be overcome. The findings revealed that the highest frequency of reported improvements (138 in total) was found to be pertaining to the method's set of key parameters.

Keywords: Project controls, Critical success factors, Shortcomings, Enhancements, Metrics.

1 INTRODUCTION

The construction industry is huge, unpredictable, and needs remarkable capital expenditures. However, client satisfaction during project execution is threatened due to cost overrun and/or schedule slippage (Kaliba *et al.* 2009). To this end, successful projects have all common characteristics; they have a well-defined scope of work, achievable schedule, and a realistic budget. Although these characteristics are the core of a successful project, project monitoring and controls is a main critical success factor for construction projects. As such, Naderpoor and Mofid (2011) defined earned value method (EVM) as a recognized project monitoring and controls method that integrates cost, schedule and scope into a unique measurement system.

Although EVM is gaining exceptional significance within the construction industry, many obstacles have hindered its wide employability as a main project monitoring and controls method. To this effect, researchers suggested improvements to EVM, including the introduction of fuzzy indices to compute the percent complete of executed works (Moslemi-Naeni and Salehipour 2011); stochastic S-curves to account for uncertain work environments (Acebes *et al.* 2015);

computerized applications for automating the controls process (Turkan *et al.* 2013); and artificial intelligence to forecast time and cost at-completion figures before the start of the project (Chao and Chien 2009).

2 PROJECT CONTROL AS A CRITICAL SUCCESS FACTOR

The first track of review of existing literature performed by this research tackled the work related to project critical success factors published in the last three decades, precisely intended to pinpoint the significance of effective controls as a main critical success factor (CSF). The adopted studies were searched for using the "Google Scholar", "Science Direct", and "ASCE" databases search engines, using the semantics and keywords-based methods. Consequently, the search led to adopting a total number of 65 relevant studies, allowing the adoption of general headings describing the main factors contributing to success of projects. Figure 1 shows the fifteen adopted headings that were found to have citation frequencies of more than ten in the reviewed studies, shown in a decreasing order of occurrence frequency. As shown in Figure 1, "effective project monitoring and controls methods" is the heading receiving the highest citation frequency of 62, thus emphasizing the significant importance of applying suitable controls processes for achieving project success.

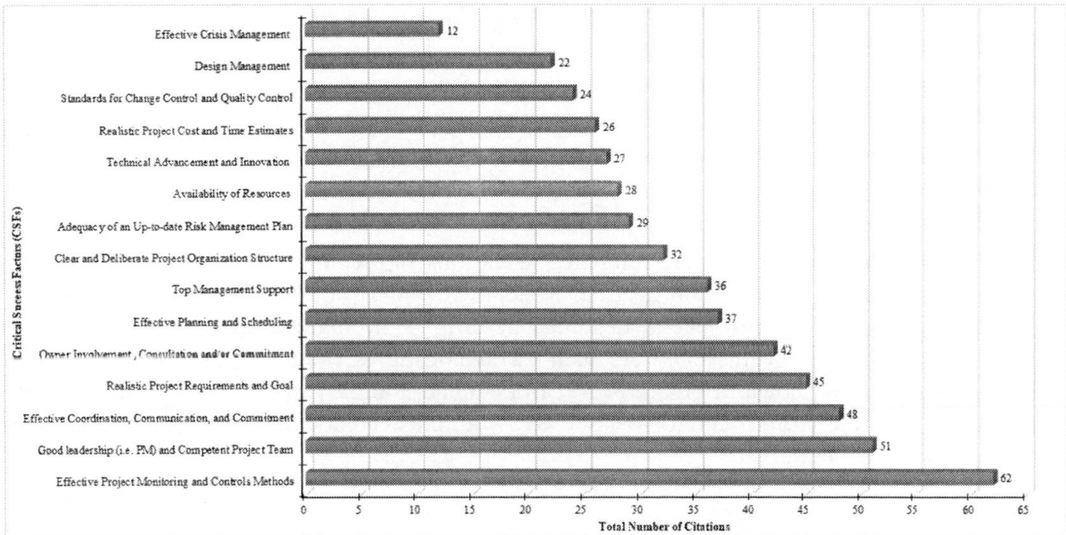

Figure 1. CSFs encountered in the reviewed literature.

3 TEMPORAL DISTRIBUTION OF EVM STUDIES

An in-depth review of the current literature related to project controls and EVM in the last three decades allowed the adoption of two sets of 56 and 230 shortlisted research studies, dealing with EVM shortcomings and proposed improvements, respectively. To this effect, the adopted EVM publications focus on (a) the main difficulties deemed to have prevented the method's applicability and (b) the proposed improvements, with the purpose of attaining a wider future employability. The filtered studies were categorized based on their years of publication, the affected EVM metrics, and the main tracks said to be related to EVM enhancements. Figure 2 shows the frequencies reported in connection with EVM limitations and improvement studies, distributed according to their years of publication. It is important to note here that the shown frequencies represent the findings of the search to date; in other words, those findings represent partial results of the on-going research. The depicted distribution clearly shows that a considerable volume of research work related to EVM limitations has been performed in the beginning of the 21st century. However, it can be safely inferred that the last three years (i.e., 2015-2017) included a substantial number of publications (33 percent) reporting on the main enhancements to the EVM methodology. This shift in the research work trend can be seen as an indication of the potential significance of encountered improvements studies in addressing EVM limitations.

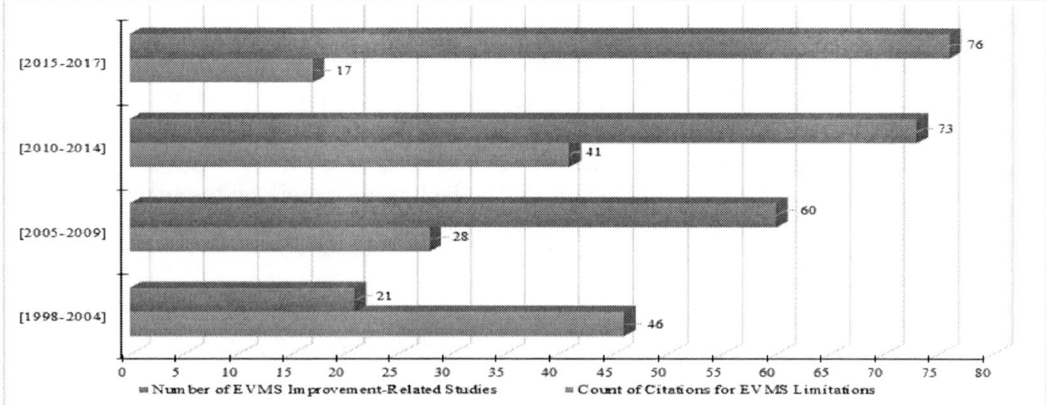

Figure 2. Temporal distribution of research studies.

4 ENCOUNTERED EVM LIMITATIONS

The 56 adopted research studies made reference to various shortcomings said to be limiting the employability of the EVM methodology. An in-depth review of all encountered shortcomings allowed their classification into general headings describing the main obstacles reported to be hindering the applicability of this tool. As such, twelve headings were generated for describing the EVM limitations, as shown in Figure 3. It can be noted that the top two frequencies, 19 and 17, of reported limitations are found to be associated with the accuracy of forecasting metrics and the quality of inputs, respectively.

5 IMPROVEMENT STUDIES

Principally, the application of EVM entails a static reference point, represented by the project baseline schedule and the budget-at-completion, coupled with the process of frequently monitoring and controlling the project performance during execution. Monitoring performance related to time and cost is done by (a) comparing the three pillars of EVM: planned value (PV), actual costs (AC) and earned value (EV), (b) generating cost and schedule variances and performance indexes, and (c) forecasting the expected project at-completion duration and cost.

As such, Figure 4 shows the breakdown of the filtered studies according to the three sets of EVM' metrics, including: key parameters, performance measures, and forecasting parameters. It is found that about 15 percent (36) of the improvement studies do not deal with a specific set of EVM' metrics, and these are classified in the "general category", as shown in Figure 4. Examples include those improvements proposed for tackling human- and process-related EVM shortcomings, such as the introduction of (a) an educational game to support and teach EVM concepts (von Wangenheim *et al.* 2012), (b) web-based database management systems to ease the application of EVM for project controls by generating project status reports (Cheung *et al.* 2004, Li *et al.* 2006), and (c) models to improve the acceptance and performance of EVM (Kim *et al.* 2003).

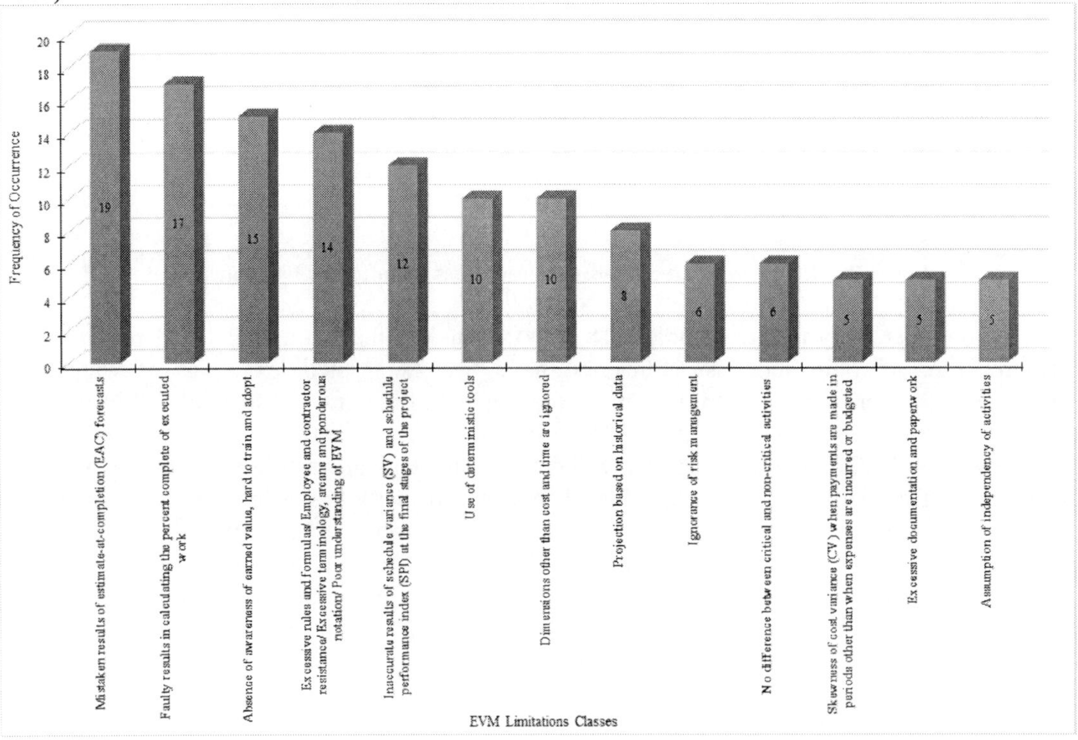

Figure 3. Frequency of reported EVM limitations classes.

Furthermore, it can be inferred that the studies dealing with EVM key parameters contributed to about 60 percent (138) of the reported enhancement studies. To this effect, six out of the 138 studies are mainly concerned with the generation of empirical S-curves in order to provide early estimates for decision-making support; examples of enhancement-related studies included the introduction of the use of neural networks and polynomial functions (Chao and Chien 2009), regression analysis models (Blyth and Kaka 2006) and analytic parametrization (Cioffi 2005). Similarly, six out of the 138 studies were related to the introduction of novel approaches for the generation of construction S-curve used for project controls, by introducing fuzzy S-curves (Ponz-Tienda *et al.* 2012), progress-based S-curves (Barraza *et al.* 2004), and optimal S-curves (Lo 2007). A number of studies (seven) were concerned with the EV/AC prediction before and during project execution, as proactive actions to advance favorable performance outcomes (Chen 2014, Chen *et al.* 2016). Moreover, 111 out of the 138 studies were dealing with the automation of data collection and progress tracking in construction sites for effective calculation of the EV of executed works, aided by information management technologies, such as those by El-Omari and Moselhi (2011), Han *et al.* (2017), and Zhu *et al.* (2017). Likewise, eight studies targeted the calculation of the EV figure by proposing modifications such as the effective EV (Lipke 2004), the quality EV (Dodson *et al.* 2015), and the customer EV (Kim *et al.* 2015), or introducing fuzzy indices to compute the EV in a more vigorous and reliable way (Moslemi-Naeni *et al.* 2011).

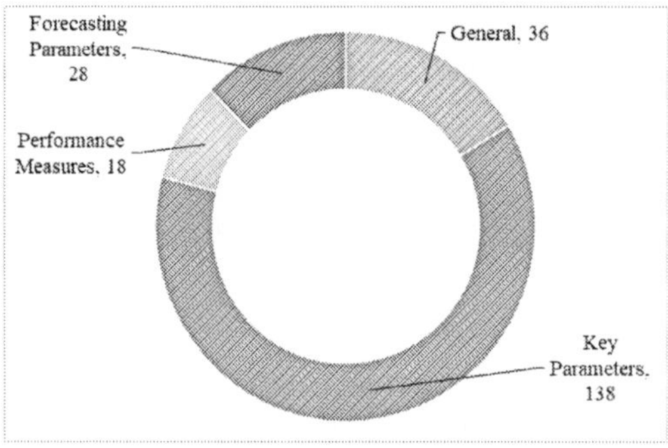

Figure 4. Number of shared improvement studies per affected EVM metrics.

To this effect, improvement studies (18 ones) related to the traditional EVM performance measures (i.e. schedule and cost indicators) have been suggested, examples of which are concerned with improvements introducing: (a) the concept of the "earned schedule" (Lipke 2003) with the purpose of shifting the focus of EVM from cost to time controls; (b) the introduction of the "weight EV" (Zhong and Wang 2011) and "the schedule forecasting indicator" (Czemplik 2014) to differentiate between critical and noncritical activities while performing EVM calculations; and (c) the formulation of a model for probabilistic evaluation of the cost performance stability of individual projects using project-specific information (Kim 2015). Lastly, in relation to the method of forecasting estimate-at-completion figures, 28 studies presented several extensions, including artificial intelligence (Cheng and Roy 2011), fuzzy logic (Moslemi-Naeni *et al.* 2011), and visualization of the performance of a project (Chou *et al.* 2010).

6 CONCLUSIONS AND FUTURE WORK

The research work reported in this paper is concerned with improvement studies aimed at addressing the EVM limitations and widening its applicability as a main project monitoring and controls tool. Thirteen general headings were eventually adopted to describe the deduced EVM limitations, with those related to the quality of inputs (faulty results in calculating the percent complete of executed work) and outputs (mistaken results of estimate-at-completion forecasts) receiving the highest occurrence frequencies of 17 and 19, respectively. To this effect, the paper investigated the distribution of the reported enhancements in the 230 reviewed studies, showing their breakdown over the main EVM sets of metrics: "key parameters", "performance indicators", and "forecasting parameters". The on-going research work by the authors involves a thorough analysis of the reported enhancements, with the aim of developing novel integrated sets of metrics for this control's tool.

References

Acebes, F., Pereda, M., Poza, D., Pajares, J., and Galán, J. M., Stochastic Earned Value Analysis Using Monte Carlo Simulation and Statistical Learning Techniques, *International Journal of Project Management*, 33(7), 1597-1609, 2015.

Barraza, G. A., Back, W. E., and Mata, F., Probabilistic Forecasting of Project Performance Using Stochastic S Curves, *Journal of Construction Engineering and Management*, 130(1), 25-32, 2004.

Blyth, K. and Kaka, A., A Novel Multiple Linear Regression Model for Forecasting S-Curves, *Engineering, Construction and Architectural Management*, 13(1), 82-95, 2006.

Chao, L.-C. and Chien, C.-F., Estimating Project S-Curves Using Polynomial Function and Neural Networks, *Journal of Construction Engineering and Management*, 135(3), 169-177, 2009.

Chen, H. L., Improving Forecasting Accuracy of Project Earned Value Metrics: Linear Modeling Approach, *Journal of Management in Engineering*, 30(2), 135-145, 2014.

Chen, H. L., Chen, W. T., and Lin, Y. L., Earned Value Project Management: Improving The Predictive Power of Planned Value, *International Journal of Project Management*, 34(1), 22-29, 2016.

Cheng, M.-Y. and Roy, A. F., Evolutionary Fuzzy Decision Model for Cash Flow Prediction Using Time-Dependent Support Vector Machines, *International Journal of Project Management*, 29(1), 56-65, 2011.

Cheung, S. O., Suen, H. C., and Cheung, K. K., PPMS: A Web-Based Construction Project Performance Monitoring System, *Automation in Construction*, 13(3), 361-376, 2004.

Chou, J.-S., Chen, H.-M., Hou, C.-C., and Lin, C.-W., Visualized EVM System for Assessing Project Performance, *Automation in Construction*, 19(5), 596-607, 2010.

Cioffi, D. F., A Tool for Managing Projects: An Analytic Parameterization of The S-Curve, *International Journal of Project Management*, 23(3), 215-222, 2005.

Czemplik, A., Application of Earned Value Method to Progress Control of Construction Projects, *Procedia Engineering*, 91, 424-428, 2014.

Dodson, M., Defavari, G., and De Carvalho, V., Quality: The Third Element of Earned Value Management, *Procedia Computer Science*, 64, 932-939, 2015.

El-Omari, S. and Moselhi, O., Integrating Automated Data Acquisition Technologies for Progress Reporting of Construction Projects, *Automation in Construction*, 20(6), 699-705, 2011.

Han, K., Degol, J., and Golparvar-Fard, M., Geometry-and Appearance-Based Reasoning of Construction Progress Monitoring, *Journal of Construction Engineering and Management*, 144(2), 04017110, 2017.

Kaliba, C., Muya, M., and Mumba, K., Cost Escalation and Schedule Delays in Road Construction Projects in Zambia, *International Journal of Project Management*, 27(5), 522-531, 2009.

Kim, B.-C., Probabilistic Evaluation of Cost Performance Stability in Earned Value Management, *Journal of Management in Engineering*, 32(1), 04015025, 2015.

Kim, E., Wells, W. G., and Duffey, M. R., A Model for Effective Implementation of Earned Value Management Methodology, *International Journal of Project Management*, 21(5), 375-382, 2003.

Kim, T., Kim, Y.-W., and Cho, H., Customer Earned Value: Performance Indicator from Flow and Value Generation View, *Journal of Management in Engineering*, 32(1), 04015017, 2015.

Li, J., Moselhi, O., and Alkass, S., Internet-Based Database Management System for Project Control, *Engineering, Construction and Architectural Management*, 13(3), 242-253, 2006.

Lipke, W., Schedule is Different, *The Measurable News*, 31-34, 2003.

Lipke, W., Connecting Earned Value to The Schedule, *The Measurable News*, 1, 6-16, 2004.

Lo, W. and Chen, Y. T., Optimization of Contractor' s S Curve, *Proc., 24th International Symposium on Automation & Robotics in Construction (ISARC 2007)*, Construction Automation Group, 417-420, Kochi, India, 2007.

Moslemi-Naeni, L. and Salehipour, A., Evaluating Fuzzy Earned Value Indices and Estimates by Applying Alpha Cuts, *Expert Systems with Applications*, 38(7), 8193-8198, 2011.

Moslemi-Naeni, L., Shadrokh, S., and Salehipour, A., A Fuzzy Approach for the Earned Value Management, *International Journal of Project Management*, 29(6), 764-772, 2011.

Naderpour, A., and Mofid, M., *Improving construction management of an educational center by applying earned value technique*, Procedia engineering, 14, 1945-1952, 2011.

Ponz-Tienda, J. L., Pellicer, E., and Yepes, V., Complete Fuzzy Scheduling and Fuzzy Earned Value Management in Construction Projects, *Journal of Zhejiang University SCIENCE A*, 13(1), 56-68, 2012.

Turkan, Y., Bosché, F., Haas, C. T., and Haas, R., Toward Automated Earned Value Tracking Using 3D Imaging Tools, *Journal of Construction Engineering and Management*, 139(4), 423-433, 2013.

Von Wangenheim, C. G., Savi, R., and Borgatto, A. F., DELIVER!–An Educational Game for Teaching Earned Value Management in Computing Courses, *Information and Software Technology*, 54(3), 286-298, 2012.

Zhong, S., and Wang, X., Improvement and Application of Earned Value Analysis in Coal Project Management, *Procedia Engineering*, 26, 1983-1989, 2011.

Zhu, Z., Ren, X., and Chen, Z., Integrated Detection and Tracking of Workforce and Equipment from Construction Jobsite Videos, *Automation in Construction*, 81, 161-171, 2017.

Responsible Design and Delivery of the Constructed Project
Edited by Abdul-Malak, M., Khoury, H., Singh, A., and Yazdani, S.

CROWD-SOURCED VISUAL DATA COLLECTION FOR MONITORING INDOOR CONSTRUCTION IN 3D

FOUAD AMER and MANI GOLPARVAR-FARD

Dept of Civil & Environmental Engineering, University of Illinois at Urbana Champaign, Urbana, United States

Complete and accurate 3D monitoring of indoor construction progress using visual data is challenging. It requires (a) capturing a large number of overlapping images, which is time-consuming and labor-intensive to collect, and (b) processing using Structure from Motion (SfM) algorithms, which can be computationally expensive. To address these inefficiencies, this paper proposes a hybrid SfM-SLAM 3D reconstruction algorithm along with a decentralized data collection workflow to map indoor construction work locations in 3D and any desired frequency. The hybrid 3D reconstruction method is composed of a pipeline of Structure from Motion (SfM) coupled with Multi-View Stereo (MVS) to generate 3D point clouds and a SLAM (Simultaneous Localization and Mapping) algorithm to register the separately formed models together. Our SfM and SLAM pipelines are built on binary Oriented FAST and Rotated BRIEF (ORB) descriptors to tightly couple these two seperate reconstruction workflows and enable fast computation. To elaborate the data capture workflow and validate the proposed method, a case study was conducted on a real-world construction site. Compared to state-of-the-art methods, our preliminary results show a decrease in both registration error and processing time, demonstrating the potential of using daily images captured by different trades coupled with weekly walkthrough videos captured by a field engineer for complete 3D visual monitoring of indoor construction operations.

Keywords: Production control, Data collection decentralization, Automation, Computer vision, Virtual construction, Structure-from-motion, Simultaneous localization and mapping.

1 INTRODUCTION

Today, capturing images and videos on construction sites using smartphones and drones has become an indispensable part of the daily workflows conducted by the field engineers and trades (Han and Golparvar-Fard 2017). The large volume of visual data has enabled 3D reconstruction techniques to be built on Structure-from-Motion (SfM) and Multi-View Stereo (MVS) algorithms and record exterior construction activities in high details and at any desired frequency (Golparvar-Fard *et al.* 2009, 2011). However, producing complete 3D as-built models for indoor construction, considering limited visibilities and fields of view of cameras, requires a large set of overlapping images and necessitates that connecting areas such as hallways and corridors be included in captured images (Kopsida *et al.* 2015). Such areas are usually textureless and colorless, and they have low lighting conditions causing SfM based reconstruction algorithms to often fail in producing compelling 3D models. Acquiring large sets of overlapping images by a single person on the jobsite -particularly in hallways and entrances where field of view of a camera is limited- can also be very time-consuming.

The opportunity is the following: indoor construction activities are mostly performed by different teams working in dispersed locations, and each team is interested in recording their own activities as proof of timely completion of their tasks or for liability purposes. We built on this *decentralized* and *crowd-sourced* data capture opportunity, and we propose a new hybrid 3D reconstruction method that (a) maps indoor construction work locations in 3D and at any desired frequency and also (b) supports routine daily reporting of activities. The proposed method uses a pipeline of SfM to register local images taken by the trades in 3D and feeds them into MVS dense reconstruction algorithm to produce dense 3D point clouds for each work area. Videos taken during weekly walkthroughs from the entirety of the site and particularly hallways and corridors are then fed into a monocular SLAM to connect the locally produced dense 3D point clouds. This approach enables project stakeholders to use their daily field reporting images and create dense 3D models of each work location at a high frequency (e.g., daily) while only needing a video walkthrough of the entire workspace to be captured on a low frequency (e.g., weekly) to link local reconstructions together. In the following, we first review the related work and present the proposed reconstructed method. Next, we present the experimental results from conducting a pilot on a real-world project. Finally, we discuss the perceived benefits and challenges of the proposed method for scaling across all project sites.

2 RELATED WORK

In the past decade, image and video-based 3D reconstruction has been the center of a large number of studies. Some of these methods use unordered site photographs together with SfM based workflows to create dense 3D point clouds (Trupp *et al.* 2004, Quiñones-Rozo *et al.* 2008, Dai and Lu 2008, Golparvar-Fard *et al.* 2009, and Golparvar-Fard *et al.* 2012). Others focus on increasing registration and 3D reconstruction accuracy and completeness by using ground control points and visual markers (Tuttas *et al.* 2014 and Karsch *et al.* 2014). The use of stereo cameras and videogrammetry along with a knowledge of *a priori* geometry to create accurate 3D as-builts for infrastructure sites has also been investigated by Fathi and Brilakis (2011), Brilakis *et al.* (2011), and Rashidi *et al.* (2015). As a result, commercial solutions such as Pix4D, Reconstruct, Autodesk RECAP, and Bentley Context Capture are now heavily used to produce highly detailed 3D as-built models of construction sites in outdoor environments. However, mapping texture-less 3D indoor environments on construction sites using images is still an open research challenge, and many solution providers encourage practitioners to use laser scanning or RGBD devices (e.g., Matterport) to conduct visual data capture in their indoor environments. While these technologies are appealing, yet they are prohibitively expensive for daily use across all construction projects.

Regardless of being captured and produced for indoors or outdoor environments, as-built 3D point clouds should be registered together and/or with BIM for any performance tracking or quality control purposes. Point-based registrations based on manually matching a set of points in both scans (Akinci *et al.* 2006, Golparvar-Fard *et al.* 2009), target-based registrations using tie points introduced to the site (Bosché *et al.* 2009), and feature-based registrations based on automatically extracting and matching distinguishable scene features (Bosché 2012) have been proposed. Furthermore, using Iterative Closest Point (ICP) (Golparvar-Fard *et al.* 2009, Bosché 2010, and Turkan *et al.* 2014) and Principal Component Analysis (PCA) refined by ICP (Kim *et al.* 2013) were investigated to improve registration accuracies and automate the process. Despite these efforts, a complete automated and feature-based registration process to tie scattered local point clouds is still needed, which is the focus of the current study.

The joint application of BIM and point clouds for automated progress tracking has also been a topic of several recent research studies. El-Omari and Moselhi (2008, 2011), Bosché (2010), Golparvar-Fard *et al.* (2012), Turkan *et al.* (2014), Tuttas *et al.* (2014), Lin and Golparvar-Fard (2016), Han *et al.* (2017), and Son *et al.* (2017) are all examples of relevant works, however, tailoring their use for indoor construction tracking is still under-investigated, and current solutions are not designed to overcome the additional complexities of mapping indoor environments. Frameworks on automatically identifying construction components in images (Roh *et al.* 2011) or assessing the state of indoor partitions (Hamledari *et al.* 2017) are also proposed but the location and orientation of the indoor images with respect to BIM is assumed to be known as a priori.

Kopsida *et al.* (2015), Hamlederi and McCabe (2016) and Han and Golparvar-Fard (2017) have specifically revealed that lack of texture, highly varying lighting conditions, the nature of indoor construction operations where progress is mostly associated with changes in wall surfaces, and the need for acquiring a large number of overlapping images can all make the process of acquiring indoor images by a single party and processing them be inefficient, time consuming, and in most extreme cases very difficult to implement. To address these inefficiencies and challenges, we propose a method that allows for crowd-sourcing the data collection process by enabling the automatic fusion of different partial 3D reconstructions at any desired frequency.

3 PROPOSED METHOD

We propose a hybrid SfM-SLAM 3D reconstruction algorithm along with a decentralized data collection workflow to map indoor construction work locations in 3D and at any desired frequency. This method builds on (Amer and Golparvar-Fard, 2018) and uses SfM and MVS algorithms --developed on (Gargallo 2016)'s OpenSfM-- to create dense 3D reconstructions from images gathered by different trades at their specific work locations on a daily basis. These point clouds are tied together via a feature based SLAM algorithm using walkthrough videos taken at a lower frequency (e.g. weekly or bi-weekly). The proposed SLAM procedure is built on ORB-SLAM2 (Mur-Artal and Tardos, 2017) and both SfM and SLAM reconstructions are performed using binary ORB features. This strategy, as demonstrated below, allows us to tackle the problem of large datasets required to map indoor construction areas by creating dense reconstructions only in the areas where they are needed. It also avoids the long computation time of the offline SfM method on large datasets and instead assigns the task of mapping connection areas such as hallways and corridors to the SLAM procedure.

To deploy the proposed method in practice, we propose a workflow which consists of the following Three phases: Planning, Monitoring, and Control. The planning phase aims at identifying the activities anticipated for the following week or two, their locations, and their assigned crews/trades. It takes place in weekly or bi-weekly meetings and it is based on 4D BIM and/or Weekly Work Plans (WWPs).

The monitoring phase consists of (1) data collection and (2) data processing. Data collection is distributed among different crews/trades, and it is composed of capturing a video of all anticipated work locations by a single person (e.g. field engineer) at a low frequency (e.g. weekly) and collecting different sets of local images of specific work locations by their respective crews/trades at any desired frequency (e.g. daily). After the data is collected, the video is used to create a sparse point cloud of the entire site (local work locations and connecting areas) using the SLAM thread of the proposed algorithm, and the local sets of images are used to create local dense point clouds of every work location separately using the SfM + MVS thread of the algorithm. The local point clouds are then registered with the global point cloud by finding matching points using a hamming distance nearest neighbor search and then calculating a rigid 3D transformation using a Random Sample Consensus (RANSAC) loop. The RANSAC loop ensures the calculated transformation is geometrically consistent with a significant number of the matches. Finally, the control phase starts by registering the locally-dense as-built map with BIM which also localizes all the collected images and video frames with respect to BIM. This registration can be done manually or using surveyed visual tags. The resulting alignment would allow different 2D image-based and 3D point-cloud based progress assessment methods to be implemented on top of the proposed algorithm.

4 CASE STUDY AND PRELIMINARY RESULTS

The proposed method was tested on a real-world project. The test aimed at mapping the work done in two different and scattered locations (L1 and L2) and registering them together and within BIM. A walkthrough video of the jobsite was captured and was later processed to create a sparse 3D reconstruction of the entire jobsite (L1, L2, and the connecting areas). Additionally, two different sets of images of both work locations were gathered separately and were later processed to create dense reconstructions of both local work locations (L1 and L2) (Figures 1d and 1e). The created local maps were then registered with the global point cloud automatically using ORB feature matching which results in the reconstruction of a locally dense point cloud of the entire site (Figures 1a and 1b). The resulting point cloud is then manually registered with BIM using identifiable common points (Figure 1c). The registration accuracies are illustrated in Table 1, and the performance of the ORB-based SfM and MVS pipeline is illustrated in Table 2.

Figure 1. Dense local point clouds aligned with the sparse point cloud and with BIM. The red dots (in b and c) show the trajectory of the walkthrough video.

Table 1. Registration accuracies.

Point Clouds	Registration Error (cm)
L1 to site map	**3.97**
L2 to site map	**2.58**
BIM to site map	31.84

Table 2. Performance of SfM and MVS using ORB.

Location	Number of Images	Images Overlap	Number of Successfully Registered Images	Sparse Reconstruction (SfM) Density	Dense Reconstruction (SfM+MVS) Density	Sparse Reconstruction time	Dense Reconstruction time	Total Reconstruction Time
L1	63	60-80%	33	11,255	359,260	19 min	22 s	~20 min
L2	42	60-80%	15	1,159	242,712	10 min	6 s	~10 min

Compared to (Amer and Golparvar-Fard 2018), where the local point clouds were reconstructed using HaHoG features (Hessian Affine Histograms of Oriented Gradients) and the registration was performed using visual tags, we notice that the number of registered images has decreased from 60 to 33 for L1 and from 41 to 15 for L2 also leading to a decrease in the densities of the generated local point clouds. However, the registration accuracy increased from an average of 7.7 cm (+ an ambiguity of 36.3 cm due to the use of tags) to an average of 3.28 cm (the average of the registration of both local point clouds in Table 1). The computation time for reconstruction also decreased from ~22 min to ~15 min.

5 CONCLUSIONS

We presented a new method to map indoor construction areas in 3D using a crowd-sourcing approach to visual data capture. The proposed method uses binary ORB features to reconstruct detailed local point clouds of scattered work locations and to reconstruct a sparse global point cloud of the entire site. The same binary features are then used to tie the different local clouds together and to create a 3D reconstruction that is only dense in work areas where construction activities need to be mapped. The experimental results revealed shorter processing times and higher registration accuracies compared to state of the art algorithms. Future work will focus on implementing a mobile version of this hybrid SfM-SLAM algorithm to provide feedback to the field engineer on the completeness of capture and on applying automated point-cloud based and image-based reasoning systems for indoor construction progress tracking purposes.

Acknowledgement

This work was supported by the National Science Foundation under grant CMMI 1446765. Any opinions, findings, and conclusions or recommendations expressed in this material are those of the author(s) and do not necessarily reflect the views of the National Science Foundation. The support of our industry partners in enabling capturing of the data and testing of the method is also appreciated.

References

Akinci, B., Boukamp, F., Gordon, C., Huber, D., Lyons, C., and Park, K. A Formalism for Utilization of Sensor Systems and Integrated Project Models for Active Construction Quality Control, *Automation in Construction*, 15(2), 124-138, 2006.

Amer, F. and Golparvar-Fard. *Decentralized Visual 3D Mapping of Scattered Work Locations for High-Frequency Tracking of Indoor Construction Activities*, Construction Research Congress, ASCE, 2018.

Bosché, F, Plane-based Registration of Construction Laser Scans with 3D/4D Building Models, *Advanced Engineering Informatics*, 26, 90–102, 2012.

Bosché, F., Automated Recognition of 3D CAD Model Objects in Laser Scans and Calculation of As-Built Dimensions for Dimensional Compliance Control in Construction, *Advanced Engineering Informatics,* 24(1), 107–118, 2010.

Bosché, F., Haas, C. T., and Akinci, B., Automated Recognition of 3D CAD Objects in Site Laser Scans for Project 3D Status Visualization and Performance Control, *ASCE, Journal of Computing in Civil Engineering,* 23(6), 311–318, 2009.

Brilakis, I., Fathi, H., and Rashidi, A, Progressive 3D Reconstruction of Infrastructure with Videogrammetry, *Automation in Construction*, 20, 884–895, 2011.

Dai, F. and Lu, M., Photo-based 3D Modeling of Construction Resources for Visualization of Operations Simulation: Case of Modeling a Precast Façade, *IEEE Winter Simulation*, 2439–2446, 2008.

El-Omari, S. and Moselhi, O., Integrating 3D Laser Scanning and Photogrammetry for Progress Measurement of Construction Work, *Automation in Construction*, 18(1), 1-9, 2008.

El-Omari, S. and Moselhi, O., Integrating Automated Data Acquisition Technologies for Progress Reporting of Construction Projects, *Automation in Construction*, 20, 699–705, 2011.

Fathi, H. and Brilakis, I., Automated Sparse 3D Point Cloud Generation of Infrastructure Using its Distinctive Visual Features, *Advanced Engineering Informatics*, 25, 760–770, 2011.

Gargallo, P., OpenSfM, 2016, retrieved from https://github.com/mapillary/OpenSfM on May 2017.

Golparvar-Fard, M., Peña-Mora, F., and Savarese, S., *Monitoring of Construction Performance Using Daily Progress Photograph Logs and 4D As-planned Models*. ASCE International Workshop on Computing in Civil Engineering, Austin, TX, USA, 2009.

Golparvar-Fard, M., Peña-Mora, F., and Savarese, S., Integrated Sequential As-Built and As-Planned Representation with D 4 AR Tools in Support of Decision-Making Tasks in the AEC/FM Industry. *ASCE Journal of CEM,* 137(12), 1099–1116, 2011.

Golparvar-Fard, M., Peña-Mora, F., and Savarese, S., Automated Progress Monitoring Using Unordered Daily Construction Photographs and IFC-Based Building Information Models. *Computing in Civil Engineering*, 29(1), 4014025, 2012.

Hamledari, H., Mccabe, B., and Davari, S., Automated Computer Vision-Based Detection of Components of Under-Construction Indoor Partitions, *Automation in Construction*, 74, 78–94, 2017.

Han, K. K. and Golparvar-Fard, M., Potential of Big Visual Data and Building Information Modeling for Construction Performance Analytics: An Exploratory Study, *Automation in Construction*, 73, 184-198, 2017.

Han, K. K., Degol, J., and Golparvar-Fard, M., Geometry- and Appearance-Based Reasoning of Construction Progress Monitoring, *ASCE Journal of Construction Engineering and Management*, 144(2), 2017.

Karsch, K., Golparvar-Fard, M., and Forsyth, D., ConstructAide: Analyzing and Visualizing Construction Sites through Photographs and Building Models, *ACM Transactions on Graphics*, 33(6), 1–11, 2014.

Kim, C., Kim, C., and Son, H., Fully Automated Registration of 3D Data to a 3D CAD Model for Project Progress Monitoring, *Automation in Construction*, 35, 587–594, 2013.

Kopsida, M., Brilakis, I., and Vela, P., *A Review of Automated Construction Progress and Inspection Methods*, Proceedings of the 32nd CIB W78, (January), 421–431, 2015.

Lin, J. and Golparvar-Fard, M., Web-Based 4D Visual Production Models for Decentralized Work Tracking and Information Communication on Construction Sites, *Construction Research Congress, ASCE,* 1731–1741, 2016.

Mur-Artal, R. and Tardos, J. D., ORB-SLAM2: An Open-Source SLAM System for Monocular, Stereo and RGB-D Cameras, *IEEE Transactions on Robotics*, 33(5), 1255-1262, 2017.

Quiñones-Rozo, C. A., Hashash, Y. M. A., and Liu, L. Y., Digital Image Reasoning for Tracking Excavation Activities, *Automation in Construction,* 17(5), 608–622, 2008 .

Rashidi, A., Brilakis, I., and Vela, P., Generating Absolute-Scale Point Cloud Data of Built Infrastructure Scenes Using a Monocular Camera Setting, *Computing in Civil Engineering*, 29(9), 4014089, 2015.

Roh, S., Aziz, Z., and Peña-Mora, F., An Object-Based 3D Walk-Through Model for Interior Construction Progress Monitoring, *Automation in Construction*, 20, 66–75, 2011.

Son, H., Kim, C., and Cho, Y. K., Automated Schedule Updates Using As-Built Data and a 4D Building Information Model, *Journal of Management in Engineering,* 33(4), 04017012-1, 2017.

Trupp, T., Marulanda, C., Hashash, Y., Liu, L., and Ghaboussi, J., Novel Technologies for Tracking Construction Progress of Deep Excavations, *Proc. Geotechnical Engineering for Transportation Projects, ASCE,* 2254–2262, 2004.

Turkan, Y., Bosché, F., Haas, C. T., and Haas, R., Tracking of Secondary and Temporary Objects in Structural Concrete Work, *Emerald, Construction Innovation: Information, Process and Management,* 14(2), 145–167, 2014.

Tuttas, S., Braun, A., Borrmann, A., and Stilla, U., Comparison of Photogrammetric Point Clouds with BIM Building Elements for Construction Progress Monitoring, *Int. Archives of the Photogrammetry, Remote Sensing and Spatial Information Sciences* - ISPRS Archives, 40(3), 341–345, 2014.

Responsible Design and Delivery of the Constructed Project
Edited by Abdul-Malak, M., Khoury, H., Singh, A., and Yazdani, S.

INFORMATION MANAGEMENT TECHNOLOGIES FOR IMPROVING EARNED VALUE QUANTIFICATION

FARAH DEMACHKIEH and MOHAMED-ASEM ABDUL-MALAK

Dept of Civil and Environmental Engineering, American University of Beirut, Beirut, Lebanon

Real-time progress tracking of construction activities is critical for successful project monitoring and control. As such, the application of modern information technologies for progress tracking are proclaimed to overcome the limitations of manual approaches and hence contribute to the automated acquisition of onsite data for the computation of the earned value of progressed work. To this end, this paper presents a thorough review of 111 studies dealing with automated data collection and progress tracking on construction sites. The researched technologies were classified into four categories, namely: information technology, geo-spatial, imaging, and augmented reality (AR). It was found that research dealing with progress tracking using imaging technologies has been speedily growing over the years. The findings also revealed that the highest frequency (57) of reported technologies is found to be related to the AR category, with 74 percent of these studies having been published in the last eight years. Moreover, 63 percent of the studies were concerned with all kinds of project activities, with more specific areas of application reported to include concrete and steel elements, MEP elements, indoor and outdoor elements. The encountered technologies are reported to decrease labor hours and the time needed for progress monitoring, and automated data collection on construction activities progress is said to facilitate the generation of status reports and permit the early detection of deviations from the as-planned status.

Keywords: Monitoring and control, Progress estimation, Project tracking, Data collection, Construction sites.

1 INTRODUCTION

Construction projects are nowadays more challenging in terms of time and budget constraints, thereby calling for devising better control and monitoring platforms (Chen *et al.* 2012). To this effect, and due to market pressure and global competition in the construction industry, a critical need for advanced management information systems had emerged. This need gave rise to the implementation of Earned Value Management (EVM) to limit projects' cost and duration overruns (Hunter *et al.* 2014). According to Naderpour and Mofid (2011), EVM is a recognized project management methodology that integrates cost, schedule, and scope into a unique measurement system, providing an early warning signal of project deviations from the plan, thus creating a chance to initiate timely corrective actions.

However, numerous researchers and practitioners have proclaimed that the main shortcoming associated with EVM is the difficulty in obtaining the percent complete of executed work. Ruskin (2004) considered "eyeball estimates" as personal and bias estimates of the percent complete of the executed works and are thus exposed to intended or unintentional biases.

Consequently, they are hard to repeat, and they may comprise huge errors (Ruskin 2004). Likewise, Humphreys and Visitacion (2009) assumed that using EVM as a standard practice may be an encumbrance, mainly since collecting significant data and information in construction sites needs a lot of effort by the project team. Moreover, Omar and Nehdi (2016) consider that construction progress tracking is not a simple task and involves challenges due to the large amounts of information in construction sites, mainly related to a diversity of functions, for instance, "scheduling, construction methods, cost management, resources, quality control, and change order management".

In recent years, construction information management has significantly benefited from developments in information and communications technology (ICT), by enhancing the speed of information flow and improving the effectiveness of information communication. As such, existing ICT has shown great potentials in enhancing on-site collection and retrieval of data (Chen and Kamara 2011).

2 SCOPE AND METHODOLOGY

The aim of this research is to explore the modern information management technologies reported in the current literature as contributing to the automated collection, retrieval, and transfer of on-site data for effective calculation of the earned value of executed works. The adopted methodology is as follows: (a) review the array of current research studies dealing with automated data collection and progress tracking; (b) cluster the various encountered technologies into general groups; and (c) address the benefits of applying those data acquisition technologies in construction sites. To this effect, a thorough in-depth review of the current literature dealing with data acquisition and progress tracking, published in the last two decades, has been carried out using online libraries and databases and based on the semantics and keywords-based methods. Many combinations of keywords, such as "data acquisition," "progress tracking", "automated data collection", "construction monitoring", "information technologies", and "progress measurement", were used to retrieve the relevant studies. The search resulted in filtering a set consisting of 111 relevant studies tackling the automation of data collection and progress tracking in construction sites, 62 percent of which were published over the last eight years. The adopted research studies were evaluated and categorized in relevant groups, allowing the identification of the main ICT areas proposed for use on construction sites for achieving more accurate earned value computations.

3 RESULTS AND ANALYSIS

3.1 Temporal Distribution of Research Studies

The 111 research studies were ordered and grouped according to their years of publication, as shown in Figure 1. It can be noticed that starting 2005 there seems to be a jump in the number of encountered studies, with the two middle ranges reflecting an average of seven publications per year. The last eight years (i.e., 2010-2017) encompassed a large number of publications (around 11 per year), showing the rapid evolvement of research tackling automated data acquisition technologies that have the potential to be used on construction sites.

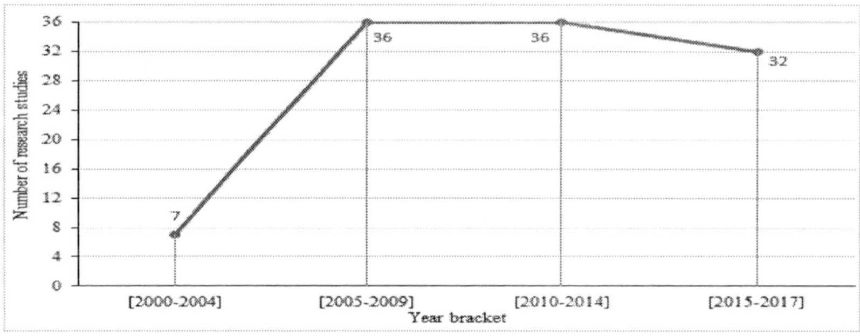

Figure 1. Temporal distribution of research studies.

3.2 General Screening and Findings

Figure 2 provides a general classification of the encountered relevant technologies for managing information on construction sites, classified according to four categories, mainly "enhanced IT", "geo-spatial", "imaging" and "augmented reality", following the classification of Omar and Nehdi (2016), along with the frequencies deduced in connection with each sub-class. To this effect, the various IT-based technologies included multimedia tools (digital camera and video) (Bohn and Teizer 2009), voice-based tools (Cheng *et al.* 2017), email and short message services (SMS) (Hegazy *et al.* 2014), and mobile computing tools (El-Omari and Moselhi 2011). Those inexpensive tools are said to enable the daily automated progress tracking of construction activities, leading to better schedule and cost control through enhanced communication.

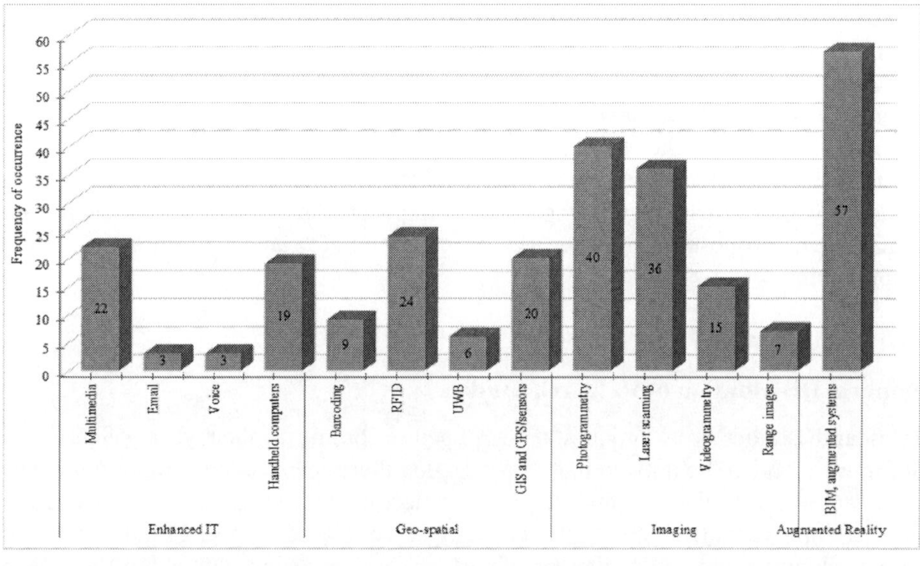

Figure 2. Frequencies of technologies for construction site data acquisition.

Moreover, Geo-spatial-based tools, including barcoding, radio frequency identification (RFID) (Khoury *et al.* 2015), ultra-wide band (UWB) tags (Shahi *et al.* 2014), geographic information systems (GIS) (Bansal and Pal 2009), and global positioning systems (GPS) (Behnam *et al.* 2016) are used to visualize on-site construction objects. Example applications include real-time 3D material, labor and equipment tracking. Furthermore, progress tracking based on imaging technologies has been focusing on using digital images to produce 3D information about various objects on a construction site in order to be used in project controls (Golparvar-Fard *et al.* 2011, Golparvar-Fard *et al.* 2012). Finally, augmented reality (AR), defined as "the combination of real and virtual scenes" (Wang *et al.* 2014), is mainly used for the comparison of different project status (Omar and Nehdi 2016). The most efficient AR application that is successively employed on construction sites is BIM (Han *et al.* 2017).

That being said, it can be noted that the highest frequency (i.e., 57) of reported technologies is found to be related to the AR applications group. This is explained by the fact that AR has a great potential to address a plethora of challenges throughout the project execution, by providing a "radical shift in human–computer interaction" (Omar and Nehdi 2016), thus receiving growing attention by researchers and industry practitioners. With an enduring technological evolvement of these applications, AR applications are progressively becoming more cost-effective with improved capabilities to deliver exhaustive data on various project activities. The evolvement of these data acquisition tools was further investigated, by providing the temporal distribution of the corresponding research work in the 111 reviewed studies. As shown in Figure 3, research on progress tracking using imaging technologies has been speedily growing over the years. Moreover, it can be seen that 42 of the 57 studies (74 percent) that employ AR applications for automated data collection were published in the last eight years (i.e., during 2010-2017). However, although work pertaining to AR technologies for construction projects has noticeably increased in the most recent years, as revealed in Figure 3, these applications are said to be still in the research phase and their complete capabilities have not yet been fully realized (Omar and Nehdi 2016).

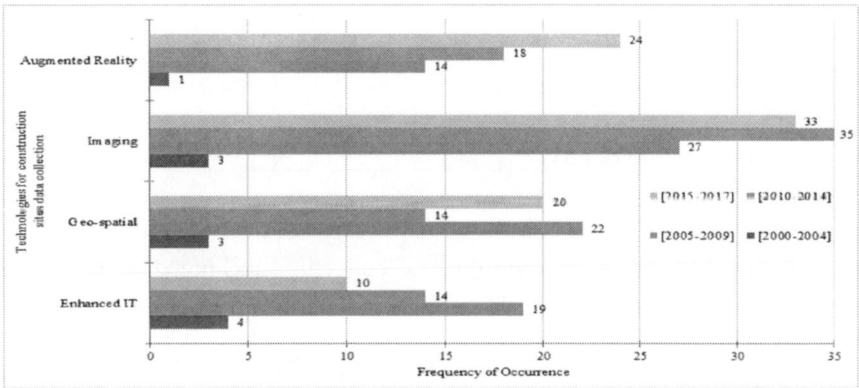

Figure 3. Temporal distribution of encountered technologies.

3.3 Areas of Application in Construction

A more detailed track of analysis tackled the significance of the above-mentioned findings from the perspective of the various areas of applications investigated as part of the reviewed published work. To this effect, Table 1 shows the breakdown of the filtered studies by the areas of application that were the subjects of investigation. It was found that 69 studies (63 percent) out of the examined studies (111, in total) were concerned with all kinds of project activities, whereas only one percent dealt with off-site construction activities. The more specific applications areas are found to include concrete and steel elements, MEP elements, indoor elements, and outdoor elements, among other similarly common on-site activities.

3.4 Benefits of Data Acquisition Technologies

Current methods for progress tracking on construction sites are mostly manual, labor-intensive, and time-consuming, and they are often based on abstruse and unreliable rules (Golparvar-Fard et *al.* 2011, Montaser and Moselhi 2015, Turkan *et al.* 2013). A considerable number of advanced automated data acquisition technologies, offering real-time on-site progress tracking, have been identified in this study. The encountered technologies are reported to significantly decrease labor hours and the time needed for progress monitoring. They allow this jobsite task to be achieved remotely and help in keeping track of the material used on construction site. It is found that real-time automated data collection on construction activities progress facilitates the generation of status reports and permits the early detection of deviations between the as-planned and current project statuses, thus allowing the timely flow of information between all interested parties. To this end, espousing such a variety of automated progress-tracking technologies can offer decision makers with prompt information in order to track the project progress more efficiently. The ultimate implication is the ability to produce accurate schedule updates and forensic delay analyses, and to better conceive suitable corrective measures.

Table 1. Distribution of technologies by areas of application in construction.

Specific application in construction	Number of research studies
General (any type)	69
Workers, materials and/or equipment	15
Infrastructure/Industrial	7
Concrete elements	6
MEP elements (cylindrical pipes and pipes spools)	3
Indoor elements	3
Structural steel elements	2
Earthwork	2
Outdoor elements	2
Secondary and temporary elements	1
Off-site construction elements	1

4 CONCLUSION

This study presents a literature review of data collection technologies that are used on construction sites for progress tracking and control. The analysis presented descriptions and classification of the emerging technologies, reported to be critical to effective data collection and retrieval. The paper highlights the usefulness of these applications in terms of increasing the efficiency of the tracking and progress reporting processes with less work done and fewer assumptions taken, with the aim of early detection of deviations between the planned and actual project statuses, thereby helping decision makers to take prompt corrective measures. Undeniably, the main obstacles for adopting the proposed technologies in construction include the excessive cost of procurement and maintenance of those emerging technologies, the necessity for extensive users training and the complexity of integration of these technologies with current applications. Data acquisition technologies that have the potential to be used on construction sites are hastily developing. Thus, it is highly suggested that decision makers carefully research their availability in order to get the up-to-date applications serving the purpose of achieving better monitoring and control of construction progress.

References

Bansal, V. K., and Pal, M., Construction Schedule Review in GIS with a Navigable 3D Animation of Project Activities, *International Journal of Project Management*, 27(5), 532-542, 2009.

Behnam, A., Wickramasinghe, D. C., Ghaffar, M. A. A., Vu, T. T., Tang, Y. H., and Isa, H. B. M., Automated Progress Monitoring System for Linear Infrastructure Projects Using Satellite Remote Sensing, *Automation in Construction*, 68, 114-127, 2016.

Bohn, J. S., and Teizer, J., Benefits and Barriers of Construction Project Monitoring Using High-Resolution Automated Cameras, *Journal of Construction Engineering and Management*, 136(6), 632-640, 2009.

Chen, S.-M., Griffis, F. H., Chen, P.-H., and Chang, L.-M., Simulation and Analytical Techniques for Construction Resource Planning and Scheduling, *Automation in Construction*, 21, 99-113, 2012.

Chen, Y., and Kamara, J. M., A Framework for Using Mobile Computing for Information Management on Construction Sites, *Automation in Construction*, 20(7), 776-788, 2011.

Cheng, C.-F., Rashidi, A., Davenport, M. A., and Anderson, D. V., Activity Analysis of Construction Equipment Using Audio Signals and Support Vector Machines., *Automation in Construction*, 81, 240-253, 2017.

El-Omari, S., and Moselhi, O., Integrating Automated Data Acquisition Technologies for Progress Reporting of Construction Projects, *Automation in Construction*, 20(6), 699-705, 2011.

Golparvar-Fard, M., Pena-Mora, F., and Savarese, S., Monitoring Changes of 3D Building Elements from Unordered Photo Collections, *Proc., Computer Vision Workshops (ICCV Workshops), 2011 IEEE International Conference on*, IEEE, 249-256, 2011.

Golparvar-Fard, M., Peña-Mora, F., and Savarese, S., Automated Progress Monitoring Using Unordered Daily Construction Photographs and IFC-Based Building Information Models, *Journal of Computing in Civil Engineering*, 29(1), 04014025, 2012.

Han, K., Degol, J., and Golparvar-Fard, M., Geometry-and Appearance-Based Reasoning of Construction Progress Monitoring, *Journal of Construction Engineering and Management*, 144(2), 04017110, 2017

Hegazy, T., Abdel-Monem, M., and Atef Saad, D., Framework for Enhanced Progress Tracking and Control of Linear Projects, *Engineering, Construction and Architectural Management*, 21(1), 94-110, 2014.

Humphreys, and Visitacion, M., Mastering Project Management Fundamentals Critical for Successful Earned Value Management, 2009.

Hunter, H., Fitzgerald, R., and Barlow, D., Improved Cost Monitoring and Control Through the Earned Value Management System, *Acta Astronautica*, 93, 497-500, 2014.

Khoury, H., Chdid, D., Oueis, R., Elhajj, I., and Asmar, D., Infrastructureless Approach for Ubiquitous User Location Tracking in Construction Environments, *Automation in Construction*, 56, 47-66, 2015.

Montaser, A., and Moselhi, O., Outdoor automated data acquistion for progress reporting, 2015.

Naderpour, A., and Mofid, M., Improving Construction Management of an Educational Center by Applying Earned Value Technique, *Procedia Engineering*, 14, 1945-1952, 2011.

Omar, T., and Nehdi, M. L., Data Acquisition Technologies for Construction Progress Tracking, *Automation in Construction*, 70, 143-155, 2016.

Ruskin, A. M., Two Issues Concerning the Use of Earned Value Measurements, *Engineering Management Journal*, 16(3), 26-30, 2004.

Shahi, A., Safa, M., Haas, C. T., and West, J. S., Data Fusion Process Management for Automated Construction Progress Estimation, *Journal of Computing in Civil Engineering*, 29(6), 04014098, 2014.

Turkan, Y., Bosché, F., Haas, C. T., and Haas, R., Toward Automated Earned Value Tracking Using 3D Imaging Tools, *Journal of Construction Engineering and Management*, 139(4), 423-433, 2013.

Wang, X., Truijens, M., Hou, L., Wang, Y., and Zhou, Y., Integrating Augmented Reality with Building Information Modeling: Onsite Construction Process Controlling for Liquefied Natural Gas Industry, *Automation in Construction*, 40, 96-105, 2014.

Responsible Design and Delivery of the Constructed Project
Edited by Abdul-Malak, M., Khoury, H., Singh, A., and Yazdani, S.

THE HAWAII SUPERFERRY BUSINESS VENTURE NAVIGATES TO FAILURE

ROCCO TRAMONTANO and AMARJIT SINGH

University of Hawaii at Manoa, Honolulu, USA

At first glance, the private business venture of the Hawaii Superferry appeared to be an economic opportunity for Superferry executives, the State of Hawaii, as well as tourists and residents looking for a cheaper alternative for inter-island travel. However, upon its introduction to the Hawaii Islands, it faced stiff opposition from several public interest groups. Plagued with legal troubles and roadblocks throughout the Superferry's short lifetime, the inter-island ferry business was no longer able to stay afloat financially. The poor implementation of conflict resolution and negotiation of the parties involved appeared to be the perfect storm for what was supposed to be smooth sailing. The eagerness with which the State waived environmental permits appeared to be done with the sole aim of benefiting the Superferry corporation, upon which the Court did not look with favor. This article explores the arguments made by the defense and opposition, and presents the findings of the court.

Keywords: Conflict, Resolution, Negotiation, Maui, Kauai, Oahu, Travel, Interisland, Ferry, Environment.

1 INTRODUCTION

In 2001, venture capitalist Tim Dick cultivated the idea to provide Hawai'i residents and tourists an efficient, rapid, and cost-effective means of transportation between the Hawaiian Islands. The result of this idea was the *Alakai*, a $100 million, 349 ft. catamaran. The *Alakai* (Figure 1) was capable of transporting 282 vehicles and 866 passengers at speeds of 35 knots between the islands of Oahu, Maui, and Kauai (Segal 2007).

On Sunday August 26th, 2007, the Hawaii Superferry set sail on its maiden voyage from Honolulu Harbor to its passenger terminal on Kahului, Maui. At a low introductory cost of $5 per seat and $5 per car, the trip sold out within thirty minutes. Upon the Hawaii Superferry's arrival to Maui, it was met by angry protestors and a restraining order preventing the vessel from docking. The protest was the beginning of many legal and financial struggles for Tim Dick and his associates that led to the demise of the interisland ferry service (Hawaii Interisland Super Ferry News 2009). In addition, the company marketed a second Superferry that would provide a four-hour route (each way) from Oahu to the Big Island in the near future. The objective of this paper is to discuss the case that went to court, and explore the implementation of conflict resolution or lack thereof between the involved parties that led to the end of an economic opportunity to the State of Hawaii.

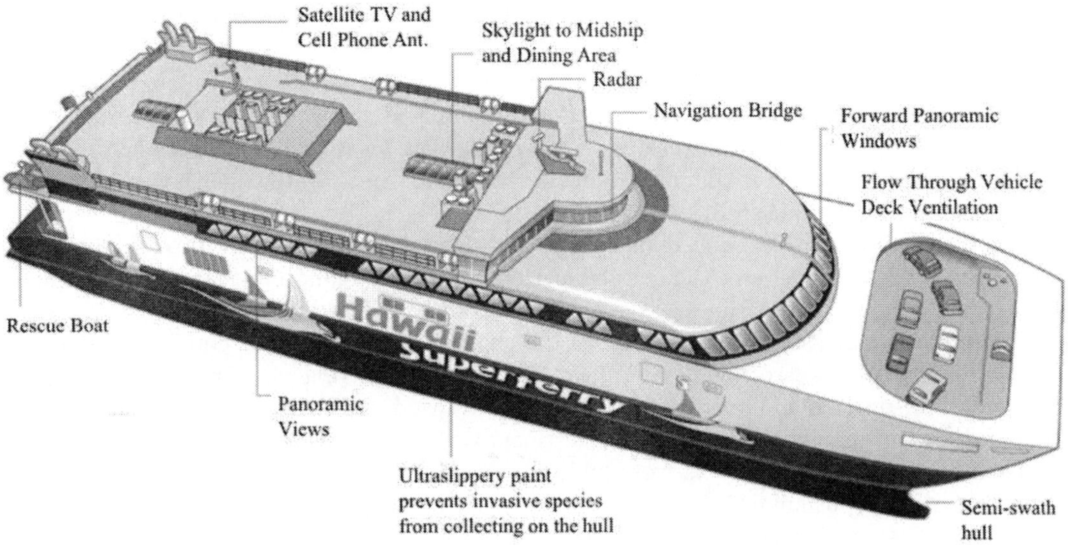

Figure 1. Graphic of the *Alakai* an interisland passenger ferry (Segal 2007).

2 SUPERFERRY TIMELINE

Tim Dick and partners secured financing through equity and federally guaranteed loans totaling $211 million from ABN Amro Bank and J. F. Lehman & Co. respectively (Schaefers 2006). In 2004, it was determined by the Hawaii State Office of Environmental Quality Control and county governments that the Hawaii Superferry venture was exempt from conducting an environmental impact study (EIS). This was under the direction of Barry Fukunaga, the Honolulu Harbor Director, who later became the interim director of the State Department of Transportation in 2007 (Niesse 2007)[1]. With this information and direction, the Hawaii Superferry began operations on August 26, 2007 without an EIS, which became the underlying reasoning for the opposition.

The launching of the Hawaii Superferry was rushed by executives knowing that legal roadblocks lay ahead. With speculation that legal troubles would delay the Hawaii Superferry debut, discounted prices of $5 were offered to over 19,000 customers to facilitate operations. On August 27, 2007, at its maiden voyage, a restraining order was issued against the *Alakai* upon arrival in Kahului, Maui, which only allowed the Hawaii Superferry to disembark passengers at its destination. On September 12, 2007, through the support of Governor Linda Lingle and a Special Session conducted by the Hawaiian Legislature, it was determined it was in the public's interest that Hawaii Superferry services were allowed to resume, beginning September 26, 2007, between Oahu and Kauai (Senate Twenty-Fourth Legislature 2007 Second Special Session State of Hawaii). However, only two days later, on September 14, 2007, services were again halted between Oahu and Maui by Judge Joseph Cardoza from the Circuit Court of the Second Judicial Circuit until hearings from all stakeholders were completed (Hawaii Interisland Super Ferry News 2009).

As ferry services continued for an extra week between Oahu and Kauai, so did protests by surfers, paddlers, and swimmers, making it impossible for the *Alakai* to dock in Nawiliwili Harbor. On September 21, trips to Kauai stopped altogether. On October 4, 2007, the Hawaii Supreme Court determined that the inter-island Superferry required an environmental assessment

[1] Mr. Fukunaga apparently consulted with the Office of Environmental Quality Control and county governments.

in order to continue operations because the corporation used $40 million in taxpayer dollars for infrastructure improvements such as harbor improvements.[2] This halt in services placed a financial burden on the Hawaii Superferry corporation as it was no longer available to generate much-needed revenue. Hawaii Superferry officials estimated approximately six weeks without revenue until they would have to relocate. The ruling was overturned on December 13, 2007 when Governor Lingle and her administration convinced the court to allow the Hawaii Superferry to continue operations while an EIS was in progress.[3]

The inter-island ferry continued operations to Maui until March 16, 2009, while facing protests organized by the Sierra Club, Kahului Harbor Coalition, and the Surfrider Foundation. During this time, Aloha Airlines, a major inter-island airline faced bankruptcy. Hawaii Superferry prices increased,[4] while still being the cheaper inter-island transportation alternative. A ruling on that date in case No. 29035 *Sierra Club v. The Department of Transportation of the State of Hawaii* rejected and overturned the law that allowed the Hawaii Superferry to continue operations while an environmental impact assessment was conducted. Assets were liquidated and auctioned off while 236 employees were laid off. The financial implications of a delay, while conducting a time-consuming EIS, while the Superferry remained idle, was too much for the corporation to stay in business (Hawaii Interisland Super Ferry News 2009).

3 HAWAII SUPERFERRY OPPOSITION

Attorney Isaac Hall represented the Sierra Club, Maui Tomorrow, and the Kahului Harbor Coalition in their prosecution against State of Hawaii's Department of Transportation. The main argument in their prosecution was that the new ferry venture was operating illegally without an EIS and that the State of Hawaii was wrong in allowing an exemption to Chapter 343 of State law (Hawaii Revised Statutes Chapter 343 2009). The environmental concern was that the vessel would be traveling through protected marine mammal habitats and sanctuaries. In addition, the private venture was using $40 million in taxpayer dollars and state land to fund infrastructure and special barge requirements. These "special privileges" that were not available to others was seen as an unethical alliance between the State and the Hawaii Superferry (*Sierra Club v. The State of Hawaii Department of Transportation* 2009).

Protestors were concerned that the private business venture was allowed without due diligence due to money-driven politics. Local opponents of the Hawaii Superferry feared that the Superferry corporation would be the sole benefactor of the state taxpayers $40 million and State land that was provided, and that they were entitled to know the business impact to the environment through an environmental impact assessment and statement. Other concerns included increased congestion and traffic due to disembarking vehicles, the spread of invasive species from island to island, and altering seafloor topography from dredging (if any was required) that would affect surf nearby breaks (Niesse 2007, Anforth and Singh 2012).

4 HAWAII SUPERFERRY DEFENSE

A service that seemed to be a valuable asset to the citizens of Hawaii, the State created an exemption to requiring an EIS through a new law known as "Act 2" from the Second Special session held by the State of Hawaii Senate Twenty-Fourth Legislature (2007). Act 2 allowed the Governor to pass an executive order that exempted large ferry vessels such as the Superferry from

[2] Operations to Kauai remained suspended because protestors would not allow the Superferry to dock.
[3] Linda Lingle passed an Executive Order, 07-10 in November 4, 2007 allowing the Superferry to operate.
[4] Owing to demand and supply, as well as because Superferry needed revenue.

having to abide by Chapter 343 of the Hawaii Revised Statutes, which required an EIS for such operations. Harbor and other infrastructure improvements were also included within the scope of this law (State of Hawaii Executive Order 07-10 2007). The Hawaii Superferry brought a valid argument that an EIS wasn't required by other large vessels such as Norwegian Cruise Lines, or shipping companies such as Matson or Young Brothers, which navigate the ocean and use harbors in the same manner. Additionally, the Hawaii Superferry was equipped with the latest sonar equipment for marine mammal avoidance, a technology that was not used by older Matson, Young Brothers, and Norwegian Cruise line vessels (Anderson 2007).

To mitigate the concerns of opposing parties and citizens, the Hawaii Superferry placed a number of self-regulations in place to reduce their environmental impact. To reduce the chances of potentially harming marine mammals, two marine mammal lookouts were hired for transits and traffic routes were altered. To reduce the likelihood of spreading invasive species from island to island, vehicles were inspected, and mud-caked vehicles were not allowed on the vessel. In addition, as an example of environmental stewardship, environmentally friendly plates were used for meals, implemented a recycling program for customers to partake, and provided public service announcements regarding the dangers of invasive species (Hawaii Superferry 2009).

The Hawaii Superferry also argued that their services would provide many benefits to the State. The venture would provide a new, cheaper transportation alternative that would generate much-needed revenue for a state that had an outstanding debt of $3.99 billion,[5] while providing approximately 240 jobs (State of Hawaii - Department of Budget and Finance 2007). In addition, bringing a vehicle over to another island via the ferry would not change the number of cars on the road than if a tourist was to rent a vehicle. This option would also alleviate another cost for residents who already owned vehicles that were traveling outer-island. Superferry executives would also make the ferry available at no cost to the state during times of emergency. A valuable asset that would be necessary in the event of a natural disaster that would shut airports down, with the ability to transport emergency goods and services (Anderson 2007).

5 FINAL RULING AND EFFECT

The final ruling in *Sierra Club v. The Department of Transportation of the State of Hawaii* determined that Act 2 of the Second Special Legislative Session violated a constitutional mandate. Due to the nature of the law, Act 2 appeared to have passed for (i) the sole purpose of allowing the Superferry to operate without an EIS (Chapter 343, Hawaii Statutes), in order to speed up its debut, and (ii) showing favoritism to one specific company by granting them harbor improvements. Act 2 was passed with a 21-month time frame, such that other ventures could not be given a fair chance to compete. The likelihood of another entity offering services like the Superferry within the applicable time frame of Act 2 was realistically impossible. The ruling in this court case was enough for the Superferry executives to throw in the towel. The company no longer had enough funds to keep the vessels in an operable condition while an environmental impact assessment was conducted. The stonewalling of what appeared to be a potential economic opportunity for the State of Hawaii by special interest groups was successful and forced the company into bankruptcy.

6 CONFLICT AND ALTERNATIVE DISPUTE RESOLUTION

According to the Hawaii Superferry, in order to be successful, $10 million was required to build special barges needed for loading and offloading the ferry, and another $30 million was required

[5] $8 billion in 2015; $6.8 billion in 2017

for pier, harbor, and other infrastructure improvements. The Superferry made clear that funding from the State of Hawaii was their only option, and that without this the transportation service would not come to Hawaii. State officials determined that the value the ferry service would provide in regards to tax revenue, boosting local economy, and providing an alternative mode of transportation between islands was worth shelling out the $40 million. With no other alternatives available to the state, their worse alternative to the negotiated agreement (WATNA) was no superferry servies to the islands. If the State of Hawaii did not aid in rushing the Superferry through Act 2, alternatives such as other private business ventures may have come to fruition.

What both parties involved did not realize at the time was that the $40 million stipulation put in place by Hawaii Superferry was used as a primary argument for their opponents as to why the Act 2 was unconstitutional. What would have avoided this situation was if Superferry executives found their own source of funding for infrastructure improvements and included a self-contained ramp in the initial design of the vessel. An option that few knew about was the ability to include a self-contained ramp within the design of the Superferry, which would have avoided the need of the special barges that were state funded. This design was incorporated in the construction of the second non-commissioned Superferry that would have been used for transits between the Big Island. The self-contained ramp option may have been withheld by Superferry executives to see how much money they could save in their venture by getting the State of Hawaii to pay for it. In the eyes of the State, there was no alternative option but to provide funding for the special barges to onload and offload passengers. In this scenario, their WATNA (worst alternative to a negotiated agreement), had they not provided funding, was the possibility of the Superferry not coming to fruition. As the only inter-island ferry service and the cheapest travel alternative, the Superferry would have recouped its costs over time and avoided the legal headache had they sourced proper funding and made the wiser investments (Anforth and Singh 2012).

6.1 The Public – The Main Stakeholder

Additionally, all stakeholders should have been involved during the planning and development phase of the Superferry. A significant stakeholder that was overlooked during this phase was the the public at the vessel's destination. With more public involvement in the planning and development phase, Superferry executives and state officials may have learned earlier in the timeline that an environmental impact assessment and statement was important to the public, even though it was deemed not to be necessary. An EIS could have been conducted while the vessel was under construction, rather than when already in operation, thus eliminating and/or reducing the fears of those concerned. The Superferry did try to remedy environmental concerns through self-regulating policies and went above what was required in attempt to ease apprehensions. However, this may have been too little too late.

6.2 Hard Bargaining by the Opposition

Despite all attempts to mitigate conflict by the private business venture, its opponents appeared to have a hard-bargaining position. Mainly, the opposition refused to find compromise through the Superferry's efforts to accommodate concerns, and no middle ground was found. In their classic negotiation book, Fisher *et al.* (2011) state that "the soft negotiating game emphasizes the importance of building and maintaining a relationship." Maintaining a positive relationship with the opposition was important to the Superferry since those are the people that they hoped to one day provide a service to. Evidence of their soft position was equivalent to concessions made to cultivate a relationship. The opposition continued to have a close-minded position, especially in regards to the fact that the Superferry would be operating like any other vessel in the ocean (if not safer due to their self-imposed regulations and the latest technology used onboard). Fisher *et al.* (2011) go on to say that "[I]n positional bargaining…the negotiating game is biased in favor of the hard player. The process will produce an agreement, although it may not be a wise one." Even if an environmental impact statement was conducted or private funds were used for the entire operation, it appears the opposition would have still been against the Superferry as it affected their way of life.

7 CONCLUSION

The Hawaii Superferry had the potential of offering a cost effective, alternative mode of travel from Oahu to Maui, Kauai, and the Big Island. Although the private business venture was a for-profit business, it provided a service to the State of Hawaii that would have benefited the majority of its residents and tourists who would have used the service. By taking shortcuts and rushing its debut to the islands, the Superferry eliminated many of its alternatives that would have probably contributed towards their success. A major misstep in the planning was the lack of inclusion of opposing parties. Playing devil's advocate in its beginning stages could have addressed the concerns of the opposition while in the company's infancy, before operation costs were incurred.

Small, yet powerful, opposition groups took a hard bargaining position, which eventually led the Hawaii Superferry to bankruptcy. In addition, the erroneous decision by the State of Hawaii that allowed the Hawaii Superferry to operate without an EIS, obtain special exemption through the legislature for Superferry operation, and partially fund a private business venture exclusively, added fuel to the fire of the opposition. With no intention of changing its position against the Hawaii Superferry, protestors used the errors of the state and the Superferry to their advantage in order to preserve their way of life. With a hard bargaining position by their opponents, dispute resolution was impossible at the point conflict was introduced. With their backs to the wall, the Hawaii Superferry had no choice but to abandon ship.

References

Anderson, H. H., In Defense of the Superferry, *Honolulu Star Bulletin.* September 3, 2007.

Anforth, K., and Singh, A., Superferry Plies and Dies, J. of Legal Affairs and Dispute Resolution in Engineering and Construction, *Journal of Legal Affairs and Dispute Resolution in Engineering and Construction* Vol. (4), No. 4, November 1, 2012.

Fisher, R., Ury, W., and Patton, B., *Getting to Yes*, Penguin Books, New York, USA, 2011.

Hawaii Executive Order N. 07-10. Honolulu, Hawaii, November 4, 2007.

Hawaii Interisland Super Ferry News, The Latest Super Ferry News, 2009. Retrieved from http://www.hawaiiinterislandsuperferry.com/hawaiisuperferrynews.html on February 15, 2018.

Hawaii Revised Statutes Chapter 343, Environmental Impact Statements. Vol 6 § 343 (2009).

Hawaii Superferry, *Hahalua News*, Hahalua, March-May: 58-63, 2009.

Niesse, M., Opponents Attempt to Stop Superferry – Legislators Might Seek an Environmental Review Wanted by the Boat's Opponents, *Honolulu Star Bulletin*, March 5, 2007.

Schaefers, A., Superferry Buoyed by Founder's Drive, *Honolulu Star Bulletin,* January 4, 2006.

Segal, D., Alakai Weighs Anchor, *Honolulu Star Bulletin,* August 26, 2007.

Senate Twenty-Fourth Legislature 2007 Second Special Session State of Hawaii. "Environmental Law: Transportation: Ferry Operations." Hawaii, October 23, 2007.

Sierra Club v. the Department of Transportation of the State of Hawaii [2009] 290356 115 H. 299, 167 P.3d 292 Supreme Court of Hawaii.

State of Hawaii - Department of Budget and Finance, Statement of Total Outstanding Indebtedness of the State of Hawaii as of July 1, 2007. Retrieved from http://budget.hawaii.gov/wp-content/uploads/2012/11/debtout2007.pdf on February 15, 2018.

Responsible Design and Delivery of the Constructed Project
Edited by Abdul-Malak, M., Khoury, H., Singh, A., and Yazdani, S.

INFLUENCE OF COLLABORATIVE BEHAVIOR ON CLAIM NEGOTIATION

RITA AWWAD[1], STEPHANIE ATALLAH[1], and CAROL MENASSA[2]

[1] *Dept of Civil Engineering, Lebanese American University, Byblos, Lebanon*
[2]*Dept of Civil and Environmental Engineering, University of Michigan, Ann Arbor, US*

The high cost incurred by the resolution of conflicts is largely affected by the existing adversarial nature of the construction industry along with the use of non-efficient dispute resolution methods in construction projects. This paper studies opinion dynamics in the negotiation of construction disputes while trying to understand the behavior and extremism of each contractual party. The developed model uses an agent-based approach to show how each agent's attitude can influence the negotiation process when solving a dispute. It can also be used to highlight the importance of alternative dispute resolution (ADR) methods and the use of a mediator in helping parties initiate negotiation and decrease the number of negotiation cycles needed to converge. The results showed that negotiation is not only affected by the attitude and character of the agents involved but it is also influenced by the delivery method of the project and the level of intensity of each agent. It was found that when the project is delivered through an Integrated Project Delivery (IPD) method, parties are more flexible and cooperative and will reach agreement within few negotiation cycles.

Keywords: Dispute resolution, Opinion dynamics, Relative agreement, Project delivery, Litigation, Social sciences, Construction management.

1 INTRODUCTION

Complex construction projects often result in complex disputes between the different contractual parties which most of the time arise from poorly prepared contracts, inadequate planning and lack of communication. One of the main factors that lead to the success of a project is the way the different parties approach their disputes and conflicts. Unfortunately, with the existing adversarial nature of the industry along with the use of non-efficient dispute resolution methods, this objective is rarely achieved.

Due to the complex structure of contracting firms and construction projects in addition to different attitudes involved, conflicts and disputes occur regularly during the entire life cycle of largescale projects. If these disputes are not addressed quickly and in an effective manner, the collaborative mode of the participants might be affected which will create an adversarial environment in which progress of the project will be largely affected. With that being said, better methodologies must be found to solve construction conflicts and promote collaborative behaviors among parties.

Social sciences have played an important role in understanding the different types of relationships among people. Many papers in the literature have targeted the subject of opinion dynamics and interaction among different agents (Deffuant *et al.* 2002, Lorenz 2008, Meadows and Cliff 2012, Stephan and Menassa 2014). However, very few studies discussed the impact of social behavior and opinion dynamics on the effectiveness of the construction negotiation process (Long Chen 2012, Yuan and Ma 2012, Azar and Menassa 2013).

This paper uses the principles of the Relative Agreement theory to simulate the negotiation process between the contractor and owner agents when trying to resolve a dispute. The developed model can largely contribute to the construction industry through the analysis of the impact of each party's attitude and behavior on the negotiation process. This model also helps understand the impact of a certain project delivery method on decreasing negotiation time and establishing better communication between parties. In addition, this model can also be used to demonstrate how using a certain alternative dispute method can play a huge role in decreasing negotiation time and efforts.

2 METHODOLOGY

2.1 Relative Agreement

The developed agent-based model is based on the relative agreement (RA) model, an extension of the Bounded Confidence (BC) model where random pair interactions occur among the agents to influence each other's opinions. The main difference between the RA and BC models is that the RA model takes into account the uncertainty in opinion dynamics. In the initial RA model defined by Deffuant *et al.* (2002), each agent is characterized by two variables: its opinion X_i and its uncertainty U_i, both being real numbers. The opinions are drawn from a uniform distribution ranging between [-1, 1] and the uncertainty ranges between [0, 2]. The process of changing opinions is a process of continuous opinion dynamics. Around each agent's opinion lies an area of confidence in which he regards all opinions in this region as relevant and all others as irrelevant.

An agent interaction is calculated by first calculating the relative agreement between agents i and j by taking the overlap between the two agents' bounds h_{ij}, given by Eq. (1) (Deffuant *et al.* 2002):

$$h_{ij} = \min(x_i + u_i, x_j + u_j) - \max(x_i - u_i, x_j - u_j) \qquad (1)$$

Subtracting the size of the non-overlapping part, the total agreement between the two agents, Eq. (2):

$$h_{ij} - (2 * u_i - h_{ij}) = 2 * (h_{ij} - u_i) \qquad (2)$$

After calculating the relative agreement between the two agents h_{ij}; if $h_{ij} > u_i$, Xj and Uj are updated as shown in Eq. (3) and Eq. (4):

$$X_j = x_j + \mu\left[\left(\frac{h_{ij}}{u_i} - 1\right)(x_i - x_j)\right] \qquad (3)$$

$$U_j = u_j + \mu\left[\left(\frac{h_{ij}}{u_i} - 1\right)(u_i - u_j)\right] \qquad (4)$$

Where μ is defined as a constant parameter responsible for controlling the speed of population convergence.

2.2 Extremism and Moderation

Agents are considered extreme in their opinion when they are more confident and less willing to make compromises and negotiate, thus their uncertainty is lower. In fact, most people who have extreme opinions tend to be less convinced. On the other hand, people who have moderate initial opinions are considered more flexible and more willing to collaborate and negotiate with the other party. Therefore, moderate parties have a higher variability or uncertainty when they interact with the other agent (Meadows and Cliff 2012).

2.3 Project Delivery Methods

In each construction project, the contractual relationships between the owner, architect/engineer (A/E), contractor and management services are defined by a certain project delivery method. Recent project delivery methods (PDM) focus on establishing a structure of collaboration and cooperative behavior while eliminating the adversarial nature of traditional methods. The four typical project delivery methods shown in Table 1 are chosen to reflect the implications of each method on negotiations in the construction industry.

Table 1. Most frequent Project Delivery Methods.

Delivery Method	Implication
Design-Bid-Build (DBB)	Promotes extreme behaviors and lack of cooperation between the agents. When a traditional DM is chosen for a project, the gap between each agent's opinions on time extension or money compensation is relatively high and the variability range of each agent is narrow or almost zero which dictates extreme behaviors.
Construction Management at-Risk (CM at-Risk)	The agents are flexible and demonstrate moderate behavior where they are willing to negotiate and interact. Therefore, the gap between each agent's opinions is much smaller than the case of the DBB and the variability of each agent's opinion has a wider range expressing a willingness to cooperate.
Design-Build (DB)	The communication between agents and collaboration is enhanced through an early involvement of both parties which leads to a moderate type of agents involved.
Integrated Project Delivery (IPD)	The IPD method allows agents to cooperate from the very beginning and agree on potential causes of conflicts. This project delivery method marks the peak of collaboration and willingness to resolve disputes. The gap between the agents' opinions is very small and their opinions' variability ranges are very wide.

3 EXPERIMENTAL MODEL

3.1 Model Initialization

When the model is launched, the user will need to make few selections on the simulation window. First, the user needs to choose the project delivery method. The choice of the PDM will influence three different parameters and variables:

- The time convergence factor: This parameter indicates the speed of convergence between the agents based on the level of communication between them. Higher values of μ will speed up the negotiation between the agents and help them reach an agreement at a faster rate.
- Owner's opinion: The owner's opinion and the gap between his opinion and that of the contractor are largely influenced by the PDM. The owner's opinion is drawn from a uniform distribution with different boundaries that are in function of the contractor's opinion. The higher the level of communication and interaction between the agents, the smaller the gap between the agents' opinions.
- Agents' variability: The variability is an indication of the agent's attitude: character, negotiation preference and approach to problem solving. When the project delivery method provides an opportunity for the agents to better interact and exchange their opinions in an open setting, their variability will increase and they will be more open to hear the other agent's opinion.

After selecting the project delivery method, the user will have to select the dispute scenario. Several claim scenarios were drawn from AIA claim classification and mainly fall under two categories: excusable and/or compensable. According to AIA clause 8.3, circumstances beyond the contractor's control can result in an excusable delay, justifying an extension of the contract time. Excusable delays may be further classified as compensable or non-compensable. If the delay is deemed compensable the party will be entitled to additional compensation for the costs of delay, as well as additional time for contract performance. However, it is possible for a delay to be compensable without extending the contract time. Seven different dispute scenarios were used for the current version of the model. However, this list can be further extended to include other relevant scenarios. When the user selects the dispute scenario, the agent-based model generates the following:

- Classification of claim: If a claim classifies as excusable, agents will be negotiating time extension in terms of days. On the other hand, if a claim classifies as compensable, agents will negotiate money compensation, in US $.
- Severity level: Each agent whether it is the owner or the contractor has a high or low severity level that expresses his attitude towards a specific dispute scenario. If the dispute is caused by the owner, the severity level of the contractor will be high since he is directly affected by the actions of the owner while the severity level of the latter is low since he is the one causing the dispute. In cases where none of the parties are directly responsible for the dispute, for example during unexpected weather conditions, both agents' severity level is low. Having a high or low severity level will also affect the gap between the agents' opinions.

The user then needs to input the amount of days or money sought by the contractor agent in the simulation window. Several values were tested for both time and money to come up with results and test the logic behind the model. Values can also be chosen based on case studies of actual disputes that have occurred, to compare with actual data (Marzouk and Moamen 2009).

3.2 Negotiation Cycle

Each agent's opinion is drawn from a uniform distribution with varying boundaries. These boundaries depend on two main factors: the severity level of each agent and the project delivery method. The logic behind choosing each factor is related to the gap between the agents' opinions.

In fact, the more open and cooperative the agents are, the smaller the gap is. The agents' attitude is largely influenced by the nature of relation between them. These factors were found through trial and error analysis.

3.3 Convergence Check

After each negotiation cycle, the model checks for convergence. The latter is defined as the difference in agents' opinions that should not exceed 5%. When the owner's opinion falls within 5% of that of the contractor, the convergence and agreement between parties is reached. This specific difference in opinions was sought to be adequate after conducting a sensitivity analysis.

4 EXPERIMENTAL RESULTS AND DISCUSSION

The results obtained confirm the logic behind negotiations among cooperative and extreme parties. It was found that when the project delivery method is a Design Bid Build, parties tend to have a high gap between their opinions with low variability and thus they do not overlap in their opinions. Therefore, no negotiation cycle is initiated, and they will need further assistance in the form of an ADR method to initiate negotiation. As for the IPD delivery method, parties are more flexible and cooperative and that is confirmed through their close opinions and relatively high variabilities (AIA 2008). The parties will reach agreement within few negotiation cycles. In fact, in the case of excusable claims, results show that it takes one negotiation cycle to converge while it might take two negotiation cycles in the case of compensable claims to reach agreement. This can be explained by the fact that agents express higher severity levels, typically the contractor, especially when it comes to monetary compensation. Therefore, the gap between the agents' opinions will be bigger and thus it will take them more time to settle. As for other project delivery methods such as a CM at risk or a Design Build project, parties will sometimes converge but it will take them more negotiation cycles, typically four to five, than when having an IPD method.

The analysis of the results obtained also show how an extreme agent can bring a moderate agent to converge towards his opinion. In fact, when the variability of one agent is extremely low, the other agent's opinions are updated to minimize the gap. The results show that when the owner for example expresses very little variability in his opinion, it is the contractor that changes his opinion and tries to make the settlement. The latter proves the theory behind extremism and uncooperative behavior where the agent showing an extreme behavior is less willing to cooperate and negotiate with the other agent (Zhang *et al.* 2016). It is also interesting to note how the convergence time factor can influence the number of negotiation rounds that the agents go through before reaching settlement. In fact, when the project is delivered through a Design Bid Build type of contract, the speed of negotiations is much slower given the large gap in agents' opinions and the lack of communication between parties. For a Design Build project, agents will have to go through more negotiation rounds than the case where the agents negotiate under an IPD type of project. The latter is explained by the fact that for a design build, the time of convergence is expressed by a factor of 0.4 while for an IPD project, the time of convergence of 0.5 is higher and thus will bring the agents to converge at a much faster rate.

5 CONCLUSION

The presented agent-based model is a simulation of negotiations between contractual parties in construction projects. The results obtained show how the behavior and attitude of a party can largely affect the negotiation process and outcome. Not only is negotiation affected by the attitude and character of the agents involved but it is also influenced by the delivery method of the project and the level of intensity of the agent. Therefore, a successful negotiation requires a combination of factors: the flexibility brought by the delivery method such as the case of an IPD method, a low severity level expressed by the parties and a willingness to negotiate and cooperate expressed through non-extreme behaviors. Future work will address how having an ADR method in the case of negotiation failure encourages ongoing and timely dispute resolution and prevents from leaving disputes pending till the end of the project. The presence of these key elements will guarantee the attainment of a settlement within few negotiation cycles and will leave all parties satisfied.

References

Azar, E. and Menassa, C. C., Framework to Evaluate Energy-Saving Potential from Occupancy Interventions in Typical Commercial Buildings in the United States, *Journal of Computing in Civil Engineering*, 28(1), 63-78, 2013.

Deffuant, G., Amblard, F., Weisbuch, G., and Faure, T., How Can Extremism Prevail? A Study Based on the Relative Agreement Interaction Model, *Journal of Artificial Societies and Social Simulation*, 5(4), 2002.

Long Chen, H., Empirical Behavioral Analysis of Project Contractors' Supply-Chain Payment Terms, *Supply Chain Management: An International Journal*, 17(3), 277-289, 2012.

Lorenz, J., Fostering Consensus in Multidimensional Continuous Opinion Dynamics Under Bounded Confidence, *Managing Complexity: Insights, Concepts, Applications*, 321-334, 2008.

Marzouk, M. and Moamen, M., A., Framework for Estimating Negotiation Amounts in Construction Projects, *Construction Innovation*, 9(2), 133-148, 2009.

Meadows, M. and Cliff, D., Reexamining the Relative Agreement Model of Opinion Dynamics, *Journal of Artificial Societies and Social Simulation*, 15(4), 4, 2012.

Stephan, K., and Menassa, C. C., Modeling the Effect of Building Stakeholder Interactions on Value Perception of Sustainable Retrofits, *Journal of Computing in Civil Engineering*, 29(4), B4014006, 2014.

Yuan, H. and Ma, H., Game Analysis in the Construction Claim Negotiations, *Procedia Engineering*, 28, 586-593, 2012.

Zhang, S., Zhang, S., Gao, Y., and Ding, X., Contractual Governance: Effects of Risk Allocation on Contractors' Cooperative Behavior in Construction Projects, *Journal of Construction Engineering and Management*, 142(6), 04016005, 2016.

Responsible Design and Delivery of the Constructed Project
Edited by Abdul-Malak, M., Khoury, H., Singh, A., and Yazdani, S.

ALTERNATIVE MEDIATION OPERABILITY MECHANISMS UNDER STANDARD CONSTRUCTION CONTRACT CONDITIONS

MOHAMMAD BARAKAT, MOHAMED-ASEM ABDUL-MALAK, and HIAM KHOURY

Civil and Environmental Engineering, American University of Beirut, Beirut, Lebanon

The use of alternative dispute resolution (ADR) techniques has been on the rise, particularly as part of the contemporary, staged-type mechanisms forming part of the pertinent conditions of the construction contract. Among the several ADR methods in use, mediation is the method widely found to be explicitly or implicitly incorporated in any such mechanisms, with its use being often stipulated as condition precedent to ultimately resorting to arbitration. This paper investigates the varying statuses of mediation as prescribed under three American-based standard contract conditions, in comparison with its implied status under international such conditions. The within and in-between comparative analyses revealed distinct differences and some resemblances in respect of several concerned attributes including: (1) the staging (sequence step) of the mediation process within the involved claim/dispute resolution timelines, (2) the condition(s) for triggering such a process, (3) the provisions governing its length, and (4) the manner with which it may be ended. The paper concludes with inferences as to the characteristics that may be hypothetically viewed as more conducive to the effective employment of such an ADR technique in resolving disputes.

Keywords: AIA, ConsensusDocs, EJCDC, FIDIC, Claims, Disputes, ADR, Resolution.

1 INTRODUCTION

Claims and disputes are almost inevitable in construction projects, and they must be dealt with according to the mechanisms prescribed within the adopted contract conditions. Construction claims may escalate into disputes between the contracting parties, potentially impacting their performances and the successful delivery of the construction project. To avoid such consequences, most standardized forms of contract conditions enforce, or make room for, an alternative dispute resolution (ADR) technique to afford the parties the chance to reach agreement prior to, alternatively, referring the matter to arbitration. Mediation outweighs other ADR techniques, in that it is a well-structured process that the parties may use in endeavoring to reach agreement in respect of the matter being in dispute (Zlatanska and Fawehimni 2016). However, mediation is not always the end result (Goodkind 1988). That is; the mediator is not in charge of the outcome, and his responsibility is merely to administer a process intended for seeking a resolution (Sgubini *et al.* 2004). The mediation process has proven to reduce the cost of resolving disputes as compared to resorting to arbitration (Goodkind 1988). In addition, it serves maintaining a good relationship between the parties, conducive of establishing future business collaborations (Goodkind 1988). Although mediation is a commonly stipulated ADR process, even the different standard forms of contract conditions reveal operational variations as to how the mediation mechanism is to be administered.

2 SCOPE AND METHODOLOGY

This research work sheds light on "mediation" as an ADR mechanism that may be prescribed as part of the claim/dispute progression timeline. The aim is to identify the attributes corresponding to the administration of such a process. The adopted methodology involved: (1) reviewing the clauses that prescribe the mediation mechanism under four different standard conditions of contracts, namely those by the American Institute of Architects (AIA), ConsensusDocs, the Engineers Joint Contract Documents Committee (EJCDC), and the International Federation of Consulting Engineers (FIDIC); (2) extracting its staging sequence, triggering conditions, governing time bars, and manners with which it may be ended; and (3) conducting an in-depth comparative analysis to highlight the potential differences and similarities among the examined mediation mechanisms.

3 MEDIATION SEQUENCES

The aforementioned standardized conditions were examined to check the sequencing of the mediation stage within the corresponding claims/disputes timelines. The extracted timelines are as illustrated in Figure 1.

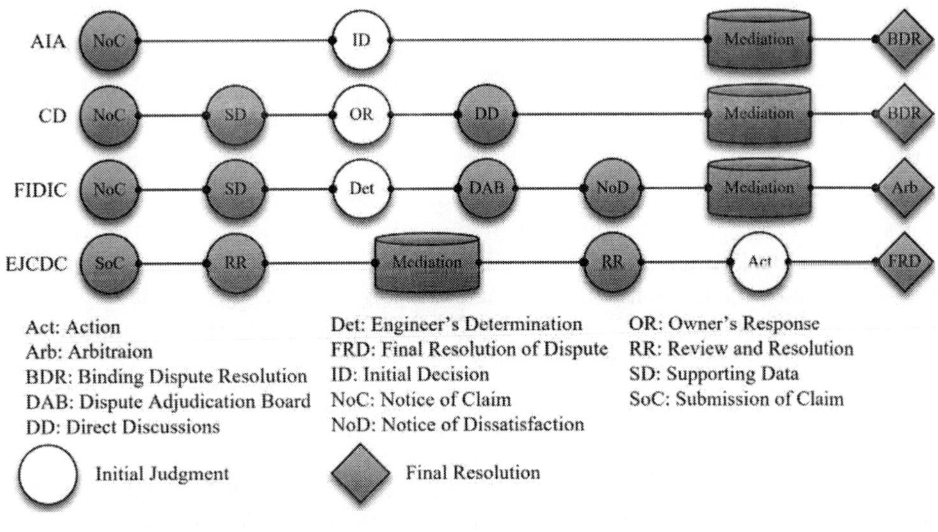

Figure 1. Mediation sequencing.

It can be seen that the mediation process is specified under EJCDC as an option nested as part of the "review and response" step, and prior to the issuing of an (initial) action by the other party (owner). Unlike the case of EJCDC, the mediation processes under AIA, ConsensusDocs, and FIDIC are found to be staged at the end of the claim/dispute timeline. That is, the mediation process is interjected with after the rendering of an initial judgement in respect of the matter in demand, and just prior to the initiation of arbitration. This sequence step of the mediation process gives the contracting parties the chance to endeavor to amicably resolve the matter in dispute prior to alternatively proceeding to arbitration. It is to be also noted that only FIDIC has two possible actions preceding the triggering of the amicable settlement stage, during which mediation can normally be entertained. More peculiar observations for each of the examined standards are further deduced in the following section.

4 OPERATIONAL VARIATIONS OF MEDIATION MECHANISMS

Different standard forms of contract conditions are found to incorporate dissimilar attributes that govern how the mediation process, when opted for, can be operated. These attributes have been analyzed for each examined standard and deductions made as discussed below.

4.1 AIA Attributes

According to "Sub-article 15.2 Mediation", the parties must refer claims to mediation, which is viewed as a condition precedent to being eligible to resort to arbitration afterwards (AIA 2017). The underlying mediation process is presented in Figure 2.

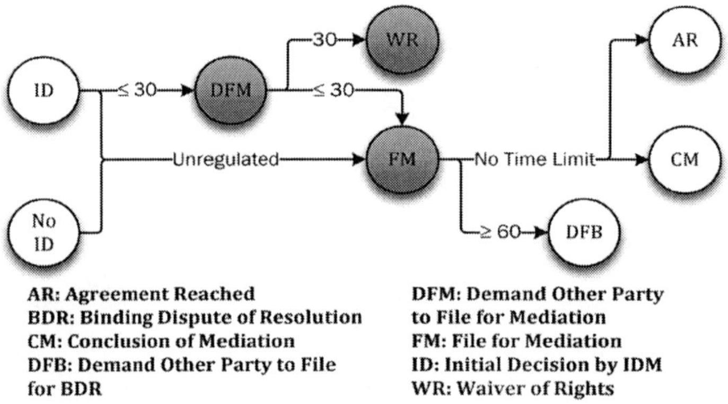

Figure 2. AIA mediation's mechanism.

Initially, claims must be submitted to the other party and to the initial decision maker (IDM), whereby the latter must render an initial decision specifying whether: (a) the claim is accepted or rejected – whether fully or partially –, or (b) the IDM is unable to resolve it. The initial decision acts as a condition precedent to mediation unless it is not rendered within 30 days from the day the IDM receives the notice of claim. In the case of rendering an initial decision, the owner or the contractor is able, within 30 days, to demand the other party to file for mediation. If the demanded party fails to file for mediation within this time bar, both parties then waive their rights to pursue mediation and binding dispute resolution in regard to the rendered initial decision. On the other hand, either party may file for mediation at any time if: (1) neither party filed a demand for the other party to file for mediation, or (2) if mediation was triggered upon the lack of an initial decision. Once mediation is filed, the parties shall endeavor to reach agreement, which could take place at any point in time. If mediation is deemed not successful, the mediator has the option to conclude it at any time. However, the option of reaching agreement or concluding mediation, after 60 days of mediation initiation, is contingent upon mediation not having been already terminated. In other words, either party may, after 60 days, issue a demand for the other party to file for binding dispute resolution, which automatically terminates the mediation process. On the other hand, the binding dispute resolution provision allows either party to file for arbitration at any time within the mediation process, but no earlier than concurrently with filing for mediation. In that case, arbitration would automatically be initiated unless agreement is reached prior to the expiry of the 60-day period of mediation.

4.2 ConsensusDocs Attributes

Claims arising on projects that adopt the ConsensusDocs contract conditions must be submitted to the owner (ConsensusDocs 2017). Subsequently, the owner shall render a response, which, if not satisfied with, can cause the contractor to initiate discussions. The discussions are triggered upon the mutual agreement of both parties in order to seek a resolution. If agreement is not reached through discussions, then the claim must either be referred to mitigation or mediation. If mediation is chosen, then the claim proceeds according to the mechanism set forth in "Sub-article 12.4 Mediation", which is depicted in Figure 3.

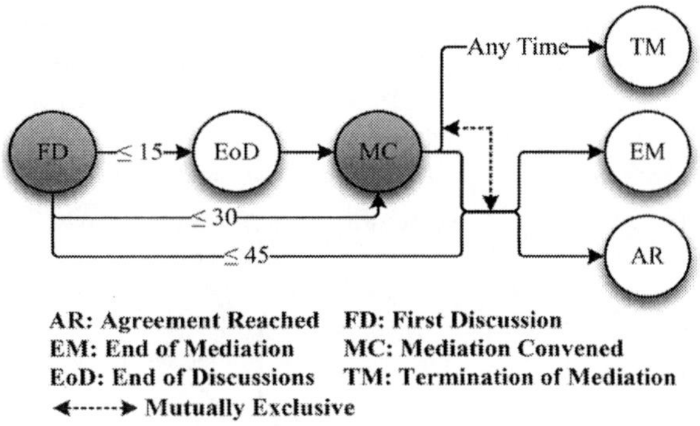

Figure 3. ConsensusDocs mediation's mechanism.

Upon the end of discussions, which can last up to a maximum of 15 days, mediation shall be convened within thirty days from the date of first discussions. Mediation shall end within 45 days from the date of first discussions, with or without agreement being reached. To be noted is that either of parties, the owner or contractor, may directly terminate mediation after the first session.

4.3 EJCDC Attributes

Another widely endorsed standard form of contract conditions is that by EJCDC. Under these conditions, parties may mutually agree to refer matters in disputes to mediation, as stated in "Provision 12.01.D Mediation" (EJCDC 2013). The mediation mechanism can be triggered at any time within the ninety-day period allotted for the review and resolution stage, which is triggered upon the submission of a claim and its supporting documents. The corresponding mediation mechanism can be summarized as shown in Figure 4.

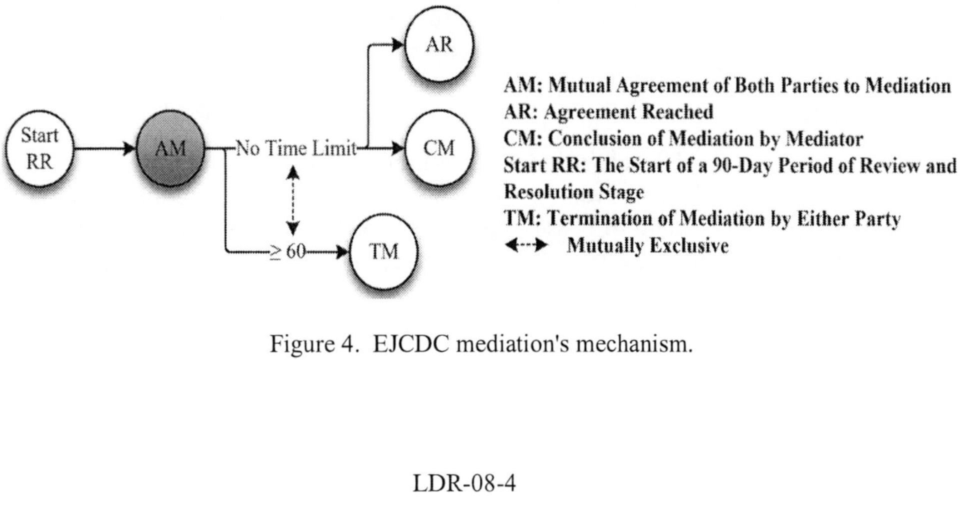

Figure 4. EJCDC mediation's mechanism.

Once mediation is initiated, the claim review and response process shall be stayed, while the parties endeavor to reach agreement. However, the mediator can conclude the mediation at any time if the process is deemed unsuccessful. The options of reaching agreement and concluding mediation are the only ones available within the first 60 days. Beyond this period, a third option avails, where either party may unilaterally terminate mediation at any time.

4.4 FIDIC Attributes

Moving to the international type of standard contract conditions, the contracting parties on a project, where the FIDIC conditions of contract are adopted, have the option of amicably resolving an arisen dispute prior to referring it to arbitration. According to "Sub-clause 20.5 Amicable Settlement" and upon the issue of a notice of dissatisfaction by either party in respect of a rendered adjudication decision, the parties are afforded a period of 55 days, during which the reaching of an amicable settlement may be attempted (FIDIC 2000). To this end, mediation is the ADR technique likely to be adopted at such an advanced stage of the claim/dispute progression. The corresponding mechanism can be illustrated as presented in Figure 5.

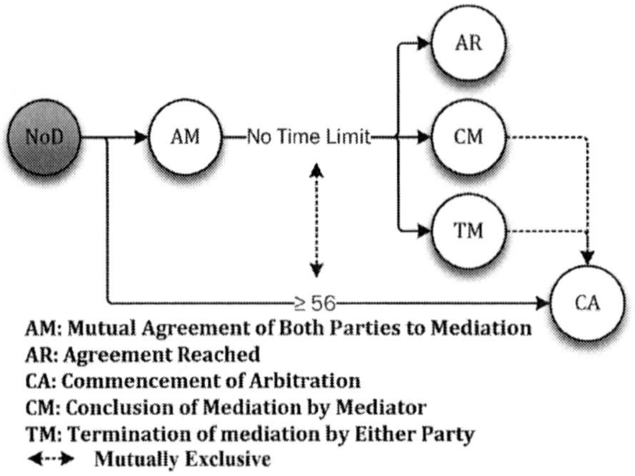

Figure 5. FIDIC mediation's mechanism.

As it can be inferred, the amicable settlement period is triggered upon the issuing of a notice of dissatisfaction by either party. Subsequently, the parties may mutually agree to mediation. Once the mediation process is initiated, agreement could be reached at any time. However, this option is contingent upon the mediation process not having already been concluded by the mediator or terminated by either party at any time. Either party may commence arbitration on or after the 56th day from the start of the amicable settlement period, unless the parties otherwise agree to an extension of the said period.

5 DISCUSSION

Although each of the discussed standard forms of contract conditions either integrates or allows for the adoption of a mediation mechanism within its corresponding claim/dispute progression timeline, each mechanism is found to possess unique properties. Firstly, the mediation process under the AIA and ConsensusDocs forms is found to be a condition precedent to the binding dispute resolution stage. Secondly, mediation is stated to be optional under the EJCDC conditions, where the parties need to mutually agree to its adoption. Thirdly, the mediation process is not explicitly referred to under the FIDIC conditions. However, parties adopting this standard may, if mutually agreed, resort to any one or more of the available ADR techniques, including, but not limited to, facilitation, conciliation, and/or mediation. Fourthly, all prescribed mediation stages have similar ending mechanisms. That is, the mediator can conclude the mediation process at any time if deemed unsuccessful, or either party has the discretion of alternatively terminating the process. For the latter option, parties adopting the AIA or EJCDC conditions may terminate mediation after the expiry of a 60-day period, whereas under the FIDIC and ConsensusDocs standards, either party may terminate mediation at any time. Finally, the specified mediation steps are found to be assigned varying time periods. Under the AIA and EJCDC conditions, mediation has a time limit of 60 days, which is extendable provided that neither party exercises termination of the process. Similarly, mediation has a duration figure of 55 days under the FIDIC standard, which is extendable upon mutual agreement of both parties. As for the ConsensusDocs, the duration of the mediation process could range between a minimum of 15 days and a theoretical maximum of 45 days.

6 CONCLUSION

The claim/dispute progression timelines of the four examined standard conditions are found to involve mediation mechanisms that have different attributes, in regard to the stage's location, initiation and ending triggers, and assigned time limits. The paper examined these mechanisms and highlighted their corresponding similarities and differences. The performed comparative analysis revealed that the ConsensusDocs standard lies on one end of the scale, where it implements mediation as a condition precedent to arbitration and enforces it automatically without the need to be filed for or mutually agreed upon. On the other end comes the FIDIC conditions, which do not explicitly refer to mediation; rather, make room for this and/or other ADR techniques to be employed as part of the stipulated amicable settlement period. In-between, the mediation process under the AIA standard comes right after that of ConsensusDocs, wherein mediation is considered as a condition precedent to the binding dispute resolution stage, but requiring that it be initiated by one of the parties. Finally, the mechanism of mediation under EJCDC is believed to be falling in the third place, prior to that of FIDIC, with mediation specified as an option whose employment is contingent upon the mutual agreement of the contracting parties.

References

AIA (American Institute of Architects), *General Conditions of the Contract for Construction*, A201, Washington, DC., 2017.

ConsensusDocs, *Standard Agreement and General Conditions between Owner and Contractor*, Arlington, VA., 2017.

EJCDC (Engineers' Joint Contract Document Committee), *Standard General Conditions of the Construction Contract*, 2013.

FIDIC (International Federation for Consulting Engineers), *The FIDIC Contracts Guide,* Thomas Telford Ltd, Geneva, 2000.

Goodkind, D. R., Mediation of Construction Disputes. *Journal of Performance of Constructed Facilities,* 2(1), 9-12, 1988.

Sgubini, A., Prieditis, M., and Marighetto, A., Arbitration, Mediation and Conciliation: Differences and Similarities from an International and Italian Business Perspective, *Bridge Mediation LCC.*, 2004.

Zlatanska, E., and Fawehinmi, F., Mediation and Conciliation: In Pursuit of Clarity, *International Journal of Arbitration, Mediation and Dispute Management*, 2016.

Responsible Design and Delivery of the Constructed Project
Edited by Abdul-Malak, M., Khoury, H., Singh, A., and Yazdani, S.

MEDIATION TACTICS AND EFFECTIVENESS IN DISPUTE RESOLUTION

MOHAMMAD HASAN SENAN, NADINE ALZAGHRINI, and ISSAM SROUR

Dept of Civil and Environmental Engineering, American University of Beirut, Beirut, Lebanon

Dispute occurrence on construction projects is almost inevitable. If not addressed adequately, disputes can cause contract termination and loss of benefits for all parties involved. Among various dispute resolution methods, mediation is gaining popularity in the construction industry as it satisfies the desired degree of control of both parties among other factors (e.g., time, cost). Mediation is a cooperative problem-solving method that benefits from the expertise of a neutral third party known as the mediator. Well-defined tactics such as creating trust atmosphere, tackling the core of dispute, and using advanced integrative approach for solution determination are essential to reduce the gap between disputants and ensure effective discussions. The aim of this paper is two-fold. First, it investigates common mediator tactics and relates them to the likelihood of having a successful mediation. The San-Ramos Channel case study is used to illustrate the efficiency of mediator tactics. Second, this paper takes a closer look into the use of mediation in Lebanon as an example of a country with minimal regulations on mediation. Barriers for an effective implementation of mediation such as the role and the influence of the engineer on the construction process and the mediator's lack of technical expertise are discussed and recommendations are derived.

Keywords: ADR, Construction, Disputants, Informal settlement, Lebanon, Mediator.

1 INTRODUCTION

Construction projects are recognized for their considerable costs and long duration spans. Even the most elaborate planning often fails to capture all eventualities in a project, thus leading to disputes among various project stakeholders. Contract incompleteness along with technical and human related factors are at the root of these construction disputes. These factor groups encompass uncertainty and risk, collaborative conflict, deficiency, ambiguity and inconsistency of contracts along with affective conflicts and opportunistic behavior (Cheung and Pang 2013). When a similar incident triggers a claim, the concerned party submits a written notice in accordance with relevant contractual clauses and within stated time frames. If all contractual and administrative processes for filing a notice are met, the engineer (i.e., the contract administrator) turns to set a determination on the issue. The disagreement escalates to a dispute when any party is dissatisfied with the engineer's judgment and the parties fail to solve the issue in a practical, well-timed, and mutually acceptable manner (Love *et al.* 2010).

Taking the case to court is the norm for conflict resolution and is known as litigation. Increased costs and the required time periods added to the reckoning of relationships encouraged parties to seek amicable settlements (Haugen and Singh 2015). These approaches provide an alternative to the adversarial pattern sensed in court and are referred to as Alternative Dispute

Resolution (ADR) techniques. Mediation stands out as a quick technique that aims to solve the dispute amicably through fruitful discussions managed by a neutral third party known as the mediator. Reducing the gap between the parties and ensuring effective communication are dependent on the strategies and tactics used by the mediator, which vary according to the complexities of the cases faced. This paper identifies the common tactics used by mediators to help disputants reach a sustainable agreement. It also discusses the current status-quo of mediation in solving disputes in the Lebanese construction sector.

2 METHODOLOGY

Mediation is distinguished by the presence of a neutral third party whose tactics are essential for the settlement's success. The first goal of this paper is to identify the common strategies used to ensure a sustainable settlement. A literature review combined with an analysis of a case study published by the U.S. Army Corps' of Engineers are used to highlight the strategies used by mediators.

The second objective of this paper is to evaluate the status-quo of dispute settlements in the Lebanese construction sector. To serve this purpose, a senior mediation consultant at the Lebanese Center for Mediation and Arbitration (LAMC) with four years of experience and a contract administrator with 20 years of experience were interviewed. The former is an accredited mediator and the latter is a chartered arbitrator. The interviews focused on the choice of ADR in the Lebanese construction industry, along with the perceived barriers to the implementation of mediation. The interview questions are attached in the Appendix.

3 MEDIATOR TACTICS

The ability to read and evaluate evidence and claim particulars is a key characteristic of the engineer, expert, or judge who issues a fair and binding decision. Likewise, the skills and attributes of the mediator are piers to a successful and efficient mediation.

Codes of conducts and scholars require mediators to be impartial and neutral. Although used interchangeably, neutrality implies not taking sides whereas impartiality requires treating all sides equally. Other norms include mainly competence, disclosure of any conflict of interests with the parties, and ensuring party self-determination (Shapira 2016). Mediator competence implies a thorough comprehension of conflicts, negotiation, and communications principles along with an understanding of the mediation process and regulations. Disclosure of conflict of interests proves the intentions of the mediator not to favor or diminish the interests of a party with whom they might have a personal or past/existing/future business relationship. The intensity, duration, and frequency of such relationships are critical factors to determining whether the neutral party must withdraw or continue with the mediation (Kovach 2014). These attributes qualify the mediator to commence with the mediation sessions to help the parties, using useful tactics, build their own agreement and determine what is acceptable to them.

3.1 Key Mediator Tactics

The presence of a qualified mediator as neutral party is essential if the parties in dispute are not able to reach an agreement themselves. The mediator, who's responsible for controlling the session, maintaining a friendly atmosphere, and reducing the gap between the two parties, uses various tactics to reach an amicable settlement. The mediator aims in the session to create a trust atmosphere between the parties and tackle the core of the dispute in order to solve the problem using an advanced integrative approach.

3.1.1 *Creating an atmosphere of trust*

The mediator, from the beginning of the mediation, works on creating an atmosphere of trust with both sides in order to help them reach an agreement. In his opening statement, the mediator explains to the audience the importance of mediation as a technique where parties can collaborate to find common grounds rather than have a solution imposed on them by an arbitrator or judge.

The mediator also uses his opening statement as a chance to gain the trust of the parties by detailing her/his experience in mediation and the success rate reached in mediations. Furthermore, the mediator is ethically required to disclose any past professional or personal relationships with any of the parties, if they exist. In this case, continuing the mediation is subject to obtaining the parties' consent. Disclosing relationships prior to mediation is important to gain parties' trust and guarantee a stable progress and a permanent solution.

In addition, the parties and/or the mediator can ask for a caucus meeting. A caucus meeting is a private meeting between the mediator (or assistant mediator) and a party. The goal of this meeting is to learn privately the concerns, objectives, and other relevant information (e.g., hidden interests, walk-away points), reduce the level of tension, and allow room for option generation (Twomey 2006). The trust established between the mediator and the parties due to the level of openness and transparency is fundamental to better guide the discussions and reduce the gap between the parties in dispute.

3.1.2 *Tackling the core of dispute*

The mediator's goal is to assist the parties in reaching a solution that will be abided by and therefore is considered sustainable. Thus, the mediator should determine the hidden interests, fears, and agendas of the parties in the opening statements, questions phase, and in caucus meetings. This process begins immediately after the opening statements of the parties. The mediator checks all the available and relevant contractual clauses and any other reachable data about the parties' firms. Based on a clear understanding of this data, the mediator prepares a list of open-ended questions that urges the party to explain its position for both the mediator and the other party. Any shown evidence or revealed information in the mediation sessions should remain confidential and cannot be referred to in court (Twomey 2006). The core of the dispute is then made clear paving the way for the mediator to start the problem resolving phase.

3.1.3 *Using advanced integrative approach*

Tension is often present in disputes. An experienced mediator transforms a tense debate into a joint problem-solving discussion which pushes the parties towards proposing new thoughts rather than splitting the pie. This technique allows for a quiet discussion to take place taking into consideration the objectives, insecurities, and interests of each party. Hence, the process of finding a solution is made easier since each party finds itself proposing what it finds acceptable while observing the reaction and ideas of the other party. While being in control of the discussion, the mediator examines the progress reached compared to the beginning of the mediation session, as s/he allows the parties to propose options and ideas that will later be crafted into an agreement.

This joint problem-resolving discussion is controlled by the mediator who pitches in every now and then in order to ensure that a friendly atmosphere remains while the discussion continues on the right track. The mediator may also intervene to show the progress reached by recapitulating what has been agreed on. The main goal of the mediation session is to reach an acceptable agreement. However, the mediator may also look for improving the parties' ability to reach solutions together for future challenges through these joint problem-solving discussions.

3.2 The San Ramos Bypass Channel Case Study

The San Ramos Bypass Channel is a case study published by the U.S. Army Corps' of Engineers. It illustrates how the ADR techniques can be used to resolve disputes between a public entity and private corporations.

The project dates back to 1986 when the Corps' decided to build a 5-mile long channel that runs through Walnut Creek city. The contractor, Brutoco, excavated a portion of the channel and stockpiled the material above high-pressured petroleum pipelines which raised concerns that the area may collapse, rupturing thus the pipeline. Additionally, the Corps' believed that slopes were cut steeper than what was designed and ordered Brutoco to build an earthen buttress.

The dispute was on assigning the responsibility for the added costs and delays of building the buttress. The Corps' put the full responsibility on Brutoco for choosing an inadequate area to place extra material. Brutoco defended itself by claiming that the Corps' authorized using that area for backfilling.

Brutoco filed a $3.1million total cost claim when no agreement was reached in negotiations on the modification order sum. The Corps' resorted to mediation in order to solve the dispute quickly and at a reduced cost. Brutoco accepted mediation with similar hopes for securing faster payments.

The mediator used the opening statements of both parties as a chance to discover the major issues in dispute according to each party since Brutoco did not submit a brief. The parties' trust in the mediator's capabilities and possibility of reaching a settlement increased when he recapped and tackled the core issue of the dispute and the points both parties agree to as well. However, the mediator was aware of the high level of tension between the parties that resulted from improper communication during project lifecycle. Hence, he had no option other than having separate "caucus" meetings with the parties to investigate further about their own risk analysis, interests, and possible settlement packages. The mediator was able, through the established trust, to discuss the weaknesses and strengths of each party's case and reduce the gap. He also motivated both parties to reach a solution not only through explaining thoroughly the implications of taking the case to litigation on cost and time, but also through proposing a deadline to reach an agreement.

The mediator's integrative approach, coupled with the intention of both Brutoco and the Corps' to reach an agreement, resulted in innovative solutions proposed by both parties. Those solutions helped in settling the main dispute in addition to other minor disputes. The mediator led the discussion through transferring those possible settlement packages while shuttling back and forth between the parties. These indirect fruitful discussions between the parties resulted in an agreement on a $1.15 million figure.

The agreement allowed the mediator to bring the parties together to draft and sign the agreement. In their evaluation, the Corps' and Brutoco praised the mediator's skills and tactics that made reaching the settlement and avoiding litigation possible (Susskind *et al.* 1992).

4 DISPUTE RESOLUTION IN LEBANON

To understand the applicability of mediation to countries with minimal regulations governing ADR, the case of Lebanon was considered. The Lebanese Arbitration and Mediation Center (LAMC), founded in 1995, is specialized in resolving disputes. Mediation has been integrated within LAMC since 2012 through a cooperation agreement signed between the Chamber of

Commerce, Industry, and Agriculture of Beirut and Mount Lebanon and the International Finance Corporation, a member of the World Bank Group.

As such, it can be inferred that mediation is a newly-introduced ADR to the Lebanese market and is not widely used yet. According to the interviewed mediation consultant, only one case has been directed to mediation and ended being resolved using arbitration. In 2014, the center passed a law to the Lebanese Parliament to incorporate mediation into the local legal system and create a statutory framework governing mediation procedures. To date, the law has not been adopted.

4.1 Barriers to Mediation Implementation in Lebanon

Recognizing common norms and practices in Lebanon is essential to understand the dispute resolution landscape. According to the interviewed contract administrator, FIDIC 1989 and 1992-reprinted contracts are the most widely used standard forms of contracts in Lebanon. Clause 67 of these contracts encourages the engineer to issue a decision should a dispute arise. Amicable settlement and arbitration can then follow if the issue is not resolved. Unfortunately, many project owners who use FIDIC contracts remove the clauses relating to new methods for clash resolution such as the dispute adjudication board, and rely on the traditional method for dispute resolution through engineer's determination and arbitration.

This strong attachment to traditional clash settlement methods is attributed by the interviewees to the grandiose role of the engineer, inherited before contracts were introduced to the Lebanese construction sector. The engineer, who is often the designer and the client representative, must authorize all changes even those coming from technical consultants. In the early 1990s, a megaproject called "Solidere" aiming at reconstructing the capital city's central business district was the first project to introduce international contracts in Lebanon. Contractors, who were used to a well-defined relationship with the designer, failed to detect initially the importance of these contract documents. Soon after, competition increased with the appearance of new contracting companies and profit margins began to drop. As a result, contractors were no longer able to blindly commit to every decision requested by the engineer. At this stage, contractors realized the value of the documents signed early on in the project and got a clearer understanding of their rights and duties as displayed in the contract.

Despite this progress, innovative approaches such as mediation and other forms of ADR techniques are still not implemented in Lebanon for several reasons, according to the mediation consultant. First, engineers remain the "rulers" of the projects and the point of reference for every major decision. As such, they prevent mediators from entering the construction industry by enhancing informal settlements on and off the site, thereby refuting a need for another third neutral party. Contractors, on the other hand, needed to conform to this decision as they wanted to maintain positive relationships with the engineer to avoid designers lobbying. Black-listed contractors endured great losses as they were no longer able to attain important projects. Furthermore, mediators in Lebanon are typically lawyers who do not possess the technical expertise needed and thus could not satisfy the needs of the parties. In contrast, designers such as the firm for which the interviewed contract administrator works developed a new contract administration department, composed of technical experts knowledgeable with dispute resolution methods and tactics. As such, mediation did not seem to offer any added benefits.

5 CONCLUSIONS AND RECOMMENDATIONS

Mediation is gaining popularity as a non-binding process that allows the disputants to collaborate and craft their own solutions. This amicable settlement is made possible thanks to the assistance of a mediator, a neutral third party who seeks to create an atmosphere of trust and to tackle the core of disputes with an integrative approach for dispute resolution.

Settling disputes through mediation is uncommon in developing countries like Lebanon despite the launch of a mediation center. Enhancing mediation in Lebanon requires several steps. First, mediation must be better introduced and advertised to decision makers. The engineers drafting contracts have to be convinced to add a clause on resolution through mediation as a final resort before arbitration. Second, the panel of accredited mediators must be limited to reputable persons in the construction industry, known for their technical expertise and impartiality throughout the years. This is an essential step for earning the trust of the stakeholders in the process. These mediators should be familiar with substantive laws, procedural requirements as well as drafting a correct and a binding settlement agreement.

Appendix: Interview Questions

1. What methods of dispute resolution are common in the contracts adopted in Lebanon?
2. What is the process commonly used for informal settlements?
3. How familiar are you with the Lebanese mediation center?
4. What are the main barriers for choosing mediation in Lebanon?
5. What are the main causes for mediation failure in Lebanon?
6. How can mediation be enhanced in the local construction market?
7. What qualifications does a person need to be assigned as a mediator?

References

Cheung, S., and Pang, K., Anatomy of Construction Disputes, *Journal of Construction Engineering and Management*, 139(1), 15-23, January, 2013.

Haugen, T., and Singh, A., Dispute Resolution Strategy Selection, *Journal of Legal Affairs and Dispute Resolution in Engineering and Construction*, 7(3), 05014004, April, 2015.

Kovach, K., Mediation in a Nutshell, West Academic Publishing, New Mexico, 2014.

Love, P., Davis, P., Ellis, J., and On Cheung, S., Dispute Causation: Identification of Pathogenic Influences in Construction, *Engineering, Construction and Architectural Management*, 17(4), 404-423, February, 2010.

Shapira, O., A Theory of Mediators' Ethics: Foundations, Rationale, and Application, Cambridge University Press, Cambridge, 2016.

Susskind, L., Babbitt, E., and Hoffer, D., Alternative Dispute Resolution Series: Brutoco Engineering and Construction Inc. Case Study #7, Institute for Water Resources, January, 1992. Retrieved from www.iwr.usace.army.mil/Portals/70/docs/iwrreports/92-ADR-CS-7.pdf on November 15, 2017.

Twomey, R. F., Mediation and its Merits as an Alternative Method of Employer-Employee Dispute Resolution, *Competition Forum*, 4(2), 414, July, 2006.

Responsible Design and Delivery of the Constructed Project
Edited by Abdul-Malak, M., Khoury, H., Singh, A., and Yazdani, S.

INVESTIGATING SITE-RELATED NOTICES IN CONSTRUCTION CONTRACTS

SALAM KHALIFE and MOHAMED-ASEM ABDUL-MALAK

Dept of Civil and Environmental Engineering, American University of Beirut, Beirut, Lebanon

Notices issued under the construction contract serve several purposes, the most significant of which is to facilitate the progression of the works. This is done through establishing necessary communications among the project participants by acting in accordance with the callings of the contract conditions. Through an in-depth examination of the notices stipulated under the standard conditions for the construction contract issued by the International Federation of Consulting Engineers (FIDIC), it was found that 43 out of 93 encountered notices deal with issues related to activities taking place on the construction site. The purpose of this paper is to present a detailed analysis of these notices and highlight their operability properties. The analysis focuses on the various characteristics pertaining to this vital group of notices, namely: their types, the participants involved in their administration, their related time bars, and the concerned aspects of the site activities. Results of the performed analysis are twofold: (1) the majority of these notices are found to be of the structured (complex composition) type; and (2) the contractor has the larger share in triggering their associated sub-clauses, and to do so, for the majority of these notices, within numerically stipulated time bars. The main contribution of the presented work is in providing a focused reference that can help contract administration practitioners, dealing with site-related matters, improve the current state of practice.

Keywords: Works, Claims, Execution, Performance, Testing, Quality, Measurement.

1 INTRODUCTION

Construction sites are characterized by being dynamic work environments. A vast number of daily activities and tasks are performed by project participants driven by having to meet tight time schedules, stringent specifications and fixed budgets. The complexity of the work, which is usually associated with such conditions, is further increased due to the need to be fully compliant with the requirements made under the conditions of the construction contract. Notice provisions, along with other provisions described in the construction contract, shall be referred to almost on daily basis during the construction phase. Contract administrators attempt to well comprehend the circumstances that warrant the need to issue notices in order to fulfill the contractual obligations and protect the parties' rights. However, the complexity of site-related works, the insufficient communication between site workforce and contract administrators, and the shallow understanding of the related provisions impede the due performance of the parties under the contact.

2 LITERATURE REVIEW

Construction contracts have been gradually developing, and their content is becoming more sophisticated by hosting new provisions (Thomas 2001). The provisions included in the contract stipulate the procedures for communicating about certain events or situations, specifically those that might affect the project's cost and time (Thomas *et al.* 1990, Aibinu 2009). As construction contracts are argued to have a major impact on project success (Von Branconi and Loch 2004), it is thus important that clauses within the contract are clearly expressed to avoid disputes resulting from misunderstandings (Eggleston 2004). Additionally, it is argued that the behaivor of the parties on a project is also shaped through the contract stipulations (Von Branconi and Loch 2004). Consequently, the parties shall be well aware of the requirements incorporated under the contract clauses in order to understand their rights, fulfill their obligations, and avoid being in any breach of the contract.

During the execution of the works, numerous occasions and events require that reference be made to the contract to figure out what actions are to be taken. In fact, professionals are often faced with the need to interpret construction contracts (Thomas 2009). Contractual issues, including compliance with the contract and clause-related issues, are major difficulties encountered during work execution (Sertyesilisik 2010). As such, there is a prime need to explore all kind of contractual obligations related to the work taking place on site, specifically in relation to notice provisions. It is important that practitioners fully grasp all circumstances and situations triggering the need to issue notices, which form a major means of communication during the construction process and facilitate the proper progression of the works.

3 RESEARCH SCOPE AND METHODOLOGY

The research reported in this paper focuses on the notices concerned with the activities taking place on construction sites and the notices' various characteristics. The research methodology included: (1) identifying and categorizing the notices found to be connected to site-related activities and (2) further analyzing these identified notices with respect to their types, the party issuing them, the time-bar governing their issuance, and their related prerequisite notices or actions.

4 SITE-RELATED NOTICES UNDER THE 1999 FIDIC CONDITIONS

An in-depth review of the FIDIC standard conditions for the construction contract (FIDIC 1999) revealed that 43 out of the total of 93 identified notices are found to be associated with site-related activities (Abdul-Malak and Khalife 2017a). These were classified into eight site sub-aspects including: accessibility, conditions, material and equipment, execution of works, inspections/tests, acceptance of works, measuring works, and taking over. These classifications show that dealing with site-related notices is spread along the project's construction phase, starting with the possession of site and up to the taking over of the built facility, then continuing through the defects notification period that leads to the final completion and contract close-out process.

The sub-clauses calling for site-related notices to be issued are shown in Figure 1, per each of the identified site aspects. As it can be seen, the highest number of notices (17 ones) is that corresponding to the execution of works. These are found to mainly connected to: Clauses 4 and 7pertaining to execution methods and remedial work, Clause 11 related to the defects notification period, Clauses 15 and 16 dealing with termination and suspension, Clause 17 concerned with employer's risks, and, finally, Clause 19 tackling force majeure issues. The second highest number of notices (10 ones) is that concerned with inspections and testing. These notices are issued whenever parts of the works are ready for examination and for the purpose of the tests on completion and the repetition of tests after remedying defects. Other notices are distributed among the other aspects, as clarified under Figure 1. To be noted is that the asterisk indicates that two notices are called for under the concerned sub-clause.

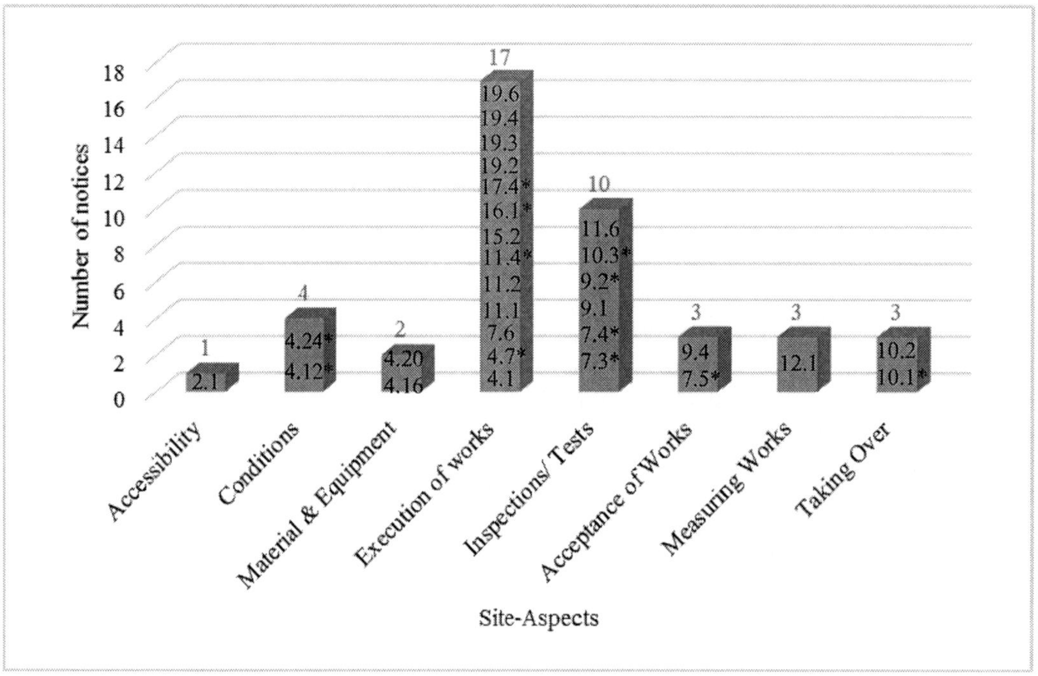

Figure 1. Number of notices per each site aspect with their corresponding sub-clauses.

4.1 Types of Site-Related Notices

Notices serve manifold purposes when called to be issued under the conditions of contracts. In addition to their main purpose of enabling communications about certain events, notices are served in order to bring attention concerning the progress of work, alert the other party about some serious events that need prompt actions, claim in respect of certain rights given under the contract, and/or object regarding a taken action, among other relevant purposes. Accordingly, the types of notices connected with site activities were deduced, as shown in Figure 2.

Attention notices are found to be the most common followed by the claim notices with 15 and 14 cases identified, respectively. The majority of the attention notices are concerned with the inspections and tests aspect (6 notices) and the work execution aspect (6 notices). Other attention notices are issued whenever there is shortage or defects in free-issue material, discovery of fossils, and measurements to be attended. As for the claim notices, these are applicable to all the site aspects except for those dealing with material and equipment and measuring works. The four information notices are found to be in connection with: material and equipment, whenever there is transport of goods; inspections and tests, whenever there is intention to attend tests or when the engineer does not require to inspect or to have a test performed; and execution of works, with respect to notifying about the items of reference. As for the three causation/alarm notices, these precede notices of claims in connection with unforeseeable physical conditions, employer's risks, and force majeure events. The five other notice types are found to be related to the following site aspects: certification notices, concerned with the taking over of the works; objection notices, issued in connection with inaccuracy in measuring the works; plant and/or materials rejection notices, related to the acceptance of works; and suspension and termination notices, concerned with execution of works.

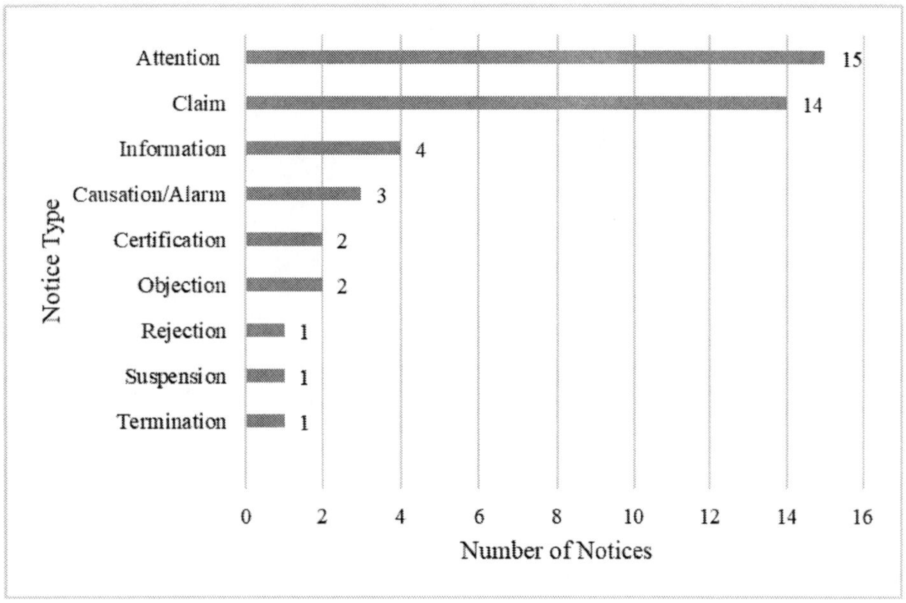

Figure 2. Site-related notices' types.

4.2 Parties Responsible for Issuing Site-Related Notices

In this section, the number of site-related notices issued by each of the involved parties is investigated, with the purpose of identifying the respective responsibilities in regards to acting in reaction to site-related events. The frequency of notices pertaining to each party is as shown in Figure 3. The contractor is found to be responsible for issuing notices in 24 instances, the employer in four instances, and the engineer in eight instances. Other cases refer to either the employer or engineer and to either party (the contractor or employer), with four and three cases, respectively. By way of elaborating, the following are the events/aspects relating to each party:

- Contractor: eight cases are related to notices of claim, two to materials and equipment, two to site conditions, five to inspections and tests, three to execution of works, two to measuring of works, and two to the taking over aspect;
- Employer: one notice is related to of termination and three to remedying defects;
- Engineer: five notices pertain to inspections/testing, one to attendance of measurements, one to acceptance/rejection of works, and one to specifying items of reference;
- Employer or engineer: four notices may be issued in association with Sub-Clause 2.5 Employer's Claims as a result of: rejection and retesting, failure to pass tests on completion, failure to comply with instructions, and failure to remedy defects or damage;

- Party: three notices, linked to force majeure events, may be issued by either party, firstly, whenever a party is or will be prevented from performing its obligation under the contract due to a force majeure event, secondly, upon the ceasing of the force majeure, and, thirdly, when the event otherwise continued for more 84 days or for multiple periods totaling more than 140 days thereby warranting a notice of termination.

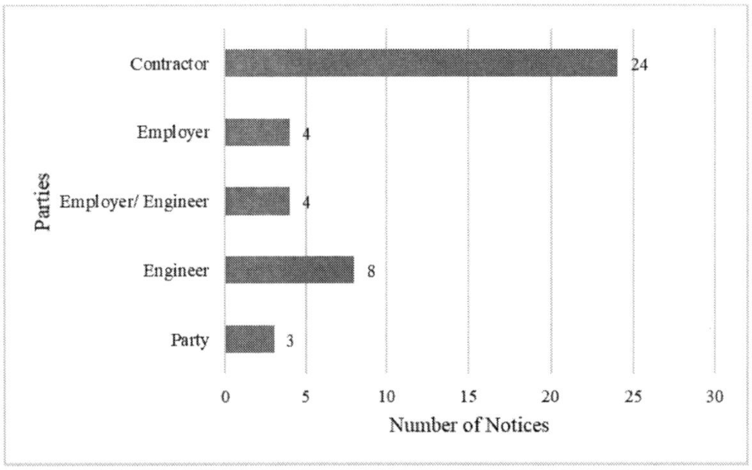

Figure 3. Frequency of site-related notices issued by each party.

4.3 Time Bars for Issuing Site-Related Notices

One of the important characteristics associated with notices is that related to the time stipulations encountered under the concerned sub-clauses and governing their issuance. Time stipulations could be numerical, where a certain period is specified for the notice issuance, or qualitative, where various terminologies, such as promptly or as soon as practicable, are used. On the other hand, some notices are found with no time bars specified in connection with their issuance. The frequency for each of these three categories is presented in Figure 4.

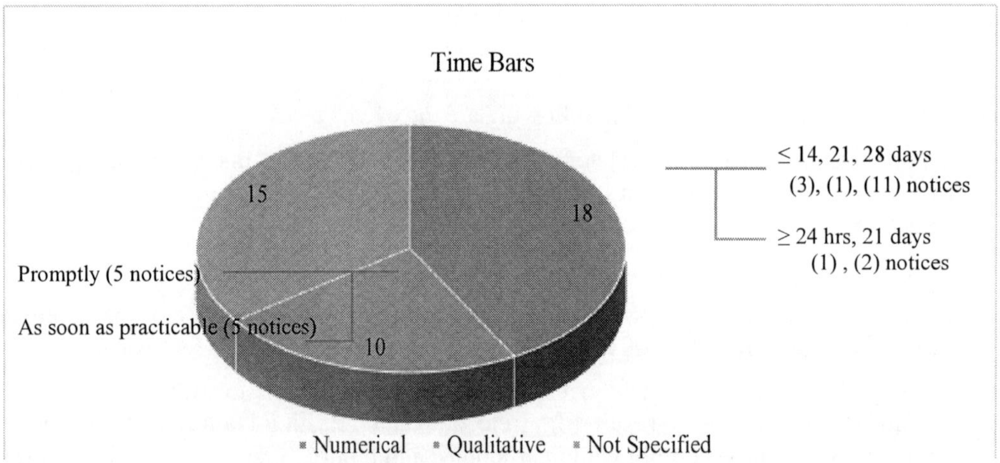

Figure 4. Number of site-related notices for each of the time-bar categories.

Eighteen notices were found to be numerically bounded by time stipulations, either by setting a minimum threshold ($\geq$ 24 hrs. and $\geq$ 21 days) or a maximum threshold ($\leq$ 14, 21, or 28 days). Of these 18 notices, 15 are issued by the contractor, two by the engineer, and one by either party.

4.4 Structured and Non-Structured Site-Related Notices

Notices can be classified as structured or non-structured depending on whether they are connected to a series of events (such as prerequisite actions and/or other notices) and/or time-bar stipulations (Abdul-Malak and Khalife 2017b). Accordingly, the site-related notices were analyzed to filter the ones that are justifiably said to be structured and the others which are not. The results showed that 32 out of the 43 notices (equivalent to 74.4 percent) were found to be structured. Such a high number of site-related notices being classified as structured notices is an indication that the scant understanding of such notices and their provisions is likely to lead to detrimental effects on the concerned aspects of the works.

5 CONCLUSION

Site-related notices pertain to different aspects of the works and events taking place on the construction site. The wide array of identified notices is found to be related to events likely to happen as early as the possession of site stage and up to the final completion and contract close out stage. Various types of site-related notices were deduced, which reflect the different purposes these notices serve, other than the customary purpose of claiming for extra time or money by a notice of claim. The participants (owner, contractor, and engineer) on a project share the responsibilities for issuing these notices; yet, the analysis revealed that the contractor has the largest share. In addition, the findings showed that the vast majority of site-related notices have to be issued within specific time period stipulations. In addition, most of the site-related notices are structured notices. Accordingly, contractual obligations, represented by serving notices whenever it is required to do so under the conditions of the contract, should be a major concern to all concerned, and more particularly to practitioners working with the contractor's team on site. The main incentives are to avoid any breach to the contract and protect the parties' rights.

References

Abdul-Malak, M. A., and Khalife, S., Classification and Analysis of Notice Requirements for Construction Contract Administration, *J. Leg. Aff. Dispute Resolut. Eng. Const.*, ASCE, 9(3), 04517016, 2017a.

Abdul-Malak, M. A., and Khalife, S., Models for the Administration of Structured Construction Contract Notices, *J. Leg. Aff. Dispute Resolut. Eng. Const.*, ASCE, 9(3), 04517017, 2017b.

Aibinu, A. A., Contractual Approach for Facilitating the Resolution of Dispute ver a Contractor's Failure to Comply with Time Limit for Notice of Delays, *J.L..A. Dispute Resolut. Eng.C.*,1(3),137-146, 2009.

Eggleston, B., *Limited Damages and Extensions of Time in Construction Contract*, Oxford, U.K., 2004.

FIDIC, International Federation for Consulting Engineers, *Conditions of Contract for Construction*, Thomas Telford, Geneva, 1999.

Sertyesilisik, B., Investigating on Particular Contractual Issues in Construction, *J. Leg. Aff. Dispute Resolut. Eng. Const.*, ASCE, 2(4), 218-227, 2010.

Thomas, R., Smith, G. R., and Wright, D. E., Resolving Disputes over Contract Notice Requirements. *Journal of construction engineering and management*, ASCE, 116(4), 738-755, 1990.

Thomas, R., *Construction Contract Claim,* Palgrave Macmillan, New York, 2001.

Thomas, R., Teaching Construction Contracts: Mutual Learning Experience, *J. Leg. Aff. Dispute Resolut. Eng. Const.*, ASCE, 1(2), 97-104, 2009.

Von Branconi, C., and Loch, C., Contracting for Major Projects: Eight Business Levers for Top Management, *International Journal of Project Management*, 22(2), 119-130, 2004.

Responsible Design and Delivery of the Constructed Project
Edited by Abdul-Malak, M., Khoury, H., Singh, A., and Yazdani, S.

ADMINISTERING CONDITION PRECEDENT NOTICES IN CONSTRUCTION CONTRACTS

SALAM KHALIFE and MOHAMED-ASEM ABDUL-MALAK

Dept of Civil and Environmental Engineering, American University of Beirut, Beirut, Lebanon

Notices represent a main area of concern for engineering and architecture professionals involved in construction contract administration, as the failure to satisfy their requirements can jeopardize the concerned party's performance under the contract. This is specifically true of those notices that are stated or inferred to be condition precedent to the preservation or the establishment of the rights of either party. A screening of all instances stipulating the issuance of notices under the standard conditions for the construction contract issued by the International Federation of Consulting Engineers (FIDIC) revealed that 41 out of the 93 encountered instances require condition-precedent notices to be issued for preserving or establishing relevant rights. This paper presents a detailed analysis of the three classes that were found to encompass these notices. As such, it sheds light on the distribution of the responsibilities for issuing them between the contractor, on one hand, and the owner or the Engineer on his behalf, on the other hand. It further discusses possible compositions of their structures, highlighting the time bars as well as the prerequisite events that are likely to be involved. The merit of the work lies in mapping these notices, with the aim of helping contract administration practitioners properly deal with them such that the contractual risks associated with their requirements can be avoided.

Keywords: Administration, Claims, Disputes, Rights, Waiver.

1 INTRODUCTION

The concept of condition precedency has long been used as a contractual tool in contract clauses to create "an obligation to be performed by one party to the contract, upon the performance of which a further obligation arises on the side of the other party" (Beck 1975). Likewise, in construction contracts, the condition precedent principle is employed in different clauses either to establish rights or to preserve them. In various standard forms of contract conditions, the notice of a claim is expressly stated as a condition precedent to the contractor's entitlement for an extension of time or extra compensation (Aibinu 2009). The non-compliance with the requirements of such a notice results in the lapse of the claim (Jaeger and Hök 2010). Enforcing strict compliance of notice provisions has been a controversial issue. Lal (2002) argued that allocating the risk of the cost of delay on the contractor is 'perfectly reasonable', and that making the notice of a claim a condition precedent is commercially sensible and fair. On the contrary, negative consequences of enforcing strict compliance were discussed by Spratt (2010), tackling the increase in costs for all parties as a result of strict compliance. To this end, it was stated that whenever changes are made, and there is a short period to realize the full impact and issue a notice for compensation, the contractor is likely to overestimate the cost of the change as a result of uncertainty (Spratt 2010). On the other hand, condition-precedent notices are usually accompanied with time-bar stipulations governing the timely fulfillment of their requirements. Such stipulations enable the owner to investigate the event and its impacts, in order to verify the claim accordingly (Lim 2012). A review of a number of cases dealing with notice requirements revealed that courts react differently when determining the operability of condition-precedent requirements and the enforceability of time-bar clauses (Lal 2002, Clark 2008, Aibinu 2009, Glover 2011). Therefore, it is in the interest of all parties to carefully comply with notice requirements, specifically those deemed to operate as condition precedent (Lim 2012).

2 RESEARCH SCOPE AND METHODOLOGY

A classification study of the notices specified under the FIDIC's standard conditions for the construction contract (FIDIC 1999) revealed that 41 out of a total of 93 encoutered instances calling for notices express a state of condition precedency for right entitlement (Abdul-Malak and Khalife 2017a). The aim of the work presented in this paper is to highlight the aspects of the works concerned with these notices and discuss the condition-precedent classes govening their triggering. The followed methodology included: (1) identifying the condition precedent notices and their relevant provisions, (2) deducing the type (class) of inherent condition precedency, (3) investigating the responsibilities of the concerned parties towards their issuance, and (4) shedding light on the potential complexity that characterizes their administration.

3 CONDITION PRECEDENT NOTICES

A comprehensive review of the twenty clauses found under the FIDIC conditions of contract unveiled a wide group of notices recognized to be condition precedent notices. These were identified and their provisions filtered out for analysis. Based on the language used in the relevant sub-clauses, three classes of condition-precedent notices were deduced: "explicit language for establishing a right", "jointly in conjunction with a specific sub-clause", and "implicit language for establishing a right" (Abdul-Malak and Khalife 2017b). For each mentioned class, the list of sub-clauses expressing condition precedency (whether explicitly, jointly, or implicitly) and the respective rights to be inherently established are identified. The notices serving these conditions, the party issuing them, and their prerequisites including the associated time periods are thereafter identified.

The above-mentioned characteristics for the first class of condition-precedent notices, where the sub-clauses explicitly state the need to serve the notices for certain respective rights to be established, are shown in Table 1. The right to claim entitlement by the contractor under Sub-Clause 20.1 is based upon serving this notice in reference to one of the 15 listed sub-clauses (see Figure 1) within 28 days after the contractor (CNT) "became aware or should have become aware" of the event (E) giving rise to the claim. The composition of this type of notice is illustrated in Figure 2 (Case 1). The prerequisites (P) for a number of these notices are as follows: first-tier notices under Sub-Clauses 1.9, 4.12, 4.24, and 16.1; instructions needed according to Sub-Clauses 7.4, 8.9, and 17.4; notice of force majeure under Sub-Clause 19.2; and the taking-over certificate as a prerequisite for notice under Sub-Clause10.3. The other notices, served in accordance with Sub-Clauses 2.1, 4.7, 10.2, 13.7, and 18.1, do not have any prerequisites. The subsequent action (S) is to provide full supporting particulars for the claim within a 42-day period.

The second explicitly mentioned case is the right by either party to refer a matter to the Dispute Adjudication Board (DAB) as a result of a dispute arising in connection with any certificate, determination, instruction, opinion or valuation of the engineer (ENG), referred to under the FIDIC conditions as forming part of the personnel of the employer (EMP). Preserving this right is conditional on the engineer serving in many such instances a determination notice (within 42 days in the case of claims raised pursuant to Sub-Clause 20.1).

Table 1. Explicit language for establishing right.

Entitlement	Relevant Sub-Clause	Condition Precedent Notice	Party Issuing Notice	Complexity (Prerequisites; Time Bars)
Right to entitlement under Sub-Clause 20.1	1.9, 2.1, 4.7, 4.12, 4.24, 7.4, 8.9, 10.2, 10.3, 13.7, 16.1, 17.4, 18.1, 19.4, 20.1	"Claim Notice" under Sub-Clause 20.1	Contractor	Structured (Case 1, in Figure 2), ≤ 28 days
Right to referral to adjudication (DAB) in connection with the time-barred determination	20.4	"Determination Notice" under Sub-Clause 3.5	Engineer	Structured, ≤ 42 days for determination in response to 20.1; no stipulation in response to other Sub-Clauses
Right to referral to arbitration	20.4	"Dissatisfaction Notice" under Sub-Clause 20.4	Party	Structured (Case 2, in Figure 2), ≤ 28 days
Right to be indemnified by the other party against an infringement claim	17.5	"Notice of a Claim Alleging an Infringement" under Sub-Clause 17.5	Party	Non-structured, ≤ 28 days
Right to terminate	19.6	"Notice of Force Majeure" under Sub-Clause 19.2	Party	Non-structured, ≤ 14 days

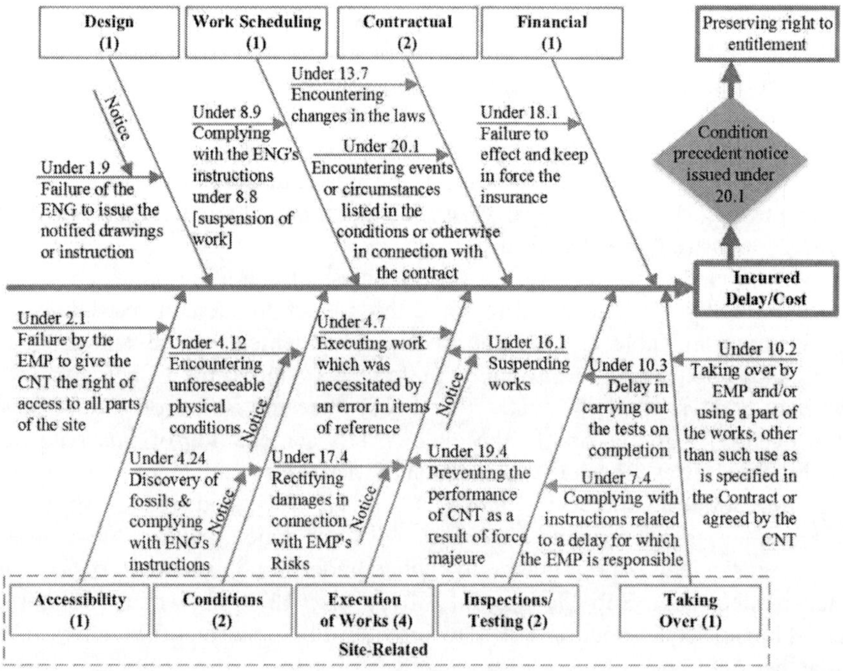

Figure 1. Circumstances for the condition-precedent notice under Sub-Clause 20.1.

The third case is in connection with establishing the right for referring a dispute to arbitration. Either party has the right to do so, provided that a notice of dissatisfaction has been issued within a period of 28 days after receiving the DAB's decision or after the expiry of the 84-day period after the matter has been referred to adjudication. Case 2 in Figure 2 shows the composition of the structure governing the issuance of this notice. The prerequisite represents the action of dispute referral to adjudication, followed by this condition-precedent notice being served within 28 days from the issuance date of the DAB decision or the expiry of the given 84-day period with no decision having been given. The subsequent action represents the possibility of commencing arbitration on or after the 56th day from issuing the condition-precedent notice. The forth situation is related to receiving an infringement claim. Either party has the right to be indemnified by the other party a gainst this claim through issuing a notice within 28 days; otherwise, the concerned party "shall be deemed to have waived any right to indemnity under this Sub-Clause" (FIDIC 1999, Sub-Clause 17.5). Finally, the fifth case in which explicit language is used to formulate a condition precedent notice is the one related to the right to terminate the contract as a result of a force majeure condition. This right is preserved whenever a notice of force majeure was served under Sub-Clause 19.2. This is clearly stated under Sub-Clause 19.6, calling for this notice to be served within 14 days from being aware (or should have been aware) of the event constituting a force majeure.

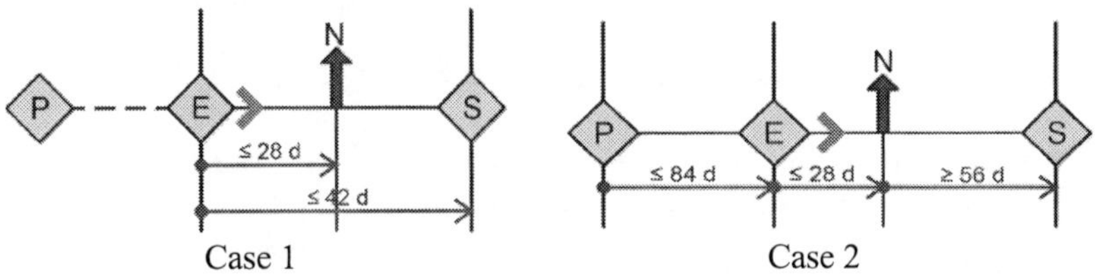

Figure 2. The composition of notices under Sub-Clauses 20.1 (Case 1) and 20.4 (Case 2).

As for the second class, where condition precedency is implicitly expressed under certain sub-clauses, the 12 instances referring to this class are detailed in Table 2. The owner's right to claim for any payment and/or the extension of the defects notification period is established in association with the notice issued under Sub-Clause 2.5, in reference to any of the identified sub-clauses (see Figure 3). The different circumstances, which lead to this case, are elaborated in Figure 3. Another case included in the second class is the right to object to an interim payment certificate formulated based on records asserted to be inaccurate by the contractor. As such, the contractor shall issue a notice of inaccuracy within 14 days after being requested to examine the records; otherwise, they shall be accepted as accurate.

Table 2. Implicit language for establishing right.

Entitlement	**Relevant Sub-Clause**	**Condition Precedent Notice**	**Party Issuing Notice**	**Complexity (Prerequisites; Time Bars)**
Right to entitlement under Sub-Clause 2.5	2.5, 7.5, 7.6, 8.6, 8.7, 9.4, 11.3, 11.4, 15.4, 18.1, 18.2	"Employer's Claim Notice" under Sub-Clause 2.5	Employer or Engineer	Structured, time period: as governed by applicable laws
Right to object to interim payment certificate due to inaccurate measurements	12.1	"Notice of Inaccuracy" under Sub-Clause 2.1	Contractor	Structured, ≤ 14 days

The third class of condition precedent notices includes 9 notices found to be jointly in conjunction with Sub-Clause 2.5 or Sub-Clause 20.1 as shown in Table 3. These notices are to be administered jointly in conjunction with the claim notices, and – as such – they shall be served before issuing further notices under Sub-Clauses 2.5 or 20.1.

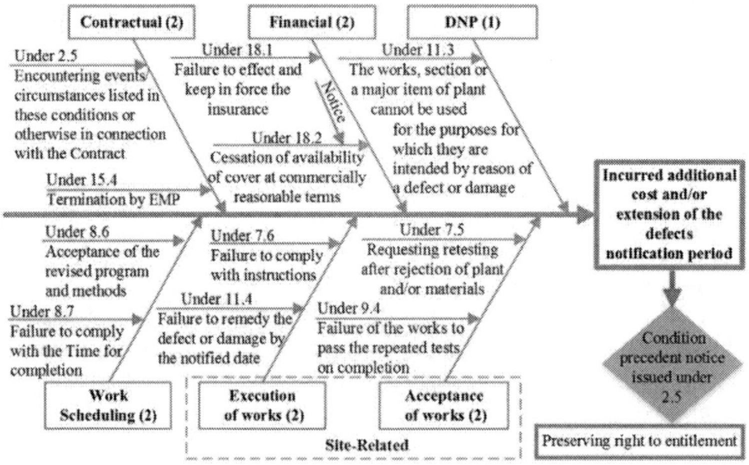

Figure 3. Circumstances for the condition precedent notice under Sub-Clause 2.5.

The first-tier notices, in conjunction with Sub-Clause 20.1, are represented in red arrows in Figure 2. To be noted is that all the listed notices have no time stipulation associated with their issuance except for the extension of time notice under Sub-Clause 8.4 where the notice needs to be issued within 28 days as per Sub-Clause 20.1.

Table 3. Condition Precedent notices under the third condition precedent class.

Entitlement and Relevant Sub-Clause	Condition Precedent Notice	Party Issuing Notice	Complexity (Prerequisites; Time Bars)
Jointly in conjunction with Sub-Clause 2.5	"Rejection of Plant and/or Materials Notice" under Sub-Clause 7.5	Engineer	Non-structured, time period not specified
	"Notice of Ceasing Availability of Cover at Commercially Reasonable terms" under Sub-Clause 18.2	Contractor	Non-structured, time period not specified
Jointly in conjunction with Sub-Clause 20.1	First-tier notices under Sub-Clauses 1.9, 4.12, 4.24, 16.1, 17.4	Contractor	Notices under Sub-Clauses 1.9 and 16.1 are structured,
	"Notice of Probable Future Events" under Sub-Clause 8.3	Contractor	Non-structured, Promptly
	"Extension of Time Notice" under Sub-Clause 8.4	Contractor	Structured, As soon as practicable and ≤ 28 days

4 NOTICE ISSUANCE RESPONSIBILITIES

The frequencies of instances calling for the issuance of condition-precedent notices by each party are as shown in Figure 4. The contractor has the highest number of condition-precedent notices to deal with, followed by the employer (or the engineer on his behalf). The engineer has two types of notices to be served as condition precedent: the determination notice and the rejection of plants and/or materials notice. However, the frequency of issuing a determination notice is expected to be dominant throughout the course of construction. That is, there are 40 sub-clauses that refer to the determination notice, and they cover multiple aspects of the works. It can therefore be concluded that the contractor and the engineer have a prime responsibility in regards to serving the condition-precedent type of notices. Furthermore, when filtering the number of structured notices versus those that are non-structured, the ratio was found to be 36:5, respectively. The high number of structured notices indicates the need to understand the different compositions associated with these condition-precedent notices.

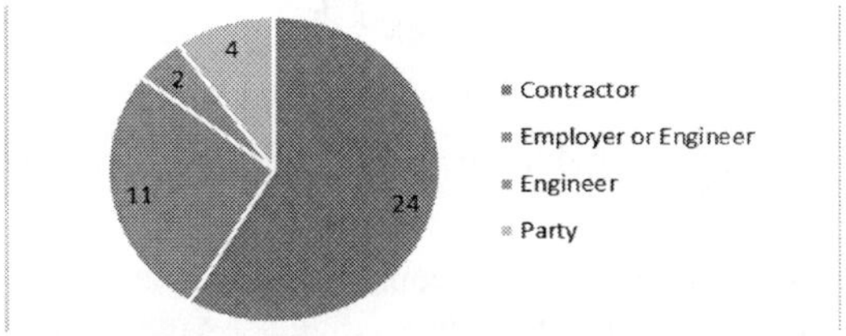

Figure 4. Number of instances calling for the issuance of condition-precedent notices by each party.

5 CONCLUDING REMARKS

The importance of identifying condition-precedent notices and the situations warranting their issuance is obviously very critical, in order for the parties to preserve their rights for certain entitlements. The condition-precedent notices, identified in reference to the 20 instances in which explicit language is used, were all found to have specific time bars governing their issuance. The findings revealed that the contractor carries the larger burden of having to administer such notices, with more than half of the notices called for being his responsibility to serve. Finally, the results showed that the vast majority of condition-precedent notices are structured, and parties shall therefore be well aware of the events, prerequisites, and time bars forming parts of their respective structures.

References

Abdul-Malak, M. A., and Khalife, S., Classification and Analysis of Notice Requirements for Construction Contract Administration, *J. Leg. Aff. Dispute Resolut. Eng. Const.*, ASCE, 9(3), 04517016, 2017a.

Abdul-Malak, M. A., and Khalife, S., Models for the Administration of Structured Construction Contract Notices, *J. Leg. Aff. Dispute Resolut. Eng. Const.*, ASCE, 9(3), 04517017, 2017b.

Aibinu, A. A., Contractual Approach for Facilitating the Resolution of Dispute over a Contractor's Failure to Comply with Time Limit for Notice of Delays, *J. Leg. Aff. Dispute Resolut. Eng. Const.*, ASCE, 1(3),137-146, 2009.

Beck, A., The Doctrine of Substantial Performance: Conditions and Conditions Precedent, *The Modern Law Review,* 38(4), 413-428, 1975.

Clark, R. P., Notice Provisions: Are Norices a Condition Precedent to Entitlement to Extensions of Time?, Contract Journal, 22, 2008.

FIDIC, *Conditions of Contract for Construction,* Thomas Telford, Geneva, 1999.

Glover, J., Time Bars: The Continued Rise of Time Bars?, *Annual Review,* 15-17, 2011.

Jaeger, A., and Hök, G., *FIDIC — A Guide for Practitioners, Springer* Berlin, Germany, 2010.

Lal, H., Extensions of Time: The Conflict between the Prevention Principle and Notice Requirements as a Condition Precedent, Society of Construction Law, London, 2002.

Lim, T., Essence of Time in Construction Contracts, *Australasian Journal of Construction Economics and Building,* 9(2), 1-6, 2012.

Spratt, P., Strict Compliance with Construction Contract Notice Provisions: Detrimental to Contractors and Taxpayers, *Public Contract Law Journal,* 40(4), 912-933, 2010.

Responsible Design and Delivery of the Constructed Project
Edited by Abdul-Malak, M., Khoury, H., Singh, A., and Yazdani, S.

EVALUATION OF SIMULATION TECHNIQUES FOR MODELING PROGRESSION OF CONSTRUCTION CLAIMS

MOHAMMAD BARAKAT, HIAM KHOURY, and MOHAMED-ASEM ABDUL-MALAK

Civil and Environmental Engineering, American University of Beirut, Beirut, Lebanon

The unstructured and dynamic nature of construction projects and the on-site work complexities have been inevitably leading to claims. These evolve according to a staged mechanism set forth in the adopted conditions of contracts. More specifically, claims might progress expeditiously or drag depending on the nature of the applied mechanism and the behavior and interaction among contractual parties. As such, this complex problem of claim progression, which entails a lot of parameters and variables, is addressed in detail in this paper by resorting to three simulation techniques namely: (1) Discrete-Event Simulation (DES), (2) Agent-Based Modeling (ABM), and (3) System Dynamics (SD). The purpose behind this study is two-fold: (1) capturing and visualizing, through three different simulation models, the dynamic and interaction among the different entities as claims are progressing and defining the weak links hindering the efficiency improvement of such a process, and (2) comparing DES, ABM, and SD simulation approaches, using choose-by-advantage technique, and evaluating the advantages and drawbacks of each when studying the progression of claims. Results of all approaches are presented and analyzed followed by a discussion of the effectiveness of each simulation technique and the potential applicability of a hybrid approach in modeling the progression of claims.

Keywords: Discrete-event simulation, Agent-based modeling, System dynamics, Disputes.

1 INTRODUCTION

Claims have become almost inevitable through the course of construction projects. This has typically urged both owners and contractors to adopt effective claim/dispute mechanisms with the aim of resolving arising conflicts. Some mechanisms might be expeditious while others might involve unregulated periods thereby dragging the claims for a long period of time. As such, there is a need to simulate the adopted claim process in order to allow contractual parties to predict the different possible outcomes and take remedial actions when needed. In this case, three simulation techniques could prove promising in simulating the aforementioned process, namely; (1) discrete-event simulation (DES), (2) system dynamics (SD), and (3) agent-based modeling (ABM) (Maidstone 2012). Although several studies compared the applicability of these methods in various fields (Brailsford and Hilton 2001, Tako and Robinson 2009a, Tako and Robinson 2009b, Siebers *et al.* 2010, Maidstone 2012), and some have adopted these tools in dispute resolution (El-Adaway and Kandil 2009, Menassa and Peña Mora 2010), none has evaluated their feasibility in modeling the progression of claims in particular. Therefore, this paper aims at evaluating these three techniques in modeling the progression of claims and identifying the most suitable one.

2 SCOPE AND METHODOLOGY

This research study uses three simulation techniques to model the progression of construction claims. The adopted methodology involves: (1) defining the generic claim/dispute timeline, (2) creating DES, SD, and ABM models to simulate the progression of claims along the timeline, and (3) evaluating the applicability of every simulation method in replicating real-life scenarios.

3 CLAIM/DISPUTE TIMELINE

Claims arising within construction projects are addressed according to the mechanism set forth in the claim/dispute administration and resolution provisions. Based on relevant standard conditions currently in use, the mechanism typically consists of three main modules: (1) the disclosure of claim, (2) the judgment(s), and (3) the final resolution of dispute, as shown in Figure 1.

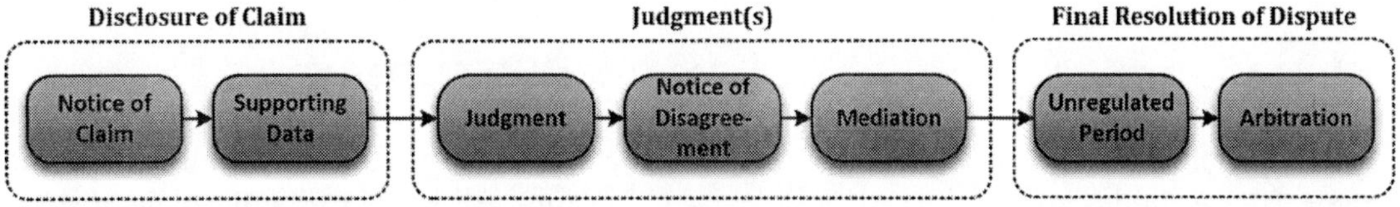

Figure 1. Claim timeline.

Starting with the disclosure of claim module, the claimant has to submit a notice of claim (NoC) to the other party to notify him of the arising issue. In addition, the claimant has to provide supporting data (SD) to justify the eligibility and/or quantum of the claim. In this study, it was assumed that the time-bars of these two stages are sequential whereby the time-bar of the supporting data stage is not activated unless and until the notice of claim is submitted. However, upon submitting the supporting data, the claim moves to the next module whereby an engineering professional shall act impartially to render a judgment. The rendered judgment is binding upon both parties with the possibility of turning final if not revoked within a specified period. In this case, either party shall submit a notice of disagreement (NoD) in order to revoke the judgment and initiate mediation. If the mediation process is not initiated, then the judgment becomes final and both parties waive their rights in pursuing mediation or arbitration. It is worth mentioning that this judgment acts as a condition precedent to mediation unless it was not issued in the first place within a specified period, allowing thereby either party to unilaterally initiate mediation at any time. When mediation is pursued, both parties shall endeavor to reach an agreement. If the mediation is deemed unsuccessful, the mediator can conclude the mediation at any time. Moreover, either party may terminate the process after a specified period. The end of the mediation process signals the start of the final resolution of dispute module. Within this module, either party can refer the matter in dispute to arbitration at any time, thereby dragging the claim within an unregulated period, as shown in Figure 1.

4 SIMULATION TECHNIQUES

The following subsections model the claim mechanism using each of the DES, SD, and ABM simulation paradigms within the anylogic environment (Borshchev and Filippov 2004).

4.1 Discrete-Event Simulation

DES is one of the most popular simulation techniques used to model a process (Maidstone 2012). This technique adopts the top-down modeling approach whereby it focuses on modeling in detail a system and not entities. Furthermore, the entities flowing through the process are passive and they cannot act independently since their behavior is affected by that of the system. Figure 2 depicts the anylogic model simulating the progression of claims using DES.

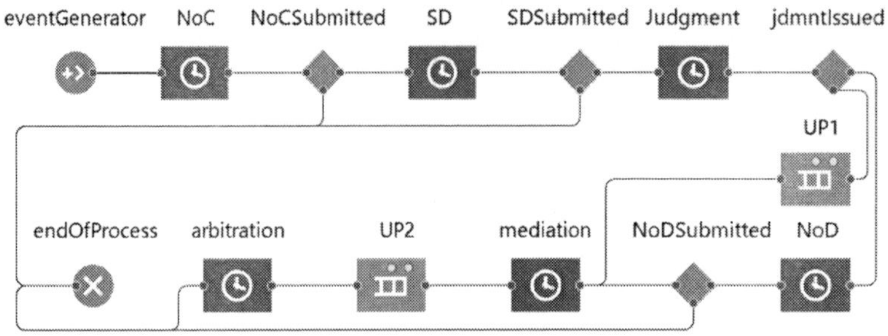

Figure 2. Simulating claim/dispute timeline using DES.

The model starting point is the source "eventGenerator" that generates the entities into the system. These entities are the events that give rise to a claim. All states within the claim/dispute timeline were modeled as tasks (i.e. dark blue rectangular shapes) except for the two instances of unregulated period. Once an entity flows into a task, an action is taken. For illustration, the event triggers the claimant to submit a notice of claim upon entering the "NoC" task. Upon exiting the "NoC" task, the claim entity is faced with two options through a "NoCSubmitted" select-output (i.e. diamond shape). In this case, if a notice of claim is not submitted within the specified period, the process ends, thereby shifting the claim entity to the "endOfProcess" sink where it leaves the system; otherwise, the claim entity moves to the supporting data stage (i.e. "SD" task). The claim entity proceeds through the claim/dispute timeline as explained in Section 3 and can possibly face two instances of unregulated periods (UP), modeled as queues in Figure 3. The first occurs when a judgment is not rendered within the stipulated time, then allowing contractual parties to initiate mediation at any time. This moves the claim entity through the "UP" queue to reach the "mediation" task. On the other hand, the second occurs when either party may refer the dispute to arbitration at any time. In this case, the claim entity stays in the second "UP" queue for an unregulated period of time before it moves to the "arbitration" task.

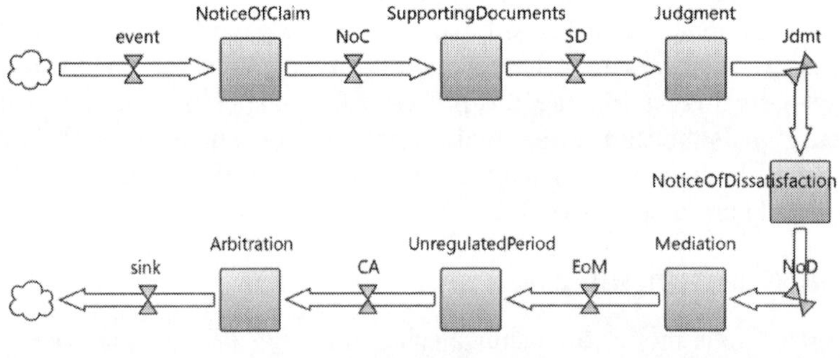

Figure 3. Simulating claim/dispute mechanism using SD.

4.2 System Dynamics

SD is another popular simulation technique used to model a system. Figure 4 shows the SD model of the claim/dispute mechanism. Similar to DES, a top-down modeling approach is adopted, but the system is modeled as a series of tanks or stocks (i.e. rectangular shapes) connected by pipes or flows (i.e. arrows), and entities are viewed as a continuous quantity flowing through this system. The rates of flow are controlled by valves, and accordingly the time spent in each stock or system state is modelled by fixing the rates of inflow and outflow (Brailsford and Hilton 2001).

In this case, the stages of the claim/dispute mechanism were modeled as the stocks, and the transitions between the stages were modeled as flows. Within the SD model, the claim entities flow from one stock to the other based on the flow rates set through direct and indirect variables. This approach allows modelers and practitioners to identify causal loops that show the impact of the system's variables on the flow of entities.

4.3 Agent-Based Modeling

Unlike DES and SD, ABM adopts the bottom-up modeling approach and aims at mainly modeling agents and the interactions among them (Maidstone 2012), and while focusing on both passive and active entities. This technique relies on state-charts to model the different agent states. For illustration, Figure 5 shows the state-chart of the claim agent, in particular.

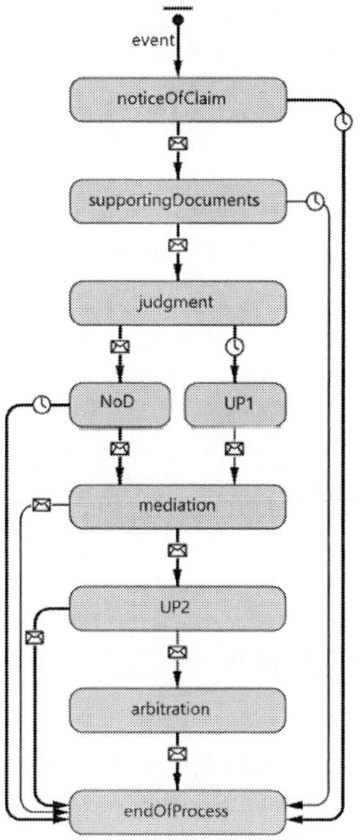

Figure 4. Simulating claim progression using ABM.

The various stages within the claim/dispute timeline were modeled as states that are connected using different types of transitions. For illustration, the claim agent transitions from the "noticeOfClaim" state to the "supportingDocuments" state upon receiving a message from the claimant agent indicating that the notice was submitted. On the other hand, the "noticeOfClaim" state is connected to the "endOfProcess" state using a time transition and not a message transition. In this case, if the time transition stipulated period expires prior to receiving the message of submitting the notice, the claim moves automatically to the "endOfProcess" state.

It is worth mentioning that other active agents such as the contracting and other involved parties (i.e. owner, contractor, judgment engineering professional, mediator, etc.) are part of the ABM model, and their interactions throughout the claim/dispute timeline contribute greatly to the overall emerging system behavior.

5 DISCUSSION

In this section, the choose-by-advantage decision making tool (Suhr 1999) was used to determine the most suitable simulation method for modeling the progression of claims, as shown in Table 1.

Table 1. Choose by advantage of the three simulation techniques.

Factors	DES		SD		ABM	
Nature of Problem (Applicability)	Process-Oriented **Highly Applicable**	85	System-Oriented Least Applicable	0	Individual-Based Moderately Applicable	60
Method's Goals vs. Problem's Goals (Similarity)	Comparison of Scenarios, predictions, and/or optimization More Similar	45	Understanding feedback dynamics and long-term system behavior Less Similar	0	Examining the system behavior that emerges from the interactions of agents **Most Similar**	100
Modeling (Ease)	1) Stochastic 2) Tasks and queues 3) Transition to the next stage is time-based Difficult	20	1) Deterministic 2) Stocks and flows 3) Transition to the next stock depends on flow rates Most Difficult	0	1) Stochastic 2) Agents and States 3) Transitions to the next state are time-based and/or action-based **Moderate**	70
Total		**150**		**0**		**230**

Starting with the nature of the problem, the claim/dispute mechanism is a process that has a queueing network. As such, the DES method fits in that regard and is highly applicable in modeling the progression of claims. However, the ABM method is considered moderately applicable (Siebers *et al.* 2010) since the mechanism is modeled as a state-chart of the claim agent. The SD technique is considered the least applicable as the aim behind modeling the progression of claims consists of focusing on the entities rather than on the system (Tako and Robinson 2009a).

On the other hand, comparing method's goals to that of the modeled problem reveals different results (Table 1). In this case, the DES goals are considered to be more similar to the problem's goals than the SD goals. However, the ABM goals are considered to be the most similar ones because this technique examines the behavior of the system that cannot be predicted automatically and emerges from the various agents' interactions (Klügl and Bazzan 2012).

Moving to the ease of modeling, SD is considered the most difficult. SD models tend to be deterministic, which is not representative of the case of claims progression. Furthermore, stages are modeled as stocks, where the flow rate between the stocks cannot be easily evaluated. In the case of DES, modeling is considered difficult as the progression of claims is modeled in a stochastic fashion due to the variability of time needed to take the corresponding action at each stage. Furthermore, the transitions between the stages are mainly time-dependent and not action-based moving the claim entity to the next stage upon an action being taken by the concerned party. It can be argued that the time needed for an action to be taken can be modeled as the time needed for the claim entity to move to the next task, which is not easy to model. Similar to DES, ABM models are stochastic in nature (Maidstone 2012), but the transitions between the states are not only time-based but also action-based whereby a claim moves to the next state based on a taken action, which facilitates modeling the problem at hand. As such, the ease of modeling level of the ABM method is considerate moderate. Consequently, the points assigned to each factor, based on their corresponding advantage, showed that the ABM approach is the most suitable one to model the progression of claims with a total of 230 points.

6 CONCLUSION

Attempting the simulation of the claim/dispute mechanism can greatly help the contracting parties in predicting the effectiveness of the adopted mechanism and its potential applicability when resolving arising conflicts. Therefore, in this paper, three different simulation techniques, namely DES, SD, and ABM, were evaluated in modeling the progression of claims throughout the claim/dispute timeline. As such, a generic timeline was first generated for use under each simulation method environment. Although all three techniques present their own pros and cons, the ABM one outweighed the other two, scoring the highest using the choose-by-advantage decision making tool. It was found that the ABM method is moderately applicable with respect to the nature of the problem, has the most similar goals to that of the studied model, and is moderately easy to model.

References

Brailsford, S. C., and Hilton, N. A., A Comparison of Discrete Event Simulation and System Dynamics for Modelling Health Care Systems, *Proceedings of the 26th meeting of the ORAHS Working Group 2000*, 18–39, Glasgow, Scotland, 2001.

Borshchev, A., and Filippov, A., Anylogic—Multi-Paradigm Simulation for Business, Engineering and Research, In *The 6th IIE Annual Simulation Solutions Conference* 150, 45, March, 2004.

El-Adaway, I. H., and Kandil, A. A., Multiagent System for Construction Dispute Resolution (MAS COR), *Journal of Construction Engineering and Management*, *136*(3), 303-315, 2009.

Klügl, F., and Bazzan, A. L., Agent-Based Modeling and Simulation. *AI Magazine*, *33*(3), 29, 2012.

Maidstone, R., Discrete Event Simulation, System Dynamics and Agent Based Simulation: Discussion and Comparison, *System*, *1*(6), 2012.

Menassa, C., and Peña Mora, F., Hybrid Model Incorporating Real Options with Process Centric and System Dynamics Modeling to Assess Value of Investments in Alternative Dispute Resolution Techniques, *Journal of Computing in Civil Engineering*, *24*(5), 414-429, 2010.

Siebers, P. O., Macal, C. M., Garnett, J., Buxton, D., and Pidd, M., Discrete-Event Simulation is Dead, Long Live Agent-Based Simulation!, *Journal of Simulation*, *4*(3), 204-210, 2010.

Suhr, J., *The Choosing by Advantages Decision Making System*, Greenwood Publishing Group, United States, 1999.

Tako, A. A., and Robinson, S., Comparing Discrete-Event Simulation and System Dynamics: Users' Perceptions, *Journal of the Operational Research Society*, *60*(3), 296-312, 2009a.

Tako, A. A., and Robinson, S., Comparing Model Development in Discrete Event Simulation and System Dynamics, *Proceedings of the Winter Simulation Conference*, 979-991, December, 2009b.

Responsible Design and Delivery of the Constructed Project
Edited by Abdul-Malak, M., Khoury, H., Singh, A., and Yazdani, S.

ADMINISTRATING THE CONSTRUCTION CONTRACT "OR" MANAGING THE RELATIONSHIP?

MOHAMED-ASEM U. ABDUL-MALAK

Dept of Civil and Environmental Engineering, American University of Beirut, Beirut, Lebanon

The construction contract conditions shall ideally be drafted in such a manner that provides for a balanced allocation of risks between the parties to any such contract. The project management teams, inclusive of contract administration professionals forming part of such teams, are then faced with decisions that are to be made on day-to-day basis in respect of observing the called-for contractual requirements. Those decisions pertaining to the initiation of claims and contributing to their progression are of the most difficult to administer and put in place, owing to such actions being at times viewed as detrimental to the relationship between the parties, both at the project operational as well as the corporate levels. This paper discusses the factors that come into play when such critical decisions get to be evaluated. As such, it presents a holistic perspective of the kinds of pressure expected to be prevailing over the involved decision-making process. The argued considerations are informed by a number of practical dispute resolution cases and the scenarios that have emanated under them. The paper concludes by providing guidelines that can help draw the line between the necessity of acting in fulfillment of the concerned contract callings and the process of preserving the relationship between the parties to the contract.

Keywords: Claims progression, Risks, Adversarial relations, Disputes, Resolution.

1 INTRODUCTION

The relationship between the owner of a construction project and the contractor selected to perform the works is legally formalized through the execution of a contract. The signature of such a contract represents the agreement of both parties to a set of respective obligations, rights, responsibilities, duties, and liabilities. Unfortunately, contractors often fail to adequately assess the inherent contractual risks and do not come to realize the undesirability of such risks until during the course of construction. In addition, the participants involved in administering these construction contracts are not only likely to be different from those who contributed to the contract negotiation and signature stage but also less knowledgeable about how the encountered risks have been originally assessed. This is aggravated by claims that are inevitably bound to surface, and their administration and potential progression into disputes can adversely affect the working relationship among the concerned participants, at the operational (project) level, as well as between the parties to the contract, at the management (corporate) level.

Raised claims are not to be expected to diminish with time if left unresolved (DiDonato 1993, Harmon 2001). On the contrary, claims evolve into disputes and could lead to negatively impacting the progress of the project (Harmon 2003), in addition to adversely affecting the communication and relationship of the contracting parties (Jones 1994, Conlin *et al.* 1996, Mix 1997, Arditi *et al.* 1998, Steen and MacPherson 2000, Harmon 2003, Ng *et al.* 2007, Cheung and Pang 2012, Ofori 2013).

2 SCOPE AND METHODOLGY

The work in this paper aims at highlighting the kinds of pressure that are likely to act on contract administration professionals who are faced on a day-to-day basis with claims/disputes administration decisions. These involve the initiation or notification of claims, the substantiation of the raised claims, the response regarding determinations or decisions given by third-party professionals, as well as the pursuit of their amicable resolution before their progression into litigation or arbitration (as the case may be) becomes more detrimental to the working relationships, both at the operational and management levels.

The work methodology involves (1) constructing the context under which the contracting business operates, by focusing on the Middle East and North Africa (MENA) region; (2) highlighting the pressures that can act on the various gates of the phases forming the claim/dispute evolvement timeline; (3) identifying the resulting potential outcomes of such raised claims being either tabled, grouped at a late stage under a global claim, or intimidatingly closed or dropped; and (4) proposing a set of strategies that can assist in achieving a balance between, on the one hand, the criticality of clearly observing the concerned contract provisions and, on the other, the preservation of a minimum degree of healthy communication and cooperation between concerned participants at both operational and management levels.

3 CONSTRUCTION CONTRACTING BUSINESS CULTURE

The construction business in the MENA region can be characterized as a dynamic environment in which unconventional contracting structures can be expected to incessantly transpire. In such structures, a hybrid mix of participants can turn out to be engaged in contractual settings that impose constraints on how various paired contracting parties pursue their respective interests under the corresponding contracts in place. Project developers, contractors, and subcontractors of various strengths, sizes and territories of operation coexist in this construction-vibrant region, with a sizeable portion of these organizations being from outside the region. This is owed to the continuous interest in the scale of project development activities and associated investments that are planned for this region over the many years to come.

Project development organizations, both public and private ones, range in terms of their level of experience in the market and their preparedness to manage complex construction-related contracts. Yet, any such organization, with or without the help of professional advisors, ends up engaging contracting organizations that range from being of a shallow structure that is in place for the sole purpose of subletting the whole of the contracted work to being of a very sophisticated structure that employs its own construction workforce in addition to relying – in accordance with a contractually-stipulated extent – on subletting practices.

On the other hand, it is not uncommon that medium- to large-scale general contracting organizations accept to act as subcontractors to other similarly-sized local or regional contracting companies and definitely to the larger international ones. These and other engagement scenarios can be inferred from the general contracting and subcontracting links depicted in Figure 1. As it can be seen, matching is possible between a developer of any size and complexity and a general contractor of any size and operation territory, whereas subcontracting can take place within any general contractors' group or to a higher- or a lower-tier group.

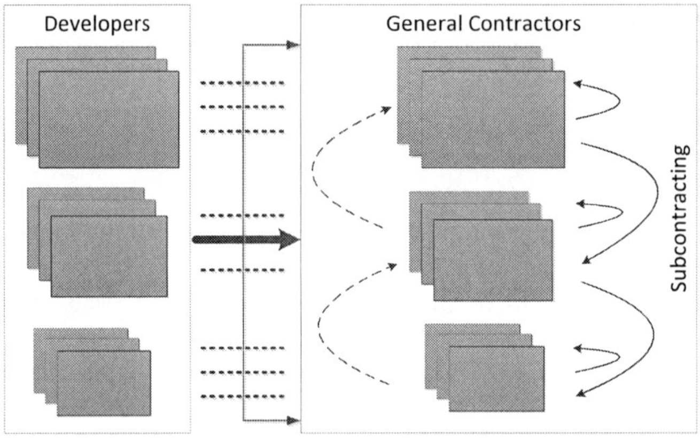

Figure 1. Developers and contractors engagement possibilities.

4 GATE-BASED ADMINISTRATION OF CLAIMS/DISPUTES

Traditionally, contracts incorporated a two-step mechanism for the resolution of disputes (Jessup and Jessup 1963), under which the design professional first determines the eligibility of a contractor's claim before either party making a referral of the dispute to binding adjudication. This approach was adopted based on the fact that the construction industry members believed that the designer is the most suitable party to assess the validity of a proclaimed change based on the intent of his design (Stipanowich 1997). However, in exercising such a quasi-adjudicatory role, the objectivity of the design professional is questioned due to conflict of interest (Cheeks 2003).

The trend in today's complex projects is for the administration of construction claims and disputes to be handled along a multi-step process. Such a process still involves an initial assessment by an engineering professional (but not necessarily the design one), but it also prescribes other alternative methods for approaching the resolution of disputes before finally resorting to litigation or arbitration. Under such mechanisms, contractors' contract administrators are expected to act at various successive gates, in a way that allows for the initiation of claims and for further documentation and debate to take place in respect of the raised claims (Abdul-Malak and Abdulhai 2017). A typical such process for the administration of claims and disputes can be schematically configured, as shown in Figure 2 (FIDIC 1999), in reference to the standard general conditions that are prescribed by the International Federation of Consulting Engineers (FIDIC).

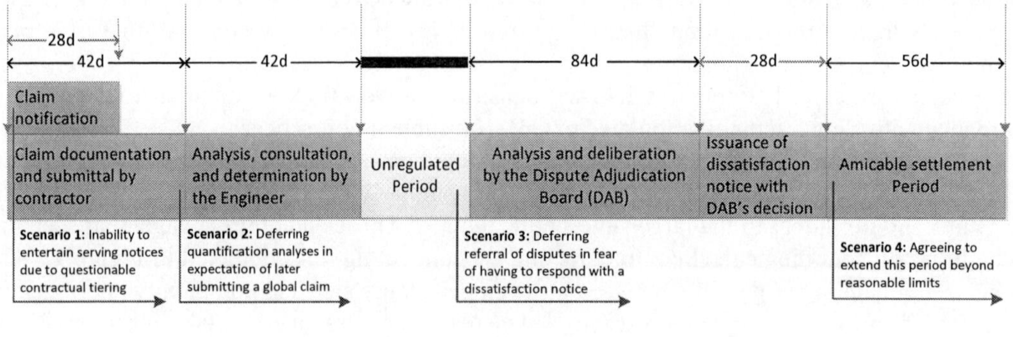

Figure 2. Gated claims/disputes administration stages (Based on FIDIC 1999).

Not only do contractors have to successively act in accordance with such requirements, but their actions are also to be rendered in certain instances within rigid time frames, such as those pertaining to the issuance of a claim notice and a notice of dissatisfaction with an adjudication decision. To this effect, failure to act in a timely fashion on these notices leads to the loss of right for the pursuit of claim and for the referral of dispute to arbitration, respectively. Several reasons have been found to prevail, which can stand in the way of the concerned contractors' contract administrators being able to act in accordance with the contract. To this end, not acting or deferring a warranted action is normally exercised in fear of damaging the relation with the projects' owners. The following scenarios (also laid onto the FIDIC's stages under Figure 1) highlight a number of such instances, to which the author has had firsthand exposure as part of dispute resolution advising roles rendered in connection with the cited cases:

- **Scenario 1.** On a community-type project involving the construction by the government of individual housing units that cater for the needs of early- to middle-career employees, serving a notice for a major claim was hindered by the fact that the general contractor had accepted to enter into a backstage subcontract whose scope was to construct the whole of the works in a shadowing capacity on behalf of the party that secured and signed the general contract with the owner. The claim was concerned with the repercussions of major delays to the works caused by the inability of the government (owner) to timely release to the eligible contracting party the entry visa quota needed for the procurement of construction labor workforce. The conditions that had facilitated arranging for such an engagement structure were behind the contracting contract signatory opting to refrain from raising claims against this governmental client.
- **Scenario 2.** Practices involving the issuance by operational project staff of notices for alleged additional time and/or compensation that were not followed by the submission of supporting particulars in due times have been mistakenly adopted as a way of not escalating the tension with the owner's management level during the course of construction. Such practices were found to be persistently detrimental as they prevented the proper administration of the construction's time schedules and eventually led to the submission of global analyses for requested time extensions, which were more difficult to sell and resolve, given their inherent underlying weaknesses.
- **Scenario 3.** Disagreement discussions concerning the determinations given by the contract engineer have been allowed to drag by deferring the placement of requests for adjudication decisions, in fear that a subsequent notice of dissatisfaction issued in response to any such decision could be highly damaging to the relationship with a high-profile client. This is owed to such a notice being a prerequisite to the establishment of right for referring a dispute in question to arbitration. The ramifications have been in the form of these unresolved claims (whose issued determinations have been disagreed with) becoming more and more complex, particularly when they were time-extension related. Making the referral of those pending claims as disputes to adjudication was acted upon only after the owner had started levying liquidated damages, a quite few months before the then-expected completion date for the works.
- **Scenario 4.** Following the notification of dissatisfaction with the issued adjudication decision, the mutual agreement to extending the amicable settlement period, until either party notifies the other that this stage is considered to have expired, allowed the said period to drag to about ten times the 56-day duration originally assigned to it under the contract. This unreasonably long (on-going) extension has given room for another non-

binding adjudicative process to be initiated, which seems to stand a slim likelihood of being conclusive. Tolerating the length of and the cost associated with such an excessively elongated period for attempting the achievement of an amicable resolution has been in return for deferring the initiation of arbitration, a process the contractor is still adamant not to be wishing to resort to. That is, agreeing to the employment of an alternative dispute resolution process (albeit being non-binding) during the amicable settlement stage is all part of the desire to preserve (or, to say the least, not to cause further damage to) the relationship with this repeat-business client.

The previously mentioned scenarios represent varying forms of claims mishandling through either (a) exercising discretion by not taking an action or deferring a warranted action, as in scenarios 1 and 3 or (b) not satisfying the requirements of submitting detailed particulars in a timely fashion and – instead – speculating on the future success of a global claim, as in scenario 2. While the situation described under scenario 4 cannot be classified as a case of mishandling, together with the situations underlying the first three scenarios have surfaced by way of purportedly aiding the preservation of the relationship with the respective projects' owners.

5 RESPONSIBLE CONTRACT ADMINISTRATION STRATEGIES

The scenarios discussed in the previous section reflect the actions (or the deferral or lack of) that general contractors are found to opt to exercise on construction projects when attempting to weigh in on maintaining a good relation with owners and contract engineers during the course of construction. In this section, strategies that contractors may alternatively embrace in order to more responsibly act or react in accordance with the contract are offered. These can be summarized as follows:

- **Act as being compliant.** The owner is the drafter of the contract; as such, contractors shall assume credit for being compliant with their contracts, and for always acting in accordance with, and within the boundaries of, their contracts' callings.
- **Make explicit reference to contract clauses.** In every instance a written communication is to issued, an explicit reference to the clause requiring the contractor to act or react, as the case may be, shall be made.
- **Pave the way for claim notices.** A close observance of all types of notifications called for under the contract conditions is to be maintained. This can be expected to provide the grounds, and subsequently pave the way, for issuing notices of claims.
- **Do not personalize issued objections.** In expressing objections to determinations, opinions, or decisions given by the contract engineer or adjudicator, contractors shall critique the argument underlying such rendered judgements and not the individual or his/her thinking process. Same as in the case of placing a claim (demand) forward, saying "no" to the engineer's or adjudicator's judgment demands of the contractor to put forward detailed particulars that counter-argue the arguments forming the bases for such judgments.
- **Do not underestimate a claim's lifespan.** By deferring those actions that are otherwise warranted, by way of buying time or postponing the escalation of tension, contractors end up acting at times when many of those staff who have had firsthand knowledge of the claim in question are already away and may not be possible to continue to rely on them in supporting the dispute resolution process. More often than none, contractors come to painfully realize that the lifespan of the claim/dispute outstretches that of the contract's

construction phase if not that of the construction and defects notification phases combined.

6 CONCLUDING REMARKS

It is well established that there is no exact science that underlies the way with which contract administrators approach the administration of construction contracts in practice. A potentially governing factor can be the degree of autonomy operational contract administration staff get to enjoy on one project versus the other. Based on the discussed case-based inferences, it can as such be understood that multiple pressures may overarch the decision-making milestones that spread along the claim/dispute tracking and resolution timeline. To this end, it remains evident that the proper documentation for, and the expeditious tracking of, claims can help better control the length of their lifespan and minimize the chance of their evolvement into disputes. By allowing the resolution of claims/disputes to drag can only be expected to contribute to increasing the level of adversarialism of the relationship between the contracting parties.

References

Abdul-Malak, M. U., and Abdulhai, T. A., Conceptualization of The Contractor's Project Management Group Dynamics in Claims Initiation and Documentation Evolution, *Journal of Legal Affairs and Dispute Resolution in Engineering and Construction,* 9(3), 04517014, 2017.

Arditi, D., Oksay, F. E., and Tokdemir, O. B., Predicting the Outcome of Construction Litigation Using Neutral Networks, *Comput. Aided Civ. Infrastruct. Eng.,* 13, 75–81, 1998.

Cheeks, J. R., Multistep Dispute Resolution in Design and Construction Industry, *Journal of Professional Issues in Engineering Education and Practice,* 129(2), 84-91, 2003.

Cheung, S. O., and Pang, K. H. Y., Anatomy of Construction Disputes, *Journal of Construction Engineering and Management,* 139(1), 15-23, 2012.

Conlin, J., Lanford, D. A., and Kennedy, P., *The Sources, Causes, and Effects of Construction Disputes: a Research Project,* CIB Rep. 0254-4083, Construction Industry Board, London, 1996.

DiDonato, L., *Alternate Methods to Resolve Contract Disputes: Approaches to Effective Prevention,* American Association of Cost Engineers, Morgantown, W. Va., 1993.

FIDIC, (International Federation for Consulting Engineers), *Conditions of Contract for Construction,* 1st Ed., Thomas Telford Ltd, Geneva, 1999.

Harmon, K. M. J., *Pseudo Arbitration Clauses in New York City Construction Contracts,* Construction Briefings, July, 2001.

Harmon, K. M. J., Dispute Review Boards and Construction Conflicts: Attitudes and Opinions of Construction Industry Members, *Ph.D. dissertation,* Nova Southeastern Univ., Fort Lauderdale, FL, 2003.

Jessup, W. E., Jr., and Jessup, W. E., *Law and Specifications for Engineers and Scientists*, Prentice-Hall, 28–29, 1963.

Jones, S. R., How Constructive is Construction Law?, *Construction Law Journal,* 10, 28-28, 1994.

Mix, D., ADR in The Construction Industry: Continuing the Development of a More Efficient Dispute Resolution Mechanism, *Ohio St. J. on Disp. Resol.*, 12, 463, 1997.

Ng, H., Pena-Mora, F., and Tamaki, T., Dynamic Conflict Management in Large-Scale Design and Construction Projects, *J. Manage. Eng.,* 52–66, 10.1061/(ASCE)0742-597X(2007)23:2(52), 2007.

Ofori, G., Conflict in Construction, *Construction Management and Economics,* 31(5), 497–501, 2013.

Steen, R. H., and MacPherson, R. J., Resolving Construction Disputes Out of Court Through ADR, *Journal of Property Management,* 65(5), 58, 2000.

Stipanowich, T. J., The Multi-Door Contract and Other Possibilities, *Ohio St. J. on Disp. Resol.,* 13, 303, 1997.

AUTHOR INDEX

AUTHOR INDEX